W9-CRY-839

Biomolecular Stereodynamics Volume I

Upcoming Adenine Titles of Related Interest

Structure and Dynamics of Nucleic Acids and Proteins, Eds. Enrico Clementi, International Business Machines and Ramaswamy H. Sarma, State University of New York at Albany. This volume deals with computer experiments, NMR and X-ray studies and thermodynamics of nucleic acids and proteins. A significant portion is devoted to the application of soliton theory to nucleic acids. The volume is based on the proceedings of the La Jolla gathering in the discipline, September 1982 and is expected to be out in early 1983.

Biochemical and Inorganic Perspectives in Copper Coordination Chemistry, Eds. Kenneth D. Karlin and Jon Zubieta, State University of New York at Albany. Oxygen utilization and electron transfer processes carried out by copper containing proteins have produced a great deal of interest in new coordination chemistry of copper. This volume focuses upon current trends in copper research emphasizing new structural chemistry, applications to catalysis and redox activity of copper compounds and enzymes. The volume is expected to be out in early 1983.

Proceedings of the Second SUNYA Conversation in the Discipline Biomolecular Stereodynamics
Volume I, ISBN 0-940030-00-4, Ed., Ramaswamy H. Sarma,
Adenine Press, New York, ©AAAS.

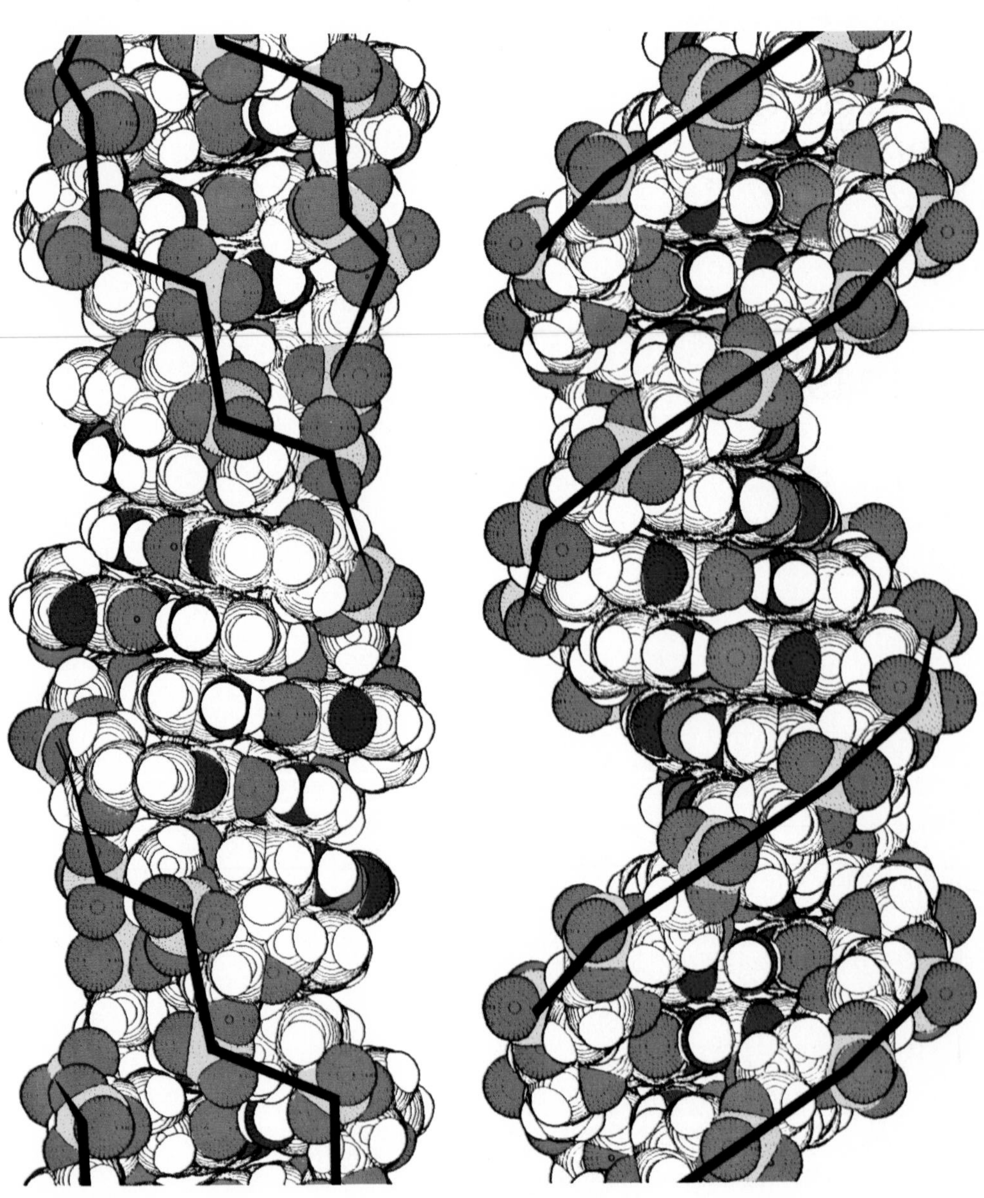

The Left-Handed Z-DNA vis-a-vis The Right-Handed B-DNA
Sponsored by
National Foundation for Cancer Research

Biomolecular Stereodynamics, Volume I

Proceedings of a symposium held at the State University of New York at Albany, 26-29 April, 1981

Edited by

Ramaswamy H. Sarma

Director
Institute of Biomolecular Stereodynamics
State University of New York at Albany

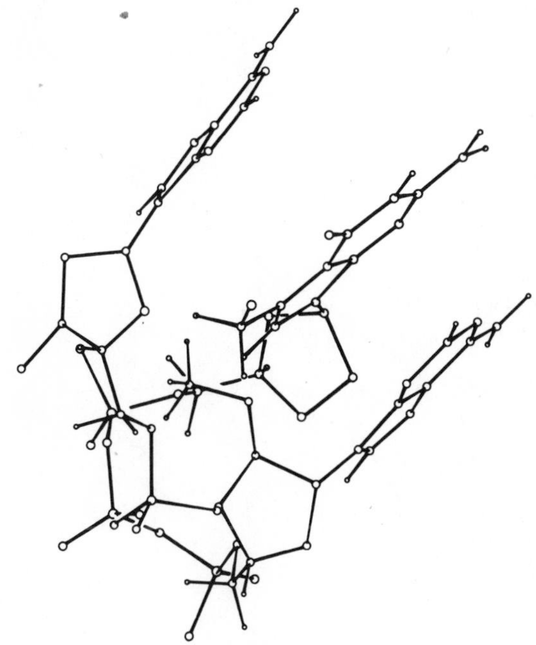

Adenine Press, New York

Adenine Press
Post Office Box 355
Guilderland, New York 12084

Cover illustration: End view of the left-handed Z-DNA by Alexander Rich, Gary J. Quigley and Andrew H.-Wang, Copyright © Alexander Rich

Library of Congress Cataloging in Publication Data
Main entry under title:

Biomolecular stereodynamics

"Proceedings of the Second SUNYA Conversation in the Discipline Biomolecular Stereodynamics held at the State University of New York at Albany, April 26-29, 1981 under the auspices of the Department of Chemistry and organized by the University's Institute of Biomolecular Stereodynamics"—Preface.
Bibliography: v. 1, p. v. 2, p.
Includes index.
1. Molecular biology—Congresses. 2. Stereology—Congresses. I. Sarma, Ramaswamy H., 1939- II. SUNYA Conversation in the Discipline Biomolecular Stereodynamics (2nd : 1981 : State University of New York at Albany) III. State University of New York at Albany. Dept. of Chemistry. IV. State University of New York at Albany. Institute of Biomolecular Stereodynamics.
QH506.B554 574.8′8 81-14867
ISBN 0-940030-00-4 (v. 1) AACR2
ISBN 0-940030-01-2 (v. 2)

Set in type in the United States by Word Management Corporation, Albany, New York.
Printed in the United States by Hamilton Printing Company, Albany, New York.

Preface

These are the Proceedings of the Second SUNYA Conversation in the Discipline Biomolecular Stereodynamics held at the State University of New York at Albany, April 26-29, 1981 under the auspices of the Department of Chemistry and organized by the University's Institute of Biomolecular Stereodynamics. In this Conversation we explored the frontiers and attempted to delineate the structural and stereodynamical subtleties of the awesome biological machinery at the molecular level. The conference was a synthesis in the discipline in which experimentalists and theoreticians in Chemistry, Physics and Biology presented papers, discussed and debated the changing structural scene in biology; and these volumes are born out of the fire and heat of this synthesis.

Volume I deals exclusively with the continuing story of DNA and RNA and opens with an extensive review, opulent with stereographics, by Richard Dickerson on the structural dynamics and conformational interconnectedness of the DNA triad—the A, B and Z spatial configurations. Then the discoverer of the left-handed Z-DNA, Alexander Rich expounds on the intimate geometric details, conformational transition and reactivities of the Z-helices. In addition Rich describes the structure of the antitumor antibiotic, daunomycin, intercalated into a right-handed double helix, and provides structural rationalizations for the physiological differences between daunomycin and the closely related adriamycin. Robert Wells reports about his success in constructing a recombinant plasmid, a fraction of which underwent salt-induced B to Z transition. The plasticity of the DNA double helix and the structures in solution *vis-a-vis* those in crystals and theory are taken up by Ramaswamy Sarma. He presents hard NMR (nuclear magnetic resonance) spectroscopic data which reveal that the structures observed in crystals are very close to those found in solution. These double helices are so well organized that when they are transferred into solution from the crystalline state, the onset of additional internal motion and freedom does not overwhelm the integrity of the molecular framework; to a large extent there is structural preservation. Nevertheless nucleic acids are capable of undergoing sequence and ionic strength dependent transitions to distinct spatial configurations. Struther Arnott and Rangaswamy Chandrasekaran, in a highly photographic essay, examines the polymorphism in fibrous polynucleotides and advocates a new classification system. They further describe the possible geometry of junctions that may exist between A and B as well as B and Z helices. V. Sasisekharan provides theoretical data which show that both right and left-handed helices can be constructed using either mono- or dinucleotides as repeating units. Rick Ornstein and Jacques Fresco present an improved theoretical formalism to obtain accurate base pair geometries and energies. This is followed by Bernard Pullman who presents detailed electrostatic molecular surface potential and accessibility maps of B and Z-forms and clearly demonstrate that the reactivity towards carcinogenic species is significantly different for both. Reactivity and structural dynamics of the globular transfer RNA molecule are described by Alberte Pullman and Alfred Redfield respectively. Particularly noteworthy is the discussion by Redfield on the care one has to pay about the assignments of NMR lines in tRNA; his NOE approach with specific deuteration has the potential of providing very accurate information about the structural dynamics of tRNA-protein interaction. Enrico Clementi and Gina Corongiu report computer experiments which combine *ab initio* potentials and Monte Carlo simulations and take into consideration essential parameters like temperature, statistical distributions, time, solvents and reaction fields, to arrive at the structure of water and counter ions around the double helix. In their model, the counter ions are present in an ordered helical array following closely the two strands of the double helix with interconnecting water filaments across the grooves. They suggest that the base pair recognition by the protein may involve a relay type mechanism. Color artwork has been extensively used by them for the translation of the results of computer experiments to molecular terms. Parthasarathy advocates a special role for water from model crystallographic studies; he suggests that the water molecule may act as a *spacer* and stabilize the helical structure of DNA, should a base be turned outside due to dynamical fluctuations or due to noncomplementary base opposition. Nadrian Seeman who is capable of giving a "local habitation and a name to airy nothing" presents the bizarre possibility that nucleic acids can exist in N-connected two and three dimensional networks using migrationally immobile and semi-mobile junctions. He further elaborates how can these junctions be used to investigate branch point migration. Together with Bruce Robinson, he presents a dynamic-structural model for this phenomenon which is in good agreement with experimental data.

The section on internal motion in the double helix opens with M. Sundaralingam and E. Westhof who from theory and crystallographic data synthesize a unified stereodynamical formalism for the interconnected motions in the double helix in which the ribose and the deoxyribose act as the *Yin* and *Yang* of the genome. Wilma Olson examines from theory the allowed backbone torsional movements and presents a stereochemically and energetically sound trajectory for B to Z transition. Hydrogen-1, phosphorus-31 and carbon-13 NMR measurements are employed by David Kearns, Phillip Hart, Randolf Rill and fluorescence emission anisotropy by Mary Barkley to explore the internal motions in the double helix. Even though important information about base pair opening, B-DNA structure, effect of sequence and coupling between backbone and base motions, has resulted from these studies, no unified motional model for the double helix has emerged because the art is new and the motional dynamics of the double helix is complicated by its multidimensional conformation space.

Neville Kallenbach and Todd Miles probe the nature of the low temperature open state using equilibrium spectroscopy and hydrogen-deuterium exchange and attempt to determine whether one is dealing with a localized open state or an extended one like a soliton. Thomas Neilson provides NMR and melting studies which clearly indicate that non base paired dangling bases provide stability to the double helix depending upon whether the direction of the stacking of the dangling bases is $3'\rightarrow5'$ or $5'\rightarrow3'$. Douglas Turner ponders the question: Why do nucleic acids form helices? Richard Blake presents an extensive review of the theory of melting and Volume I concludes with an opus by Leonard Lerman where he describes a possible way to map completely a genomic DNA taking advantage of the sequence dependence of the electrophoretic mobility of DNA in denaturing gels.

In Volume II we take up biological structures considerably more complex than nucleic acids—the integral membrane proteins, protein-nucleic acid complexes, protein dynamics and related systems. This volume opens with a philosophical review by G. N. Ramachandran on the evolution of conformational biology as an independent discipline in the last three decades. It is an intensely personal voyage by the author through the challenging years which ultimately found the solution of the collagen structure and the formulation of the (ϕ,ψ) relationships. Then Gobind Khorana elaborates his studies on the structure and orientation of the light driven proton pumping protein bacteriorhodpsin in the purple membrane. He reveals his plans to comprehend the mechanism of proton translocation by selective *in vitro* mutagenesis on the genomic DNA of the protein. He hopes to generate small changes in the amino acid sequence of the protein by this method. He has already partially purified the messenger RNA for bacteriorhodpsin and has prepared using reverse transcriptase a 80 nucleotide fragment of the genome. He is using this fragment to identify the bacteriorhodpsin gene among the cloned genomic fragments from the halobacter DNA. This is reminiscent of the days in which Khorana was deeply involved in nucleic acid structural studies and we nucleic acid people hope that he is coming home. Robert Stroud describes the structure of an acetylcholine receptor based on x-ray diffraction and electron microscopy examination and advocates a hypothesis for a dynamic mechanism of its action. James Prestegard addresses the question of membrane transport and storage and attempts to obtain possible answers using hydrogen and deuterium NMR spectroscopy. L. W. Reeves uses deuterium NMR to probe intra-chain motion, kinks and jogs in bilayer model membranes.

In the section on protein-nucleic acid interactions, Aaron Klug shows, by means of nuclease digestion, x-ray and neutron scattering studies, that in the nucleosome core particles, the DNA is wrapped around the protein core in a left-handed supercoil and the periodicity of DNA changed in the transition from free DNA in solution to DNA complexed with proteins as in the nucleosomes. Klug illustrates the structural details, the orientation of the superhelices and the relation between the histone octamer and the double helix in a series of compelling color art work. Gary Felsenfeld takes up the question of the accessibility of nucleosome DNA to high mobility group proteins and externally added histone octamers. He also marshals evidence which suggest that accessibility plays a crucial role in the formation of higher order chromatin structure. Chromatin structure is extensively discussed by Morton Bradbury and Harry Matthews. Particularly, they examine how acetylation of lysines and phosphorylations of serines and threonines of the core histone affect chromatin structure.

Olof Sundin and Alexander Varshavsky elaborate on their work on the topological problems faced by the SV40 minichromosome in the course of its replicative cycle. When the DNA has completely replicated, the chromosome then forms a catenated structure in which two relaxed circles are linked. The catenation linking numbers can go from 1 to 25. While this process is going on, each of the relaxed circles is also being supercoiled; thus two reactions are happening simultaneously: catenation linking numbers are being decreased while supercoiling is increased. These reactions are under the control of two separate enzymes. Binding of the antitumor drug *cis*-diammine-dichloroplatinum II to DNA and to the nucleosome core particle is described by Stephen Lippard. His studies revealed that the antitumor agent unwinds closed circular DNAs, shortens it up to 50% of its original length, nicks the DNA and inhibits the action of restriction endonuclease. Gunther Eichhorn explores the effects of several metal ions and polypeptides on the structure and packing of DNA and finally extends his studies to the RNA polymerase system. So far in all of these reports one has observed that the DNA maintains its double helical structure on its interactions with the protein. In contrast, Alexander McPherson, Ian Molineux, Paula Fitzgerald and Alexander Rich describe the way in which the dimer of the gene 5 protein of bacteriophage fd is able to separate the strands of the double helix and is able to bind single stranded DNAs.

Martin Karplus introduces the subject of structure and motion in proteins. He delineates the dynamics of positional fluctuations in the vicinity of a given structure and the motions involved in going from one structure to another. Karplus presents *tour de force* computer experiments in which a ligand is charged up with substantial kinetic energy and released near the binding site in the three dimensional static x-ray structure; the trajectory of the test molecule is then followed for a few picoseconds to trace the motions involved in its voyage from the binding site through the protein matrix to the solution outside and then in again. In addition Karplus considers hinge bending motions, exterior and loop motions and motions associated with activated and rare events. He elaborates how information on these can be gained by a combination of Monte Carlo and molecular dynamics simulations. David Weaver follows Karplus with a discussion of the microdomain dynamics in folding proteins. From nuclear Overhauser measurements C.M. Dobson concludes that the crystal structure is a very good description of much of the average structure of lysozyme in solution. He examines internal motions of specific groups by Overhauser effects and arrives at the extent of fast dynamical fluctuations of individual torsion angles, very close to those predicted by molecular dynamics simulations. Robert Griffin presents a detailed and lucid account of the fundamentals of solid state NMR spectroscopy and, with copious illustrations, portrays how this form of spectroscopy in carbon-13 and deuterium configurations, can be employed to follow motions in lipid bilayers, peptides and proteins. Following Griffin is Eric Oldfield who describes the use of deuterium NMR to follow protein dynamics in solution, in membranes and in the crystalline solid state. Walter Englander attempts to obtain information about protein breathing and hemoglobin allostery by a newly developed hydrogen exchange approach. His results point to an interesting kind of protein structural opening as an intermediate in the exchange process. He interprets these openings in terms of the transient, small scale, cooperative unfolding of individual segments of secondary structure in which internal H-bonds are severed and reformed to water. Mössbauer spectroscopy is employed by S.G. Cohen to study the dynamics of the iron containing core in crystals of the iron storage protein, ferritin; he concludes that the core is undergoing bounded diffusive motion within a cage.

Crystallographic data and computer modelling have been employed by William Duax and David Smith to arrive at the conformational flexibility and the dynamics of ion capture by ion transport antibiotics valinomycin, monensin and lasalocid A . Using the same approaches Vivian Cody arrives at thyroid hormone conformations and the spatial changes that accompany its interactions with the receptor. Rama Krishna and Gideon Goldstein employ a variety of powerful NMR techniques such as Overhauser effects, transfer of solvent saturation, two dimensional J resolved and spin echo correlated NMR spectroscopy to arrive at the dynamic spatial configuration of cell differentiating peptide fragments of the thymic hormone thymopoietin. David Cowburn analyses the NMR spectra of the peptide hormones oxytocin and vasopressin containing deuterium, carbon-13 and nitrogen-15 isotopes and arrives at their overall dynamic conformations. In addition, direct observation of nitrogen-15 NMR, and of the detailed exchange kinetics of amide hydrogens by NMR have been used to characterize the solvation and hydrogen-bonding properties of these peptide hormones. Michael Blumenstein attempts to unearth the rotation rates of the aromatic rings in these hormones while bound to the receptor neurophycin by using carbon-13 and fluorine-19 NMR spectroscopy. Ian Armitage using carbon-13 NMR characterizes the structure and binding of the aminoterminal glyco-octapeptide of the integral membrane protein glycophorin A. He unequivocally demonstrates that the blood group M and N antibody determinants reside within the aminoterminal glyco-octapeptide.

Binding of the glycopeptide bleomycin—an antibiotic employed in cancer chemotherapy—to the DNA double helix is discussed by Arthur Grollman. He advocates a model in which there is coordination of molecular oxygen to the metal in bleomycin, Fe(II), and preferential intercalation of the bithiozole rings between two DNA base pairs in which one chain contains GpT or GpC base sequence. J. D. Glickson reports the characterization of the binary, ternary and quarternary complexes of bleomycins with various metals, nucleic acids and molecular oxygen analogs. Philip Bolton extensively discusses the principles and applications of heteronuclear two dimensional J-resolved NMR spectroscopy to study the conformational dynamics of nucleic acid fragments and their complexs with proteins. Suse Broyde, using potential energy calculations arrives at the deformation that the carcinogenic agent N-acetoxy-2-

acetylaminofluorene and its analogs can cause to DNA. She reports five possible base displaced conformations for the drug-nucleic acid complex that can cause severe bends in the helix. Kenneth Miller presents a rigorous theoretical treatment of drug intercalation into DNA. He finds three closely related structural models which are sterically and energetically sound. In order to account for the stabilization that Mg^{++} provides to double helices, Theophile Theophanides and Moschos Polissiou examine in detail the interaction between Mg^{++} and 5′-guanosine monophosphate. They find that the metal coordinates to N7 of guanosine and causes subtle conformational variations.

The volumes conclude with the brief closing remarks delivered at the conference by Sir David Phillips. He recalls that the decade of 1965-75 was influenced by the solution of some of the formidable protein crystal structures and that in the euphoria of describing these structures, the crystallographers largely neglected the evidence of mobility present in their data. The decade, dominated by brass models of double helical DNA and protein molecules, Sir David calls, the decade of the rigid macromolecule. "There were, of course, voices crying in the wilderness; Walter Englander among them, no doubt protested like Galileo 'but they do move,' and the message going out from this meeting is that indeed they do."

In editing this book, while I have attempted to establish connections among the various contributions, I have also attempted to retain the individuality of each chapter as written by the authors. I have not removed some of the repetitious matter such as figures and definitions of nomenclature etc. A few authors have begun to use the new IUPAC-IUB nomenclature for the backbone torsions of nucleic acids; but a majority of them, including this editor, continue to use the Sundaralingam nomenclature.

The conference and hence these volumes came into being primarily because of support from various sources. The SUNY Conversation in the Discipline series started it. I thank President Vincent O'Leary of SUNYA, Dean Daniel Wulff of the College of Science and Mathematics, Prof. Harry L. Frisch, Executive Director of the Center for Biological Macromolecules and the Chairmen of the Departments of Chemistry (Prof. Larry Snyder), Biology (Prof. Leonard Lerman) and Physics (Prof. Walter Gibson) for their support. I express my gratitude to Professors Volkner Mohnan and Eugene McLaren for extending the facilities of the Atmospheric Sciences to hold the poster session. Here at the University I have received assistance from several individuals. I acknowledge with gratitude the help from Virginia Dollar, Crystal Hutchins, Mary Kantrowitz and Charles Heller of Chemistry and John Elliot of Biology and Al Dasher of the College of Science and Mathematics.

I am immensely grateful to the following institutions for their support: Bristol Laboratories, Bruker Instruments, Inc., Exxon Educational Foundation, General Electric Company, Hoffmann-LaRoche, IBM Instruments, Inc., Merck Sharp and Dohme, National Foundation for Cancer Research, Nicolet Xrd Corporation, Searle Research, Sterling Winthrop, Upjohn Company, Varian Associates and Wilmad Glass Company.

I must particularly express my deep gratitude to Dr. Roland W. Schmitt and Dr. Leroy S. Moody of General Electric Corporate Research and Development for their continuing support of these conversations in biomolecular stereodynamics.

National Foundation for Cancer Research provided the funds for the B-DNA *vis-a-vis* Z-DNA Frontispiece in color. I am grateful to Dr. Franklin Salisbury, Executive Director of the Foundation for this generosity. It may be noted that three Regional Directors of the National Foundation for Cancer Research have contributed to these volumes. They are Enrico Clementi at IBM, Bernard and Alberte Pullman at Institut de Biologie Physico-Chimique, Paris, France.

I express my gratitude to Ben Fuina, and all the people of the Word Management Corporation for their professionalism, cooperation and concern for the successful completion of typesetting and production of mechanicals in time.

I thank my friends and colleagues in the discipline who have either served in the organizing committee or have provided valuable suggestions or have physically helped me in several ways. I must particularly mention: Neville Kallenbach, Tom Krugh, Leonard S. Lerman, Suraj Manrao, Chanchal Mitra, William D. Phillips, Alfred G. Redfield, Alexander Rich, Mukti H. Sarma, Nadrian C. Seeman and Larry Snyder.

I thank the authors for contributing to these volumes and for providing the manuscripts in time.

It has been a demanding mission, but I enjoyed it and learned a great deal; hope that you will join us for the Third Conversation.

Ramaswamy H. Sarma
Albany, NY
July 21, 1981

CONTENTS

Volume I

CONTENTS

Volume II

Proceedings of the Second SUNYA Conversation in the Discipline Biomolecular Stereodynamics
Volume I, ISBN 0-940030-00-4, Ed., Ramaswamy H. Sarma,
Adenine Press, New York, ©Adenine Press

Single-Crystal X-ray Structure Analyses of the A, B, and Z Helices or One Good Turn Deserves Another

Richard E. Dickerson,* Horace R. Drew* and Ben Conner*
Norman W. Church Laboratory of Chemical Biology
California Intitute of Technology
Pasadena, California 91125

"... ille sinistrorsum, hic dextrorsum abit,
unus utrisque error, sed variis illudit partibus."

(This to the right, and that to the left hand strays,
And both are wrong, but wrong in different ways.)

—Horace, *Sermonae* (Satires) II.3

Since the first proposal of a double helical structure for DNA by Watson and Crick in 1953, based on the superb fiber diffraction photographs obtained by Rosalind Franklin and Maurice Wilkins, great advances have been made in our understanding of DNA structure by fiber methods, by Langridge, Arnott, Sundaralingam, and others whose names and achievements are documented in the two previous volumes in this series. Yet there is a certain sense in which all of this work has been "wrong": not that it was incorrect, but that it was by necessity incomplete. Fiber diffraction photographs could yield the averaged structure for a particular type of DNA polymer, but could not reveal structural details at individual base pairs along the helix. This detailed level of information became possible only with the recent advent of single-crystal x-ray analyses of short, homogeneous DNA molecules; and the x-ray work in turn was made possible only by the development of efficient triester methods of DNA synthesis.[1] In the wake of this new synthetic technology there quickly followed analyses of several DNA oligonucleotides: *p*ATAT,[2] CGCGCG,[3] CGCG in its high-salt[4,5] and low-salt[6] forms, CGCGAATTCGCG,[7-11] a doubly-intercalated complex of daunomycin with CGTACG,[12] and more recently CCGG.[13] (Throughout this paper, DNA oligonucleotides will be identified only by base sequence, and will be assumed not to have a leading 5′ phosphate unless preceded by a *p*.) These structures have begun to yield a wealth of new information about the principles of DNA double helical structure, possible sequence-dependent structural modifications, and the nature of the B-to-A and B-to-Z helix transitions.

---

*Present address: Molecular Biology Institute, University of California, Los Angeles, California 90024.

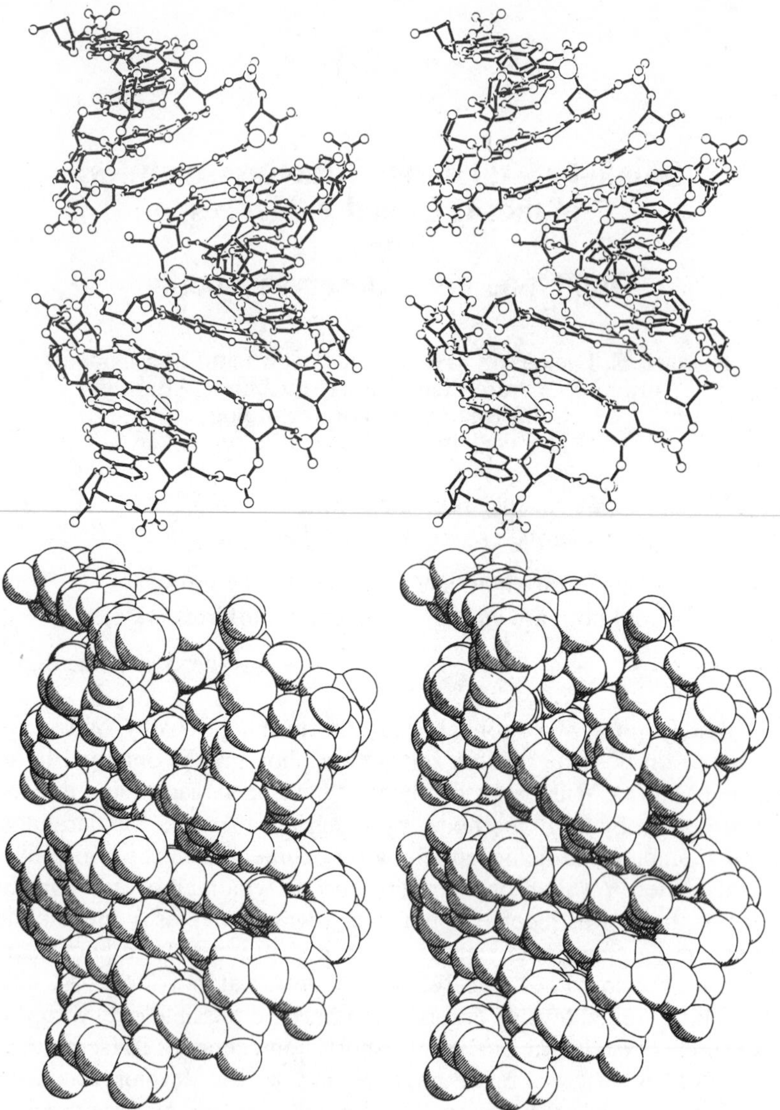

Figure 1. (a) Skeletal and (b) space-filling stereo drawings of A-DNA, constructed by stacking three identical CCGG double-stranded tetramers atop one another in proper helical registration. The structure of each tetramer is the same as in the crystal analysis, although the packing of molecules in the crystal is different. The helix has 10.7 base pairs per turn, and 2.3Åper base pair rise along the helix axis. Each cytosine on the outermost two base pairs of a tetramer has an iodine (large sphere) at its 5 position. Other atoms in order of decreasing size are: P, O, N and C. Note the depth of the major groove in (b), and the shallowness of the minor groove. From reference 13.

Figure 2. (a) Skeletal and (b) space-filling stereo drawings of B-DNA as established by the CGCGAATTCGCG dodecamer. The bend in helix axis is induced by crystal packing, and has little effect on local helix parameters. Base pairs 4 through 9 in the center of the molecule have an average of 9.8 base pairs per turn; the ends are slightly more variable. The vertical rise per base pair is 3.33Å. (a) from reference 10. *Figure, adjacent page.*

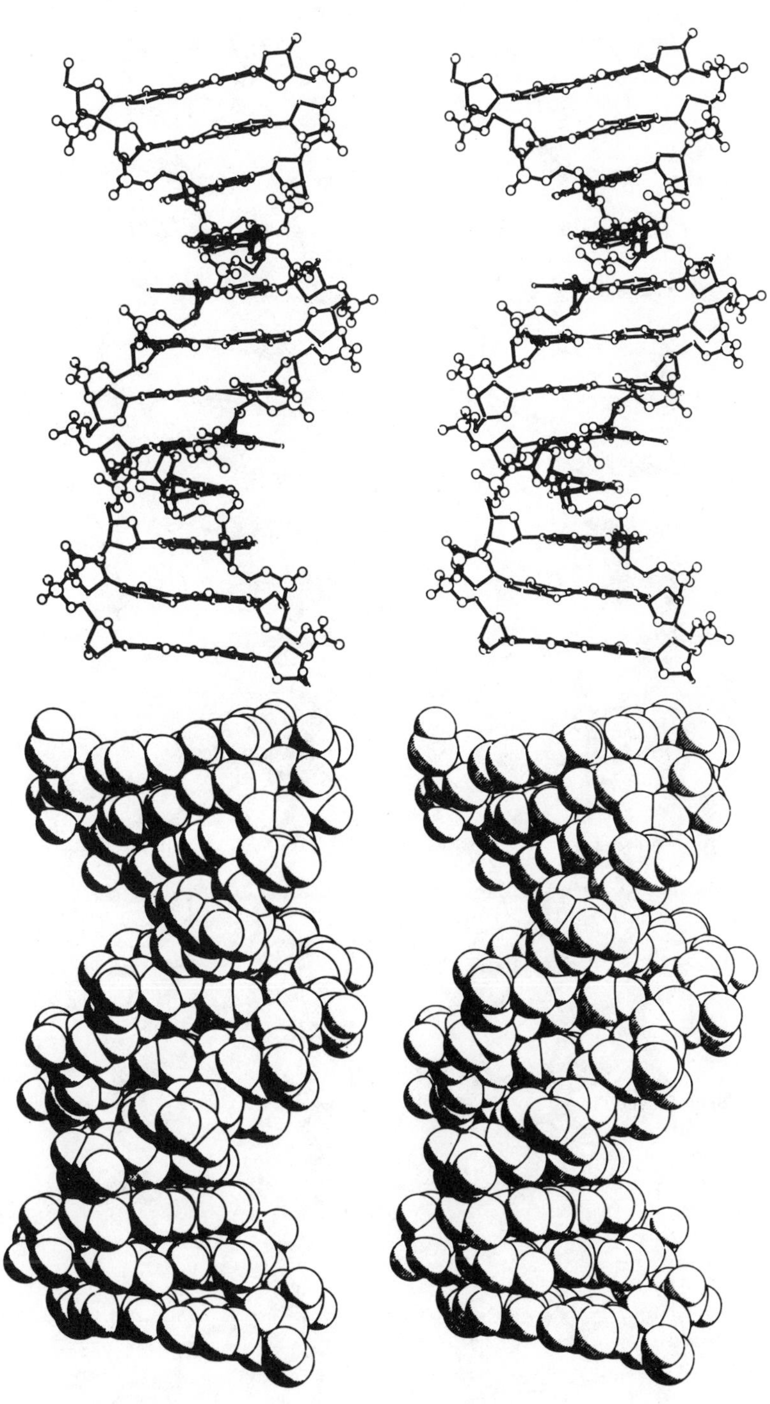

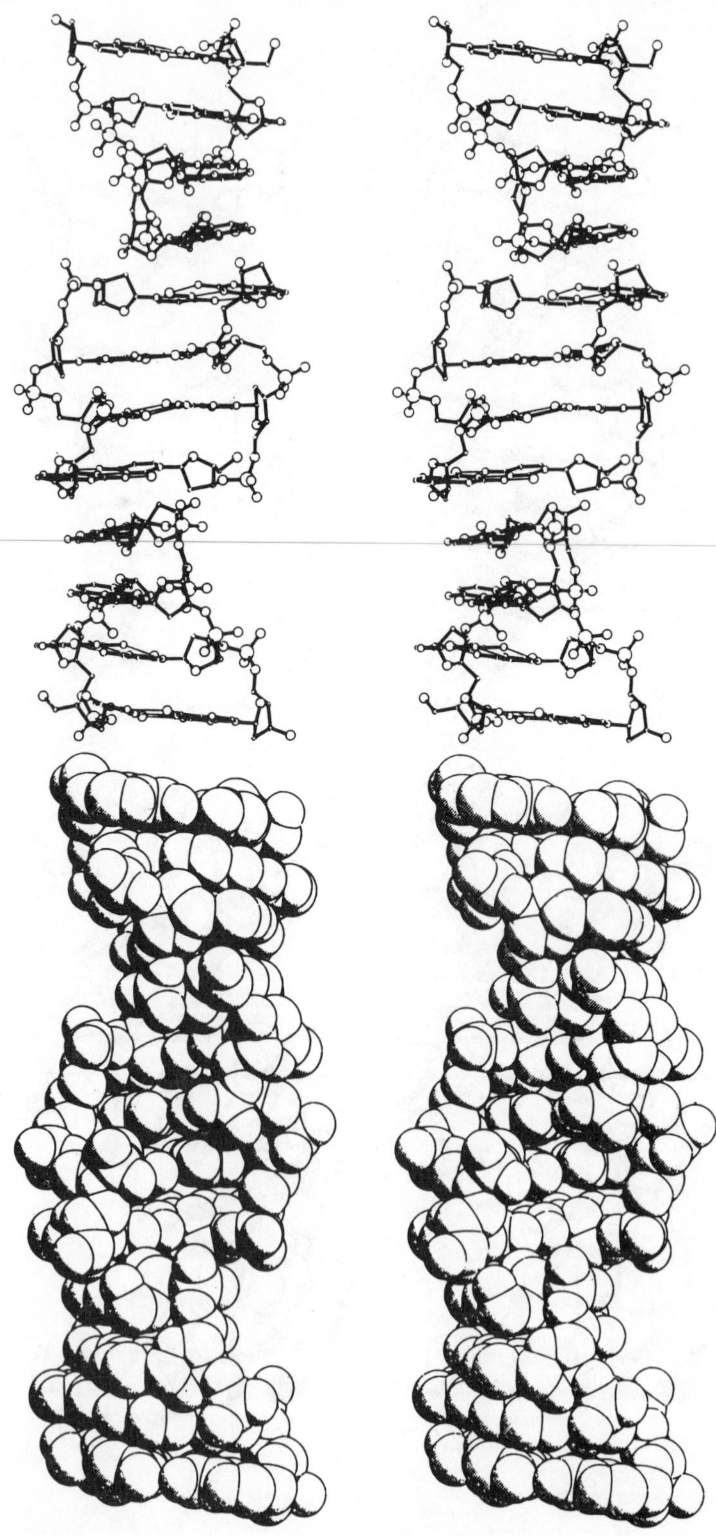

The three structure analyses carried out in this laboratory: CCGG, CGCGAA-TTCGCG, and high-salt CGCG, have turned out serendipitously to provide one example each of three known families of DNA helix: A, B and Z respectively. CGCGAATTCGCG has been refined at 1.9Å resolution to a 2σ R factor of 17.8%, CGCG has been refined at 1.5Å to R = 21%, and CCGG is being refined at 2.1Å resolution, with a current R factor of 20.5%. The three structures are depicted in ball-and-stick and space-filling representations in Figures 1-3. In each structure analysis the entire double strand, tetramer or dodecamer, has been the crystallographic asymmetric unit. For the dodecamer one complete molecule is shown. For the tetramers, three identical molecules have been stacked atop one another in such a way as to continue the helical repeat observed within the molecule, and thus to simulate a continuous helix. (The actual packing of molecules in crystals of the two tetramers is quite different.) No attempt has been made in Figures 1 and 3 to idealize the A and Z helices; each is built from molecules just as they are observed in the x-ray analysis.

Figure 3. (a) Skeletal and (b) space-filling stereo drawings of Z-DNA, constructed by stacking three identical high-salt CGCG double-stranded tetramers atop one another in proper helical registration. The structure of each tetramer is the same as in the crystal analysis, although the packing of molecules in the crystal is different. The helical repeating unit in this left-handed structure is two base pairs. There are six such per complete turn of helix, with a translational repeat along the helix axis of 7.61Å. Note the depth of the minor groove, and the fact that the "major groove" is hardly a groove at all. *Figure, adjacent page.*

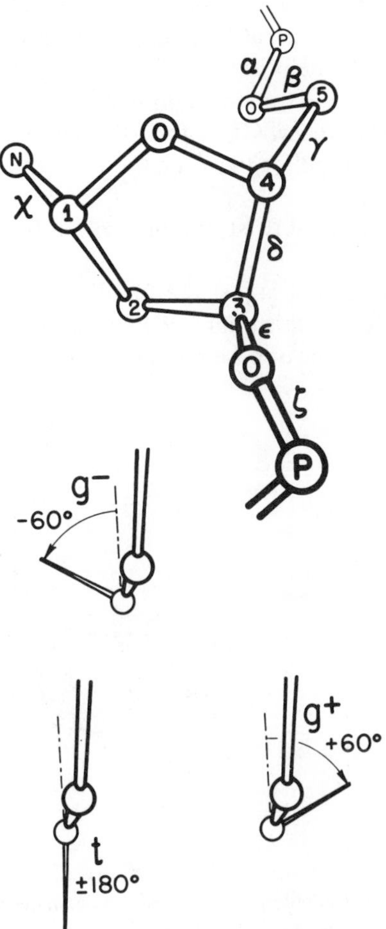

Figure 4. (a) IUB-IUPAC-recommended definition of main chain (α − ζ) and glycosyl (χ) torsion angles.[14] χ is defined by O1′-C1′-N1-C2 for pyrimidines and O1′-C1′-N9-C4 for purines. (b) Torsion angle sign conventions, and definition of the three staggered bond configurations: *gauche*$^-$, *trans,* and *gauche*$^+$. The main chain configuration in (a) is that of the B-DNA helix: g^-, *t*, g^+, *t*, *t*, g^-, and the sugar conformation is C2′-*endo*.

From examination of these three structures has come the beginning of an understanding of the relationships between the three families of helix: A, B and Z, and the transitions between them. This paper presents some of these interrelationships and our tentative conclusions about their significance.

Helix Conformation in B-DNA

More is known from single-crystal structure analysis about helix conformation in B-DNA than either of the other types, since the data include eleven independent base pair steps in the CGCGAATTCGCG dodecamer,[7-11] and one nonintercalated and unperturbed base pair step in the center of the daunomycin complex of CGTACG.[12] Sixteen parameters seem particularly useful in describing the structure of the helix:

a) Inclination of base pairs to the helix axis.
b) Displacement of base pairs away from the helix axis as viewed from above.
c) Rise per base pair along the helix axis.
d) P-P separation along one chain, and radial distance of phosphorus atoms from the helix axis.
e) The six main chain torsion angles (Figure 4).
f) Glycosyl C1′-N torsion angle χ (Figure 4).
g) Propellor twist, or the total dihedral angle between base planes in one base pair.
h) Helical twist, tilt, and roll of bases from one base pair to the next (Figure 5).

Helical twist is just the familiar rotation per base pair about the helix axis. 360° divided by this quantity gives an equivalent but frequently more graphic measure, the number of base pairs per turn of helix.

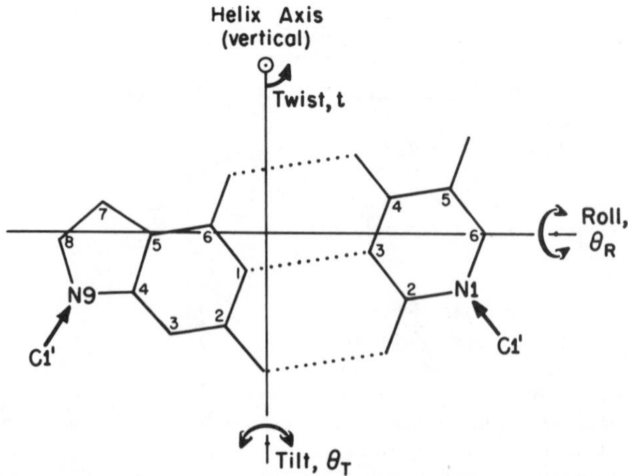

Figure 5. Definitions of the three rotational degrees of freedom of a base pair: helical twist, t, roll about the long axis, Θ_R, and tilt about the short axis, Θ_T. From reference 9.

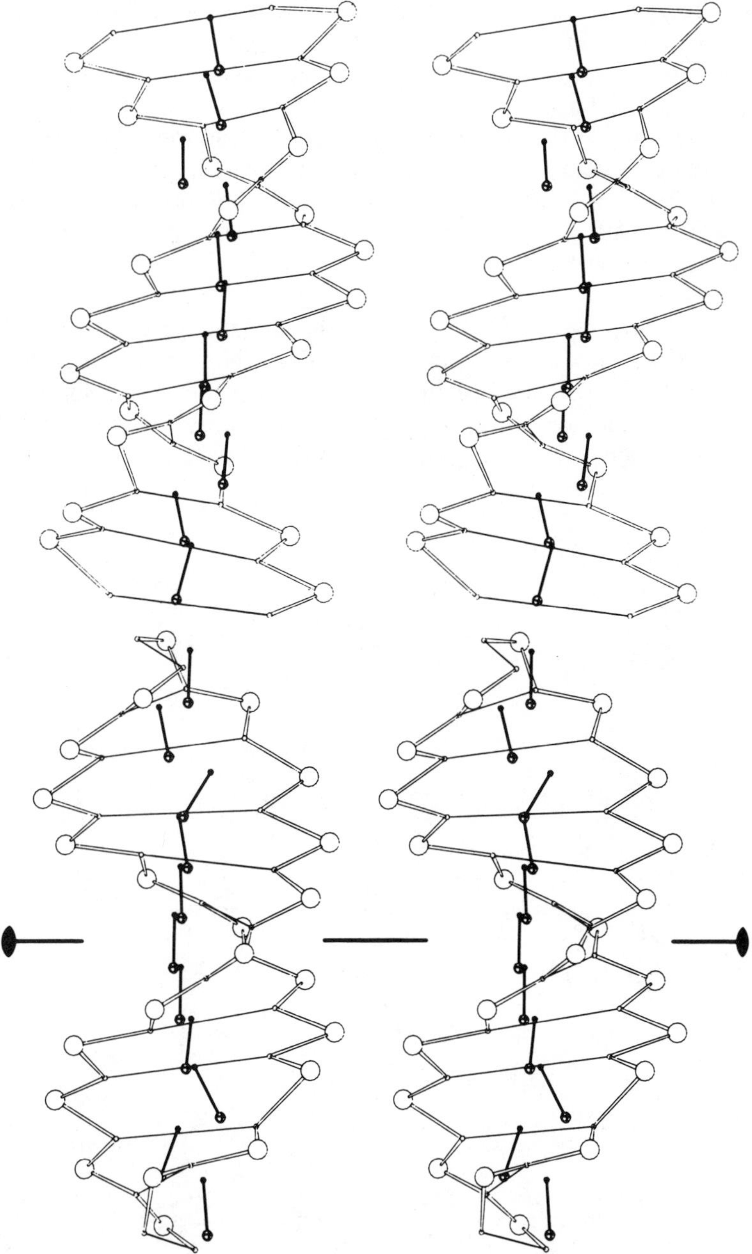

Figure 6. Local helix rotation vectors for successive base pair steps along the CGCGAATTCGCG dodecamer, plotted on a skeletal representation of the helix. Spheres are phosphorus atoms, re-entrant angles are sugar C1′ positions, and base planes are represented only by thin lines connecting C1′ atoms. If the helix axis were not bent, then the molecule would possess a lateral twofold symmetry axis, perpendicular to the page through the molecular center in (a), and horizontal in the plane of the page as indicated in (b). Note that the local helix vectors still show this twofold symmetry in spite of the axial bend. From reference 9.

The eleven individual helix rotation vectors that relate one base pair to its successor are plotted in Figure 6 on two skeletal diagrams of the dodecamer. The central five helical steps are quite regular, with their axes nearly colinear, with 9.76 base pairs per turn, and with the bases at -8° inclination away from perpendicularity to the helix axis. (The sign convention is chosen so the A helix has a positive inclination angle.) The helix axis passes between the hydrogen bonds of each base pair, as expected for B-DNA, and the mean rise per residue along the entire dodecamer is 3.33Å(0.013Å). (Numbers in parentheses following quoted parameters always are standard deviations.)

In contrast to the regularity of the center of the helix, the three outermost base pair steps on either end show a more complex behavior. The third step in from each end has its local helix vector tilted relative to the base planes and displaced into the major groove in a manner typical of A-DNA. Nevertheless, the uniform P-P separation along one chain, 6.68Å(0.23 Å), suggests that this is no more than an A-like

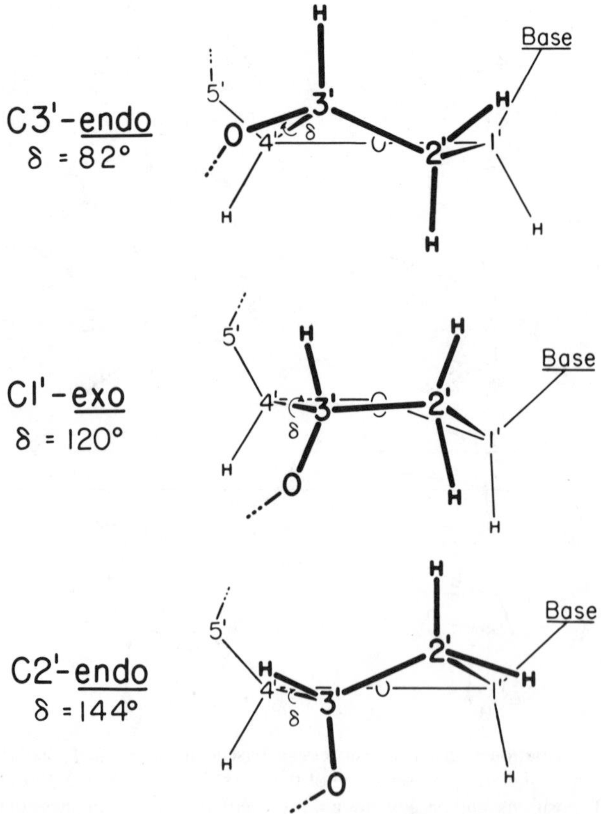

Figure 7. Relationship between deoxyribose sugar puckering and the C5′-C4′-C3′-O3′ main chain torsion angle δ. This torsion angle is a much more sensitive and easily observed measure of sugar conformation than is the appearance of the five membered ring itself. From reference 4.

variation within a B-helical framework.[9] An ideal A helix would have P-P separations of around 5.64Å.[15]

The observed local variations in helix properties at the outermost three steps must be properties of the dodecamer in solution, and not simply effects attributable to the crystalline environment. For, as Figure 6 shows, these local helix vectors reflect the twofold, end-for-end symmetry that is a consequence of the chemical identity of the two nucleotide strands. Overlapping of helix ends in the crystal produces a 19° bend that is obvious in all of the dodecamer stereos, and this bend is in such a direction that it formally removes the twofold symmetry. Yet the local twofold character of the individual rotation vectors remains. Hence the symmetrical variations in helicity illustrated in Figure 6 can only result from the particular sequence of bases themselves and not from outside influences, illustrating the strong effect that base sequence can have on helix structure.

The conformation of the five-membered sugar rings, one of the most discussed aspects of DNA helices, is most easily and most accurately measured by the C5′-C4′-C3′-O3′ main chain torsion angle δ, as illustrated in Figure 7. Examination of the final refined electron density maps of the dodecamer suggests that errors or uncertainties in main chain torsion angles are of the order of 10° or less, and this is quite sufficient to establish individual sugar conformation.[21] Main chain angle δ and the glycosyl torsion angle χ in a B helix are strongly correlated, as can be seen in Figure 8. χ and δ values are scattered along the best linear regression line: $\chi = 0.837\delta - 220°$, with a linear regression coefficient of 78%.[22] The two points for the central TpA step of the daunomycin complex (squares) fall within this same distribution.

Four aspects of the distribution in Figure 8 are structurally significant and perhaps unexpected:

a) The points are distributed over a broad range of angles, rather than being centered around the C2′-*endo* sugar conformation,
b) Torsion angles χ and δ are linearly correlated,
c) Purines (dark circles and square) tend to adopt higher values of χ and δ, closer to the classical C2′-*endo* conformation, than pyrimidines do (light circles and square).
d) Points for the purine and pyrimidine of one base pair tend to be spaced equally far to opposite sides of a common midpoint in χ and δ. This has been termed the "principle of anticorrelation".[8]

These are specific attributes of B-DNA, and not of double-helical DNA in general, and can be understood from reasonably simple structural considerations.[11]

The correlation of angles χ and δ follows from the particular geometry of the B-helix backbone, and the relative constancy of P-P distances along that backbone (Figure 9). As can be seen from Figure 2a, the C4′-C3′ bond vector is inclined

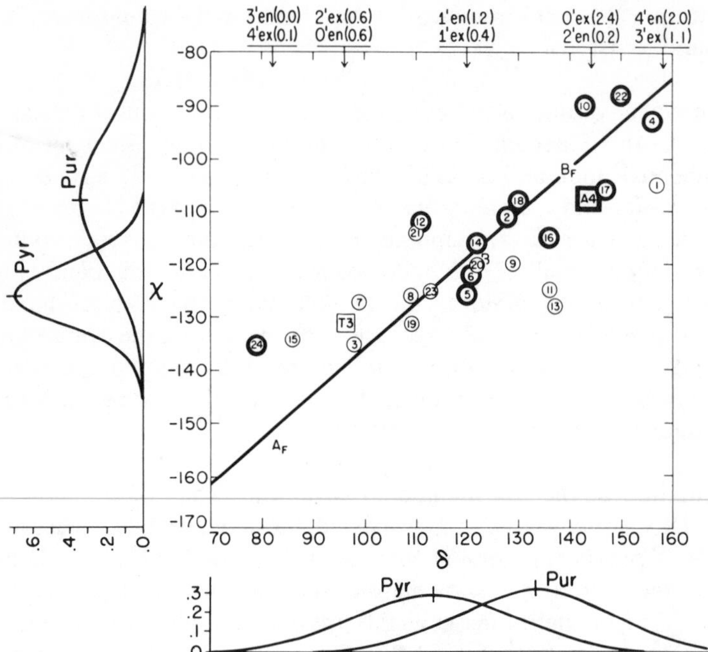

Figure 8. χ/δ torsion angle correlation plot for CGCGAATTCGCG (circles) and the central step of the daunomycin complex of CGTACG (squares). Normalized Gaussian distributions (i.e. – equal areas under the curves) have been fitted to the χ and δ values for purines and pyrimidines separately in CGCGAATTCGCG , omitting base pair 1/24 as representing an end effect. A_F and B_F represent classical A and B helix conformation as deduced from fiber diffraction data.[15] Notice the linear correlation between χ and δ, the preference of purines for higher χ and δ values than pyrimidines, and the anticorrelation within each base pair, or the tendency for points from one base pair to be spaced equally far to either side of the midpoint of the diagram. From reference 11.

Figure 9. Explanation of the correlation between χ and δ. If the O3′ to C5′ distance is relatively constant along the B helix, then a rotation of χ from (b) to (a) requires an opening-up of δ in order to continue to span the gap. See text for elaboration. From reference 11.

approximately 45° to the direction of the main sugar-phosphate backbone chain, so closing or opening torsion angle δ exerts a pull or a push, respectively, on the adjacent chain. (Had the C4′-C3′ bond been parallel to the main chain direction, then variations in δ would have had relatively little effect. This will prove to be important in Z-DNA.) Assume, in Figure 9, that the distance to be spanned between phosphorus atoms (or more immediately, between atoms O3′ and C5′) remains constant as χ and δ are changed. If the sugar ring is tilted as in Figure 9b, then the C3′-C4′ bond itself contributes toward bridging the gap, and torsion angle δ can remain relatively closed. But if the glycosyl bond is rotated to less negative values so that the sugar ring straightens up as in Figure 9a, then torsion angle δ must open in compensation in order to maintain a fixed O3′ to C5′ distance. In sum, the correlation between χ and δ in B-DNA arises because of the orientation of the C3′-C4′ bond relative to the main chain direction, and the relative constancy of P-P distances along the framework of a B helix.

The preference of pyrimidines for lower χ and δ values than purines is a consequence of tight steric contacts between sugar ring hydrogen atoms and substituents on the six-membered pyrimidine ring; contacts that are avoided by the more open geometry of bonding to the five-membered ring of purines (Figure 10). At $\chi = -120°$ as depicted in Figure 10, a marginally short contact exists between the C1′

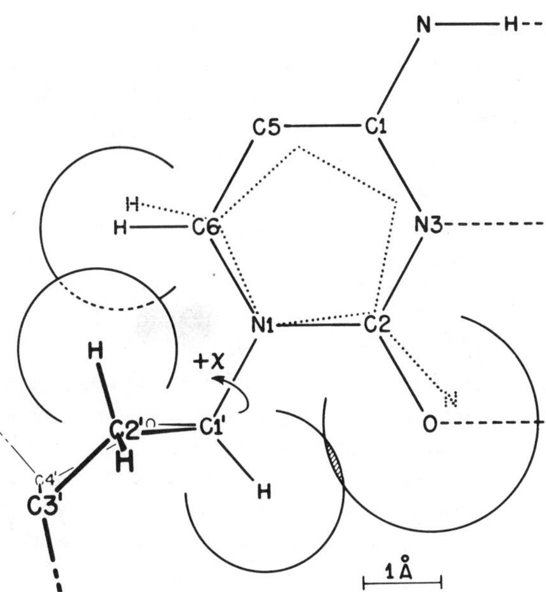

Figure 10. Explanation of the preference of pyrimidines for lower χ values. At $\chi = -120°$ as drawn here, a slightly close contact (shaded) occurs between the C1′-H on the sugar and the C2-O on the base. Rotation to higher χ values in the direction of the C2′-*endo* configuration brings about an even more severe clash between a C2′-H and the C6-H of the base. But rotation to χ lower than $-120°$ (more negative) relieves all short contacts. Purines (dotted outline) have none of these steric clashes, and can adopt the classical C2′-*endo* configuration. From reference 11.

hydrogen on the sugar and the C2 oxygen on the pyrimidine (2.2Å). Rotation to higher (less negative) χ values is prevented because it brings into contact the C2′ sugar hydrogen and the C6 pyrimidine hydrogen atom. Rotation to lower χ values, however, meets with no obstruction. Hence the pyrimidine χ values in the dodecamer are distributed in a narrow Gaussian envelope centered at $-128°$. In contrast, the extra room provided by the five-membered ring of a purine (dotted outline in Figure 10) permits purines to take up a broad distribution centered closer to the classical C2′-*endo* sugar conformation.

The principle of anticorrelation, or the tendency for members of one base pair to adopt conformations at equal distances to either side of the midpoint of Figure 8, is a consequence of the way in which sugar rings are linked to base pairs (Figure 11). The linkage is not linear; the tetrahedral bond geometry at C1′ means that the glycosyl bond, C1′-N, makes an angle with the best mean plane of the sugar ring. Hence, when sugar planes are rocked as from Figure 11a to 11b, the N end of each glycosyl bond dips down into the plane of the paper, inducing a downward motion of the upper or major groove edge of the base pair. Conversely, rocking the sugar rings in the opposite direction would raise the N ends of the glycosyl bonds out of the plane of the paper, and would lift up the major groove edge of the base pair. Because the two sugar-phosphate chains in the helix run in opposite directions, rocking both sugar rings in the same direction as viewed from outside the helix (Figure 11a to 11b, for example) means rotating the glycosyl angles in opposite direction, and rocking the sugar planes to the same extent means rotating the glycosyl angles by the same amounts. Hence the principle of anticorrelation is just an expression of the fact that a base pair and its attached sugar rings make up a linked, semi-rigid structural unit. If the base plane is tipped and the sugar rings are rocked about their glycosyl angles, they tend to rock in the same (external) direction and to the same degree.

This tandem rocking of sugar rings can be seen clearly in the stereo drawings of the eleven base pair steps of the dodecamer in Figure 12. In base pairs 1/24 and 2/23 the sugar rings are roughly perpendicular to the plane of the page. In 3/22 they are rocked forward (i.e., their near edges are moved toward the top of the drawing, as in Figure 11b), and at 4/21 they are perpendicular again. These are all alternating CG steps, and this alternation of rocking motion again becomes particularly clear at the other CG end of the dodecamer: 9/16 is vertical, 10/15 is rocked back (near edge toward the bottom of the page), 11/14 is rocked slightly forward, and 12/13 slightly back. In the AT-rich center of the helix these motions are less obvious, except that a relative rock-forward can be seen between the central steps 6/19 and 7/18.

This rocking of sugar rings has another observable consequence: Whenever the rings rock forward (as defined by Figure 11b), the major groove edge of the base pair dips into the plane of the page, and whenever the rings rock back, the major groove edge rises out of the page. As has already been mentioned, purines systematically adopt larger χ values than pyrimidines do. Careful examination of the stereos of Figure 12 shows that a consequence of this is that 5′-purine-pyrimidine-3′

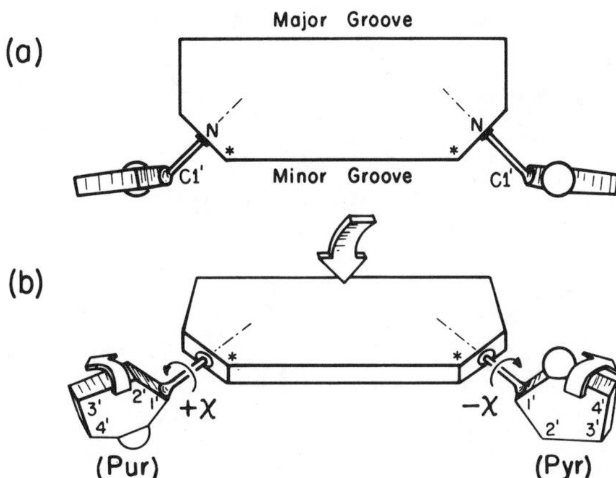

Figure 11. Linkage between base pair plane orientation, glycosyl bond angle χ, and sugar plane orientation, resulting from the fact that the glycosyl bond C1′-N is inclined at an angle to the best mean sugar plane. The principle of anticorrelation simply means that, when sugar rings rotate at one base pair, they do so in tandem as viewed from outside the helix, and to roughly the same extent. Rotation of sugar rings as at (b) also causes the upper or major groove edge of the base pair to dip into the plane of the page. See text for the consequences in terms of base roll angles. From reference 11.

steps along the sequence, such as 2/23 to 3/22 (GpC), 6/19 to 7/18 (ApT), and 10/15 to 11/14 (GpC), roll their two base pairs in such a way that the angle between them opens toward the major groove. Conversely, at pyrimidine-purine base pair steps such as the CpG steps at 3/22 to 4/21 and 9/16 to 10/15, the angle between successive base pair planes opens toward the minor groove.

This geometric effect was discovered empirically before its relationship to glycosyl bond angles was recognized.[9] Figure 13 shows the angle of relative roll of two successive base pairs about their long axes (as defined in Figure 5), plotted for the 11 base pair steps along the dodecamer. All three purine-pyrimidine steps, GpC at 2 and 10 and ApT at 6, open their roll angles toward the major groove (negative θ_R), and all of the pyrimidine-purine steps (of which only CpG are present in this helix) are more positive. We would predict this to be a general sequence-dependent property of B helices: purine-pyrimidine steps open their base toward the major major groove, pyrimidine-purine steps open toward the minor groove, and homopolymer steps (pur-pur and pyr-pyr) tend to resist rolling in either direction.[9,11] This observed or predicted behavior is illustrated for several sequences in Figure 14.

The downward trend of Figure 13 from left to right is the sole structural feature of the dodecamer that can definitely be ascribed to the 19° crystal-induced bend. As Figure 15 shows, at the upper end of the molecule (steps 1, 2, 3 ...) the bend in the axis is concave toward the major groove (M), and at the lower end (steps ...9, 10, 11) it is concave toward the minor groove (m). Pushing the top end of the helix to the left to produce the 19° bend has the effect of closing down the major groove there, or of opening up the minor groove side, making all roll angles θ_R more positive locally than they would otherwise have been in a straight helix. Conversely, pushing

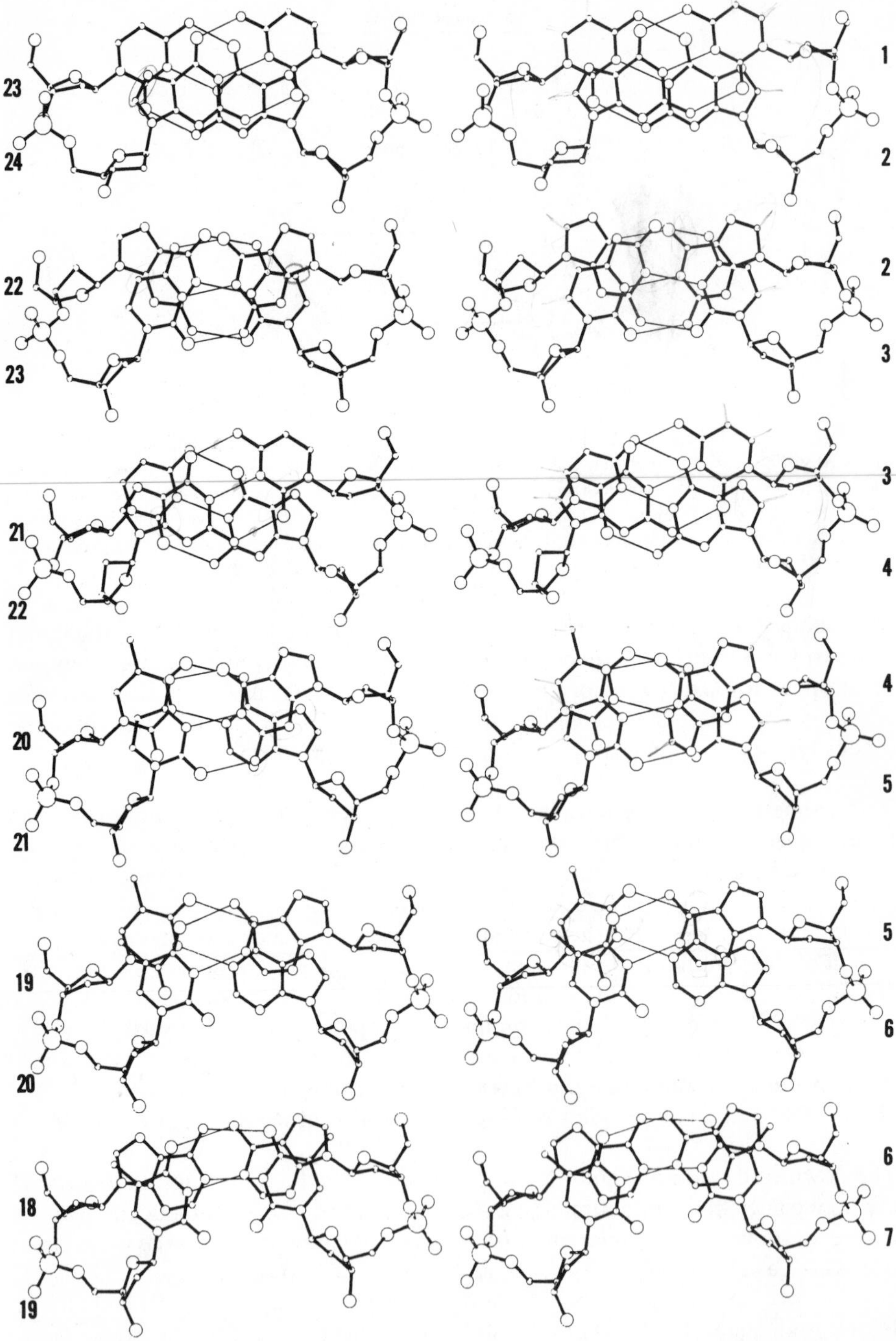

23 1

24 2

22 2

23 3

21 3

22 4

20 4

21 5

19 5

20 6

18 6

19 7

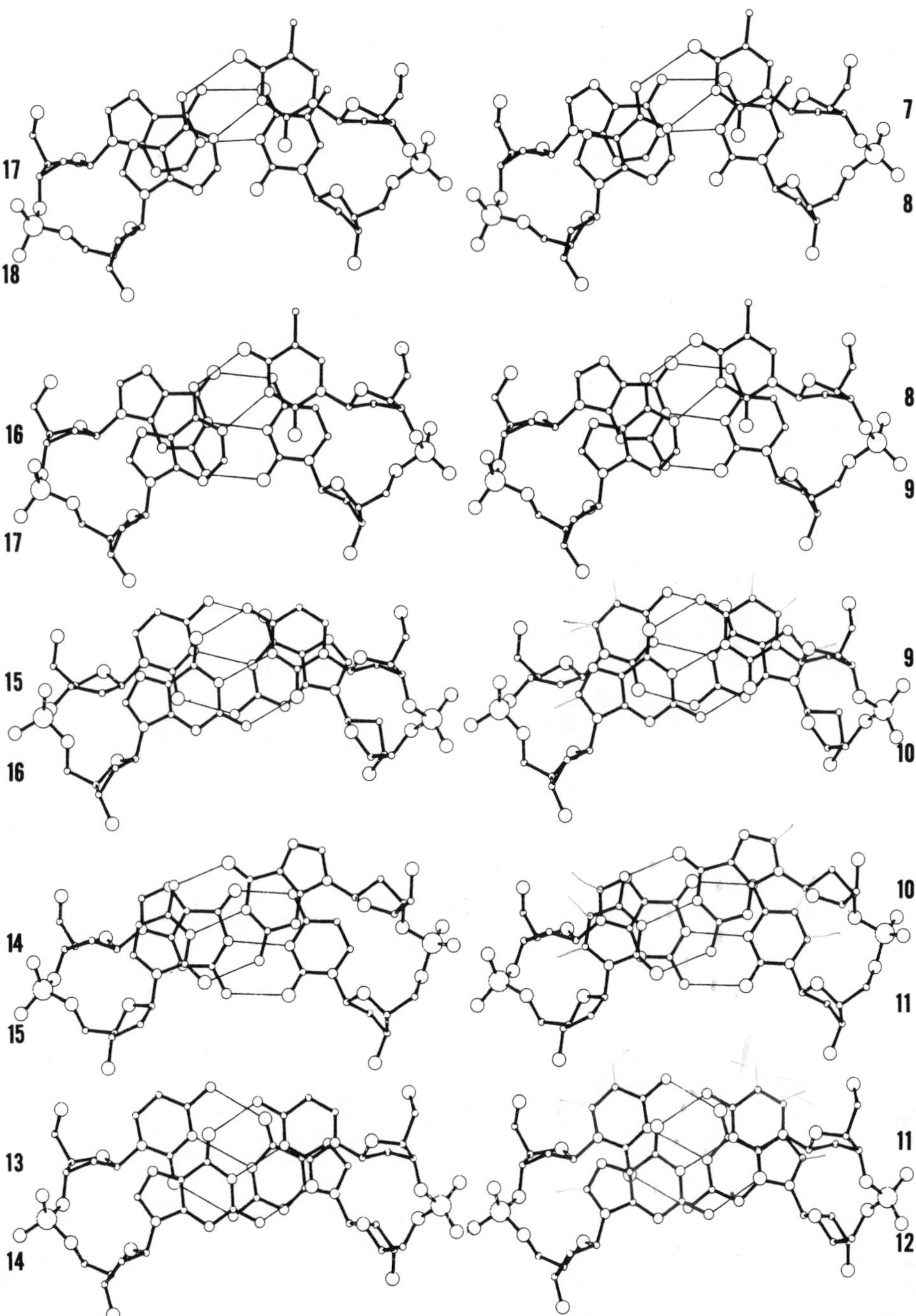

Figure 12. Stereo views of the eleven base pair steps in the CGCGAATTCGCG dodecamer. All of the geometrical relationships presented in Figures 8-12 and 14 are visible in (and in fact were derived from) these stereo views. Bases in strand 1 are numbered along the right, and those of strand 2 are numbered at the left. From reference 9.

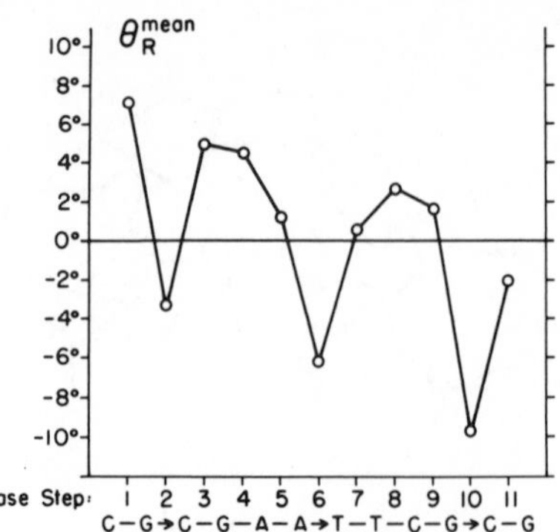

Figure 13. Behavior of the roll angle between best mean planes fitted to successive base pair steps, as a function of steps along the dodecamer helix. Note that all three purine-pyrimidine steps roll open the angle between base planes toward the major groove (negative θ_R), while the pyrimidine-purine steps tend to open toward the minor groove (positive roll angle). For an explanation of the downward trend from left to right in tems of the bend in helix axis, see text and Figure 15. From reference 9.

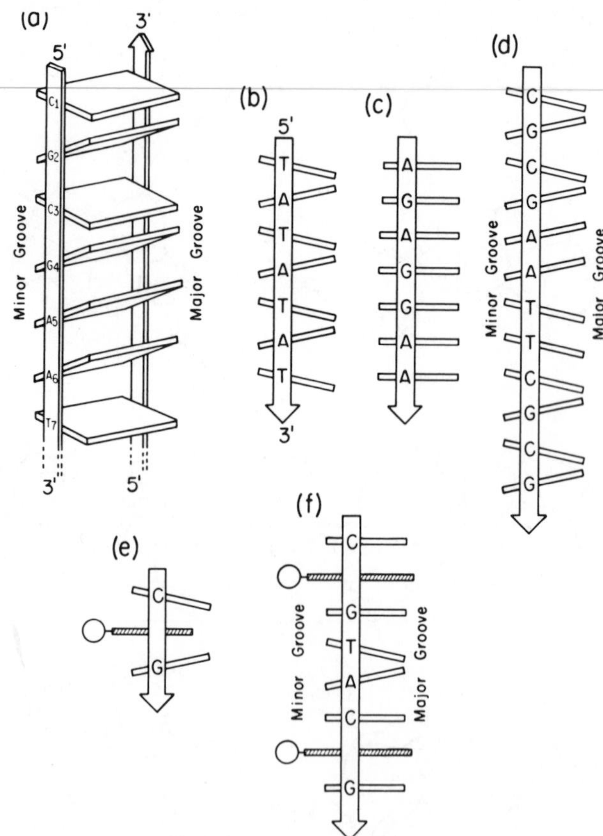

Figure 14. Unrolled-helix or Venetian blind diagrams of DNA helices, illustrating observed and expected base plane roll behavior, as deduced from Figure 13. (a) Observed behavior at the first seven steps of CGCGAATTCGCG, (b) Expected alternation of roll in an alternating purine-pyrimidine copolymer, (c) Expected parallel stacking in a homopolymer sequence, (d) The entire dodecamer, (e) Observed opening up of base pairs toward the minor groove in several intercalator complexes of r(CpG) and r(UpA), (f) Observed roll in the center of the daunomycin complex of CGTACT, and parallel base stacking around the fully-penetrating intercalator at the two ends. For discussion, see reference 11.

the lower end of the helix toward the left tends to close down all local roll angles on their minor groove side, making θ_R systematically more negative. Hence the left side of the graph in Figure 13 is raised, and the right side is depressed.

Aside from this one effect, the other local helix parameters are scarcely influenced by the 19° helix bend at all.[9] The B helix seems to be an inherently flexible object, capable of undergoing deformation in a smooth, unbroken and unkinked manner. The radius of curvature of this dodecamer in the crystal is 110Å, whereas that required for winding about a nucleosome core differs only by a factor of two: 45Å. Hence it could be imagined that nucleosomal winding, too, is carried out in a similar fashion.

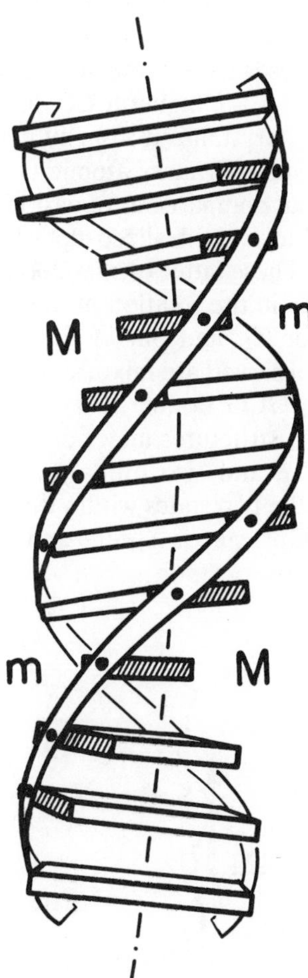

Figure 15. Ribbon-and-plank schematic of the dodecamer, to illustrate the effect that the helix axis bend has on local roll angles. At the top end, the major groove (M) is compressed, and all roll angles Θ_R are more positive than for a straight helix (see left half of Figure 13), whereas at the bottom end, the minor groove is compressed, and all roll angles are more negative (right half of Figure 13).

Helix Conformation in Z-DNA

The Z helix, which can be adopted by alternating CG polymers (and possibly mixed (G-T) (A-C) polymers also, as judged from fiber data[15,19]), is quite a different structure from B. Not only is it a left-handed helix; it has a *syn* glycosyl conformation at purines, with the sugar ring rotated toward the minor groove (*s* at left in Figure 16), in contrast with the *anti* conformation encountered at pyrimidines (*a* at right in Figure 16) and at both types of base in every other DNA double helix. This peculiar alternation of *syn* and *anti* glycosyl bond orientations gives the main chain pathway a zigzag appearance, as can be seen in Figure 17. The P-O5′-C5′-C4′-C3′-O3′-P main chain pathway past each guanine is effectively vertical (parallel to the helix axis), whereas the equivalent pathway past a cytosine is horizontal (tangential to the helix). In addition, the C4′-C3′ bond vector at guanine sugars is parallel to the overall main chain direction (Figure 18), whereas the C4′-C3′ bond at cytosine sugars is tilted at an angle of roughly 45° to the chain direction as in B-DNA. This will prove to be significant for the torsion angle behavior of the A helix.

X-ray analyses have been carried out on three Z helices: CGCG crystallized from 500 mM $MgCl_2$ ("high-salt CGCG"[4,5]) and 15 mM $MgCl_2$ ("low-salt CGCG"[6]), and CGCGCG crystallized from 10 mM $MgCl_2$ or $BaCl_2$[3,16], with and without spermine or spermidine. Primary atomic coordinates have been published only for high-salt CGCG, but comparisons of various salt forms of CGCGCG and of low-salt CGCG have led to two idealized coordinate sets for what are termed the Z_I and Z_{II} helices.[16] These latter two models are compared in Figure 19, and are seen to differ principally in the rotation of the phosphate group at the GpC step: into the minor groove for Z_I, and out of the groove for Z_{II}. The tetramers and hexamers as actually observed are mixtures of these two forms. One of the two GpC steps in the low-salt CGCG double helix is Z_I and the other is Z_{II}. In the comparison of CGCGCG structures under different salt conditions, one of the four GpC steps is always Z_{II}, and two other steps occasionally are so, with crystalline disorder between the two states within the same structure.[16] In high-salt CGCG also (Figure 20), one GpC step is Z_{II}-like and the other resembles Z_I.

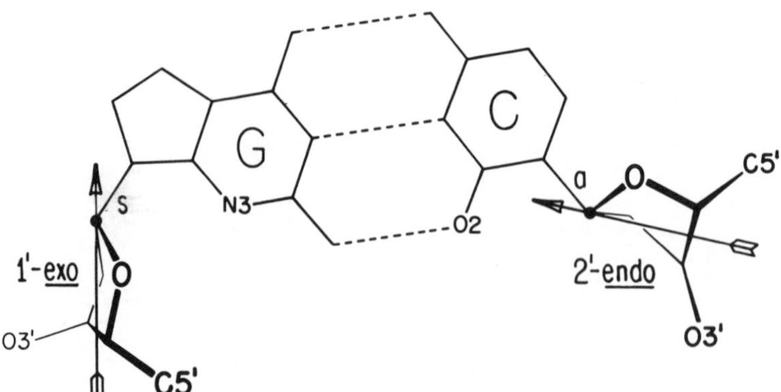

Figure 16. Comparison of guanine *syn* (s) and cytosine *anti* (a) glycosyl conformations in Z-DNA. Arrows indicate the direction of motion of the sugar ring (and base pair) if each O3′-C3′-C4′-O5′ torsion angle, δ, is closed while holding fixed the overall main chain pathway. At guanines the motion is only a rotation of one base out of the stack; at cytosines a translation of the entire base pair is required. This accounts for the difference in χ/δ behavior in Z-DNA and B-DNA.

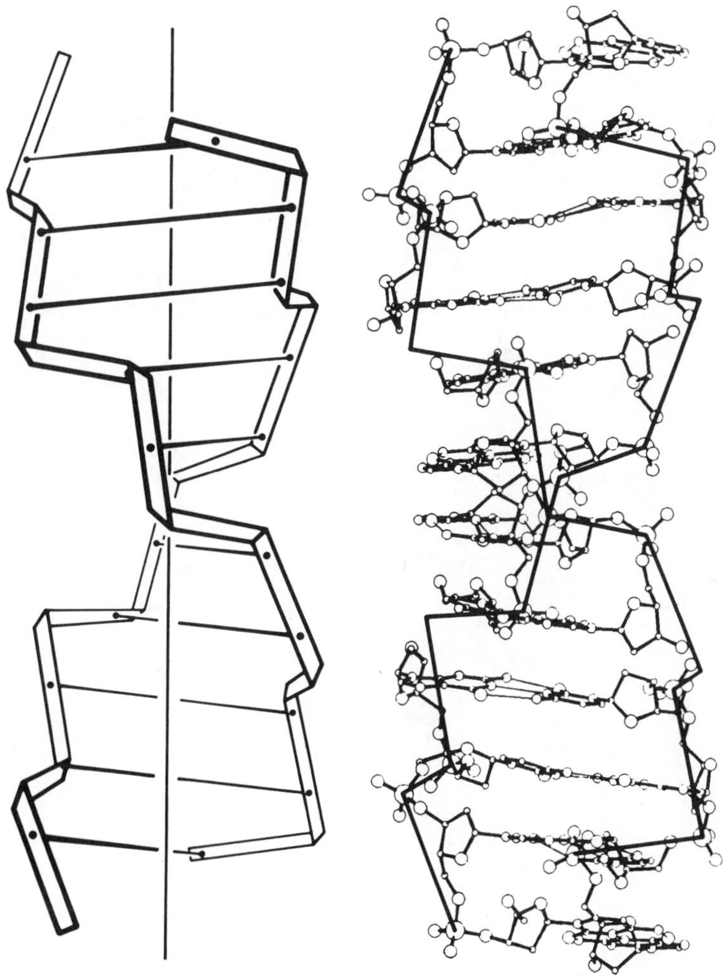

Figure 17. Stacking of three high-salt CGCG molecules to form a full turn of Z helix, with the phosphate-sugar-phosphate main chain marked by straight lines at right, and abstracted into a ribbon diagram at left. The path is vertical past guanines, and horizontal when going past cytosines. Adapted from reference 4.

In view of the intermingling of conformations within the same molecule, and even of intramolecular disorder, one can ask: are Z_I and Z_{II} genuinely distinct conformations, or are they only samplings of the continuum of conformation that are thermally accessible to a double helix of the Z type? Indeed, does this continuum of conformations also include the Z′-helix, which is somewhat different from Z in having C1′-*exo* sugars at guanines rather than C3′-*endo* (Figure 7)? (All Z helices have C2′-*endo* sugars at cytosines.)

χ and δ values are plotted against one another in Figure 21 for the eight observed bases in high-salt CGCG, and the four idealized base conformations of the Z_I and

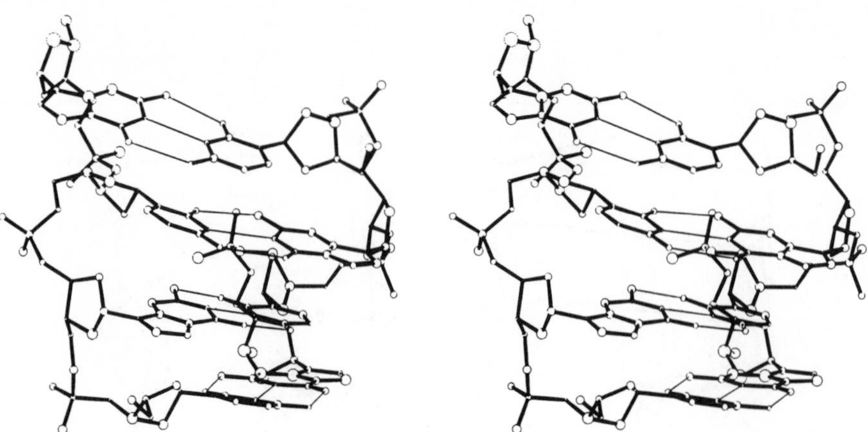

Figure 18. Structure of a single molecule of high-salt CGCG. This is a "thermal" diagram, in which the radius of each atom has been made proportional to the square root of its refined isotropic temperature factor, B. From reference 5.

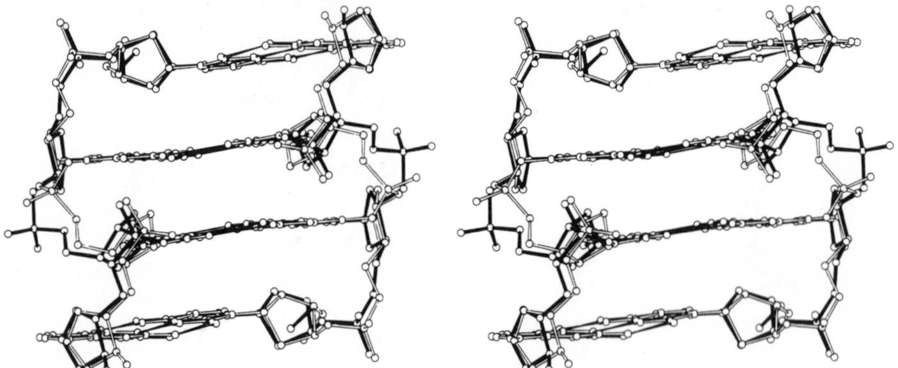

Figure 19. Superposition of CGCG tetramers generated from published coordinates for the idealized Z_I (open bonds) and Z_{II} (dark bonds) helices.[15] The principal difference is the rotation of phosphate groups at GpC steps in or out of the minor groove. From reference 5.

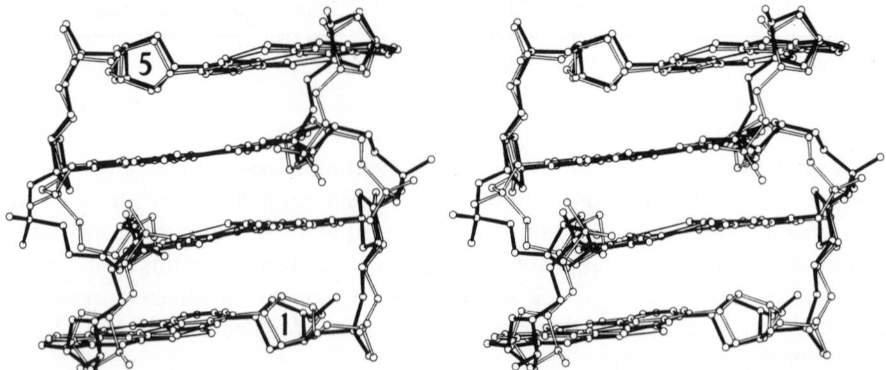

Figure 20. Superposition of the idealized Z_I tetramer (open bonds) and the actual high-salt CGCG structure (dark bonds). The GpC conformation at right in high-salt CGCG is that of Z_I, and that at left is more like Z_{II}. From reference 5.

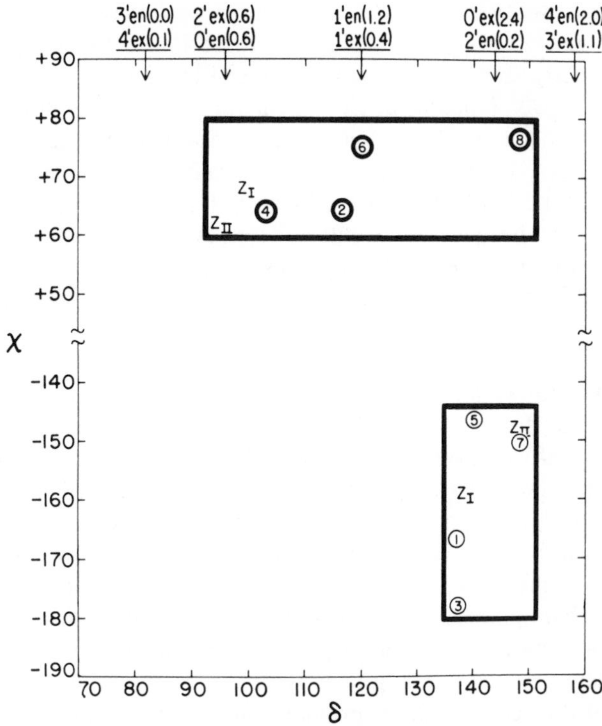

Figure 21. χ/δ torsion angle diagram for the Z helix. Heavy and light circles represent purines and pyrimidines as in Figure 8. Nubers 1-8 are bases in high-salt CGCG; Z_I and Z_{II} are idealized helices from reference 16. χ and δ are not linearly correlated in Z-DNA as they are in B-DNA. Instead, purines and pyrimidines occupy relatively restricted configurations as indicated by rectangular outlines.

Z_{II} model structures. Guanines and cytosines are observed to behave in opposite ways: for guanines (dark circles and attendant Z's) the range of χ values is restricted but the δ distribution is broad, whereas for cytosines (light circles and nearby Z's) the δ range is small but the χ variation is large. Unlike the situation in B-DNA, χ and δ values are not correlated at all.

The reasons for this quite different behavior of the Z helix can be understood by considering the special structural features of the helix. The tight contacts between the deoxyribose ring and guanine N3 produced by a *syn* glycosyl bond conformation limit χ to a small range to either side of the position in which the mean sugar plane is at right angles to the guanine ring plane (Figure 16, left). Tilting χ outside this range brings either one side of the sugar ring or the other too close to N3. In contrast, opening or closing δ at a guanine merely pushes the guanine ring back and forth along the direction of the arrow at the left side of Figure 16. If δ is decreased while the main chain is anchored in place, then the guanine ring is pushed up in the direction of the arrow; and if δ is opened up, then the guanine is pulled back in the reverse direction. This represents only a small slipping of one end of the base pair in or out of the helical stack (Figure 18), and is a relatively easy motion to accomplish. Unlike the situation in B-DNA, altering δ produces an insignificant change in overall P-P distance past a guanine, since the direction of the main chain is vertical: atom O5′ is directly above C5 at the left side of Figure 16, and a phosphorus atom lies directly below O3′. Opening or closing torsion angle δ does not alter the *vertical* separation between O3′ and C5′. Hence at guanines, χ is restricted by *syn* geometry whereas δ is relatively unrestricted. The C1′-*exo* sugar conformation seems to be favored in the interior of high-salt CGCG, away from end effects (points 2 and 6 in Figure 21), probably because of the chloride ions bound to guanine N2, and because this enables the outer atoms of the sugar ring to move as far as possible away from the guanine N3. The C2′-*exo*/C3′-*endo* conformation is reported for both the Z_I and Z_{II} helix models.

The ground rules for χ and δ at cytosines are entirely different. Now δ is restricted to a narrow range, for reasons that can be understood from Figure 16. The *anti* glycosyl bond conformation alters the direction of the arrow that represents δ-induced motion relative to a fixed main chain framework (Figure 16, right). The displacement now is a slippage of the entire base plane past its neighbors above and below *along its long axis.* Such a motion transmits the deformation to the other chain across the helix, and produces a more radical disruption of structure. Moreover, the main chain pathway past cytosines is horizontal when viewed as in Figure 16: both the P attached to O3′, and the O5′ and P attached to C5′, lie roughly in the plane of the page. Hence opening or closing δ directly affects the intrachain P-P separation, and this, as in B-DNA, provides another restriction on the free alteration of δ. As a result, the range of allowable δ values for cytosines is curtailed to the immediate vicinity of the C2′-*endo* conformation. In contrast, χ has what could be considered a "normal" range of variation, comparable with that for either purines or pyrimidines in B-DNA (Figure 8).

In summary, χ is restricted for guanines because of steric constraints of the *syn* glycosyl conformation, but δ can vary because it is relatively easy to slip one end of a base pair in or out of the stack. For cytosines, δ is restricted because it is harder to slide an entire base pair along its long axis, and because the overall intrachain P-to-P distance is not easily changed, whereas χ has a normal range of variation. The linear correlation of χ and δ encountered in B-DNA is not required in Z-DNA because of its different backbone geometry.

In view of the foregoing discussion, it appears that Z_I, Z_{II} and Z′ may not truly represent distinct conformations, but may be only statistical samplings of the full range of conformations available to the Z helix, represented to a first approximation by the outlined rectangles in Figure 21. Only more Z structures, hopefully involving A and T bases, will decide the issue.

Helix Conformation in A-DNA

Less can be said at present about A-DNA helix geometry from the CCGG structure since refinement at 2.1Å resolution is still in progress, with a current two-sigma

R-factor of 20.5%. Crystals were grown from 40% isopropyl alcohol, and this dehydration, plus the presence of a N2 amino group in the minor groove at every base pair (see next section) is probably the reason why CCGG adopts the A-helical conformation in these crystals. Every possible derivative brominated or iodinated at a single cytosine was synthesized (1-Br and 2-Br, 1-iodo and 2-iodo), but no derivative was crystallographically isomorphous with the native form or with another derivative. Hence the structure of the 1-iodo derivative, ^{I}CCGG, was solved using only anomalous scattering from iodine as a primary source of phasing information. This represents the first time that a purely anomalous scattering phase analysis has been used for oligonucleotides, and only the third time that it has been used for macromolecules, the previous analyses being of the proteins hemerythrin and crambin, by Wayne Hendrickson and coworkers.[17,18,23]

The structure of CCGG is shown in front and side views and down the helix axis in Figures 22-24. A stacking of three identical molecules along this axis has already been presented in Figure 1. Although helix parameters may change slightly as refinement continues, the helix appears to have 10.7 base pairs per turn and a rise per residue along the helix axis of 2.3Å. The bases are tilted $+19°$ away from perpendicularity to the helix axis. The two iodinated cytosines on the ends of the molecule appear to have an abnormally high rotation, that may not be typical of an ideal A-helix, but the two interior base pairs seem unperturbed.

A χ/δ torsion angle plot for this molecule is quite different from those of B-DNA (Figure 8) and Z-DNA (Figure 21). It would be premature to publish incompletely refined values, but with one exception all of them—for purines and pyrimidines alike—lie in a restricted region centered about the classical A-helix position marked by A_F in Figure 8, with a scatter of roughly $\pm 15°$ in χ and less than $\pm 10°$ in δ. The sole exception, G4, appears at the present stage of refinement to exhibit a C2′-*endo* sugar puckering, but this must be regarded as tentative until refinement is completed. This sugar ring is nearest the viewer at the lower left of Figure 24, and it is quite possible that its conformation is genuinely anomalous, since there is no continuation of helix backbone at its O3′ position to maintain the A structure.

Hydration in B-DNA, and Interconversion to the A and Z Families

Figures 1-3 show that the A, B and Z helices have quite different appearances: the A helix is short and fat, with a very deep major groove and a shallow minor groove. In some respects it resembles a cylindrically coiled ribbon rather than a helix of stacked base pairs. The B helix is thinner and more elongated, with major and minor grooves of comparable depth, although the narrowness of the minor groove gives it a sharper profile. The Z helix is still thinner and more elongated, with a cavernous minor groove and no major groove at all: the major groove edges of the base pairs are pushed to the surface of the helix.

Geometrical constraints on the backbone of the three helix types are just as different. The A helix has little room for variation in either χ or δ. In the B helix, more

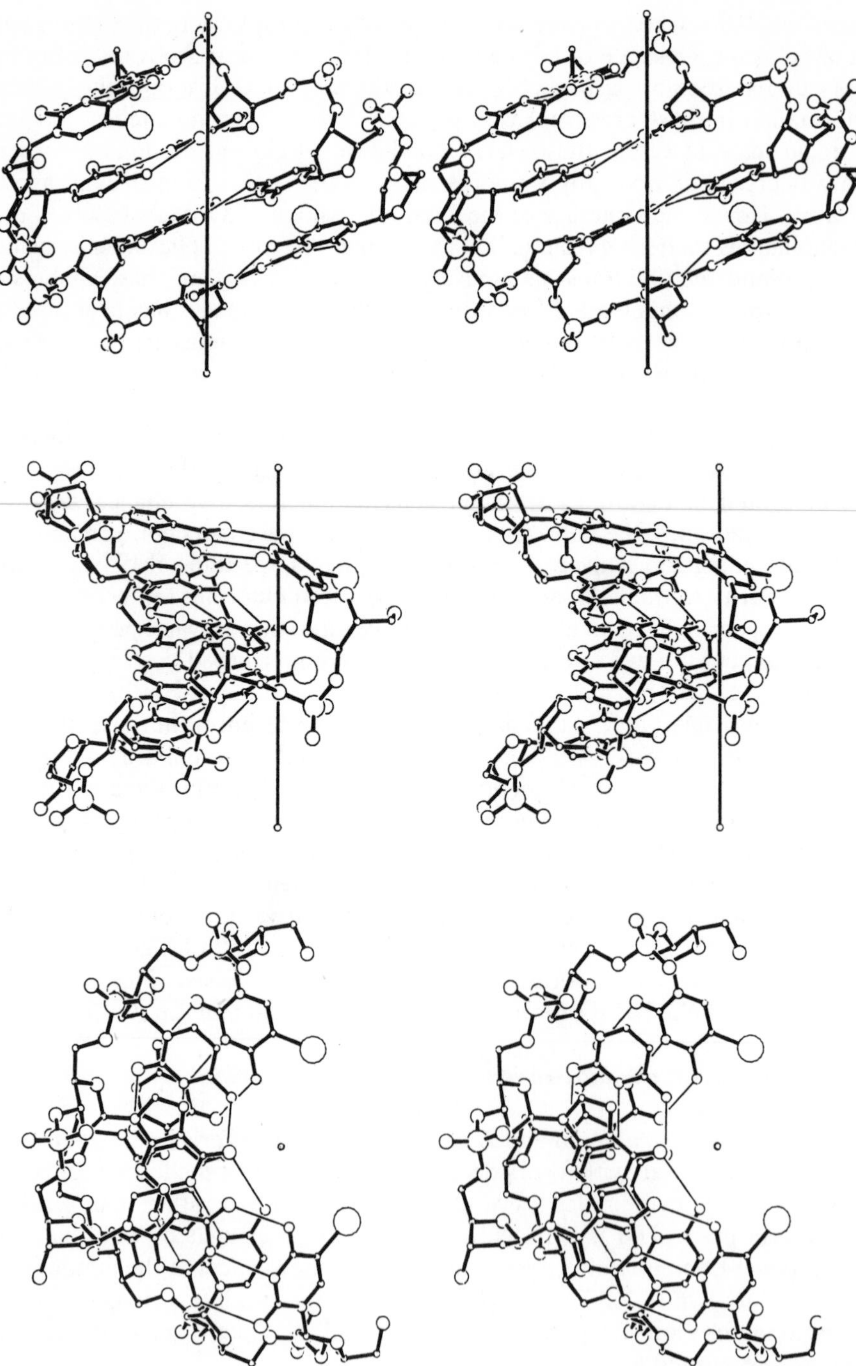

The respective legends are on next page.

motion is possible, but the changes in χ and δ are coupled and compensatory, canceling out one another's effects on the main chain pathway. In the Z helix, the two torsion angles are decoupled from one another, but each has its own restrictions when associated with one base type or the other.

One of the most interesting structural aspects of double-helical DNA is the transition from one of the above three families to another. The factors affecting such transformations have been reviewed in reference 4 (as well as many other places), and no attempt will be made to duplicate that discussion here. The interconversions are summarized in Figure 25. In brief, those conditions that favor a transition from B-DNA to another form, whether in the A or Z families, or to the C helix within the same family, seem to be conditions that lower the activity coefficient or effective concentration of the surrounding water molecules. This can be accomplished by lowering the *actual* concentration, as by drying or by replacing water with another solvent. It also can be accomplished by lowering the *effective* concentration of water molecules by tying them up with salt ions.

The hydration state observed for the CGCGAATTCGCG dodecamer has provided powerful clues as to the origin of the B to A helix transition. Water molecules were added slowly during X-ray refinement, by inspection of electron density and difference density maps, and incorporating new peaks as solvent only when they (a) appeared in both the density and difference density maps, and (b) occurred at reasonable coordination distances from other DNA or solvent atoms. In this way a network of solvent peaks was slowly built outward from the DNA molecule, and false or premature peaks hopefully were kept from influencing subsequent stages of the analysis. Rather than expressing partial occupancy in terms of reduced numbers of electrons, a standard eight electron value was used for each solvent peak, and the isotropic temperature factor, B, was refined as a measure of occupancy. The first water molecules added had B values similar to the DNA atoms to which they were bonded, and the search for solvent peaks was brought to an end when further addition would have required B values of 75 or above, equivalent to 1½ electrons or less of occupancy.

Eighty solvent peaks were identified by this process, numbered 25-104 (numbers 1-24 were used for base identification), and eight of these ultimately were reinterpreted as a linear spermine molecule spanning the major groove. Of the remaining

Figure 22. Major groove view of the A-helical tetramer CCGG, with the helix axis drawn as a vertical line. The largest two atoms, on cytosines, are the iodines used in anomalous phase analysis. Other atoms in order of descending size are: P, O, N and C. From reference 13.

Figure 23. CCGG viewed 90° to the left of Figure 22. Vertical line again is the helix axis. Note the propellor twist of base pairs. Part of the extreme twist of the two end iodocytosines may be a consequence of crystal packing. From reference 13.

Figure 24. Top view of CCGG, looking directly down the A-helix axis. Notice that this particular sequences causes all four guanines to be well stacked atop one another, while the cytosines are in a sense peripheral. From reference 13.

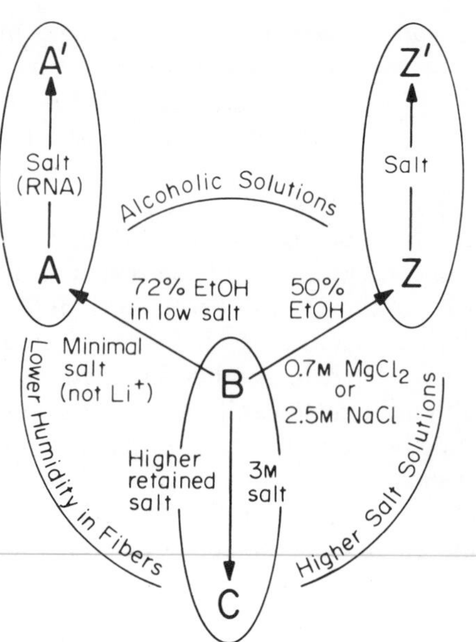

Figure 25. Salt, alcohol, and humidity-induced transformations between the three families of DNA helices. B-DNA is the high-humidity form. Low humidity or alcohol, under low salt conditions, drives B-DNA to the A form. High salt drives it to the C form which is another member of the B family. For alternating CG sequences, either alcohol or high salt can bring about a conversion to left-handed Z-DNA. From reference 4. (Note: The +2 charge ascribed to the lithium ion in reference 4 by the editors, not the authors, merely illustrates the fact that nature and *Nature* sometimes are at odds.)

72, the 50 most closely associated with the DNA molecule are shown in Figure 26. Four generalizations about these ordered solvent molecules can be drawn:

1) The solvent peaks appear to be water molecules and not magnesium ions, as judged from peak separations. This perhaps is not surprising, since in other nucleotide structures, localized divalent cations generally have been found only when they could bridge between phosphate groups in neighboring molecules, and the packing of dodecamers in the crystal provides no such opportunities.

2) Ordered hydration within the major groove appears to involve a single layer of water molecules associated with all of the polar O and N atoms on the base pairs that are not otherwise blocked by spermine. Further layers of hydration within this groove appear to be disordered, with few distinct, localized peaks.

3) Hydration of the phosphate backbone also appears to be generally disordered, with three notable exceptions: at three of the four thymines, a water molecule is immobilized in clathrate-like fashion between the 5-methyl group and a nearby phosphate. One of these "frozen" water molecules can be seen at the far right in the major groove in Figure 26b, and the other two can be seen "around the corner" of the major groove at the left.

4) Most importantly of all for this discussion, hydration of the minor groove is much more ordered, and extends for three or four shells or layers. The first two hydration shells build a zig-zag spine of water molecules down the floor of

the minor groove, and we believe that disruption of this spine is the first step in the B-to-A helix transition.

The actual locations of water molecules in the minor groove spine can be seen in Figure 27, and a linear, schematic representation is given in Figure 28. The first hydration shell water molecules form bridges between thymine O2 and adenine N3 atoms along the floor of the groove, connecting bases from adjacent base pair steps. The particular bases connected in this manner are those that are brought into closer proximity by the helical rotation. Figure 27 shows that in each strand, these are the second of the two bases contributing to the two base pairs. That is, in the two base pair steps A6/T19 and T7/A18, a water molecule connects bases T7 (the second base in strand 1) and T19 (the second base in strand 2). This particular bridge occurs at the very center of Figure 27.

These first-shell solvent molecules then are connected by a second shell, giving them each a local tetrahedral geometry, so that a zig-zag chain or spine of water molecules progresses down the minor groove. Third and fourth shell water molecules are arranged in a less regular manner, frequently bridging to a phosphate group on the rim of the minor groove. An unrolled-groove representation of the water spine is shown in Figures 28a and 28b: the former showing bond angles, and the latter giving bond lengths and temperature factors (effective occupancies). At the center of the molecule, the tetrahedral geometry around the first-shell waters is reasonably good, and the bond distances lie within an acceptable hydrogen-bond range. But the spine pulls away from the floor of the groove at either end, and dissipates in two great plumes of semi-ordered water molecules. Both the spine structure and its breakup at either end are represented in Figure 29.

Part of the reason for the destruction of the spine at both ends is the blocking of the minor groove by overlap from neighboring molecules, but we believe that the changeover from A/T to G/C base pairs at the ends of the molecule also is critical. As one moves from the center of the dodecamer helix toward either end, the spine begins to pull away from the bottom of the groove at steps G4/C21 and C9/G16, even before the overlapping ends of neighbor molecules are encountered. Figure 27 shows that the guanine N2 atoms interfere with the regular pattern of bonding to N3.

This is a fundamental difference between G/C and A/T base pairs. The former have an additional N2 amino group in the minor groove that is doubly intrusive (Figure 30): the $-NH_2$ physically displaces spine water molecules that would otherwise bind to N3, and also introduces a hydrogen donor into the groove, breaking the pattern of hydrogen bond acceptors N3 and O2 (pyrimidine). The spine of hydration should be stable in AT-rich regions and destabilized by GC sequences.

Leslie and coworkers[19] have made a systematic survey of the types of fiber diffraction patterns observed with DNA double-helical polymers of various repeating sequences. In all cases the B-helical pattern is the standard form at high humidity.

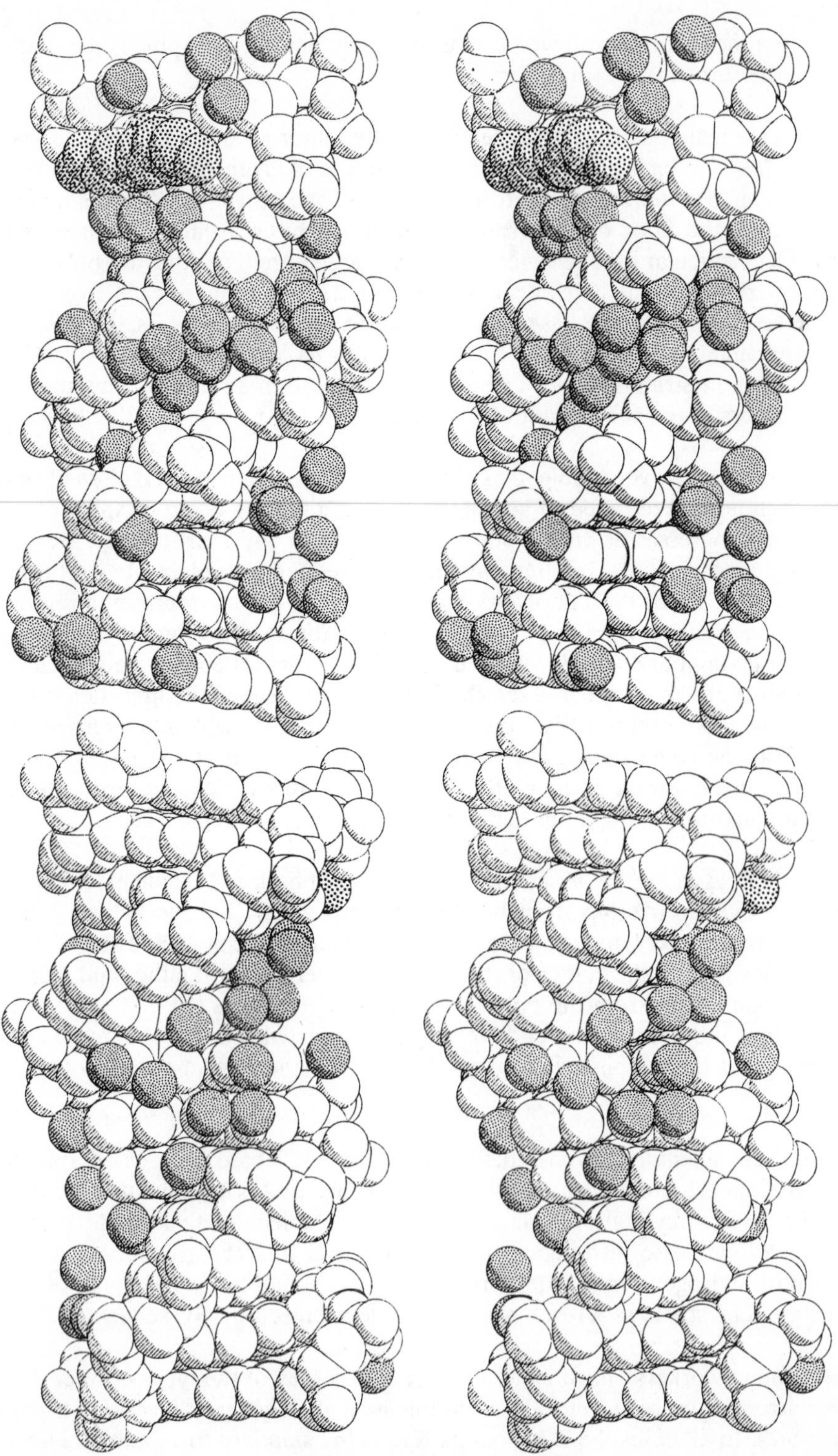

The respective legends are on next page.

Table I
Sequence Dependence of DNA Conformational Behavior

DNA Polymer	Purine N2 Amino Group Present?	Observed Helix Conformations
(A) (T)	No	B
(I) (C)	No	B
(A-I) (C-T)	No	B
(A-T) (A-T)	No	B, (A)
(I-C) (I-C)	No	B
(I-T) (A-C)	No	B
(A-A-T) A-T-T)	No	B
(A-I-T) (A-C-T)	No	B
(A-I-C) (I-C-T)	No	B
(I-I-T) (A-C-C)	No	B
(G) (C)	Yes	B, A
(A-G) (C-T)	Yes	B
(G-C) (G-C)	Yes	B, A, Z
(G-T) (A-C)	Yes	B, A, Z
(A-A-C) (G-T-T)	Yes	B, A
(A-G-T) (A-C-T)	Yes	B, A
(A-G-C) (G-C-T)	Yes	B, A
(G-A-T) (A-T-C)	Yes	B, A
(G-G-T) (A-C-C)	Yes	B, A

Notes: Adapted from Leslie *et al.*[19] The repeating unit in each polymer strand is given in conventional $5' \rightarrow 3'$ order within parentheses. "B" includes all members of the B family of helices: B, B′, C, C′, C″, D and E. "Z" includes the S helix. (A) indicates that the A form of poly(dA-dT) could not be obtained reproducibly; Pilet *et al*[20] have noted that the A conformation of this sequence is stable only over a narrow range of relative humidity.

But they found that some sequences could be converted to the A helix when the fiber was dried, and others could not. As Table I shows, with one questionable exception, every sequence that lacked N2 amino groups on purines (A and I) could not be converted from the B form to the A, and (again with one exception) every sequence that *did* possess N2 amino groups (on G's) could be converted from B to A, or to Z in the case of alternating purine-pyrimidine chains.

On the basis of the observed hydration state of the CGCGAATTCGCG dodecamer, and these fiber diffraction findings, we propose the following interpretation of the B-to-A helix transition:

1) The ordered spine of water in the minor groove of B-DNA helps to maintain the integrity of the deep and narrow minor groove, preventing it from opening

Figure 26. Stereo drawings of water molecules (finely stippled spheres) surrounding the CGCGAATT-CGCG dodecamer. The more coarsely stippled horizontal "caterpillar" is a spermine molecule, bridging the major groove at the upper end of the helix. (a) View into the minor groove. (b) Major groove view. The lack of water molecules within the minor groove at the very top and bottom of (b) arises because this part of the groove is involved with overlapping contacts with helices above and below in the crystal lattice. From reference 10.

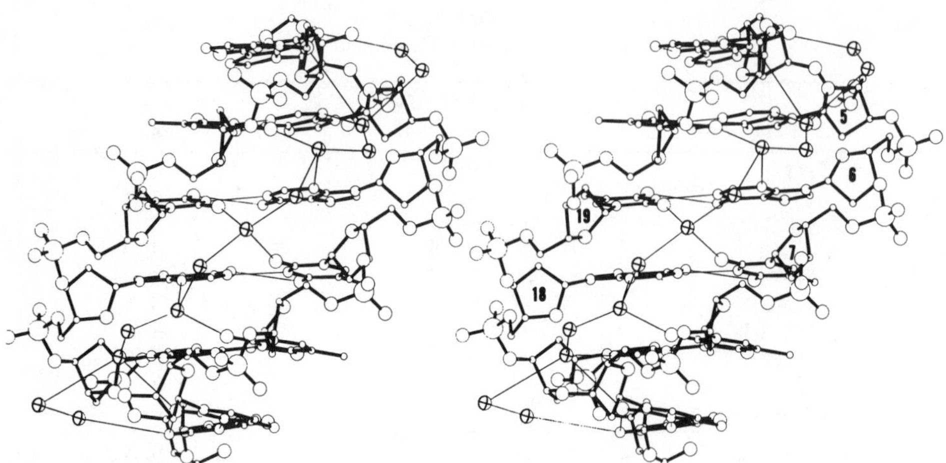

Figure 27. Stereo view of hydration in the minor groove in the AT-rich center of the dodecamer. Crossed spheres are water molecules, and a few bases are numbered for reference within their sugar rings. First hydration shell water molecules bridge O and N atoms from bases at the bottom of the minor groove, and these first-shell waters in turn are bridged by a second shell, giving them a local tetrahedral coordination. First and second-shell water molecules form a zig-zag hydration chain or spine down the minor groove. From reference 10.

up into the shallow groove associated with the A helix. (Compare Figures 1b and 2b.)

2) This spine is disrupted by N2 amino groups on purines, or by the introduction of guanine into the sequence. Hence polymers with high guanine content are easier to convert by dehydration from the B form to the A.

3) Because inosine has no N2 amino group, although it is otherwise identical to guanine, I/C base pairs behave like A/T pairs rather than like G/C. (See Table I.) Conversely, 2-amino adenine/thymine base pairs, if incorporated into a synthetic DNA polymer, should mimic G/C pairs in their disruptive effect on the spine and the B conformation.

4) The B-to-A transition is cooperative in nature not only because of the large structural changes involved, but also because disruption of the spine structure at any one point causes the ends to fray at either side, making it easier for disruption to propagate to adjacent sites.

The B-to-Z helix transition also appears to depend on water activity, and not on simple screening of phosphate charges by cation clouds (4). Providing that an alternation of purines and pyrimidines is present, the same conversion can be effected by high salt or by alcohol. High salt concentrations would tend to screen the repulsion by phosphate groups that are poised 8Å apart across the minor groove (center of Figures 3a and 3b), but alcohol would actually increase phos-

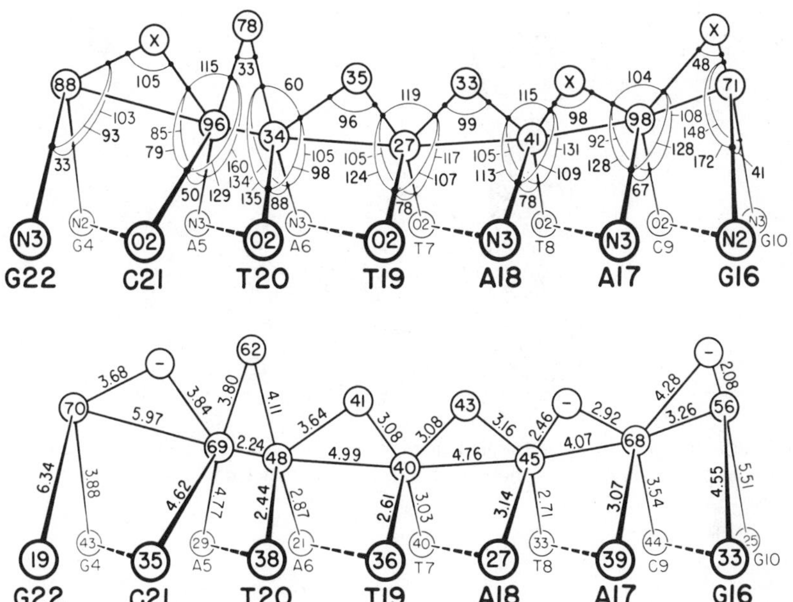

Figure 28. Unrolled-groove representation of minor groove hydration. Circles at bottom are the O or N atoms of base pairs, and the atoms labeled from left to right: 88-x-96-78-34-35 .... are the spine of water molecules. Bases are identified at the bottom of each diagram. Drawing (a) is labeled with atom identification numbers and bond angles; drawing (b) has refined isotropic temperature factors and bond lengths. Water identification numbers show the order in which they were incorporated into the structure analysis, and this corresponds generally with the order of increasing isotropic B values. Peaks with (x) and (−) were considered too low to include during analysis, but were found later to be present in the density and difference density maps when the spine structure was being examined. The spine pulls away from the floor of the minor groove at both ends; the distances are too long here to be considered bonds in any real sense. From reference 10.

phate repulsion by lowering the dielectric constant of the surrounding medium. The main effect that both high salt and alcohol have on the medium is a reduction in activity of water molecules. Apparently, when conditions are such that water is functionally in short supply, the B helix will slump into either the A or the Z, depending on sequence. The A helix effectively eliminates the minor groove; the Z helix eliminates the major. Perhaps a balance between the spine, maintaining the integrity of the minor groove, and the sheet of water hydrating the surface of the major groove, is critical in the preservation of the B structure.

Retrospect

Single-crystal x-ray structure analyses now are in hand for at least one example from each of the three families of double-helical DNA: A, B and Z. It is gratifying, and in a sense amazing, to see how well these crystal structure analyses have confirmed the earlier work based on fiber patterns. The Z helix, of course, was new; and the B helix in CGCGAATTCGCG was sufficiently long that it had its imme-

Figure 29. Interpretation of the dragon's spine found in the minor groove of B-DNA. The dragon's breath may have some connection with the dehydration necessary to induce a B-to-A helix transition.

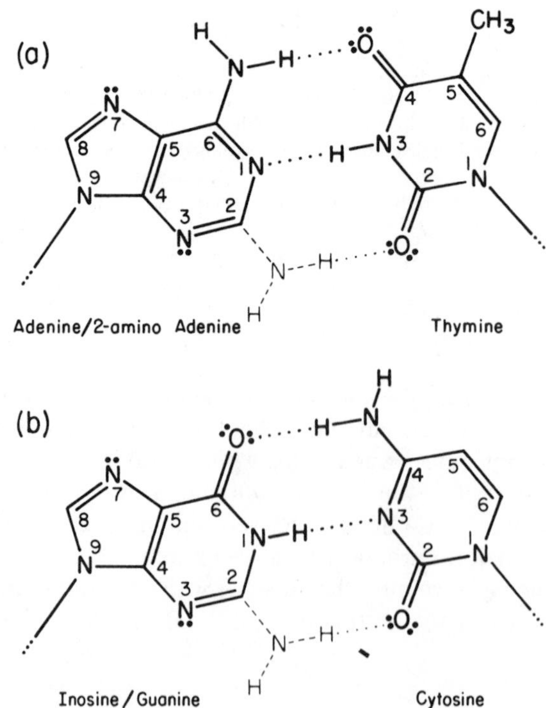

Figure 30. Influence of the 2-amino group on the geometry of the minor groove. Of the two purines separated by a slash at the left of each drawing, that to the left of the slash does not have the 2-amino group shown in faint dashes, while that to the right of the slash does have it. If the 2-amino group does produce a minor groove spine disruption that is responsible for the cooperative B-to-A helix transition, as maintained in the text, then I/C base pairs should behave like A/T (which is known to be true), and 2-amino-A/T pairs should behave like G/C (which has yet to be tested). From reference 10.

diate and unquestioned identity. But to solve the structure of CCGG, notice that its glycosyl angles and sugar geometry were those of A-DNA, determine the internal helical rotation and translation parameters, generalize these into an infinite helix, and then see the tetramer fit into a 10.7 base pair A helix like the final piece in a cosmic jigsaw puzzle, leaves one almost with a feeling of awe.

In a sense the intellectual *tour de force* in DNA structure analysis lay in the deduction of the correct structures from the small harvest of fiber diffraction data. Anyone can solve a structure, given sufficient data; it takes a greater leap of intuition to solve the structure with insufficient data. "Does this confirm or disprove Struther?" became a common laboratory query—"Struther" being a handy eponym for several independent laboratories and a quarter of a century of research. The conclusion, of course, is that the single-crystal work does indeed confirm that which was learned from fiber diffraction. But it advances the field one step further, into sequence-dependence, details of hydration states, and fine structure that only the thousands of reflections from ordered crystal diffraction patterns can provide. Nevertheless, a building is only as secure as its foundations, and in this case the foundations were well-laid.

Acknowledgements

We would like to express our appreciation to Tsunehiro Takano for his help and guidance during restrained least-squares refinement of these structures, and for his demanding standards as to what constitutes physically meaningful (as opposed to mathematically elegant) refinement of a macromolecule. Lillian Casler deserves much credit for giving tangible form to some of our less-coherently expressed ideas, in the form of the meticulous line drawings in this paper. Al Fratini and Mary Kopka have acted as continual sounding-boards, listening to our debates and pointing out some (perhaps not all) of the logical inconistencies in our arguments. Finally, Tony Wright must take credit (or blame) for finding the opening quotation from the other Horace. Financial support for this research was provided by National Institutes of Health grants GM-12121 and GM-24393, National Science Foundation grant PCM79-13959, and a grant from The Upjohn Company.

References and Footnotes

1. Itakura, K., Katagiri, N., Bahl, C.P., Wightman, R.H. and Narang, S.A, *J. Am. Chem. Soc. 97,* 7327-7332 (1975).
2. Viswamitra, M.A., Kennard, O., Jones, P.G., Sheldrick, G.M., Salisbury, S, Falvello, L. and Shakked, Z. *Nature 273,* 687-688 (1978).
3. Wang, A.H.-J., Quigley, G.J., Kolpak, F.J., Crawford, J.L., van Boom, J.H., van der Marel, G. and Rich, A. *Nature 282,* 680-686 (1979).
4. Drew, H., Takano, T., Tanaka, S., Itakura, K., and Dickerson, R.E. *Nature 286,* 567-573 (1980).
5. Drew, H.R. and Dickerson, R.E. *J. Mol. Biol. 151* (1981a).
6. Crawford, J.L., Kolpakj, F.J., Wang, A.H.-J., Quigley, G.J., van Boom, J.H., van der Marel, G. and Rich, A. *Proc. Natl. Acad. Sci. U.S. 77,* 4016-4020 (1980).
7. Wing, R., Drew, H., Takano, T., Broka, C., Tanaka, S., Itakura, K. and Dickerson, R.E. *Nature 287,* 755-758 (1980).

8. Drew, H.R., Wing, R.M., Takano, T., Broka, C., Tanaka, S., Itakura, K. and Dickerson, R.E. *Proc. Natl. Acad. Sci. U.S. 78,* 2179-2183 (1981).
9. Dickerson, R.E. and Drew, H.R. *J. Mol. Biol. 149,* 751-776 (1981).
10. Drew, H.R. and Dickerson, R.E., *J. Mol. Biol. 151,* (1981)
11. Dickerson, R.E. and Drew, H.R., *Proc. Natl. Acad. Sci. U.S.* — in press, 1981.
12. Quigley, G.J., Wang, A.H.-J., Ughetto, G., van der Marel, G., van Boom, J.H. and Rich, A., *Proc. Natl. Acad. Sci. U.S. 77,* 7204-7208 (1980).
13. Conner, B., Takano, T., Tanaka, S., Itakura, K. and Dickerson, R.E., *Nature* — submitted, 1981.
14. Altona, C., Arnott, S., Danyluk, S.S., Davies, D.B., Hruska, F.E., Klug, A., Lüdemann, H.-D., Pullman, B., Ramachandran , G.N., Rich, A. , Saenger, W., Sarma, R.H. and Sundaralingam, M. (1981). "Abbreviations and Symbols for the Description of Conformations of Polynucleotide Chains", IUB-IUPAC Joint Commission on Biochemical Nomenclature, Discussion Draft.
15. Arnott, S., Chandrasekaran, R., Birdsall, D.L., Leslie, A.G.W. and Ratliff, R.L. *Nature 283,* 743-845 (1980).
16. Wang, A.H.-J., Quigley, G.J., Kolpak, F.J., van der Marel, G., van Boom, J.H. and Rich, A., *Science 211,* 171-176 (1981).
17. Hendrickson, W.A. and Teeter, M.M., *Nature 290,* 107-113 (1981).
18. Hendrickson, W.A. and Smith, J.L. — private communication.
19. Leslie, A.G.W., Arnott, S., Chandrasekaran, R. and Ratliff, R.L., *J. Mol. Biol. 143,* 49-72 (1980).
20. Pilet, J., Blicharski, J. and Brahms, J., *Biochemistry 14.* 1869-1876 (1975).

21. This point cannot be emphasized too strongly: At less than atomic resolution, one cannot refine sugar rings sufficiently to establish ring puckering merely from the positions of the ring atoms themselves. But the extended pathway of the main chain can be established with high precision. If, as in our refinement, strong energy constraints are placed on bond lengths, somewhat more elastic constraints are placed on bond angles, and virtually no constraints at all are placed on torsion angles, then the torsion angle values must come from the (F_o-F_c) terms, or from the fitting of chain to the X-ray data. Under these circumstances, the sugar puckering can be inferred safely from observed torsion angle values. The torsion angles are the primary information; sugar conformations are secondary deductions.
22. $\chi = A\delta + B$, where

$$\sigma_\chi = \left[\frac{\sum\chi^2 - \frac{1}{N}(\sum\chi)^2}{N-1}\right]^{\frac{1}{2}}$$ and σ_δ is similarly defined.

$$A = \frac{\sum\chi\delta - \frac{1}{N}(\sum\chi)(\sum\delta)}{\sum\delta^2 - \frac{1}{N}(\sum\delta)^2}$$

$$B = \frac{1}{N}(\sum\chi - A\sum\delta)$$

R = linear regression coefficient. $= A\left(\frac{\sigma_\delta}{\sigma_\chi}\right)$

23. Seeman and coworkers have also used anomalous scattering from phosphorus to solve the structure of a dimer, r-(ApU), complexed with 9-aminoacridine, *Nature 253,* 324-326 (1975).

Proceedings of the Second SUNYA Conversation in the Discipline Biomolecular Stereodynamics
Volume I, ISBN 0-940030-00-4, Ed., Ramaswamy H. Sarma,
Adenine Press, New York, ©Adenine Press

DNA: Right-handed and Left-handed Helical Conformations

Alexander Rich, Gary J. Quigley and Andrew H.-J. Wang
Department of Biology
Massachusetts Institute of Technology
Cambridge, MA 02139

Introduction

Most of our knowledge of the detailed three-dimensional conformation of nucleic acids is obtained through x-ray diffraction studies. The earlier studies of macromolecular DNA were carried out on fibers, both of natural and synthetic origin. These studies have both advantages and disadvantages. One of the advantages was the simplicity of the experimental technique but a significant disadvantage was that there was only a limited amount of diffraction information. In general, it is impossible to solve the three-dimensional structure of a macromolecule from a fiber diffraction pattern without making a large number of simplifying assumptions as to the nature of the structure. In the case of DNA fibers, the amount of diffraction information available is quite limited and accordingly, many assumptions had to be made regarding the structure. The DNA structure was not solved in the same way that the three-dimensional structure of proteins are solved; rather the structure was guessed at, initially by Watson and Crick in 1953,[1] and it has been subjected to a series of refinement calculations in which the model was changed to make better agreement with the observed diffraction data. In any case, the diffraction data is limited both in the number of reflections and more specifically by the resolution. In no case was it possible to observe details of structure as most details had to be inferred or guessed at by using different models.

This situation has changed dramatically in recent years with the availability of oligonucleotides in quantities sufficient to carry out three-dimensional crystallization experiments. Crystals have the advantage that they usually diffract to higher resolution up to and including atomic resolution (less than 1Å resolution). In addition, there is a very large amount of diffraction data available from crystals and it is not necessary to make assumptions about models. The methods that have been developed for the macromolecular proteins can be applied to these compounds. In recent years a wealth of information has come from these studies. We are now in a position to see fine details of the B-DNA structure from the recent work of Wing *et al.*[2] This approach makes it possible to discover novel features dealing with the interaction of B-DNA with a number of molecules which are known to modify its activity. In addition, it also makes it possible to discover novel conformations of DNA. In the present article we discuss an example of each of these, one dealing with the manner in which right-handed B-DNA accomodates an anti-tumor antibiotic, daunomycin, and in the other case, an example in which a novel left-handed form of DNA was discovered.

Daunomycin and DNA

Daunomycin was the first antibiotic to be used in the treatment of leukemia in man.[3] Daunomycin (Figure 1) as well as the closely related adriamycin (14-OH daunomycin) are both widely used in the treatment of human tumors. Even though these substances are closely related to each other, they nonetheless have quite

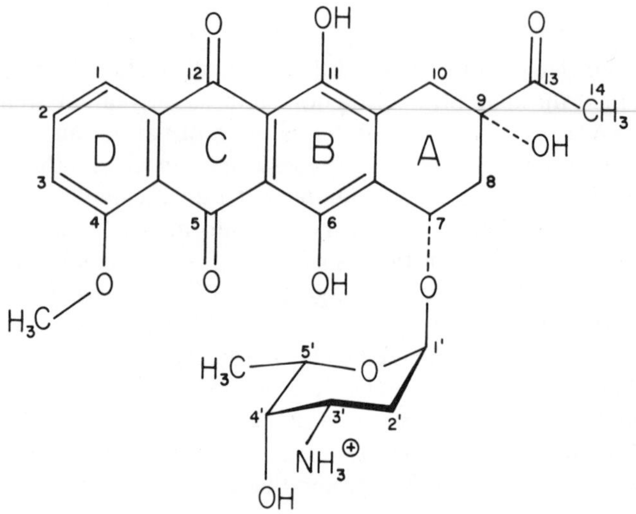

Figure 1. Daunomycin.

distinct biological activities. Daunomycin is used in the treatment of leukemias, whereas adriamycin is used in the treatment of solid tumors.[3] Many studies have been carried out concerning the manner in which these molecules interact with DNA.[4] They are known to intercalate in DNA, usually at a saturation level of one daunomycin per three base pair. Intercalation is believed to occur via the planar part of the chromophore involving the three unsaturated rings (B through D in Figure 1). Although frequent guesses were made about the manner in which these molecules interact with DNA, very little information was known up to the present study. In particular, it was not obvious why the fourth ring A in the chromophore (Figure 1) is unsaturated and has both equatorial and axial constituents. Attached to Ring A is an amino sugar. The clinical importance of these drugs is very great, and over 500 derivatives have been synthesized in an attempt to modify their biological activity.[3,4]

In an attempt to study the manner in which these drugs interacted with DNA, daunomycin was co-crystallized with a fragment of DNA, a hexanucleoside pentaphosphate d(CpGpTpApCpG). This fragment is self-complementary and it was found to form a 1:1 crystalline complex with daunomycin, producing attractive red crystals which diffracted to 1.5Å resolution. The crystals were in the space group $P4_12_12$ with dimensions a = b = 27.92Å, c = 52.89Å. The crystal structure was solved by molecular replacement and refined to an R value of 20%.[5] The structure was found to contain a double helix of right-handed B-DNA containing six base pairs, into which were intercalated two daunomycin molecules in between the deoxy CpG sequence. Upon refinement 80 water molecules appeared around the 6 base pair double-helical fragment. A two-fold rotational axis was found passing through the center of the duplex. The structure of the daunomycin oligonucleotide complex is illustrated in Figure 2 with space-filling diagrams. Figure 2a shows the structure of the DNA fragment without the daunomycin in it. It can be seen that the oligonucleotide is forming a fragment of right-handed B-DNA with the CpG segments at either end of the oligomer separated to allow the intercalation of the planar daunomycin chromophore. In Figure 2(a-c) there is a horizontal two-fold axis which lies in the plane of the paper in the middle of the molecule. In Figure 2b the chromophore is drawn and in the upper part the non-planar ring A can be seen with the 9-hydroxyl pointing down towards the middle of the molecule. At the bottom of Figure 2b ring D can be seen protruding into the major groove of the double helix with its methoxy group at the left. In Figure 2c the amino sugar is added to the chromophore and it can be seen at the top that it fits into the minor groove of the helix. It should be noted that the geometry of the daunomycin is nicely suited to intercalate into a right-handed double helix, since the manner in which the amino sugar comes off the chromophore has a handedness which fits a right-handed double helix. In Figure 2d the molecule has been rotated so that we are looking down the 2-fold axis. This shows the extent to which the amino sugars of the two daunomycin molecules fill the minor groove of the double helix with the positively charged amino group in the middle of the minor groove well separated from the two negatively charged phosphate groups on either side. The daunomycin molecule covers almost three base pairs, thereby accounting for the stoichiometry of DNA fully saturated with daunomycin.

Figure 3 is a view of the molecule looking down the helix axis which shows the manner in which the outer dCpG nucleotide sequence accomodates the shaded daunomycin molecule. It can be seen that the elongated chromophore skewers the base pair with ring D protruding well into the major groove of the molecule and ring A in the minor groove. It is worth noting that the base pair G2-C5* is displaced in the direction of the major groove relative to the base pair C1-G6*. This 1.3 Angstrom displacement is only one of the distortions associated with the binding of the drug. There is also an unwinding of the helix. Interestingly, this unwinding does not take place in the base pairs which surround the intercalator but instead, the unwinding occurs in the base pair one removed. The unwinding in the crystal structure is 8°, which is very close to the 11° which has been reported in solution.[6,7]

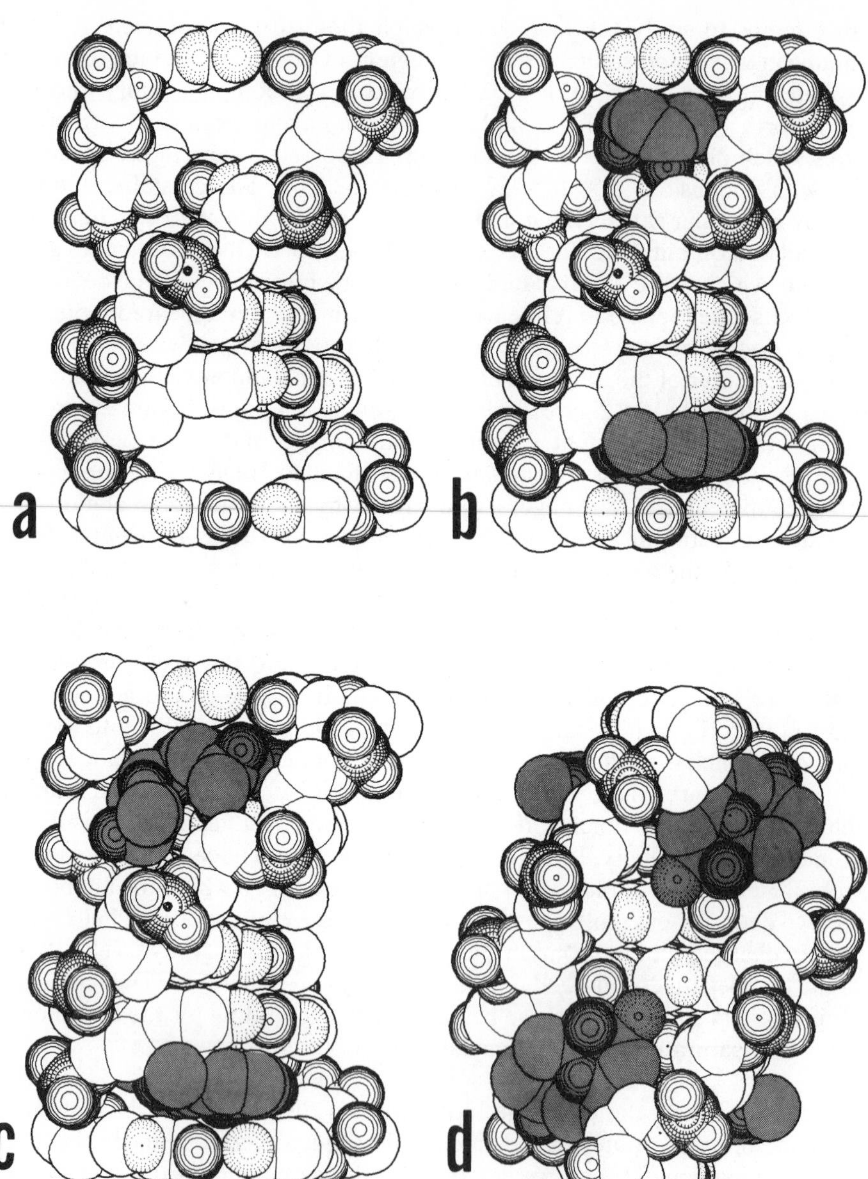

Figure 2. Space-filling drawings of the daunomycin-d (CpGpTpApCpG) complex. In a-c, the molecular 2-fold axis is horizontal and in the plane of the paper. (a) Right-handed helical hexanucleotide duplex by itself. (b) Hexanucleotide and aglycone without the acetyl group on C9. Note the way O9 projects down over G2 from the aglycone at the top. (c) Hexanucleotide and complete daunomycin. The amino sugar extends into the minor groove. (d) The complex as viewed into the minor groove down the molecular 2-fold axis perpendicular to the paper. The amino sugars and ring A fill most of the minor groove. The C4 methoxy group can be seen extending through into the major groove. Heavy stippling, daunomycin atoms; open circles, carbon; dotted circles, nitrogen; solid circles, oxygen; radial-spiked circles, phosphorus. Hydrogen atoms are not shown.

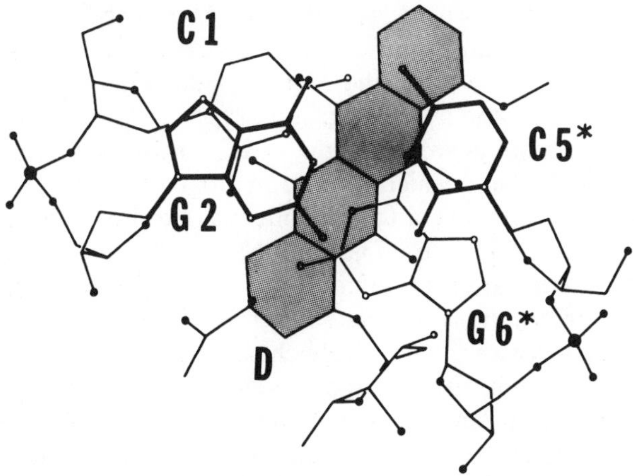

Figure 3. View of the intercalator perpendicular to the base plane. The daunomycin (D) ring system is stippled. The adjacent G2-C5* base pair closer to the reader is shown by thick lines, and the C1-G6* base pair further away is shown by thin lines. The two nucleotide backbones are different. Also note that the center of the G2-C5* base pair has moved up, toward the major groove relative to the C1-G6* pair. designates the complementary sequence.

Drugs Anchored to DNA

The detailed manner in which daunomycin interacts with the double helix is shown diagramatically in Figure 4, where we are viewing the double helix at a slight angle to the base pairs. It can be seen that there are hydrogen bonds which connect

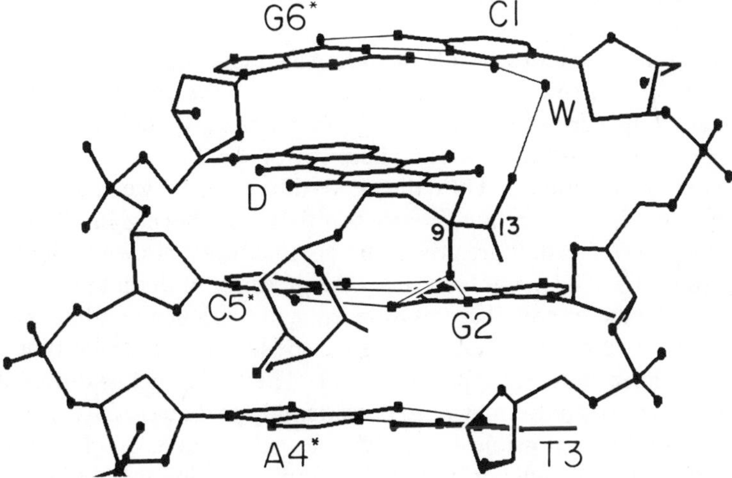

Figure 4. Diagram of daunomycin (D) intercalated into d(CpGpTpApCpG), showing intermolecular attractions. Note two hydrogen bonds between O9 of daunomycin and N2 and N3 of G2. In addition, water (W) forms a hydrogen bond bridge between O13 of daunomycin and O2 of C1. Oxygen atoms are shown as ellipses; nitrogen atoms are shown as squares.

substituents of unsaturated ring A with the base pairs above and below the intercalation site. Thus, the hydroxyl group on O9 forms a hydrogen bond by donating its hydrogen to the guanine N3 of the base guanine G2. This is a tight hydrogen bond (2.6Å). In addition, the hydroxyl group oxygen receives a hydrogen from the amino group of the same guanine forming a 2.9Å hydrogen bond. In addition, a water molecule (W) is found which forms a hydrogen bonded bridge between the 013 of the acetyl group and O2 of the cytosine residue above the intercalation site. The electron density of this water molecule indicates that it is a fully occupied site. It can be seen that association of daunomycin with the DNA double helix not only has base stacking interactions associated with the intercalation and electrostatic interactions associated with the positively charged amino sugar and the negatively charged phosphate, but it also has some highly specific hydrogen bonds which provide what is termed an "anchoring function" holding the antibiotic to the double helix.[5] Although good hydrogen bonds seem to be formed with the dCpG pair, they do not indicate great specificity. For example, if an adenine were present instead of guanine 2, the hydroxyl in position 9 could still form a hydrogen bond with N3 of adenine even though the second hydrogen bond to that hydroxyl group would not be formed. Similarly, the water molecule could form a hydrogen bond with bases other than cytosine 1, since all four bases have an electronegative atom very close to the position of the 02 of cytosine 1.

The specificity of the interactions of the anchoring function imply that these play an important role in the binding of daunomycin to the minor groove of the double helix. None of these interactions were anticipated by model-building studies, although it was realized that many of the substituents on ring A are important for binding. For example, an inversion of the two groups on the carbon C9 position of daunomycin results in a complete loss of activity.[8]

It is interesting to consider the effect of adding an additional hydroxyl group on the 14 position of daunomycin, which would convert it to the antibiotic adriamycin. If this hydroxyl group were added to the diagram in Figure 4, the hydroxyl would find a position below the carbon 14 atom where it would be within hydrogen bonding distance of O3′ of guanine G2. Thus, the major difference between daunomycin and adriamycin appears to be related to an additional anchoring function hydrogen bonding interaction. Crystals are now available of adriamycin with oligonucleotide fragments; thus, it will be possible to verify the presence of this additional hydrogen bond. It is interesting to note that this difference in hydrogen bonding is associated with considerable differences in the biological spectrum of these antibiotics as they are effective against different types of tumors. The basis for this differentiation is not clear, but this knowledge may have considerable practical utility. In view of this structure, it should be possible to modify the substituents on ring A in such a manner as to modify the anchoring function. For example, one could increase the number of hydrogen bonds holding the antibiotic to the DNA backbone, or to base pairs near the intercalating site. Because of the considerable differences in the activity spectrum of daunomycin and adriamycin, this may give rise to even further biological differences which result in some tumor cells being most sensitive to a particular derivative. This remains an active field for chemical modifications so that suggestions such as these derived from the three-dimensional structure of the complex may prove of considerable utility.

A large number of derivatives of these substances have been made,[3,8] and many of them have biological activities which would be anticipated from viewing of the structure. For example, removal of the methoxy group on ring A is associated with a slight increase of activity. Since the methoxy group does nothing but protrude into the wide groove of the double helix, it is not unreasonable that this should be the case. However, changing substituents on rings B and C may make it more difficult to intercalate and this may account for reduced activity associated with those derivatives. For workers involved in a synthetic program of drug modification, knowledge of the manner in which the drug interacts with its receptor (DNA in this case) is of great importance in developing rational drug design. This may now be pursued in the development of further modifications of daunomycin and adriamycin.

The Surprise of a Left-Handed Double Helix

Some of our earliest attempts to solve the three-dimensional structure of DNA oligonucleotides involved working with oligonucleotide fragments composed entirely of guanine and cytosine residues. This approach was carried out in order to find double helical fragments which had a greater stability. For this reason, studies were carried out on all of the self-complementary tetramers containing guanine and cytidine residues. Among those, the fragment d(CpGpCpG) produced good crystals. Accordingly, the series was extended by studying the structure of the hexamer d(CpGpCpGpCpG). It was found that this substance would crystallize in the presence of spermine (Figure 5) and $MgCl_2$ to produce an orthorhombic crystal in the space group $P2_12_12_1$ with a = 17.88Å, b = 31.55 and c = 45.58Å. This crystal diffracted to 0.9Å, and using the method of multiple isomorphous replacement it was solved and refined to an R factor of 13%.[9] The asymmetric unit of this structure was a double helix of DNA containing six base pairs plus two spermine molecules, a hydrated magnesium ion, and 62 water molecules. The structure was highly ordered and diffracted to atomic resolution, thus everything in the unit cell was visualized including all of the water molecules of solvation and ions. This was the first oligonucleotide which was analyzed at this resolution. Solution of the crystal structure revealed that it did not form the familiar right-handed B-DNA. Instead it formed a two-stranded double helix with antiparallel sugar phosphate chains and Watson-Crick base pairs between the bases, but it was twisted into a left-handed helix. Figure 6 has a van der Waals drawing showing the molecules as they are found in the crystal lattice, together with a van der Waals diagram of the right-handed B-DNA form of the same molecule. In this diagram a heavy line is drawn from phosphate to phosphate to aid in following the sugar phosphate backbone. Since the backbone has a zig-zag conformation in the left-handed form it was called "Z-DNA". From Figure 6 it can be seen that the hexanucleotide fragments have crystallized in the form of an infinite double helix. The bases are stacked upon each

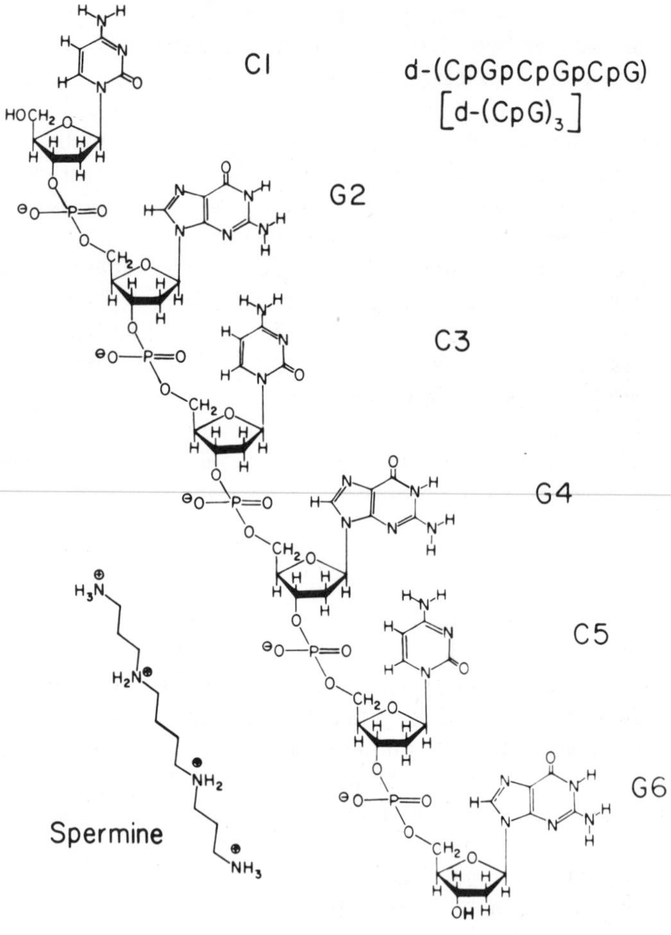

Figure 5. Chemical structure of the hexanucleotide pentaphosphate d(CpGpCpGpCpG) and spermine. Note that only the amino hydrogens of spermine are shown.

other in a continuous left-handed helical array with only every sixth phosphate group missing due to the fact that the oligonucleotide was a hexanucleotide pentaphosphate. Unlike B-DNA, Z-DNA has only one groove which is formally analogous to the minor groove of B-DNA. The atoms that form the concave major groove of B-DNA forms a convex outer wall of Z-DNA.

Cations and DNA Conformation

The existence of conformers of DNA with alternating sequences of guanine and cytidine has been known since 1972 when Pohl and Jovin[10] showed that raising the salt concentration of a solution of poly(dG-dC) resulted in a near-inversion of the circular dichroism in the solution. This inversion occurred with a midpoint of 2.7M NaCl or 0.7M $MgCl_2$. It has been shown that the Raman spectrum of the high

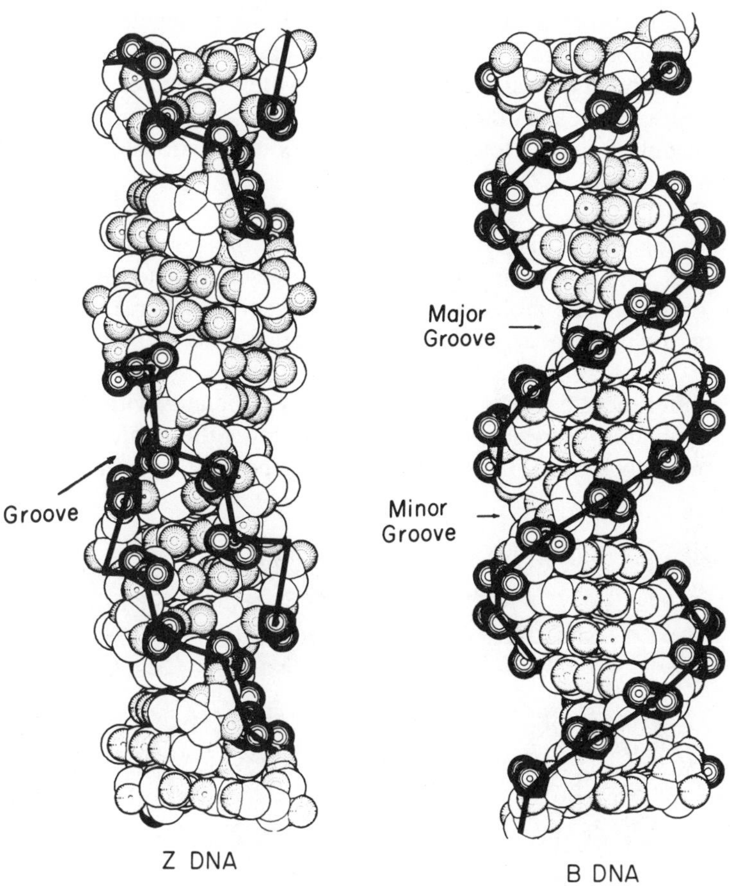

Figure 6. Van der Waals side views of Z-DNA and B-DNA. The irregularity of the Z-DNA backbone is illustrated by the heavy lines which go from phosphate to phosphate residues along the chain. This includes positions where the phosphate residues are missing in the crystal structure but would be occupied in a continuous double helix. The groove in Z-DNA is quite deep, extending to the axis of the double helix. In contrast, B-DNA has a smooth line connecting the phosphate groups and two grooves, neither one of which extends into the helix axis of the molecule.

and low salt forms of this polymer are different[11] and we have recently shown that the Raman spectrum of the Z-DNA crystals is identical to that of the high salt form.[12] The structural basis for this salt effect can be seen in Figure 6. When the Z-DNA conformation is adopted by a polymer of poly(dG-dC), the distance between phosphate groups on opposite strands is much smaller than it is in B-DNA across the minor groove of the right-handed double helix. Stabilizing the Z-DNA conformation requires a higher concentration of positive charges to shield the electrostatic repulsion between the negatively charged phosphate groups. When the phosphate-phosphate repulsion is shielded, the Z-DNA conformation is the stable one for this polymer. In low salt solutions, electrostatic repulsion between phosphate groups destabilizes the Z-DNA conformation.

In the crystal structure the molecules of the Z-DNA oligonucleotide fragments are packed in a hexagonal array (Figure 7). The molecules do not touch each other, but rather are surrounded by a sheath of hydrated water molecules which separate the oligonucleotides from each other continuously. The spermine molecules are also in this intermolecular region. However, the same conformations form in the absence of spermine molecules.[13]

Figure 8 is a schematic diagram which illustrates the form of Z-DNA. The anti-parallel chains are shown, and the serrated edge represents the position of the phosphate groups in the molecule. The helical repeat is 44.6Å. There are twelve base pairs per helical repeat, six of the dinucleotide units dCpGp.

The major differences between B- and Z-DNA is associated with the conformation of the deoxy guanosine residues. Figure 9 illustrates the conformation of deoxy-guanosine in Z-DNA and B-DNA showing two important differences. The guanine

Figure 7. Z-DNA packing. A view down the c-axis of crystals of the spermine, magnesium salt of d(CpGpCpGpCpG). The crystals show pseudo hexagonal packing with no direct contact between adjacent molecules.

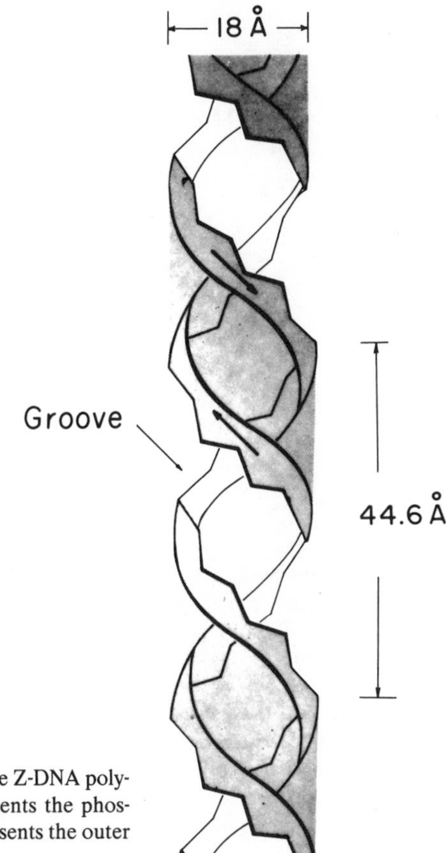

Figure 8. A schematic diagram of the Z-DNA polymer. The saw-toothed edge represents the phosphate positions. The stippling represents the outer surface.

residue can rotate about the glycosylic bond connecting it to the sugar. There are in general two major positions which this ring can occupy; *anti,* away from the sugar and *syn,* toward the sugar. Both of these conformations are equally stable, and NMR studies of model nucleotides show that they exist in approximately equal distributions in solution. It is interesting to note that the same is not true for pyrimidines, which are mostly found in the the *anti* conformation, and are only occasionally found in the *syn* conformation. In Z-DNA, all of the guanine residues are found in the *syn* conformation, while all of the cytosine residues are found in the *anti* conformation.

A second difference in conformation is associated with the pucker of the furanose ring. In Figure 9 the rings are drawn so that the plane formed by the atoms C1′, O, C4′ is perpendicular to the plane of the paper. Atoms found above this are referred to as having the *endo* conformation. It can be seen that in B-DNA the C2′ atom is endo, while in Z-DNA the C3′ atom is endo. This difference in ring pucker is associated with significant differences in the extension of the sugar phosphate chain. In the C2′-*endo* conformation, as in B-DNA, the phosphate groups are far

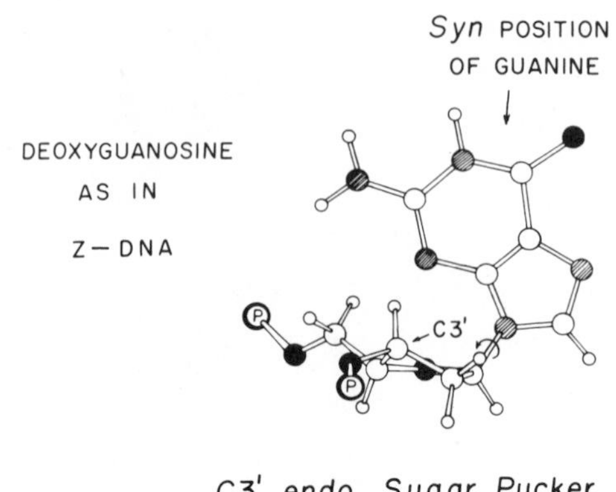

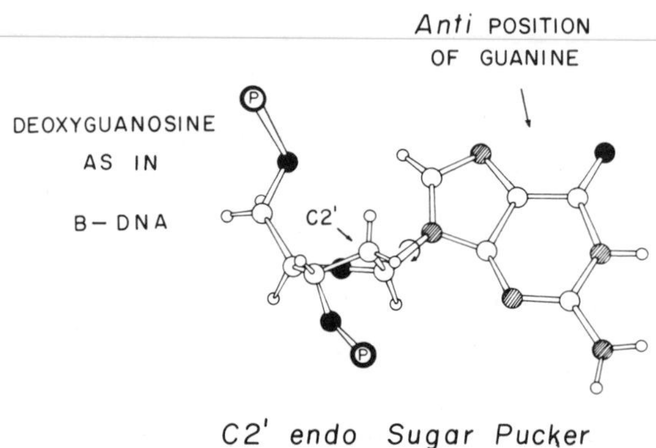

Figure 9. Conformation of deoxyguanosine in B-DNA and in Z-DNA. The sugar is oriented so that the plane defined by C1′-O1′-C4′ is horizontal. Atoms lying above this plane are in the *endo* conformation. The C3′ is *endo* in Z-DNA while in B-DNA the C2′ is *endo*. These two different ring puckers are associated with significant changes in the distance between the phosphorus atoms. In addition, Z-DNA has guanine in the *syn* position, in contrast to the *anti* position in B-DNA. A curved arrow around the glycosidic carbon-nitrogen linkage indicates the site of rotation.

apart, while in the 3′-*endo* conformation, the phosphate groups are closer together. This change in pucker represents an elastic element in the sugar phosphate backbone. Furthermore, it is an elastic element which is used extensively in naturally-occurring polynucleotides. Most RNA molecules are found with a C3′-*endo* pucker. However, in the three-dimensional structure of yeast phenylalanine transfer RNA, 8 of the 76 nucleotides are found with a C2′-*endo* conformation.[14] These are often found in positions where the ribose phosphate chain has to be extended in order to link different parts of the molecule. They are also frequently associated with the presence of intercalating units within the structure. Although C2′-*endo* conformations are usually found in DNA molecules, the energy barrier between these two conformations for deoxynucleosides is not very great. Thus it is not surprising to find an alternative conformation of DNA involving a change in the ring pucker. In

Z-DNA all of the guanosine residues have the C3′-*endo* pucker, while all of the cytidine residues have the C2′-*endo* pucker.

These changes in conformation are associated with significant differences between B- and Z-DNA. An end view of Z-DNA and B-DNA is shown in Figure 10. Z-DNA has a diameter of 18Å, in contrast to the 20Å diameter for B-DNA. There are significant differences in the position of the bases. The guanosine residues are shaded in Figure 10, and in B-DNA these are found in the center of the molecule while in Z-DNA they are located near the outer wall. It can be seen that the five-membered imidazole ring of guanine is located at the outer part of the mole-

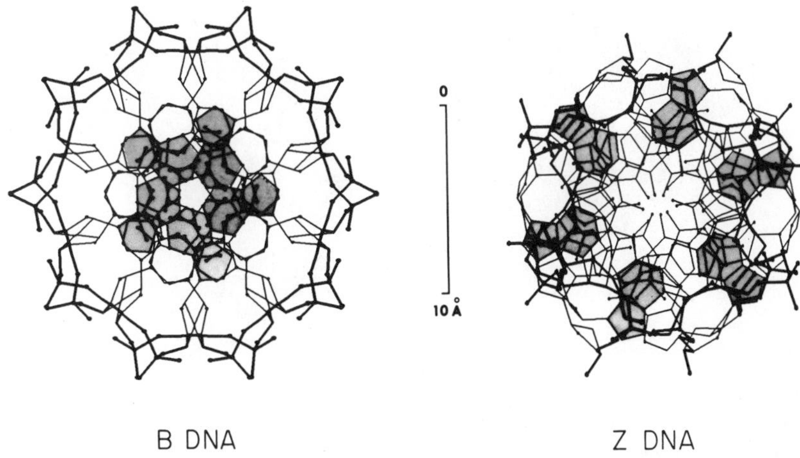

Figure 10. End views of B-DNA and Z-DNA are illustrated in which the guanine residues of one strand have been shaded. The Z-DNA figure represents a view down the complete c-axis of the crystal structure encompassing two molecules. The shaded guanine residues illustrate the approximate 6-fold symmetry. The imidazole part of the guanine residue forms a segment of the outer cylindrical wall of the molecule together with the phosphate residues. The B-DNA figure represents one full helix turn. In contrast to Z-DNA, the guanine residues in B-DNA are located closer to the center of the molecule and the phosphates are on the outside.

cule. The position of the rings can be seen more clearly in Figure 11, which shows a space-filling end view of a three base pair Z-DNA double helix as viewed down the helix axis. Three phosphate groups are seen on the left and one on the right. In Figure 11, the groove rotates clockwise as the structure comes towards the reader. This accounts for the three phosphates seen at the left side. The groove is quite deep in Z-DNA, extending almost to the axis and the edge of the groove is lined by negatively charged phosphate residues. There are considerable differences in the accessibility of different atoms in the base pairs at the upper part of Figure 11. The outer wall is formed by the atoms C8-H, N7 and O6 of guanine as well as N4-H and C5-H of cytosine. This is in marked contrast to B-DNA, where these atoms are located at the bottom of the wide groove. In addition, the *anti* conformation of the sugar in B-DNA means that the C8-H part of guanine abuts against the sugar

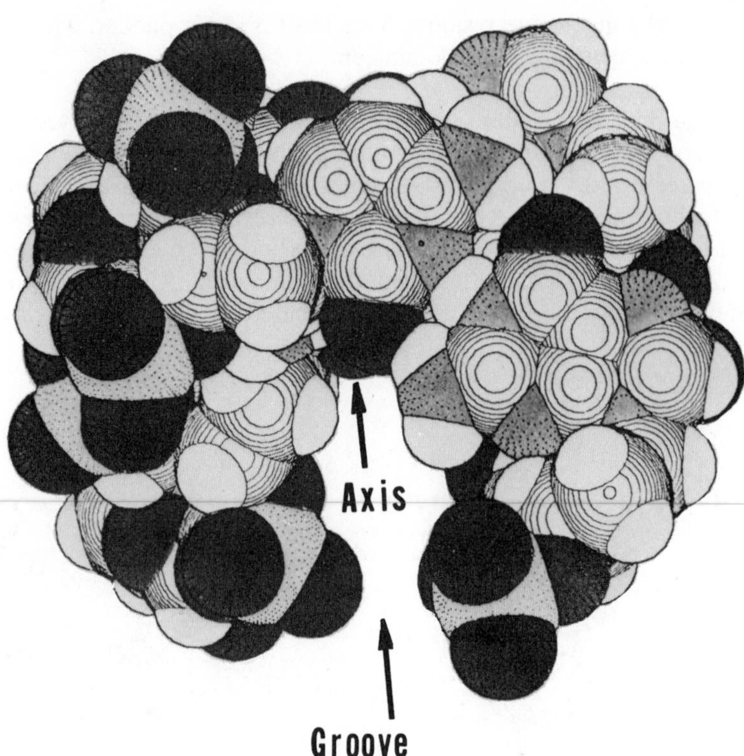

Figure 11. Van der Waals drawing of a fragment of Z-DNA as viewed down the axis of the helix. Three base pairs are shown, and the deep groove is seen which extends almost to the axis of the molecule. In these three base pairs the groove rotates clockwise toward the reader. For that reason, three phosphates are visible on the left and only one on the right. The N7 and C8 atoms of guanine are near the outer wall of the molecule. This drawing is made directly from the segment of the spermine-magnesium Z-DNA crystal which is in the Z_I conformation. The solid black dot indicates the axis of the molecule.

residue. This difference is important in understanding the chemical modifications of DNA which are found to stabilize it in the Z conformation. It should also be noted that the C5-H position of cytosine is located on the periphery of the molecule, as it can be methylated in eukaryotic systems.

Figure 12 summarizes some of the major differences between B-DNA and Z-DNA. The asymmetric unit of Z-DNA is no longer a nucleotide but rather a dinucleotide as the two nucleotides have different conformations. In addition, the part of Z-DNA containing the dCpG dinucleotide has a conformation in which there is not much rotation from one base pair to the next along the fiber axis. However, the bases are strongly sheared, as illustrated in Figure 12. Because of this shearing, the cytidine residues of one strand are stacked over that of the opposite strand. Because of this shear, the guanine residues are exposed so they are no longer stacked on a base on one side. Instead, the guanine is stacked on the O1′ atom of the sugar residue below. In B-DNA there are pseudo two-fold or dyad axes which are found in the

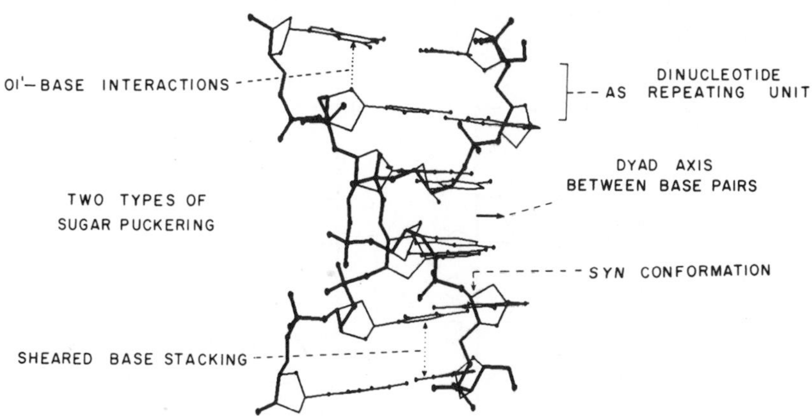

Figure 12. Schematic diagram showing a projection of the left-handed double helix of the spermine-magnesium d(CpGpCpGpCpG) hexamer. Seven structural features are found in this left-handed DNA conformation which are different from those found in right-handed B-DNA, as indicated in the diagram and discussed in detail in the text.

plane of the base pair as well as between the base pairs. In Z-DNA the only dyad which remains is that which is found between the base pairs. The difference in conformation of adjoining residues has made the dyad in the plane of the base pairs disappear. Another important difference is associated with the fact that the orientation of the bases relative to the direction of the sugar phosphate chains is different in the right- and left-handed forms of DNA. In Z-DNA the bases are flipped over, so to speak, so that they are stacked in an opposite orientation relative to what is found in B-DNA. This is illustrated in the diagram in Figure 13. On the left B-DNA is drawn as a ladder in which the guanine and cytosine bases are represented as rectangles which are white on the top and shaded below. In forming a segment of Z-DNA in the middle of the B-DNA strand, it is necessary to flip the bases over so that their shaded side is up, as shown in the right-hand side of the diagram in Figure 13. Alternate phosphate groups in Z-DNA are found at different distances from the axis due to the zig-zag array. It is only the pair of phosphate groups on opposite strands, furthest away from the axis that are far enough apart so that they can join to a segment of B-DNA. This means that the unit of Z-DNA which would be found in the center of the B-DNA segment would consist of the sequence dCpG. In the diagram in Figure 13, two units of dCpG are forming the Z conformation in the middle of a long stretch of alternating (dG-dC). The mechanism whereby this transition takes place is not known. However, it is likely that this takes place for the guanosine residues by simply going from an *anti* to a *syn* conformation. This would effectively invert the plane of the base. In the case of the cytidine residues a more complex transformation must occur in which both the cytidine base and the sugar must invert themselves. It is this inversion of the cytosine sugar residues which gives rise to the zig-zag conformation of the polynucleotide backbone.

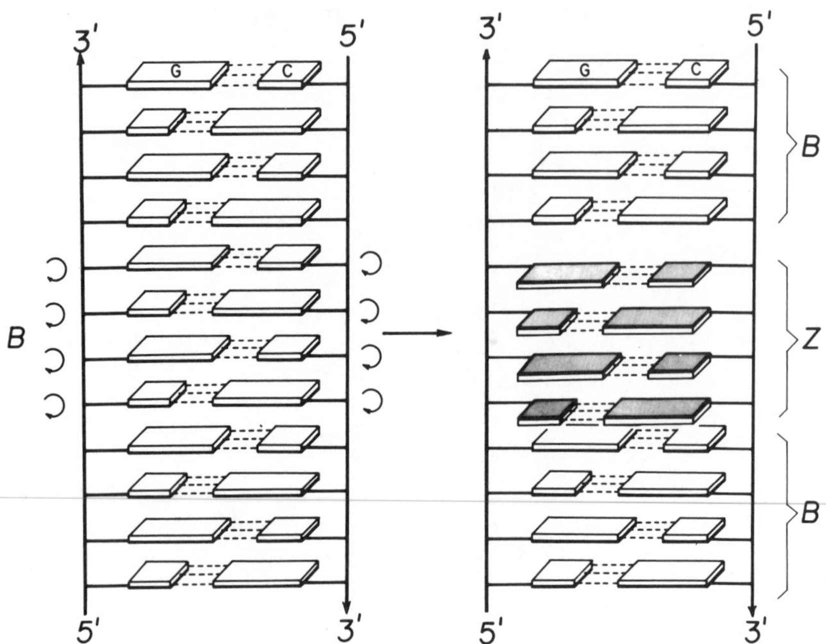

Figure 13. A diagram illustrating the change in topological relationship if a four base pair segment of B-DNA were converted into Z-DNA. This conversion could be accomplished by rotation of the bases relative to those in B-DNA. This rotation is shown diagrammatically by shading one surface of the bases. All of the dark shaded areas are at the bottom in B-DNA. In the segment of Z-DNA, however, four of them are turned upwards. The turning is indicated by the curved arrows. Rotation of the guanine residues about the glycosidic bond produces deoxyguanosine in the *syn* conformation while for dC residues, both cytosine and deoxyribose are rotated. The altered position of the Z-DNA segment is drawn to indicate that these bases will not be stacking directly on the base pairs in the B-DNA segment.

Local Changes in DNA Conformation

The Z conformation has now been found in seven different crystal structures involving both tetramers and hexamers of alternating (dC-dG) sequences[13,15,16] In studying these structures in detail, it can be seen that not all residues have the same conformation. An example of this is shown in Figure 14, in which one strand of the hexanucleotide complex has been inverted so that it is parallel to the other strand and then rotated to show the similarity in conformation. It can be seen that all of the conformations are similar to each other with the exception of one phosphate group in a GpC sequence. The phosphate group in the position G4pC5 is different than that found in the other three GpC sequences. This difference is associated with a rotation of the phosphate group so that it is in a position where it can now form a hydrogen bond with the hydrated magnesium complex found coordinated to N7 of guanine G6. We have called the majority conformation of phosphate as in G10pC11 the Z_I form, while the turned-out phosphate in the G4pC5 is called Z_{II}.[13] The majority conformation found in most crystals is Z_I, but all of the crystals show some residues with the Z_I conformation.

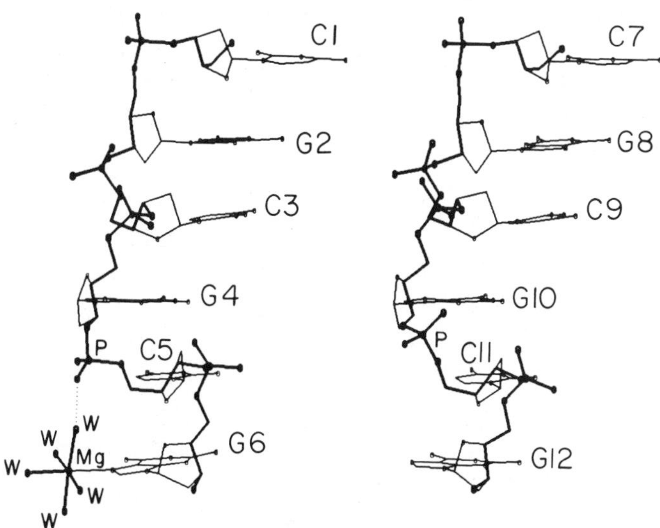

Figure 14. Diagram showing the conformation of the two independent hexanucleotide molecules in the spermine-magnesium Z-DNA crystal. In the crystal the two strands are antiparallel, but in this diagram they have been arranged in a parallel alignment to show the similarity of the two chains. The two independent chains are very similar except for the linkage G4pC5. That phosphate group has rotated in such a manner that its oxygen is forming a hydrogen bond with a water molecule (W) in the octahedral coordination shell surrounding the magnesium ion, which is complexed to N7 of guanine 6.

Conformational changes which are associated with differences in either sequence or local ionic environment are generally found in oligonucleotide structures. This is true for the Z conformation as well as the conformations of the nucleotides found in the dodecamer B-DNA fragment.[1] These pictures of DNA fragments obtained from single crystals are truer representation of the actual conformation of nucleic acids than that which is obtained by the study of fiber diffraction. Because fiber diffraction studies have so little information in them, one assumes that the structure is quite regular. However, it is likely that the fiber conformation is some type of average structure around which the actual structure may fit. The real structure will have conformational differences due to the presence of ions or proteins. This is an important consideration, since some of the biologically interesting information may be associated with sequence-dependent differences in conformation which may be recognized by the ions or proteins which bind to them. These studies further illustrate the extent to which one has to accept the interpretation of fiber x-ray diffraction studies with extreme caution, since they give only an approximate view of the molecule.

The newer view of DNA which we obtain from single crystal studies will undoubtedly transform our perception of DNA from a completely regular molecule to one which is conformationally active. The conformation responds to changes due to molecules which bind to it, as with daunomycin, or it can undergo sequence-dependent changes such as is seen in the left-handed Z-conformation. These recent studies are just the beginning of a new understanding of the many and varied structures of DNA. Because of the central role of DNA in information storage and transmission, it is likely that these structural changes will be associated with significant biological effects.

Acknowledgements

This research was supported by grants from the National Institutes of Health, National Science Foundation, National Aeronautics and Space Administration, and the American Cancer Society.

References and Footnotes

1. Watson, J.D. and F.H.C. Crick, *Nature* (London), *171,* 737 (1953).
2. Wing, R., Drew, H., Takano, T., Broka, C., Tanaka, S., Itakura, K. and Dickerson, R.E. *Nature* (London) *287,* 755-758 (1980).
3. diMarco, A., Arcamone, F. and Zinino, F. In *Antibiotics,* ed. by J.W. Corcoran and F.E. Hahn, Springer Verlag Press, Berlin, vol. 3, pp. 101-128 (1974).
4. Neidle, S. *Prog. Med. Chem. 16,* 151-220 (1979).
5. Quigley, G.J., A.H.-J. Wang, G. Ughetto, G. van der Marel, J.H. van Boom and A. Rich *Proc. Nat. Acad. Sci. USA 77,* 7204-7208 (1980).
6. Waring, M.J. *J. Mol. Biol. 54,* 247-279 (1970).
7. Wang, J.C. *J. Mol. Biol. 89,* 783-801 (1974).
8. Henry, D.W. *Cancer Treat. Rep. 63,* 845-854 (1979).
9. Wang, A.H.-J., G.J. Quigley, F.J. Kolpak, J.L. Crawford, J.H. van Boom, G. van der Marel and A. Rich *Nature 282,* 680-686 (1979).
10. Pohl, F.M. and T.M. Jovin *J. Mol. Biol. 67,* 375 (1972).
11. Pohl, F.M., A. Ranade and M. Stockburger *Biochim. Biophys. Acta 335,* 85-92 (1973).
12. Thamann, T., R.C. Lord, A.H.-J. Wang and A. Rich *Nucleic Acids Res.* (in press).
13. Wang, A.H.-J., G.J. Quigley, F.J. Kolpak, G. van der Marel, J.H. van Boom and A. Rich *Science 211,* 171-176 (1980).
14. Quigley, G.J. and A. Rich *Science 194,* 796-806 (1976).
15. Drew, H.R., T. Tanako, S. Tanaka, K. Itakura and R.E. Dickerson *Nature (London) 286,* 567 (1980).
16. Crawford, J.L., Kolpak, F.J., A.H.-J. Wang, G.J. Quigley, J.H. van Boom, G. van der Marel and A. Rich *Proc. Nat. Acad. Sci. USA 77,* 4016-4020 (1980).

Proceedings of the Second SUNYA Conversation in the Discipline Biomolecular Stereodynamics
Volume I, ISBN 0-940030-00-4, Ed., Ramaswamy H. Sarma,
Adenine Press, New York, ©Adenine Press

Structure of the DNA Double Helix in Solution, Crystals and Theory[1ab] Coming of Age of Z-DNA

Ramaswamy H. Sarma, C.K. Mitra and Mukti H. Sarma
Institute of Biomolecular Stereodynamics
State University of New York
Albany, NY 12222

I am the original fragrance of the earth
I am the life of all that lives
I am the original seed of all existences
All states of being are manifested by My energy
I am unborn and my transcendental form never deteriorates
Although I appear in so many configurations
You know not my true transcendental form

Lord Krishna in *Bhagavad-gita*

The Beginning

The blue prints of the mechanism of life are preserved in genomic DNA. Little that we know about the three dimensional dynamic solution geometry of any genomic DNA and that how the structure controls expression. Dickerson et.al.[2] conclude in their opus in this volume that the intellectual *tour de force* in DNA structure analysis lay in the deduction of correct structures from the small harvest of the fiber diffraction data. The A and B helices proposed by Arnott and Hukins[3] from fiber studies have been elegantly confirmed, except for details, by single crystal examinations.[2] However, the left-handed Z-helix of Rich[4] was never even suspected by fiber structural scientists. Theoretical investigations have led to a plethora of models for the double helix: the alternating B-DNA of Klug,[5] the propeller twist model of Levitt,[6] the vertical double helix of Olson,[7] and the right and left-handed models of Sasisekharan[8] in which the mononucleotide or the dinucleotide may form the repeating unit.

One does not know how valid are these structures derived from solid state and theoretical studies for the double helix in solution. In this article we undertake an examination of the structure of the double helix for poly(dG-dC)•poly(dG-dC), in solution and summarize the present knowledge on the structure of double helices containing G-C sequences.

Computation of Magnetic Shielding Constants

In our approach we use ^1H NMR spectroscopy and take advantage of the fact that

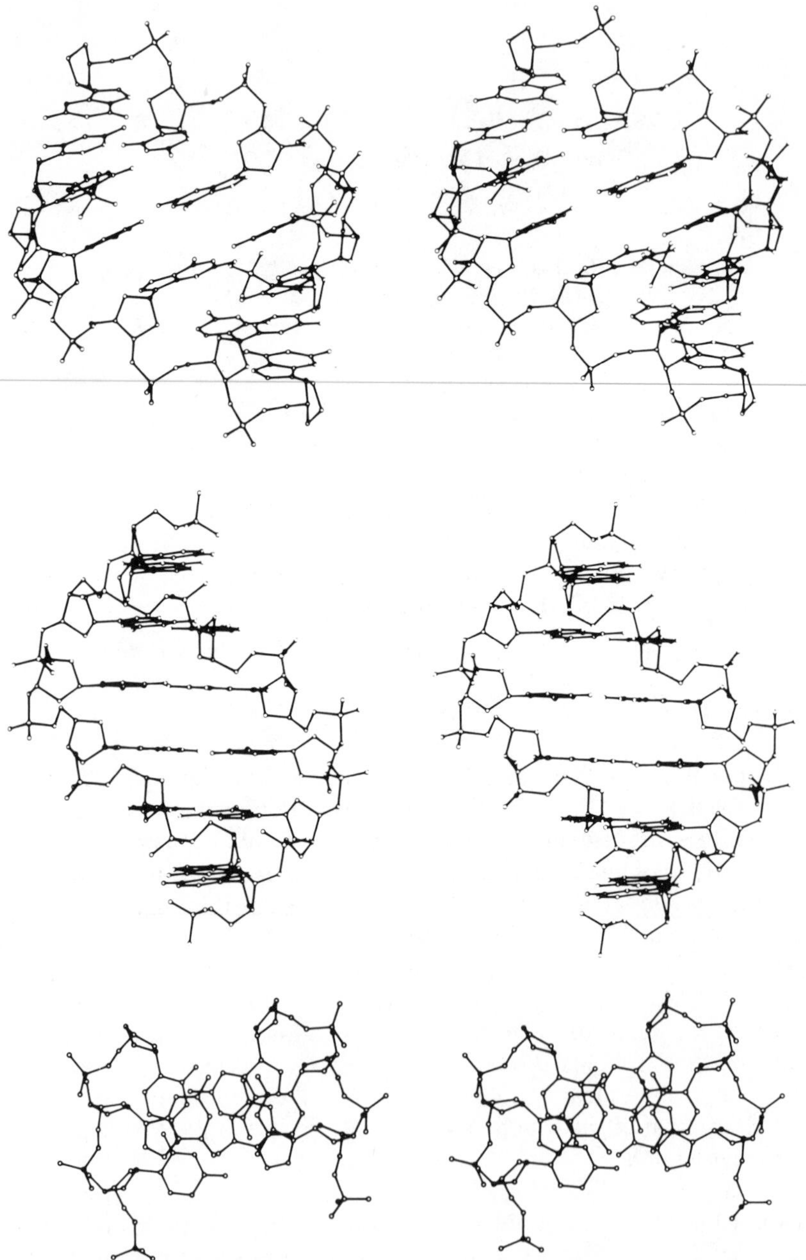

Figure 1. Stereographic perspectives of A-DNA (top), Arnott and Hukins B-DNA (middle). Geometric relationships among the bases in the latter are shown at the bottom. The structures displayed in Figs. 1-7 are for a hexamer segment of poly(dG-dC)•poly(dG-dC).

sufficient geometrical differences exist among the various proposed models. The chemical shift of a given nuclei, for example a proton, in the NMR spectrum is strongly influenced by the local geometric and chemical environments of the nuclei. To drive home the geometric differences among the diverse structures, in Figures 1 through 7, we stereographically illustrate, the structures of A-DNA,[3] the A/H B-DNA (i.e., Arnott/Hukins[3]), C-DNA,[9] D-DNA,[10] alternating B-DNA,[5] the vertical double helix,[7] Z_I-DNA,[9] Z_{II}-DNA,[10] the Levitt and Dickerson[2,13] propeller twist models. In Table I we have derived the torsion angles from the x, y and z coordinates and the nomenclature explained in Fig. 8. These illustrations are for the alternatinq G-C system. In the Dickerson model,[12,13] we have confined ourselves to the d-GAATTC stretch of the self-complementary d-CGCGAATTCGCG double helix and have replaced the A and T by G and C. Note that in the Dickerson[2,13] crystal structure it is this domain which displays significant propeller twists. An examination of Figures 1 through 7 clearly illustrates the significant geometric differences among the various helices. Even though A, B, C, D and Olson's B-DNAs are made up of a monomer repeat unit, the A and Olson's forms have ^{3}E sugar pucker, the sugar pucker of B, C, and D lie in the ^{2}E domain. In the A form the χ_{CN} is low anti and the

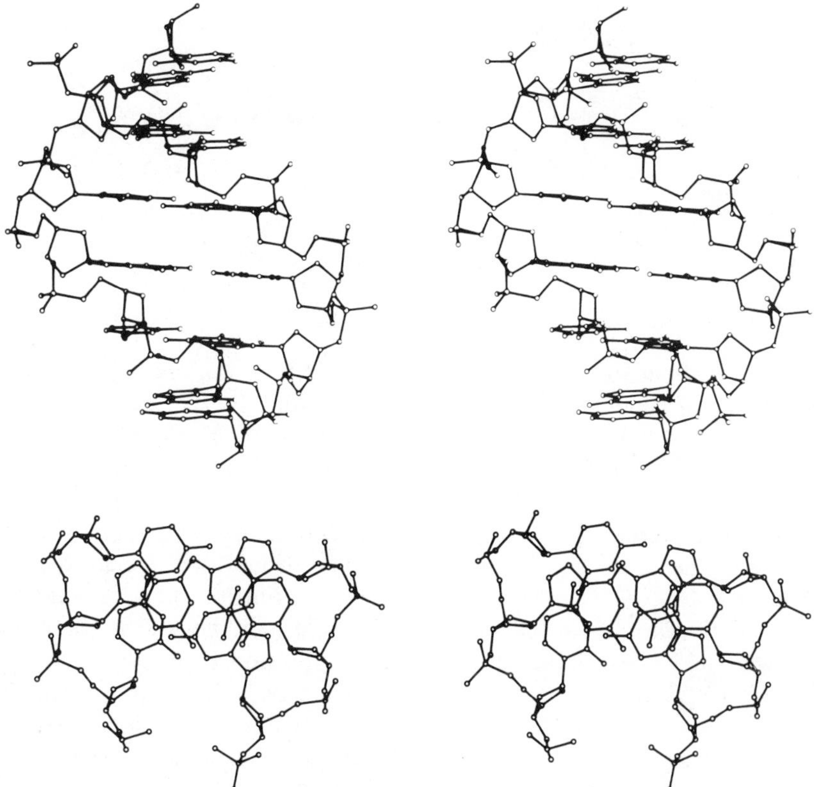

Figure 2. Stereographic perspective of C-DNA, view perpendicular to the helix axis (top); an end view of a three base pair segment is shown at the bottom.

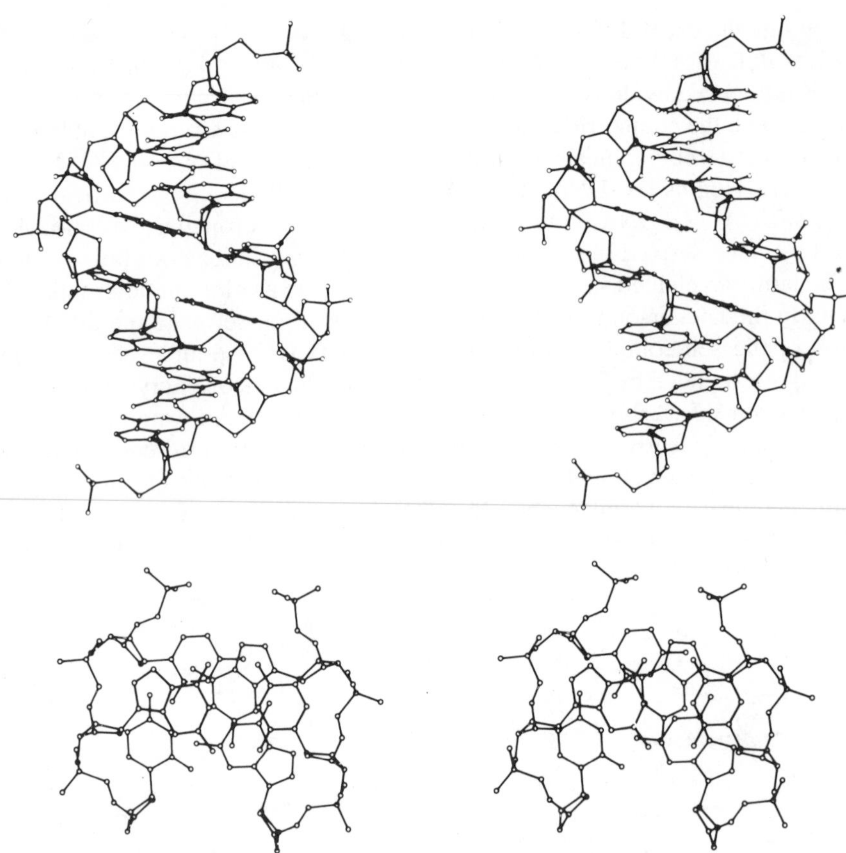

Figure 3. Stereographic perspective of D-DNA; views same as in Figure 2.

bases are tilted about 19° to the helix axis; in the Olson's form the χ_{CN} is high anti and the bases are almost parallel to the helix axis and the structure has a vacant inner core of about 35Å in diameter. The B, C, and D forms belong to a close family of structures as an examination of the torsion angles in Table I reveals.

Even though the alternating B-DNA, Z_I/Z_{II} DNA's have a repeat unit of a dinucleotide, the former generates a right-handed helix, but the latter left-handed helices; in the former the G is *anti* and in the latter it is *syn.* Further there are significant differences in the various torsion angles (Table I) between the alternating from and the Z forms, and these create morophological differences in the double helix (Figures 4, and 6). It should be noted that even though both alternating B and Z-DNA's are antiparallel, in the Z forms at each base pair there is local parallelism because the sugar direction undergoes local inversion to accommodate a *syn* dG to enter into Watson-Crick base-pairing in such a way that the glycosyl orientations are *cis.* A principal feature of the Levitt[6] and Dickerson[13] models is the propeller twist (Fig. 7) between the base pairs. An examination of Figure 4 reveals that the

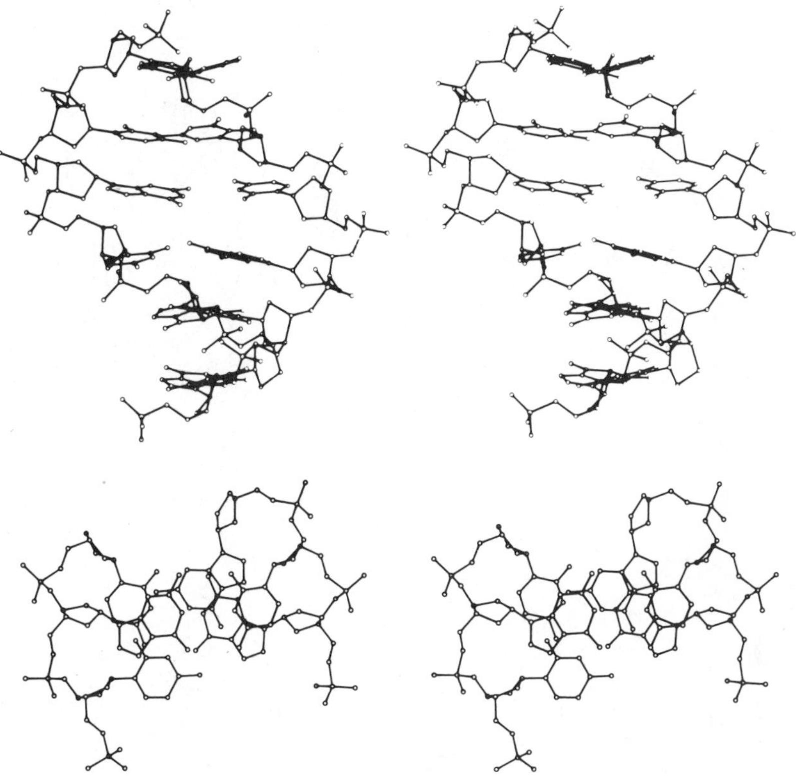

Figure 4. Stereographic perspective of Klug's alternating B-DNA; views same as in Figure 2. A stereographic examination will reveal that the base pairs are significantly propeller twisted in this model.

base pairs in Klug's alternating B-DNA model are also significantly propeller twisted, and this was not noticed by the authors.[5]

In our approach we compute from the theory of NMR spectroscopy, the relative chemical shifts of the protons, i.e., magnetic shielding constants, of the double helix from the x y and z coordinates. We have described elsewhere *in extenso* the assumptions[14,15] and principles involved in the computation of magnetic shielding constants for a double helix. However to make this presentation as self-contained as possible we provide a brief summary here. We assume that the chemical shift of the central base-paired nucleotides in an heptamer duplex are the shift of protons in any nucleotide unit in the polymer duplex and that there are no end effects. This assumption is reasonable because the polymer contains close to 100 base pairs and there is translational symmetry; the chemical shifts are not significantly affected by units beyond the third neighbor. We assume that the heptamer duplex can exist in any of the ten

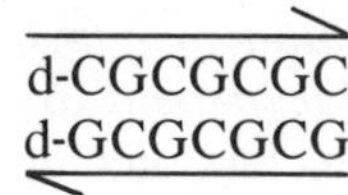

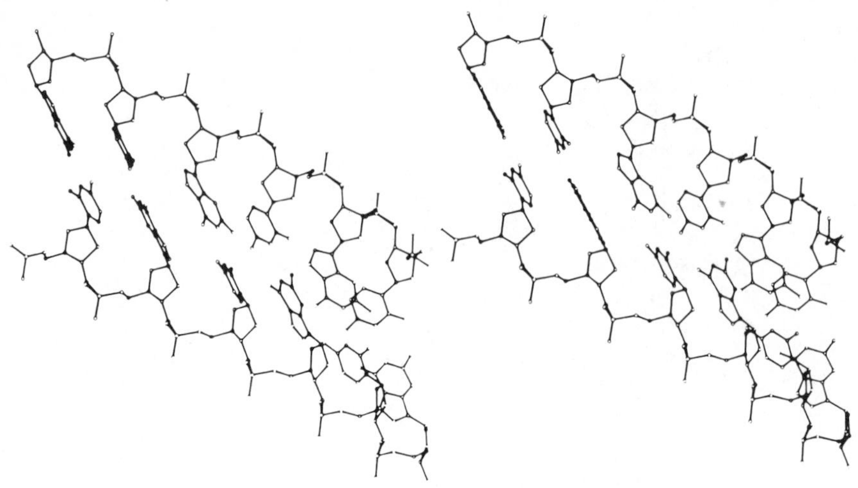

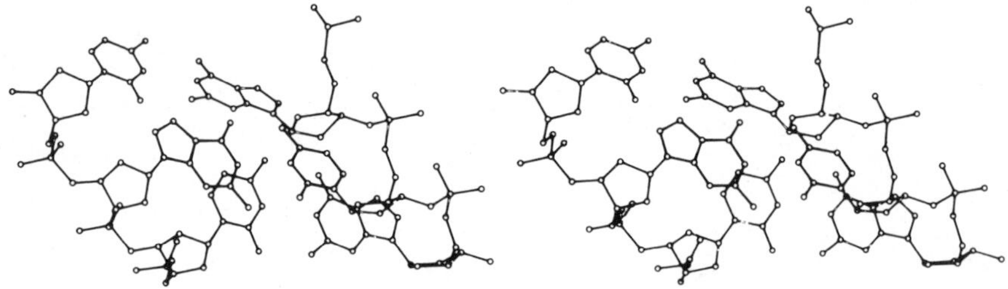

Figure 5. The vertical double helix of Olson, in stereo.

spatial configurations illustrated in Figures 1-7. In Figure 9 we have schematically drawn an heptamer segment. The chemical shifts of the central cytidine C_0 or G_0 (Figure 9) will be affected by the remaining 13 nucleotide units.[14,15] Extensive calculations in this laboratory and that of Pullman and Giessner-Prettre[16] have shown that nucleotide units as far away as the sixth neighbor can affect the shifts. However, the chemical shifts are not significantly affected by nucleotide units beyond the third neighbor.

The contribution to the chemical shifts originate from (a) ring current effect of the bases, (b) the diamagnetic component of the atomic magnetic anisotropy of the bases, and (d) the diamagnetic and paramagnetic components of the atomic magnetic anisotropy of the sugar phosphate backbone. Shielding constants for a given proton of a nucleotide unit in structures like the ones in Figures 1-7 can be computed from x y and z coordinates taking all the above contributions into account.[14,15,17-19] *The calculated shielding constant essentially provides the magnitude and direction of shielding a proton in a nucleotide unit such as C_0 (Figure 9) will experience as the unit is moved from an isolated environment to that in an organized structure like the ones in Figures 1-7.* In such a calculation one cannot include the contribution to

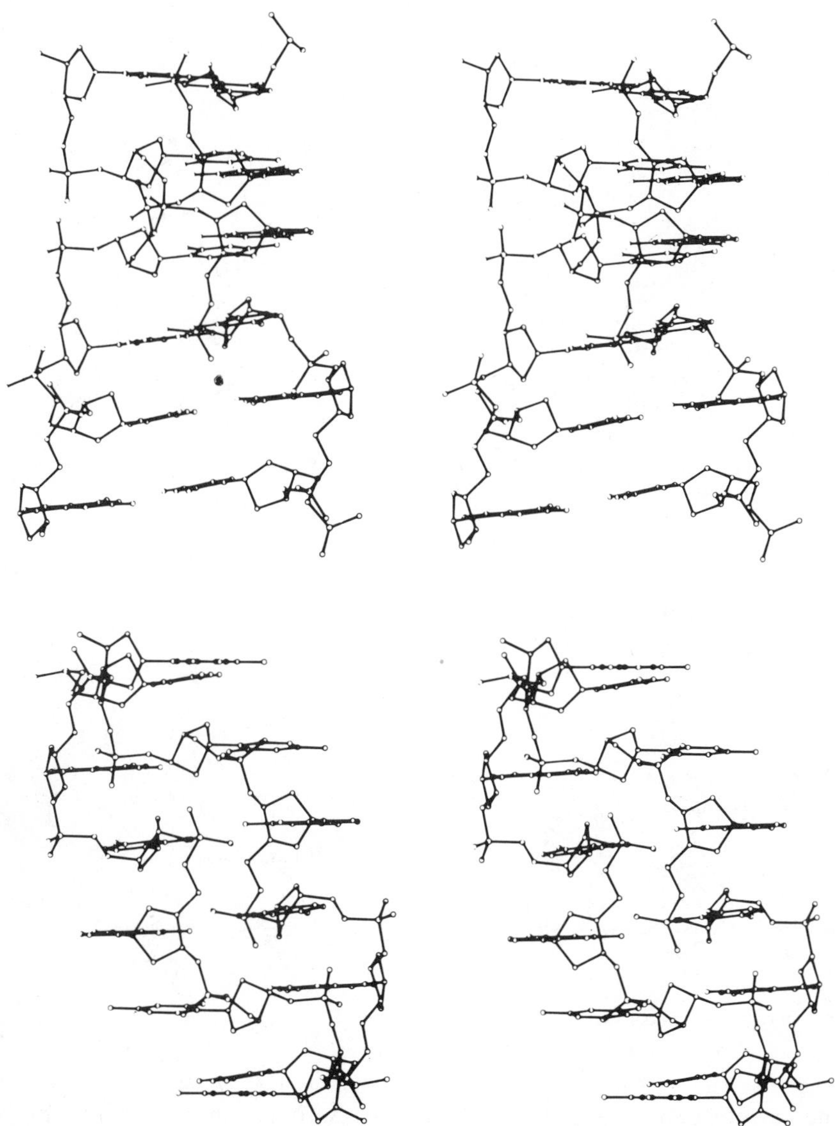

Figure 6. Stereographic views of Z_I (top) and Z_{II} (bottom) DNA's.

shielding from the parent nucleotide unit to which the proton belongs i.e., C_0 should be excluded when computing the effect on C_0 from the remaining 13 units in configurations Figure 1 through 7. G_0 should be excluded when computing the effect of the remaining thirteen units on G_0. Here one is assuming that the conformation of the isolated mononucleotide is the same as it is in the various organized structures (Fig. 1-7). This is not true and hence a calculation at a second level should be undertaken to correct for the effect of this on shifts and this is presented later.

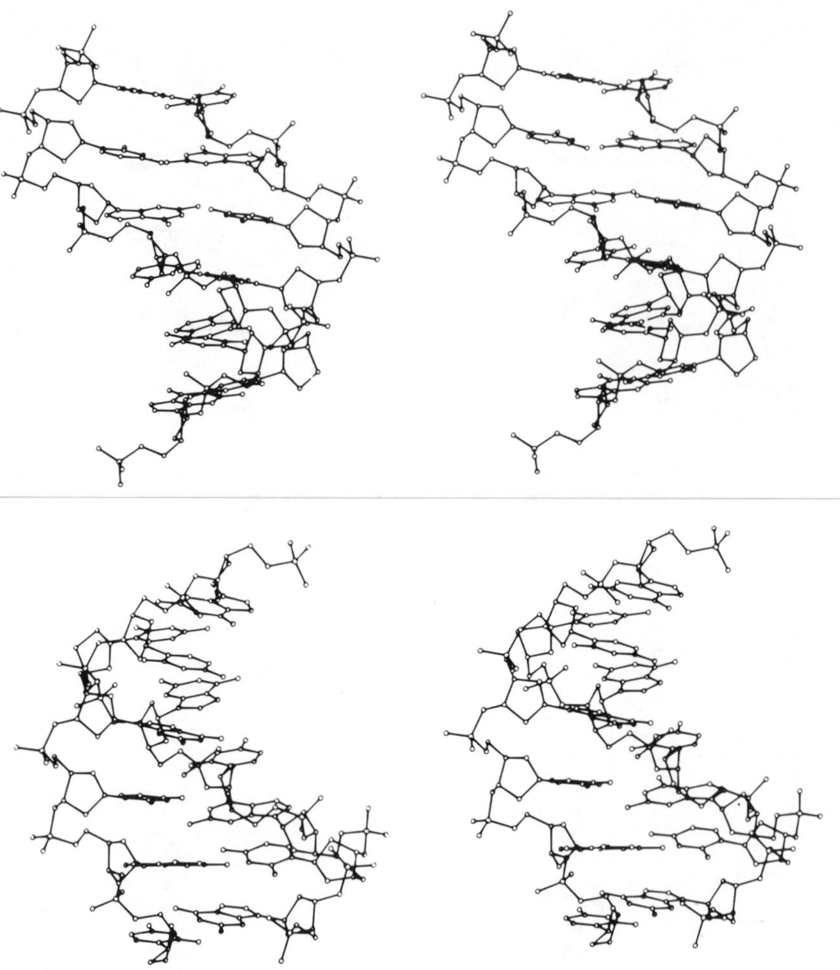

Figure 7. Stereographic perspectives of the propeller twisted models of Levitt (top) and Dickerson (bottom).

The ring current constants and the magnetic anisotropy tensor elements for the calculations were kindly provided by Pullman and Giessner-Prettre before publication.[20] Using the methods described elsewhere[14,15,17-19] and from the x y and z coordinates we have calculated the magnetic shielding constants for CH5, CH6, CH1′, GH8 and GH1′ for the central C_0-G_0 base pair in an heptamer segment (Figure 9) of poly(dG-dC)•poly(dG-dC) for the ten different spatial configurations it may display (Figure 1-7). For these computations we have taken into account contributions to shielding from ring current effects as well as effects from the diamagnetic and paramagnetic components of the atomic magnetic anisotropy. The results of the computations are presented in Table II. Before one examines Table II, we should point out that in the earlier papers in this series[14,15] where we presented for the first time the NMR methodology to handle the details of a double helix, we have presented an extensive table which indicates the magnitudes of the various factors such as ring currents and the paramagnetic and diamagnetic component of the atomic magnetic anisotropy from various nucleotide units in a double helix. In Table II, the data are presented in a considerably reduced format. Examination of the data in references 14 and 15 clearly indicate that ring current contributions alone are inadequate to rationalize NMR data and that a dominant contribution comes from the paramagnetic term.

How does one experimentally measure the calculated shifts. One has to experimentally obtain the shifts for poly(dG-dC)•poly(dG-dC) duplex in the solvent condition of interest and measure those for isolated mononucleotides i.e., same solvent conditions at extremely low concentration and high temperature. The difference in shifts between the experimental value for the monomers and the double helix should then be compared with those computed for the various spatial configurations. *However, there is a problem.* In these calculations the shielding effect of 13 neighboring nucleotide units arranged in a particular configuration on the shift of a nucleotide unit was carried out and we neglected, in this first level of calculation the effect of its own local geometry on its shifts. This would have been alright if the experimentally measured isolated mononucleotides have a geometry identical to what they have in the organized structure. In fact a plethora of NMR studies from this and other laboratories[21-26] have shown that monomers particularly the purine ones are highly flexible structure and they are in a conformational blend. The shielding data in Table II do not take into account the change in shifts that will result as a conformationally equilibrating monomer is made part of an organized structure. Hence a calculation at the second level should be made to correct the computed shielding values in Table II to reflect the true situation. The dominant factor which affects the shielding values of the observable protons is a change in χ as the monomer becomes part of the double helix. We have described in considerable detail[14] how calculation at the second level can be done and corrections can be made using experimentally determined syn $\Leftrightarrow$ anti and $^2E \Leftrightarrow {}^3E$ equilibrium for the monomers[21-26] and from the dependence of shielding constants on χ[27-29] for B- and Z-DNA. Similar calculations were performed in the present case for the various forms of DNA studied. Last time[14] we did not carry out calculations at the second level for the shifts of CH5, CH6 and CH1′. However we have done so in the present time. This was done on the basis that pyrimidine nucleotides overwhelmingly exist in *anti* conformation[21,26] and the expected average value[30-31] of χ is $\sim 50°$. The final set of shielding constants obtained are given in Table III.

Chemical shifts of isolated mononucleotides in 10 mM and 4 M salt were obtained by recording the ^{1}H NMR spectra of 3′dCMP, 3′dGMP 5′dCMP, and 5′dCMP, 8mM, 90°C using the super-conducting 270 MHz FT NMR spectrometer. For all practical purposes the shifts of the monomers were identical at 10 mM and 4 M salt at 90°C at 8 mM concentrations. The average of the shifts for CH5, CH6, CH1′, GH8 and GH1′ for the 3′ and 5′ mononucleotides were taken. The shifts for the duplex of poly(dG-dC)•poly(dG-dC) at high salt were obtained from Patel.[32] For the low salt form the data for poly(dG-dC)•poly(dG-dC), 10 mM in total nucleotides,

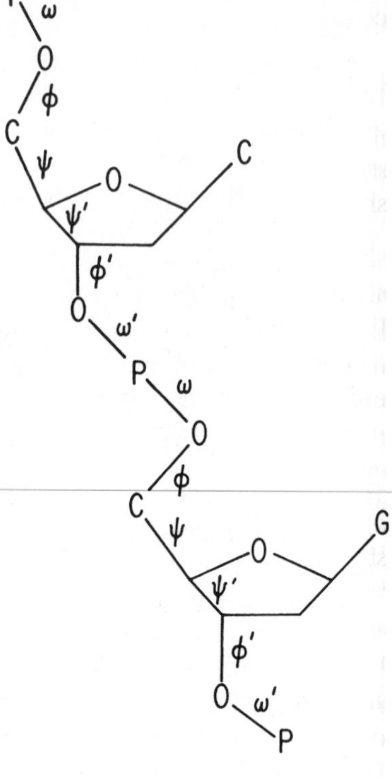

Figure 8. The nomenclature employed for the torsion angles for data in Table I. Note that for A, B, C, D and Olson's DNAs the repeat unit is a mononucleotide. For the alternating B and the Z_I and Z_{II} structures, the repeat unit is a dinucleotide. For these latter systems we have at first given the torsion angles for the pGp unit starting with ω' (P-O3-) and we have reported the angles for the pCp unit starting with again ω'.

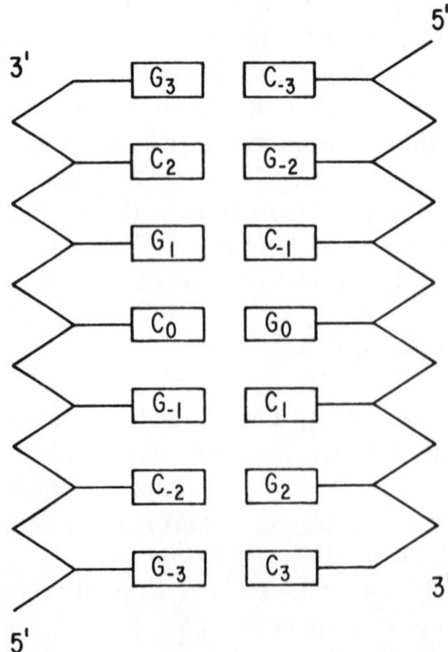

Figure 9. Schematic drawing of an heptamer segment of double stranded poly(dG-dC)•poly(dG-dC).

Table I
Torsion Angles in the Various Models (All angles in degrees). Note that $_3E \simeq {}^2E$.

Model	Backbone						Sugar pucker	Furnose					Glycosyl torsion	χ
	ω'	ϕ'	ψ'	ψ	ϕ	ω		τ_0	τ_1	τ_2	τ_3	τ_4		
A-DNA	314	178	83	46	208	275	^3E	4	334	37	324	21	anti	26
Alt-BDNA 3′pG5′p	278	192	99	59	151	300	^3E	326	17	5	334	38	anti	33
Alt-BDNA 3′pC5′p	227	200	143	65	172	293	^2E	326	42	327	15	12	anti	76
B-DNA	265	155	156	36	214	313	$_3E$	356	25	325	33	342	anti	82
C-DNA	254	161	156	37	200	321	$_3E$	356	25	325	33	341	anti	83
D-DNA	259	142	156	69	208	298	$_3E$	356	24	325	33	341	anti	83
Olson's B-DNA	268	198	86	58	178	295	^3E	0	340	30	329	20	high anti	122
Levitt-DNA	273	170	108	68	168	294	$^{O4}E\text{-}_1E$	319	36	357	338	39	anti	48
Z-DNA-I 3′pG5′p	291	256	99	190	179	48	^3E	8	340	24	340	7	syn	248
Z-DNA-I 3′pG5′p	80	266	138	56	221	223	^2E	334	35	330	17	6	anti	21
Z-DNA-II 3′pG5′p	55	181	93	157	193	92	^3E	345	358	16	334	25	syn	241
Z-DNA-II 3′pC5′	74	259	147	66	163	146	$_3E$	336	39	322	26	359	anti	33

Table II

Computed Magnetic Shielding Constants for CH5, CH6, CH1′, GH1′ for Various DNA Models. This is the computation at first level and does not take into account contribution to shielding from change in χ.

Proton	DNA Model	Total Shielding
CH5	A-DNA	1.4701
	Alt-BDNA	0.4210
	B-DNA	0.7956
	C-DNA	0.6035
	D-DNA	0.4564
	Olson's-DNA	0.0798
	Levitt	0.7081
	Dickerson	0.8381
	Z_I-DNA	0.9385
	Z_{II}-DNA	0.7544
CH6	A-DNA	0.4715
	Alt-BDNA	0.2515
	B-DNA	0.1330
	C-DNA	0.1264
	D-DNA	0.0755
	Olson's-DNA	0.6715
	Levitt	0.3421
	Dickerson	0.3604
	Z_I-DNA	0.6395
	Z_{II}-DNA	0.6227
CH1′	A-DNA	0.1407
	Alt-BDNA	0.2824
	B-DNA	0.3070
	C-DNA	0.4846
	D-DNA	0.5854
	Olson's-DNA	0.2118
	Levitt	0.1890
	Dickerson	0.2771
	Z_I-DNA	0.5154
	Z_{II}-DNA	0.5912
GH8	A-DNA	0.4606
	Alt-BDNA	0.4250
	B-DNA	0.2291
	C-DNA	0.2302
	D-DNA	0.1593
	Olson's-DNA	0.3025
	Levitt	0.4277
	Dickerson	0.4424
	Z_I-DNA	−0.0217
	Z_{II}-DNA	0.0226
GH1′	A-DNA	0.2005
	Alt-BDNA	0.2740
	B-DNA	0.4054
	C-DNA	0.5664
	D-DNA	0.6917
	Levitt	0.2304
	Dickerson	0.4149
	Z_I-DNA	0.0238
	Z_{II}-DNA	0.0485

were obtained in 3 mM NaCl buffered at neutral pH. The data were obtained at 81°C. Such a high temperature was used because at lower temperatures the resonances were too broad to measure the shifts accurately. The onset of the melting of the duplex was at 89°C so much so our structure in low salt corresponds to the form, a few degrees before melting. It is possible that our low salt temperature corresponds to the domain of high energy transient base pair breakage and breathing.[33,34] The difference in chemical shifts between the monomers and the duplex poly(dG-dC)•poly(dG-dC) ($\Delta\delta$) at high and low salt conditions are in Table III. The assignments of the various resonances employed in the present study are identical to the one employed previously.[14]

Spatial Configuration of Poly(dG-dC)•Poly(dG-dC) in High Salt Solution

A comparision of the computed shielding constants for the ten different spatial configurations of DNA with the experimental data for poly(dG-dC)•poly(dG-dC) at high salt brings out a revealing story. In view of the large number of structures and the numbers to compare, we show in Figure 10 a computer drawn histogram for effective visual comparison. We have indicated elsewhere[14] that in comparing computed shielding constants for a given structure with experimental data, one has to look for agreement individually for a collection of protons and that an agreement between theoretical computation and experimental data should be considered excellent if the difference is about 0.1 ppm and very good to fair if it is about 0.15 to 0.2 ppm. Any difference beyond 0.25 ppm is poor and unsatisfactory. Further certain protons are more crucial than others. For example in the present case CH5 whose shift is entirely dependent upon overall geometry and little affected by small internal local fluctuations should show agreement within 0.1 ppm.

Examination of the data in Table III and the histogram in Figure 10 shows that out of the five separate sites (i.e., CH5, CH6, CH1′, GH8 and GH1′) we have used in the double helix there is overall agreement between the computations and experimental observations in high salt in each case for Z_I-DNA (Figure 6); in four sites (CH5, CH6, GH8 and GH1′) the agreement is excellent (0.02 to 0.12 ppm) and in one case (CH1′) it is fair (0.22 ppm); it may be noted that the theory overestimates the shielding of CH1′ by 0.22 ppm. It is crucial to realize that a small fluctuation in the χ_{CN} of cytosine which will cause a 3° to 4° increase in the average χ_{CN} in solution of dC of Z_I-DNA (i.e., $\chi_{CN} \simeq$ 24°-25° instead of presently used 21°) will practically compensate the above and will not affect significantly $\Delta\delta$ CH6 and this will result in excellent agreement with experimentally observed data. This is because the shifts of CH6 and CH1′ are such that δCH1′ is considerably sensitive and δCH6 little sensitive when χ_{CN} is increased from 20° toward 40°.

In the case of Z_{II}-DNA (Figure 7) the data starts with a disagreement 0.3 ppm between calculated and that experimentally observed for CH5. This is a serious disagreement because the shift of CH5 is independent of local fluctuations and suggest that Z_{II}-DNA may be an untenable structure for poly(dG-dC)•poly(dG-dC) in high salt. It is true that for the Z_{II} form one notices excellent agreement for GH8

Table III
The Theoretically Computed Magnetic Shieding Constants and Those Experimentally Observed for poly(dG-dC)•poly(dG-dC) High and Low Salt Conditions

Proton	Model	Computed ($\Delta\delta$)	High salt $\Delta\delta$	Low salt $\Delta\delta$
CH5	A-DNA	1.49	1.05	0.82
	Alt-BDNA	0.46		
	B-DNA	0.85		
	C-DNA	0.65		
	D-DNA	0.51		
	Olson's-DNA	0.24		
	Z_I-DNA	0.95		
	Z_{II}-DNA	0.75		
	Levitt Propeller	0.71		
	Dickerson Propeller	0.84		
CH6	A-DNA	0.20	0.62	0.70
	Alt-BDNA	0.72		
	B-DNA	0.73		
	C-DNA	0.73		
	D-DNA	0.68		
	Olson's-DNA	1.68		
	Z_I-DNA	0.51		
	Z_{II}-DNA	0.45		
	Levitt Propeller	0.34		
	Dickerson Propeller	0.36		
CH1′	A-DNA	0.45	0.69	0.56
	AH-BDNA	0.42		
	B-DNA	0.58		
	C-DNA	0.75		
	D-DNA	0.86		
	Olson's-DNA	1.15		
	Z_I-DNA	0.91		
	Z_{II}-DNA	0.77		
	Levitt Propeller	0.19		
	Dickerson Propeller	0.28		
GH8	A-DNA	−1.12	0.30	0.27
	Alt-BDNA	−0.33		
	B-DNA	−0.15		
	C-DNA	−0.15		
	D-DNA	−0.22		
	Olson's-DNA	0.34		
	Z_I-DNA	0.18		
	Z_{II}-DNA	0.22		
	Levitt Propeller	−0.44		
	Dickerson Propeller	−0.80		
GH1′	A-DNA	−0.21	0.15	0.41
	Alt-BDNA	−0.12		
	B-DNA	0.25		
	C-DNA	0.41		
	D-DNA	0.53		
	Olson's-DNA	0.42		
	Z_I-DNA	0.17		
	Z_{II}-DNA	0.20		
	Levitt Propeller	−0.35		
	Dickerson Propeller	−0.21		

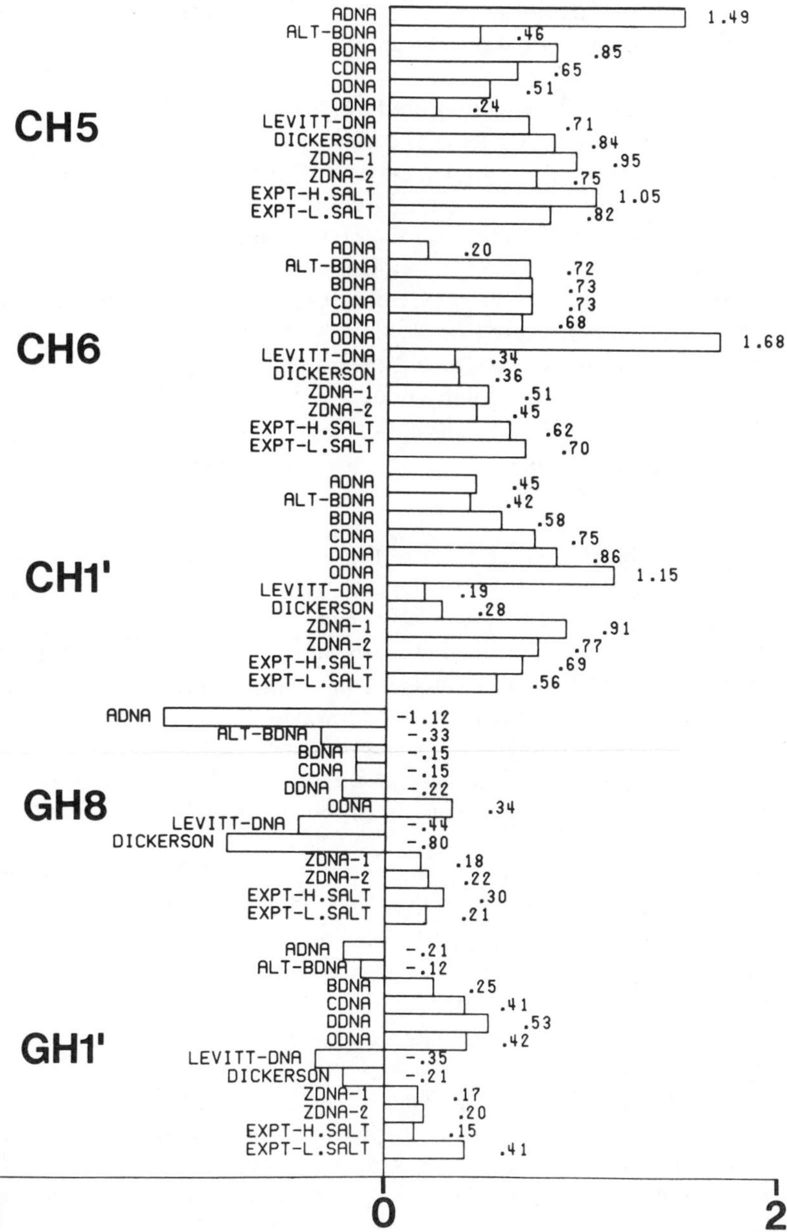

Figure 10. The histogram corresponding to Table III. The O-DNA is the vertical double helix.

and GH1′, but this is most likely accidental because both Z_I and Z_{II} forms predict the same magnitude for GH8 and GH1′. Particularly noteworthy is that the projected values for Z_{II} overestimates the shielding of CH1′ only by 0.08 ppm, but underestimates the same for CH6 by as much as 0.17 ppm. The magnitude and direction of these shifts are such that they cannot be internally compensated by a few degrees of fluctuations about the employed value of 33° for χ_{CN} of dC in Z_{II}.

The vertical double helix was advocated by Olson[7] as a model for the high salt form of poly(dG-dC)•poly(dG-dC) from theoretical considerations. Though this is an opulent and a rich piece of biological architecture (Figure 5; see Dhingra and Sarma[35] for a breathtaking space filling color stereograph of the vertical double helix), the data in Table III and Figure 10 resoundingly show that this is totally an untenable structure for poly(dG-dC)•poly(dG-dC) in high salt. The data show that the experimental $\Delta\delta$ are off by −0.81, 1.06, 0.46 and 0.27 ppm for CH5, CH6, CH1′ and GH1′ from the projected ones. The overwhelming underestimation of the shielding of CH5 is due to insufficient overlap and stacking among the bases in a given strand; the large overestimation of the shielding of CH6 and CH1′ is partly due to the high χ value of 122°, a situation in which CH6 has moved away from the magnetic anisotropy of sugar ring oxygen and CH1′ away from the same of >C=O at C2 of the heterocycle.

Even though in the alternating B-DNA[5] (Figures 4) the repeat unit is a dinucleotide like Z_I-DNA, the data (Table III, Figure 10) suggest that this cannot be the true structure for poly(dG-dC)•poly(dG-dC) in high salt. The structure predicts shielding of CH5 by 0.46 ppm, whereas the observed value is 1.05 ppm. Further the structure overestimates the shielding of CH6 by 0.1 ppm and underestimates the shielding of CH1′ by 0.27 ppm. Local fluctuations of a few degrees around the employed χ_{CN} of 76° for dC cannot internally compensate these opposite trends for CH6 and CH1′. Also notice that there is a discrepancy of −0.63 ppm for GH8 and −0.27 ppm for GH1′ between what is projected by the structure and what is observed.

Arguments similar to the above using data in Table III, Figure 10 can be used to demonstrate the classical structures like the A, B, C and D forms or the new ones like the Levitt and Dickerson propeller twist forms do not hold true for poly(dG-dC)•poly(dG-dC) in high salt.

From the above discussion it is obvious that among the ten different spatial configurations for DNA examined i.e., A. B. C. D forms of DNA, the vertical double helix, the alternating B-DNA, the $Z_I$and Z_{II}-DNAs, and the two propeller models only the left-handed Z_I-DNA correctly predicts the observed experimental data. It seems safe to conclude that Z_I-DNA is most likely the and the only spatial configuration that poly(dG-dC)•poly(dG-dC) in high salt may assume. The time averaged, fluctuation averaged solution structure is essentially identical to the solid state structure except that in the solution state fluctuations about the χ_{CN} of dC may displace this χ_{CN} to an average value of 24°-25° rather than the 21° projected from the single crystal studies of the self-complementary d-CGCGCG hexamer. It is crucial to realize that an average of 25° *does not* mean that their extreme values could be anything provided the average is 25°. This is because the shifts are not linearly dependent on the magnitude of torsion. For example if the χ of dC changes from 20° to 30°, δCH6 will be unaffected, but if it goes from 20° to 10°, δCH6 will experience a high field shift of 0.4 to 0.3 ppm! Because the conclusions are arrived at by examining the shift patterns of a fair number of protons and because the direction and magnitudes of the shifts are nonlinearly sensitive to torsional events, it is an inescapable conclusion that in these instances when we derive a single average solution structure, for example Z_I-DNA for poly(dG-dC)•poly(dG-dC) in high salt, it is a real structure with finite lifetime—it is not an average of some widely different spatial configurations.

It is necessary to point out that due to the lack of the availability of refined coordinates we have been unable to examine the S-DNA of Arnott et al.[36]

Spatial Configuration of Poly (dG-dC)•Poly(dG-dC) in Low Salt

Examination of the projected magnetic shielding constants for the various forms of DNA *vis-a-vis* the experimentally observed $\Delta\delta$ for poly(dG-dC)•poly(dG-dC) in low salt (Table III, Figure 10) provides insights regarding its structure under low salt conditions. Such comparison as was done before for the high salt form (where heavy weight was placed for the shift of CH5, the internal compensation between the shifts of CH6 and CH1′ due to fluctuations about the χ_{CN} of dC was taken into account and overall agreement for a collection of protons were sought) clearly show that A-DNA, D-DNA, the alternating B-DNA, the vertical double helix, the Z_I, Z_{II} DNA's or the propeller twisted forms of Levitt or Dickerson cannot be the true structure of poly(dG-dC)•poly(dG-dC) in low salt solutions.

In the case of the classic Arnott and Hukins[3] B-DNA, there is excellent agreement between projections and what is experimentally observed at three sites ie CH5, CH6 and CH1′; in the case of GH1′ the agreement is good to fair i.e., within 0.17 ppm; but GH8 shows a violent disagreement of up to 0.36 ppm. If one takes C-DNA the situation is somewhat same, the agreement for CH5 and CH1′ is less satisfactory than that for B-DNA, but that for GH1′ and CH6 are excellent. However one should place high weight on CH5 shift and should be able to explain discrepancy in CH1′ and CH6 due to fluctuations in the χ_{CN} of dC. Under these operational criteria C-DNA fails even though we realize that the torsion angles for the AH/B- and C forms are close (Table I) and in solution internal fluctuations may allow accessibility to both forms. It is our thesis that the time averaged fluctuation averaged structure lies close to the AH/B- form and that the large discrepancy in GH8 shifts represent a special dynamical situation which the time averaged and space averaged fiber structure does not recognize i.e., the ability of guanine to assume *syn* conformation under our experimental conditions ie at 81°C, 8°C below the onset of melting. It is possible that for very high temperature melting polymer duplexes, this temperature range may represent the high energy breathing mode in which there is transient base pair opening and some very small percentage of bubble formation.

Our observation that as we increase the temperature from 70°C to 81°C, there is detectable change in the line width suggest the onset of detectable motion or high energy breathing at about 81°C under our experimental conditions. Our data can be rationalized on the basis of an equilibrium between large populations of AH/B-DNA and very small populations of a model in which the dG is free to adopt *syn* conformation. We are unable to obtain any detailed information about the small population of the *syn* form. It could originate from a bubble in which the dG is *syn;* could be the unwound *syn* helix of Arnott.[37,38]

Examination of the χ dependence of δCH8 in purine nucleotides[27-28] clearly reveals that a change of χ from 30° to 240° can cause an upfield shift of CH8 by as much as 1.5 ppm ie even a very small percentage of the bubble in which the χ for dG is free to move from *anti* to *syn* range can explain our data. One cannot rationalize the data on the basis of an equilibrium between AH/B-DNA and Z_I-DNA, unless one is willing to assume significant populations of the latter. There are several lines of independent evidence which can be "stretched" to support our thesis. Examination of the recently solved[13] crystal structure of the self-complementary d-CGCGAATTCGCG shows that the χ_{CN} for dG varies from as low as 60° to as high as 93° in the double helix;[31] in the crystal structure of double stranded d-CGCG and d-CGCGCG, the dG in *syn;*[11,39] these suggest that even in the organized double helix the χ_{CN} about dG can vary considerably; extensive solution studies of mononucleotides and single stranded oligonucleotides which are akin to bubbles have clearly shown[25] that unlike the pyrimidine systems, the χ_{CN} of purine systems show considerable flexibility so much so they exist as syn ⇔ anti equilibrium.

Are the G•C Pairs Propeller Twisted in Low Salt?

Recently Crothers and coworkers[40] have concluded that the base pairs in low salt poly(dG-dC)•poly(dG-dC) are not flat but propeller twisted and the twist is smaller than what they reported for calf thymus DNA.[41] This does not at least *prima facie* agree with our conclusion that the low salt form of poly(dG-dC)•poly(dG-dC) is the classic Arnott and Hukins' B-DNA in which the bases are flat. Before we present our explanations for these two apparently different findings from two separate spectroscopic methods, we want to examine available information in the literature.

Examination of the crystal structure of the self-complementary d-CGCGAATT-CGCG[2,13] clearly illustrate that the AATT region is prominently propeller twisted and that as one moves toward the ends in the GCG and CGC regions propeller twists becomes very small (See Figs. 2b and 3b in reference 50). The crystal structure of the miniature double helix GpC clearly shows[42,43] that the bases are flat. These observations along with the fact that there are three H bonds in GC pairs suggest that GC pairs may have an intrinsic tendency to be flat.

However Crothers and coworkers do not see a pronounced propeller twist in the GC systems. The data in Table III make it vividly clear that Levitt and Dickerson A•T propeller models are totally untenable for the GC series. Comparison of the computed magnetic shielding constants with experimental data for the low salt form shows that among the *five* sites examined only in *one,* these propeller models agree. For example for CH6 Levitt and Dickerson models predict shielding of 0.35 ppm where as what is observed is 0.70 ppm; at CH1′, the models predict 0.19 and 0.18 ppm respectively, but what is observed is 0.56 ppm. In the case of GH8 and GH1′ these models are off by 0.6 to over 1 ppm. There exists no internal motional mechanism in DNA to correct for these outrageous discrepancies. *Obviously the Levitt and Dickerson A•*T propeller twist models are untenable for poly(dG-dC)•poly(dG-dC) in low salt.

An examination of the energy minimized (*a la* Levitt) alternating B-DNA of Klug[5] in stereo (Fig. 4) will reveal that here the (G•C) base pairs are indeed propeller twisted. This interesting aspect of the structure has not been discussed by Klug and coworkers.[5] It was pointed out earlier that the data in Table III do not support the model of Klug et.al. for G•C series either in low or high salt.

Poly(dG-dC)•poly(dG-dC) in low salt can essentially maintain an organized structure of AH/B-DNA and could have a small twist between base planes of ≃ 10°, distributed +5° for G and −5° for C (or −5° for G and +5° for C) from the AH/B-DNA base plane and there could be interconversion ie (+5°) G-C (−5°) ⇔ (−5°) G-C (+5°). In the NMR, especially at the employed 81°C, it will appear as flat base pairs. It is possible that electric dichroism[40] is able to pick up such nuances and subtleties to which NMR with its long time scale is transparent. Obviously the total twist should be in the neighborhood of 10° and should be symmetrically distributed between the base pairs from the plane of AH/B-DNA. If it is larger, the non-linear and directional dependence of shifts *vis-a-vis* twist will reflect in the NMR data.

Effect of Methylation on the Sense of Helix in dG-dC Systems.

Recently Rich and coworkers have discovered that 7-methylation of the G in poly(dG-dC)•poly(dG-dC) under low salt physiological conditions will promote the transition from the right-handed B to the left-handed Z-form.[44] It is important to note that in Z-DNA the sugar ring of dG is in ^{3}E conformation and it has been demonstrated by Kim and Sarma[45] that 7-methylation of Gp promotes the equilibrium ^{2}E ⇔ ^{3}E to the right so much that 7-methyl 5′GMP is the only purine mononucleotide that is known to exist in solution with a preference for ^{3}E sugar pucker. Even though this effect on sugar pucker may play some role in the promotion of B ⇔ Z transition, a more important effect may originate from electrostatic forces as has been pointed out by Rich.[44] The 7-methylation makes the guanine a cation, thereby reducing the effective charge on the DNA molecule. Since the phosphates are closer together in Z-DNA than in B-DNA, electrostatic rationalizations for this effect are self-evident. In this connection it is interesting to note that Felsenfeld and coworkers[46,54] discovered that methylation of the 5-position of cytosine in dG-dC systems, with a touch of Mg^{++}, will cause B → Z transition. Rich[44] has suggested that this effect comes from the fact that the 5-methyl group of cytosine is able to stabilize the Z-form by its interaction with the hydrophobic domains of Z-form and that the geometries of B-DNA do not allow such interaction.

Effect of Cross-Linking Ligands on the B → Z Transition

Mercado and Tomaz[47] have shown that mitomycin very specifically interacts with dG-dC sequences and that mitomycin cross-linked poly(dG-dC)•poly(dG-dC) has a CD very similar to that in high salt solutions. Lippard has shown that the monofunctional chlorodiethylenetriamineplatinum (II) interacts with poly(dG-dC)•poly(dG-dC) and inverts the CD spectrum.[48] These are indications that drugs and related ligands may express their biological activity by promoting B→Z transition. In fact Pullman,[49] elsewhere in this volume, provides detailed electrostatic computations on B-DNA and Z-DNA and the data suggest that the Z-form is considerably more accessible and reactive toward carcinogenic species.

Plasticity of the DNA Double Helix and Coming of Age of Z-DNA

It is exciting to note that a DNA of dG-dC•dG-dC sequence can take up a left-handed Z_I-spatial configuration in one set of conditions and in another set of conditions it can undergo so much local structural alterations that the very morphology and handedness display dramatic changes—a change from a thinner helix devoid of major grooves to a thicker helix with distinct major and minor grooves—the AH/B-DNA. (See Dickerson et. al.[2] for a scaled display in stereo of the dimensions of A, B and Z-forms). Elsewhere in the volume we[50] demonstrate that in *solution* the AATT stretches of the double helix display pronounced propeller twists *a la* Levitt, Crothers and Dickerson,[6,13,40,41] and that the AATT can induce the nearest neighbor GC's to assume propeller shapes.

Sarma, Ikehara and their coworkers[51] recently demonstrated that changing the χ_{CN} from 80° to 120° in a double helix changes the handedness from right to left *a la* Sundaralingam.[52] These results along with what we documented in this opus clearly suggest the rich plasticity in the structure of the DNA double helix and its ability to assume sequence and ionic strength dependent distinct spatial configurations and it is likely that this plasticity plays a crucial role in the control of genomic expression.

When the Z-DNA came on the scene in late 1979, there was considerable skepticism whether this was simply an artifact of crystallization conditions or a true structure with biological relevance. What is abundantly clear, from the large number of articles in this volume, is that the Z-DNA has weathered the rigorous scrutiny and has come of age. Wells and coworkers elsewhere in this volume[53] demonstrate that both the B and Z forms coexist in a recombinant plasmid. From his antibody studies Rich[44,55] has shown that the Z-form is present in the interband regions of Drosophilia polytene chromosomes. The observation that 5-methylation of cytosine in poly(dG-dC)•poly(dG-dC) promotes B to Z transition[46,54] may have bearing on control of expression because it has been suggested that such methylation in eukaryotes is related to genomic expression. Under these conditions, one cannot but speculate whether the left-handed helix may be playing a role in this control. It is also possible that the Z-helix is involved in carcinogenic processes. For example, the interaction of the N-acetoxy derivative of the hepatic carcinogen acetylaminofluorene with poly(dG-dC)•poly(dG-dC) results in random modification of the double helix[56,59] with characteristics similar to Z-DNA. These are the beginnings of the indications that the left-handed Z-helix may indeed have important biological functions, but, as has been pointed out by Stephen Neidle,[60] one indeed keenly awaits firm evidence that any biological function actually does involve a switch of helicity from B to Z.

So far experimental studies have shown that the double helix exists in three distinct, conformationally active spatial configurations—the A, B, and Z forms. It is likely that this triad controls genomic expression by adopting structural variations within the allowed conformation space of each family. Due to the multidimensionality of its conformation space, it is also possible that there exist structurally distinct DNA double helices not discovered by man yet. It has been inscribed in the scriptures:

Although I appear in so many configurations
You know not my true transcendental form

Lord Krishna in *Bhagavad-gita*

References and Footnotes

1. a. Papers I and II in this series are references 14 and 15.
   b. The authors are deeply indebted to Professor Bernard Pullman and Dr. Claude Giessner-Prettre for providing them with the tensor elements to compute the contribution to shielding from the paramagnetic and diamagnetic components of the atomic magnetic anisotropy and the ring current constants to compute the contribution to shielding from ring current effects. We are indebted to Professor Alexander Rich for providing us the coordinates of Z_I-DNA and Z_{II}-DNA and to Professor Wilma K. Olson for the coordinates of the vertical double helix. We thank Professors M. Levitt and R. E. Dickerson for making available the coordinates of their respective propeller twist forms. We thank Professor Donald M. Crothers for communicating his results before publication and Dr. D.J. Patel for our collaborative work on d-CGCGAATTCGCG. This research was supported by grants from National Cancer Institute of NIH (CA12462) and from National Science Foundation (PCM7822531). The NMR measurements were conducted at the North East NSF NMR facility at Yale.

2. Dickerson, R.E., Drew, H., and Conner, B., in *Biomolecular Stereodynamics, Volume I,* Ed., Sarma, R.H., Adenine Press, N.Y., p. 1 (1981).
3. Arnott, S. and Hukins, D.W.L., *Biochem. Biophys. Res. Commun, 47,* 1504 (1972).
4. Rich, A., Quigley, G.J., and Wang, A. H-J., in *Biomolecular Stereodynamics, Volume I,* Ed., Sarma, R.H., Adenine Press, N.Y., p. 35 (1981).
5. Klug, A., Jack, A., Viswamitra, M.A., Kennard, O., Shakked, Z., and Steitz, T.A., *J. Mol. Biol. 131,* 669 (1979).
6. Levitt, M., *Proc. Natl. Acad. Sci., USA 75,* 640 (1978).
7. Olson, W.K., *Proc. Natl. Acad. Sci., USA 74,* 1775 (1977).
8. Sasisekharan, V., Bansal, M., Brahmachari, S.K., and Gupta, G., in *Biomolecular Stereodynamics, Volume I,* Ed., Sarma, R.H., Adenine Press, N.Y., p. 123 (1981).
9. Marvin, D.A., Spencer, M., Wilkins, M.F.H., and Hamilton, L.D., *J. Mol. Biol. 3,* 547 (1961).
10. Arnott, S., Chandrasekaran, B., Hukins, D.W.L., Smith, P.J.C., and Watts, L., *J. Mol. Biol. 88,* 523 (1974).
11. Wang, A.H-J., Quigley, G.J., Kolpak, F.J., Crawford, J.L., van Boom, J.H., van der Marel, G., and Rich, A., *Nature* 282, 680 (1979).

12. Crawford, J.L., Kolpak, F.J., Wang, A. H-J., Quigley, G.J., van Boom, J.H., van der Marel, G., and Rich, A., *Proc. Natl. Acad. Sci., USA. 77,* 4016 (1980).
13. Wing, R., Drew, H., Takano, T., Broka, C., Tanaka, S., Itakura, K., and Dickerson, R.E., *Nature 287,* 755 (1980).
14. Mitra, C.K., Sarma, M.H., and Sarma, R.H., *Biochemistry 20,* 2036 (1981).
15. Mitra, C.K., Sarma, M.H., and Sarma, R.H., *J. Am. Chem. Soc.* (in press).
16. Pullman, B., and Giessner-Prettre, C. (private communication to RHS).
17. Mitra, C., Sarma, R.H., Giessner-Prettre, C., and Pullman, B., *Internatl. J. Quant. Chem. QBS 7,* 39 (1980).
18. Mitra, C.K., Dhingra, M.M., and Sarma, R.H., *Biopolymers 19,* 1435 (1980).
19. Cheng, D.M., Danyluk, S.S., Dhingra, M.M., Ezra, F.S., MacCoss, M., Mitra, C.K., and Sarma, R.H., *Biochemistry 19,* 2491 (1980).
20. Prado, F.R., Giessner-Prettre, C., and Pullman, B., *J. Mol. Str.* (in press).
21. Lee, C.H., Ezra, F.S., Kondo, N.S., Sarma, R.H., and Danyluk, S.S., *Biochemistry 16,* 3627 (1976).
22. Ezra, F.S., Lee, C.H., Kondo, N.S., Danyluk, S.S. and Sarma, R.H., *Biochemistry 17,* 1977 (1977).
23. Cheng, D.M., and Sarma, R.H., *J. Am. Chem. Soc. 99,* 7333 (1977).
24. Son, T-D, Guschlbauer, W., and Gueron, M., *J. Amer. Chem. Soc. 94,* 7903 (1972).
25. Davies, D.B., *Prog. NMR Spectros. 12,* 135 (1978).
26. Davies, D.B. *Studia Biophysica 55,* 29 (1976).
27. Prado, F.R., Giessner Prettre, C., and Pullman, B., *J. Theor. Biol. 74,* 259 (1978).
28. Giessner, C., and Pullman, B., in *Nuclear Magnetic Resonance Spectroscopy in Molecular Biology,* Pullman, B., Ed., D. Reidel Publishing Co. 147 (1978).
29. Giessner-Prettre, C., and Pullman, B., *J. Theor. Biol. 65,* 189 (1977).
30. Sundaralingam, M. *Jerusalem Symposia on Quantum Chemistry and Biochemistry 5,* 417 (1973).
31. We have derived the torsion angles for the self-complementary d-CGCGAATTCGCT (13) from the xyz coordinates. The average of the χ_{CN} of the dG's vary from a low value of 60° to a high value of 93° in the double helix.
32. Patel, D.J., in *Stereodynamics of Molecular Systems,* R.H. Sarma, Ed., Pergamon Press 397 (1979).
33. Kallenbach, N.R., Mandel, C., and Englander, S.W., in *Nucleic Acid Geometry and Dynamics,* Ed. R.H. Sarma, Pergamon Press, 233 (1980).
34. Nakanishi, M. and Tsuboi, M., *J. Mol. Biol. 124,* 61 (1978).
35. Dhingra, M.M. and Sarma, R.H. *Internatl. J. Quant. Chem. QBS 6* 131, (1979).
36. Arnott, S., Chandrasekaran, R., Birdsall, D.C., Leslie, A.G.W. and Ratliff, R.L., *Nature 283,* 743 (1980).
37. Arnott, S., Chandrasekaran, R., Bond, P.J., Birdsall, D.L., Leslie, A.G.W. and Puigjaner, L.C. in *Structural Aspects of Recognition and Assembly in Biological Macromolecules,* Eds., M. Balaban, J. Sussman, W. Traub, and A. Yonath; Balaban 1SS, Rehovot *2,* 487 (1981).
38. Arnott, S., Bond, P.J., and Chandrasekaran, R., *Nature 287,* 561 (1980).
39. Drew, H., Takano, T., Tanaka, S., Itakura, K., and Dickerson, R.E. *Nature 286,* 567 (1980).
40. Wu, H.M., Dattagupta, N., and Crothers, D.M., private communication to RHS from DMC.
41. Hogan, M., Dattagupta, N., and Crothers, D.M., *Biochemistry 18,* 280 (1979), and private comunications to RHS from DMC.
42. Seeman, N.C., Rosenberg, J.M., Suddath, F.L., Kim, J.J.P., and Rich, A., *J. Mol. Biol. 104,* 109 (1976).
43. Hingerty, B., Subramanian, E., Stellman, S.D., Sato, T., Broyde, S. B., and Langridge, R., *Acta Cryst. B32,* 2998 (1976).
44. Alexander Rich at the inaugural address: Second SUNYA Conversation in the Discipline Biomolecular Stereodynamics, SUNY, Albany, April 26-29, 1981. The written article by Rich and coworkers in the volume does not contain this information; but they are in the audio and video tapes of the conference.
45. Kim, C.H., and Sarma, R.H., *J. Am. Chem. Soc. 100,* 1571 (1978).
46. Gary Felsenfeld in the delivered lecture at the Second SUNYA Conversation in the Discipline Biomolecular Stereodynamics, SUNY, Albany, April 26-29, 1981.
47. Mercado, C.M., and Tomaz, M., *Biochemistry 16,* 2040 (1977).

48. Lippard, S.J., in *Biomolecular Stereodynamics, Volume II,* Sarma, R.H., Ed. Adenine Press, N.Y., p. 165 (1981).
49. Pullman, B., in *Biomolecular Stereodynamics, Volume I,* Sarma, R.H., Ed. Adenine Press, N.Y., p. 163 (1981).
50. Sarma, R.H., Wagner, B.J., and Mitra, C.K., in *Biomolecular Stereodynamics, Volume I,* Sarma, R.H., Adenine Press, N.Y., p. 89 (1981).
51. Dhingra, M.M., Sarma, R.H., Uesuji, S., Shida, T., and Ikehara, M., *Biochemistry* (in press).
52. Sundaralingam, M., and Yathindra, N., *International J. Quant. Chem. QBS 4,* 285 (1977).
53. Wells, R.D., Klysik, J., Stirdivant, S.M., Hart, P., and Larson, J.E. in *Biomolecular Stereodynamics, Volume I,* Adenine Press, N.Y., p. 77 (1981).
54. Behe, M. and Felsenfeld, G. *Proc. Natl. Acad. Sci. USA 78,* 1919 (1981).
55. Lafer, E.M., Moller, A., Nordheim, A., Stollar, B.D. and Rich, A., *Proc. Natl. Acad. Sci. USA 78,* 3546 (1981).
56. Sage, E. and Leng, M., *Proc. Natl. Acad. Sci. USA 77,* 4597 (1980).
57. Sage, E. and Leng, M., *Nucleic Acids Research 9,* 1241 (1981).
58. Santella, R.M., Grunberger, D., Weinsteing, I.B. and Rich, A., *Proc. Natl. Acad. Sci. USA 78,* 1451 (1981).
59. Broyde, S., Higerty, B., and Stellman, S., in *Biomolecular Stereodynamics Volume II,* Ed. Sarma, R.H., Adenine Press, N.Y., p. 455 (1981).
60. Neidle, S., *Nature 292,* 292 (1981).

Proceedings of the Second SUNYA Conversation in the Discipline Biomolecular Stereodynamics
Volume I, ISBN 0-940030-00-4, Ed., Ramaswamy H. Sarma,
Adenine Press, New York, ©Adenine Press

Left-Handed DNA in Restriction Fragments and a Recombinant Plasmid

R. D. Wells, J. Klysik, S. M. Stirdivant
J. Larson and P.A. Hart
Department of Biochemistry and School of Pharmacy
University of Wisconsin—Madison
Madison, Wisconsin 53706

Introduction

The role of DNA structure in gene regulation has been reviewed recently.[1,2] The fundamental concept in these studies is that DNA properties and conformation have an intimate role in gene regulation and that DNA does not have the same conformation throughout its entire length. Instead, DNA has interesting structural features at certain loci; one may think of DNA with respect to conformations as "microheterogeneous." A corollary to this notion is that DNA structure is not unalterable but may be perturbed when certain proteins are bound. We have described elsewhere[1] different types of DNA conformations and their properties including that of several types of non-DNA B conformations as well as cruciforms.[3,4] One of the most dramatically unusual DNA conformations, the Z-family, has been described recently (reviewed in 1).

Left-Handed DNA

The notion of left-handed helices is not new. In 1970, Wells *et al.*[5] showed that $(dI\text{-}dC)_n$ adopts an unusual conformation which was interpreted as a left-handed double helix on the basis of CD and X-ray studies on oriented fibers. It was proposed[5] that these types of structures are recognition sites for regulatory proteins. Moreover, the conformational feasibility of left-handed helices has been reported (reviewed in 1).

A substantial contribution to this area of research was made when Rich *et al.*[10,11] and Dickerson *et al.*[12] determined the crystal structures of small oligonucleotides containing (dC-dG) residues. This work revealed a very unusual conformation. The helical symmetry is left-handed and the distribution of nucleotides along the helix is not as even as in B-DNA. However, the base pairs are in an anti-parallel arrangement with Watson-Crick hydrogen bonds. Whereas all common DNA structures described so far require the nucleotides in the *anti* conformation for the base with respect to the deoxyribose, all guanosine residues in this structure exhibit the *syn* conformation (reviewed in 1). In addition, the oligomers have unusual deoxyribose puckers. The result is a left-handed helix with several unusual parameters which the authors call the Z-form. Rich *et al.* conclude that this conformation corresponds to the high salt form of $(dG\text{-}dC)_n$ as revealed by circular dichroism and Raman spectroscopy (reviewed in 1). Furthermore, Arnott *et al.* (reviewed in 1) have interpreted X-ray diffraction patterns on several DNA polymers and conclude, in agreement with Rich *et al.*, that the repeating purine-pyrimidine sequences may adopt left-handed structures.

^{31}P NMR measurements performed on oligomers of (dG-dC) in high concentrations of salt show two distinct peaks of roughly the same intensity for the phosphate residues (reviewed in 1). This observation is consistent with the interpretation that the alternating phosphodiester bonds may adopt different conformations. Recently Sarma and coworkers[13] computed the shielding constants for poly(dG-dC)•poly(dG-dC) from the x, y, z coordinates of the left-handed Z-DNA[11] and the right-handed B-DNA of Arnott and Hukins[14] and compared the derived NMR data with the experimental NMR data in high salt. The comparison unequivocally revealed that poly(dG-dC) • poly(dG-dC) in high salt solution exists in the left-handed Z-DNA conformation.

Z-type DNA in a Natural Genome

These crystallographic and solution studies on small oligonucleotides and synthetic DNA's containing dC-dG sequences were quite revealing. However, the important question regarding gene regulation is the relevance of these observations to natural DNA and their possible implications. We therefore decided to construct and study a recombinant DNA in which poly dG-dC was cloned in a plasmid DNA.[6] Generation and characterization of an appropriate clone would enable a number of studies on important questions regarding Z-DNA structure including the following: (a) can Z-DNA exist between flanking segments of B-DNA, (b) What length of (dG-dC) is required to provide a stable Z-conformation. (c) What is the nature and length of the junction between the Z-DNA segments and the flanking B-regions. (d) To what extent does the left-handed Z-type structure influence the biological and physiochemical properties of the neighboring sequence of random DNA (B structure) or vice versa. (e) Since a Z-type conformation is left-handed, does a B→Z transition influence the tertiary structure (i.e., supercoil) of a plasmid molecule. Moreover, will the presence of supercoiling alter the environmental conditions necessary to stabilize a Z-type conformation. (f) Do some biological molecules, such as proteins or small ligands (drugs or carcinogens), specifically interact with the Z-form.

The generation of appropriate clones is outlined in Figure 1. The 95 bp *lac* fragment was ligated with dC-dG oligomers generated by *Fnu* DII digestion of the high-molecular weight polymer. This ligation product was then joined with a filled in *Bam* HI site of pBR322 DNA. Since the *Fnu* DII products have a 5′-C, this should regenerate the *Bam* HI sites between the insert and the vector. Transformation of competent cells and screening for blue colonies revealed the presence of the *lac*

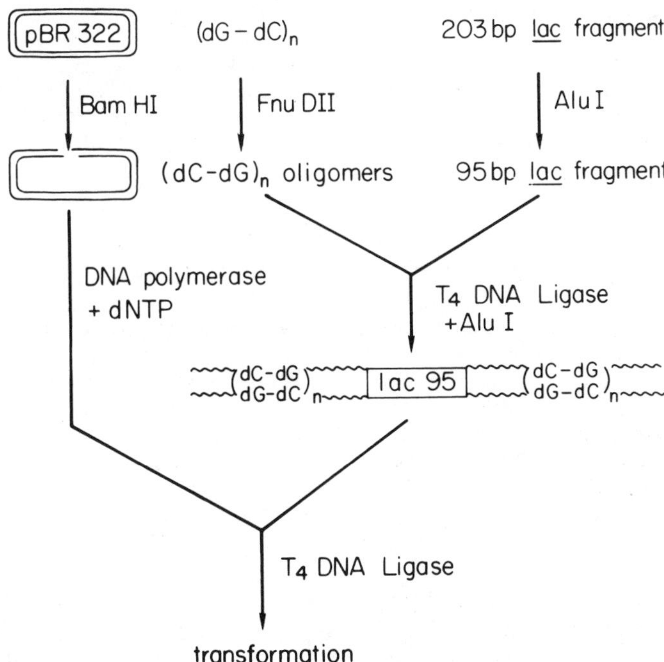

Figure 1. Strategy of cloning of dC-dG tracts flanking the *lac* 95 base pair control region.

repressor binding site.[7] DNA restriction mapping and DNA sequencing studies were performed to totally characterize the inserts.

Figure 2 illustrates two of the molecules which were employed in this study. The recombinant plasmid (pRW751) was constructed which contained an insert 157 bp in length. A full description of the preparation and characterization of this family of recombinant DNAs will be described elsewhere (Klysik and Wells, manuscript in preparation). Digestion of this plasmid DNA by *Bam* HI generates the 157 bp insert and the vector DNA which are easily separated in milligram quantities by high pressure liquid chromatography on RPC-5.[8] Three other related DNA fragments were also prepared and characterized.[6]

Z-DNA in the 157 bp Fragment

Figure 3 shows the CD spectra for the 157 bp fragment in 1.0 M and 5.0 M NaCl. In addition, the calculated spectra are shown under identical conditions for the mixture of the appropriate molar ratio of $(dC\text{-}dG)_n \bullet (dC\text{-}dG)_n$ and the purified 95 bp fragment. In 1.0 M NaCl, the 157 bp fragment shows a typical spectrum observed for DNA fragments.[9] Moreover, there is very close agreement between the observed spectrum and the spectrum calculated for the component parts.

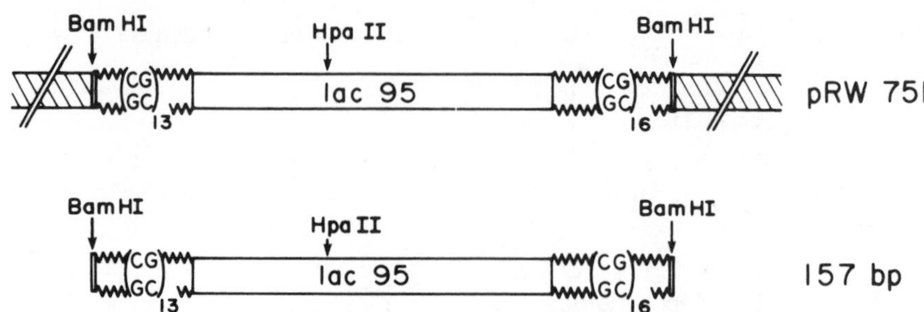

Figure 2. Two DNAs used in these studies. pRW751 is a recombinant plasmid derived from pBR322 which contains a 157 bp insert into the "filled-in" Bam HI site. The cloning procedure, preparation of plasmid DNA, and preparation and characterization of DNA restriction fragments has been described.[6] The 157 bp fragment was sequenced by the procedure of Maxam and Gilbert. The DNA sequence revealed the presence of a perfectly alternating dC-dG sequence shown in this figure along with the presence of the *lac* sequence. The localization of the 26 bp and 32 bp (dC-dG) arms relative to the *lac* 95 bp fragment is precisely known. However, the orientation of the entire 157 bp insert with respect to the vector is not known. pBR322 segments are designated with cross hatching.

However, in 5.0 M NaCl, a dramatic change is observed in the spectrum with negative CD bands at 293 and 245 nm and a positive band at 268 nm. This spectrum agrees well in shape with the calculated spectrum but is lower in amplitude at the peaks at 293 and 268 nm (but not at 245 nm). This general shape in high salt is a result of a marked change in the CD spectrum for $(dG\text{-}dC)_n \bullet (dG\text{-}dC)_n$. The inset in Figure 3 shows the cooperative nature of the change in CD at 295 and 250 nm as expected for the salt-induced B→Z change in conformation in the (dC-dG) arms. In contrast, no cooperative change is observed under similar conditions for the 95 bp *lac* fragment alone. The midpoint of this transition (~3.6 M) is substantially higher than for $(dG\text{-}dC)_n \bullet (dG\text{-}dC)_n$ (~2.5 M), presumably due to the junction with the *lac* sequence.

To confirm that the inversion in the CD spectrum was due to formation of a Z-type conformation in the dC-dG ends of the 157 bp fragment, ^{31}P-NMR studies were performed. Figure 4 shows the spectra observed in 0.5 M and in 5.0 M NaCl. A single symmetrical peak is observed in the lower salt whereas a second resonance, which is 1.45 ppm downfield from the main resonance, was observed in 5.0 M salt. This result is clearly consistent with previous observations, which documented a splitting of similar magnitude in the phosphorous spectrum of the high salt form of oligomers of $(dG\text{-}dC)_n \bullet (dG\text{-}dC)_n$.

Integration of the peaks in Figure 4 indicates that approximately 20% of the phosphorous resonances are in the second small peak which was shifted 1.45 ppm. Since the (dG-dC) segment comprises approximately 37% of the 157 bp fragment, this suggests that virtually all of the (dC-dG) tracts have converted to a Z-type conformation. That the NMR studies indicate that a somewhat larger percentage of the (dC-dG) sequences undergo the transition than deduced from the CD measure-

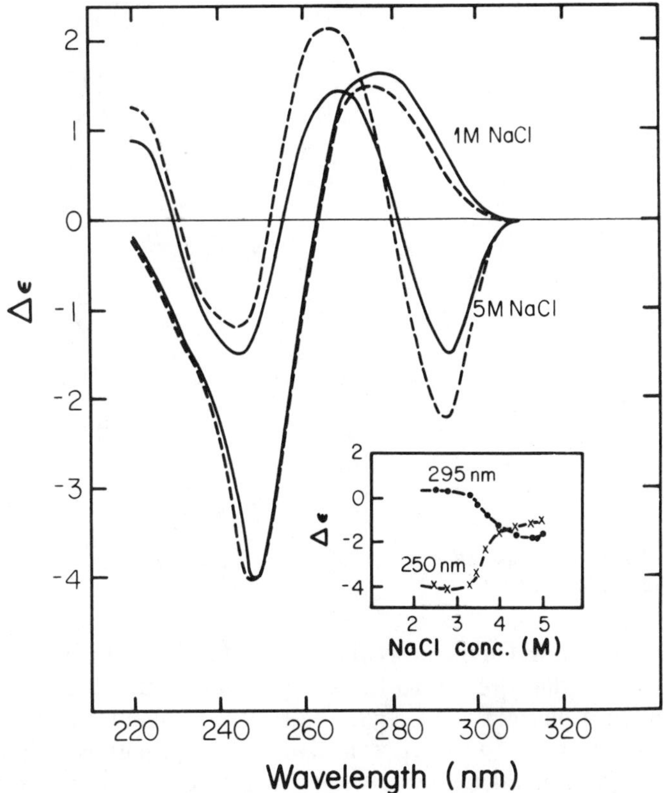

Figure 3. Circular dichroism spectra of 157 bp fragment. The solid lines show the spectra of the 157 bp fragment in 1.0 and 5.0 m NaCl. The dotted lines show the calculated spectra for a mixture of the *lac* 95 bp fragment and $(dG\text{-}dC)_n \bullet (dG\text{-}dC)_n$ at the same molar ratio as in the 157 bp fragment. The 1.0 and the 5.0 M NaCl spectra for the *lac* 95 bp fragment as well as that for the $(dG\text{-}dC)_n \bullet (dG\text{-}dC)_n$ were taken separately. A solution of $(dG\text{-}dC)_n \bullet (dG\text{-}dC)_n$ was equilibrated for 4 hours at room temperature before the spectra were determined. The extinction coefficients used were the following: *lac* 95 bp fragment, 6.5×10^3; $(dG\text{-}dC)_n \bullet (dG\text{-}dC)_n$, 7.1×10^3; the 157 bp fragment, 5.7×10^3. The inset in this figure shows the influence of salt concentration on the CD at 295 and 250 nm. The spectra at the different salt concentrations were performed by diluting the 5.0 M NaCl solution of the 157 bp fragment with 10 mM cacodylate buffer (pH 7.2), 0.1 mM EDTA. After each dilution, the sample was equilibrated for 0.5 hours at room temperature before the CD spectra was determined. The equilibration time for this fragment was actually very short; 95% of the transition was obtained after 7 minutes.

ments, may be due to the fact that different conformational features of the molecule are being investigated.

Laser Raman spectroscopy studies[15] are in complete agreement with these results.

Z-DNA in A Supercoiled Plasmid

It was of interest to determine if a Z-type conformation can exist within a covalently

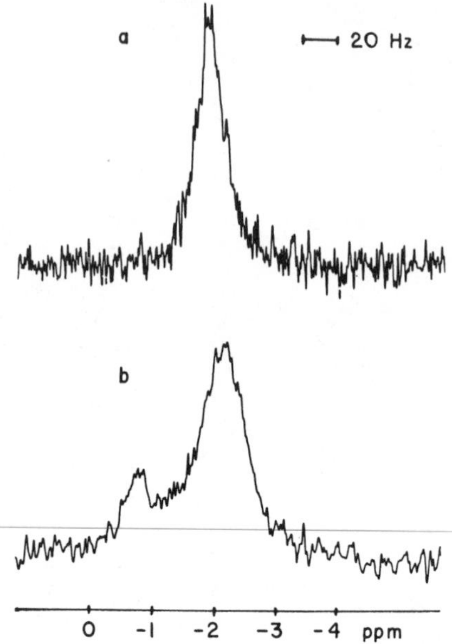

Figure 4. 36.44 MHz [31]-P-NMR spectra of the 157 bp fragment in 0.5 M and 5.0 M NaCl.

closed and circular DNA. On the basis of previous studies on supercoiled DNA, it is logical that a B→Z transition in a (dC-dG) segment should alter the number of supercoil turns in a plasmid DNA since the supercoil number is dictated by the specific linking difference which is sensitive to conformational changes in the primary helix. The general behavior which might be expected is shown in the cartoon in Figure 5.

pRW751 was a suitable molecule for studying this question. The total length of (dC-dG) sequences in pRW751 is 58 bp or 1.3% of the length of the total recombinant plasmid (4516 bp). The unwinding of one turn of primary helix removes one right-hand supercoil. In addition, each turn of left-handed primary helix which is induced will also remove one right-hand supercoil. Thus, we can calculate that a complete transition of 58 bp of (dC-dG) from B→Z would theoretically remove 10.3 supercoils from pRW751 (58 bp/10.4 bp per turn + 58 bp/12 bp per turn = 10.3).

We have established conditions for running agarose gels in NaCl solutions up to 5.0 M (Figure 6) to determine the number of supercoil turns in a DNA. pRW451 served as a control for these electrophoretic experiments since it is identical to pRW751 except that it contains a 174 bp insert (formerly a *Hha* I fragment of pBR322 DNA which contains 54% G+C) instead of the 157 bp insert in pRW751. This 174 bp insert does not contain unusual sequence features nor stretches of (dC-dG) sequence.

Figure 6 shows the gel electrophoretic patterns observed for pRW451 and pRW751 in 3.0 M and 4.5 M NaCl. In 3.0 M, the partially relaxed family of topoisomers for both DNAs comigrate (lanes A and B). For both DNAs under these electrophoresis

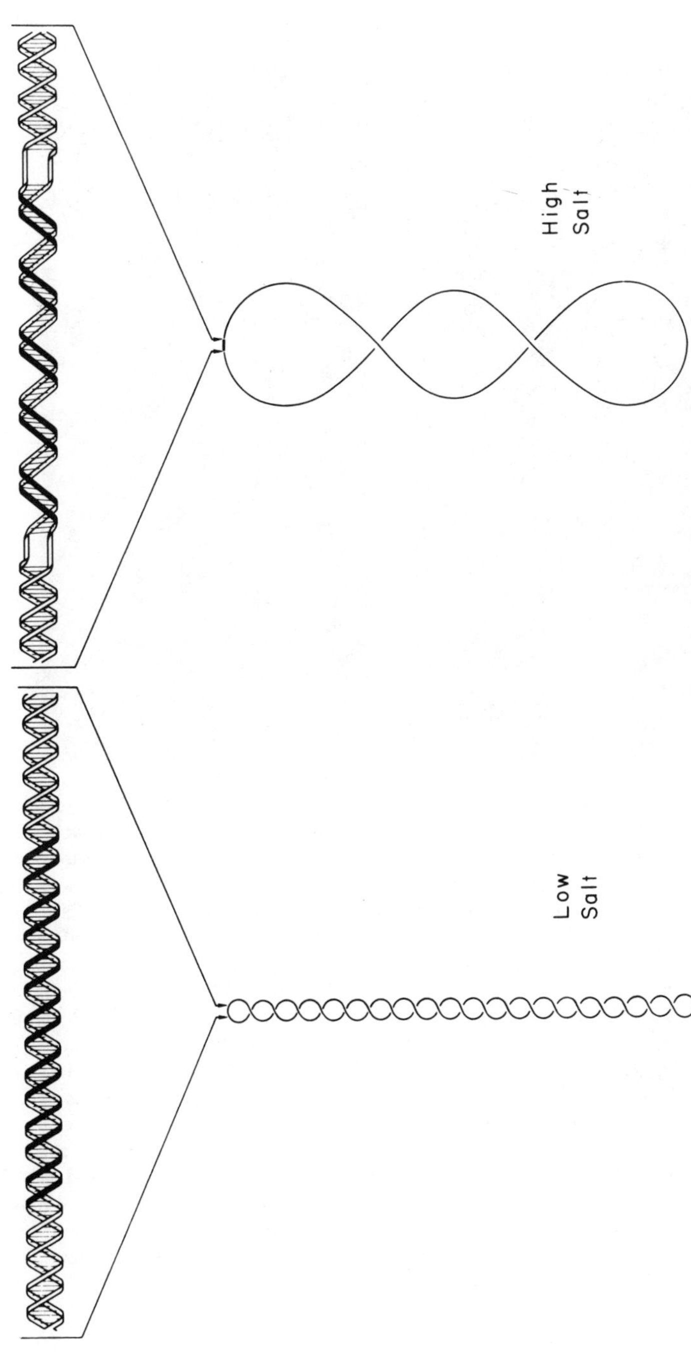

Figure 5. Cartoon depicting capacity of B→Z transition in small segment of recombinant plasmid to unwind supercoiled DNA. The left panel shows the supercoiled DNA with the entire primary helix in a right-handed B-conformation. In the right panel (high salt), a small segment (1.3%) of the recombinant plasmid is converted to the left-handed Z-conformtion thereby partially relaxing the supercoiled DNA.

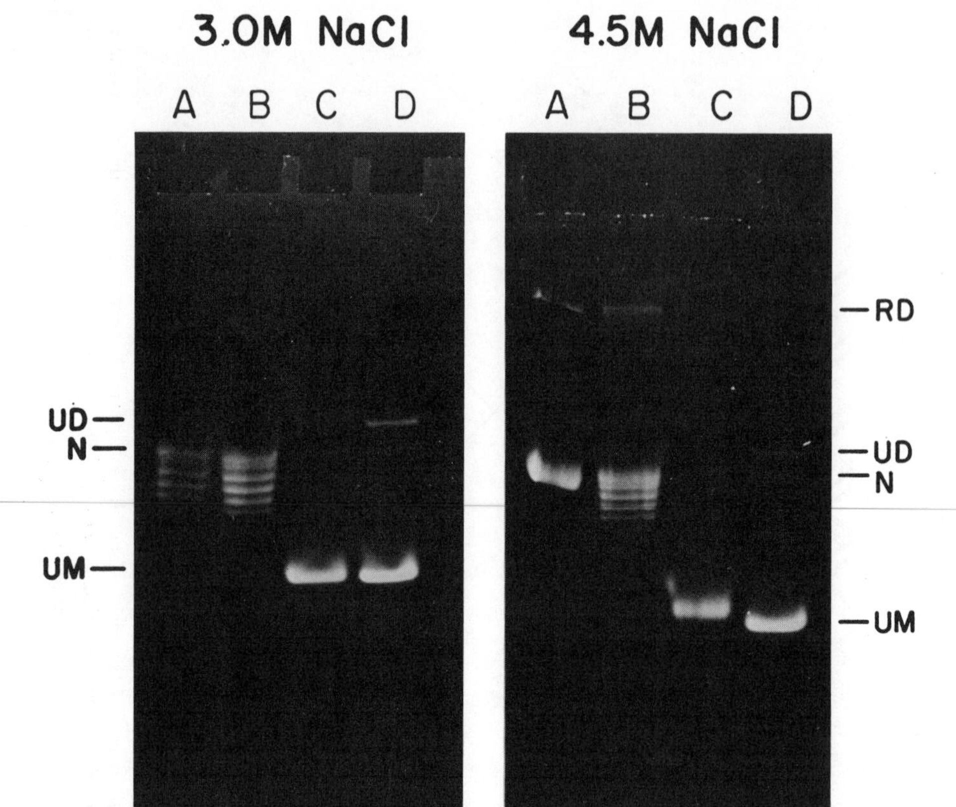

Figure 6. High salt gels on supercoiled plasmids. Monomeric supercoiled pRW751 was separated from higher forms (dimer, tetramer, etc.) by HPLC on RPC-5 using a 1.5 mm x 20 cm column at 43°C. Elution was with a 0.65 M to 0.8 M KCl gradient using the general techniques described previously. pRW451 contains the 174 bp *Hha* I fragment from pBR322 cloned into the "filled-in" *Bam* HI site of pBR322 (G.D. Staffeld and R.D. Wells, unpublished). pRW451 was isolated in a similar manner. Partially relaxed pRW751 and pRW451 were prepared by treatment with wheat germ topoisomerase (gift of R.R. Burgess). The unrelaxed and partially relaxed plasmids were electrophoresed through 1% agarose gels with a buffer of 80 mM Tris-HCl, 40 mM sodium acetate, 4.4 mM disodium EDTA, 0.525 M acetic acid, all at pH 8.3, plus either 3.0 or 4.5 M NaCl. The gels were 8 cm x 14 cm x 1 mm. The DNA samples were equilibrated in a loading buffer identical to the electrophoresis buffer but with an identical molarity of $NaClO_4$ substituted for NaCl and 0.05% bromophenol blue was added. The gels were electrophoresed for 36 hours at 25 volts for the 4.5 M NaCl gel and 20 volts for the 3M NaCl gel. The gels were desalted by soaking in water for 10 minutes and then stained with 100μg/ml ethidium bromide. The abbreviations are: UM, unrelaxed monomer; N, nicked monomer; UD, unrelaxed dimer; RD, relaxed dimer. The following samples were electrophoresed in the designated lanes; (A) partially relaxed pRW751, (B) partially relaxed pRW451, (c) unrelaxed pRW 751, (D) unrelaxed pRW451. Resolution of fully relaxed plasmid from supercoiled DNA containing +1 or −1 supercoils is difficult under these conditions. Similarly, the logarithmic nature of topoisomer migration makes impossible the resolution of individual bands for topoisomers containing ~ -10 supercoils.

conditions, the family ranges from approximately five right-handed supercoil turns to fully relaxed DNA (the exact number of supercoil twists is somewhat uncertain due to our inability to resolve topoisomers near the fully relaxed position, but we believe that this number is accurate within at least ±1). This experiment was performed with plasmids which were partially relaxed with topoisomerase in order to obtain better resolution of the family of topoisomers and thus enhance our ability to detect smaller changes in supercoil number. Lanes C and D show that the unrelaxed plasmids comigrate in 3.0 M NaCl.

When similar studies were done in 4.5 M NaCl, no change was observed for the migration pattern of the partially relaxed control plasmid (pRW451) (lane B). However, lane A shows that all of the topoisomers for pRW751 comigrated under these conditions at the fully relaxed position. Thus, a loss of as many as approximately six supercoil turns were observed, but there was no apparent generation of left-handed supercoil twists.

This striking result clearly demonstrates that a salt-induced structural change occurred in pRW751 which had a dramatic influence on the supercoil number. This result is consistent with a conformational transition from B- to a Z-type structure for at least a portion of the (dC-dG) tracts in pRW751.

A small difference was observed in the migration of pRW451 and pRW751 at 4.5 M NaCl (lanes C and D). Based on our experience with topoisomer migration patterns, we believe this difference is equivalent to a loss of 4-8 supercoil turns. However, the relatively poor resolution of topoisomers of high superhelical density makes difficult a precise quantitation, especially under these electrophoretic conditions.

Electrophoretic studies were performed on the partially relaxed plasmids at several salt concentrations between 3.5 and 4.0 M in addition to those shown in Figure 6 to determine the degree of cooperativity. At salt concentrations of 3.5 M to at least 4.0 M, the topoisomers with more supercoil turns decreased in supercoil number at lower salt concentrations than topoisomers with fewer supercoil turns. This suggests that supercoiling contributes a favorable free energy to the B- to Z-type transition.

Thus, the studies with both the partially relaxed and unrelaxed plasmids indicate an unwinding of ~5 supercoil turns (conceivably not fewer than 4 and not more than 8) in 4.5 M NaCl. In the calculations described above, if the entire (dC-dG) segment in pRW751 converted from a B→Z conformation, we should have observed a loss of 10.3 supercoil turns. Since this large an effect was not observed, the entire length of the (dC-dG) segments may not be in a Z-type structure, as predicted from the CD results.

Junction between B- and Z-type Conformations

The studies described above on supercoil plasmids demonstrate that a junction must exist between a left-handed Z-type conformation and B-DNA-type structures. This junction was also studied with three DNA restriction fragments which are related to the 157 bp fragment shown in Figure 2. These results suggest[6] that a junction region exists between the Z structure and the B conformation which is approximately 11 bp in length. The structure of this intermediate region (approximately one turn of helix) is unknown at present.

Conclusions

Circular dichroism and ^{31}P-NMR studies on fragments show that the (dC-dG) segments undergo a cooperative transition to a left-handed conformation in the presence of high salt. The natural sequences from the *lac* operon and pBR322 do not appear to convert to a left-handed conformation. Figure 7 depicts the general conclusions from this study. It seems that a junction exists between the left-handed conformation and B-DNA of approximately 11 bp in all cases which is in neither a Z-type nor a B-conformation but must be intermediate in nature.

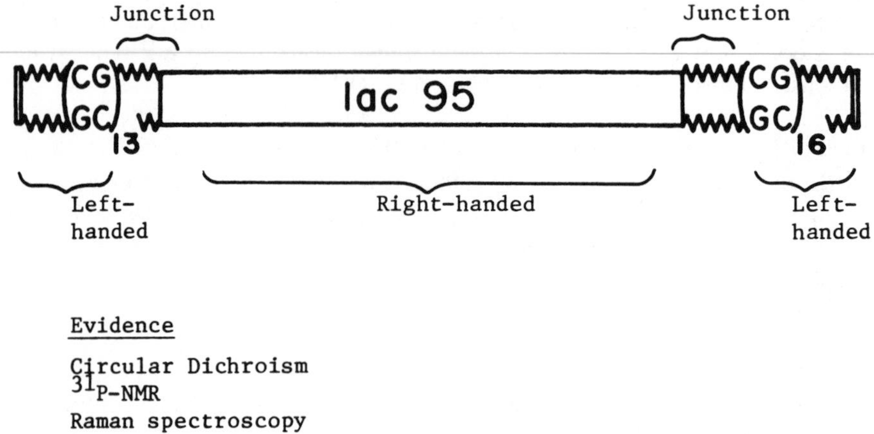

Evidence

Circular Dichroism
^{31}P-NMR
Raman spectroscopy

Figure 7. Summary of conformations in regions of the 157 bp fragment in high salt.

Moreover, electrophoretic analyses of plasmids containing the (dG-dC) sequence demonstrate a cooperative transition from a right-handed conformation to a left-handed Z-type structure. The presence of the (dG-dC) segment in the amount of 1.3% of the recombinant plasmid is sufficient to profoundly influence the supercoil properties of the DNA. These observations are consistent with prior determinations on the extreme flexibility of the DNA chain (reviewed in 1).

The demonstration that different segments of one DNA chain can coexist with conformations as different as a right-handed B-helix and a left-handed Z-type structure may have far reaching implications concerning the role of DNA structure in regulatory processes (reviewed in 1).

Acknowledgement

This work was supported by grants from the N.I.H. (CA 20279) and the N.S.F. (PCM 77-15033 and PCM 77-19927).

References and Footnotes

1. Stirdivant, S.M., Wells, R.D., Goodman, T.C., Hillen, W., Horn, G.T., Klein, R.D., Larson, J.E., Muller, U.R., Neuendorf, S.K., and Panayotatos, N., *Progress in Nucleic Acid Research and Molecular Biology 24*, 167-267 (1980).
2. Wartell, R.M., Wells, R.D., Blakesley, R.W., Burd, J.F., Chan, H.W., Dodgson, J.B., Hardies, S.C., Horn, G.T., Jensen, K.F., Larson, J., Nes, I.F., and Selsing, E., *Crit. Rev. Biochem. 4*, 305 (1977).
3. Wells, R.D., Panayotatos, N., *Nature 289*, 466-470 (1981).
4. Lilley, D.M.J., *Proc. Natl. Acad. Sci. USA 77*, 6468-6472.
5. Wells, R.D., Mitsui, Y., Langridge, R., Shortle, B.E., Cantor, C.R., Grant, R.C., and Kodama, M., *Nature 228*, 1166 (1970).
6. Wells, R.D., Klysik, J., Stirdivant, S.M., Larson, J.E., Hart, P.A., *Nature 290*, 672 (1981).
7. Wells, R.D., Hardies, S.C., Patient, R.K., Klein, R.D., Ho, F., and Reznikoff, W.S., *Journal of Biological Chemistry 254*, 5535-5541 (1979).
8. Wells, R.D., Hillen, W., Klein, R.D., *Biochemistry*, in the press (1981).
9. Wells, R.D., and Hillen, W., *Nucleic Acids Research 8*, 5427-5444 (1980).
10. Rich, A., Wang, A. H-J., and Quigley, G.J. in *Biomolecular Stereodynamics*, Volume 1, Ed., Sarma, R.H., Adenine Press, New York (in press).
11. Wang, A.H-J., Quigley, G.J., Kolpak, F.J., Crawford, J.L., van Boom, J.H., van der Marel, G., and Rich, A., *Nature 282* 680-686 (1979).
12. Drew, H., Takano, T., Tanaka, S., Itakura, K., and Dickerson, R.E., *Nature 286*, 567-571 (1980).
13. Mitra, C.K., Sarma, M.H., and Sarma, R.H., *Biochemistry 20*, 2036-2041 (1981).
14. Arnott, S., and Hukins, D.W.L. *Biochem. Biophysics. Res. Commun. 47*, 1504-1509 (1972).
15. Wartell, R.M., Klysik, J., Hillen, W., and Wells, R.D., in *Collected Abstracts, Second SUNYA Conversation in the Discipline Biomolecular Stereodynamics*, State University of New York at Albany, April 26-29, 1981.

Proceedings of the Second SUNYA Conversation in the Discipline Biomolecular Stereodynamics Volume I, ISBN 0-940030-00-4, Ed., Ramaswamy H. Sarma, Adenine Press, New York, ©Adenine Press

Propeller Twisted Adenine-Thymine Base Pairs in the DNA Double Helix in Solution[1] Solution Spatial Configuration of the AATT Domain in the Double Helices of d-GGAATTCC and d-CGCGAATTCGCG

Ramaswamy H. Sarma, B.J. Wagner and C.K. Mitra
Institute of Biomolecular Stereodynamics
State University of New York
Albany, NY 12222

Introduction

Nuclear magnetic resonance studies of DNA double helices of dG-dC sequence have shown[2-4] that the double helix under *solution conditions* takes up the left-handed Z-configuration in high salt and overwhelmingly the classic Arnott-Hukins B-form in low salt. Very little is known about the solution structural details of DNA double helices of dA-dT sequence.

Recent single crystal structural studies[5-8] showed that the self-complementary d-CGCGAATTCGCG adopts a right-handed propeller twist structure. One of the conspicuous features of this structure is the pronounced propeller twist of the AATT stretches with very small such twists for the G•C pairs. This observation in the solid state along with the report of Crothers[9] and Hogan et al.[10] that in calf-thymus DNA which is AT rich, the base pairs are propeller twisted, strongly suggest that the A•T pairs of the DNA double helix in solution may be in general propeller twisted. In order to check this thesis we have undertaken a detailed study of the spatial configuration of the double-stranded AATT stretch in the self-complementary d-GGAATTCC. We examine whether the structure of the AATT domain in solution corresponds to that of AH/B-DNA (Arnott/Hukins)[11] and alternating B-DNA[12] or does it correspond to that of the propeller Levitt[13] and Dickerson models.[5-8] We also examine the solution spatial configuration of the dodecamer d-CGCGAATTCGCG *vis-a-vis* the Arnott-Hukins B-form and the Dickerson propeller twist model. We have not included in this study the models such as A, C, D, Z_I, Z_{II} and the vertical double helix because our previous study[2-4] has clearly demonstrated that for the low salt from of poly(dG-dC)•poly(dG-dC) these structures are not applicable and it is likely that the above models may not be true for the low salt from of A•T double helices.

Computation of Magnetic Shielding Constants

As we have elaborated elsewhere in this volume,[4] we use the x-ray or theoretically predicted x, y and z coordinates to derive the experimentally observable NMR parameters. In the beginning we will assume that the double helix of d-GGAATTCC in solution may exist in the AH/B-DNA, the alternating B-DNA, the Levitt or the Dickerson propeller twist froms. In the original x, y and z coordinates of the above models, appropriate substitution of the bases was made to generate the coordinates of the double helix of the self-complementary d-GGAATTCC. To generate the Dickerson model of d-GGAATTCC the coordinates from the central eight nucleotide segments of d-CGCGAATTCGCG were used and appropriate base substitutions were made. In Figures 1 and 2 we illustrate in stereo the four structural models for the double helix of d-GGAATTCC. It can easily be seen that all the base pairs in AH/B-DNA (Figure 1) are flat and those in the Levitt Model (Figure 2) are propeller twisted. In the case of the Dickerson model (Figure 2), the central AATT stretch shows pronounced propeller twists, compared to the G•C pairs. It is important to note that in Klug's alternating B-DNA (Figure 1), the base pairs are significantly propeller twisted—this will become clear if one examines the structure in stereo. The authors[12] have not mentioned about this feature of their structure. In Figure 3 we illustrate the AH/B- and the Dickerson model for the double helix of the dodecamer d-CGCGAATTCGCG. The x, y and z coordinates were used to compute the magnetic shielding constants of AH2, AH8, TMe, TH6 and TNH3 as elaborated by Mitra et al.[2,3] In our computations we take into account the contribution to shielding from (a) ring current effects (b) the diamagnetic component of the atomic magnetic anisotropy and (c) the paramagnetic component of atomic magnetic anisotropy. For deriving the magnetic shielding constants we have used the paramagnetic and diamagnetic tensor elements as well as the ring current constants of Prado, Giessner-Prettre and Pullman.[14]

The magnetic shielding constants for the AATT double helical region in d-GGAATTCC for AH/B-DNA, alternating B-DNA, the Levitt and Dickerson propeller twist models are summarized in Table I. From these shielding constants for each spatial configuration, one will know, the extent to which a given proton in a base pair is shielded *vis-a-vis* the same proton in another base pair of the same kind. For example, according to Table I, in AH/B-DNA the shielding constants for AH2 of base pair 3 is 1.33 ppm, that of base pair 4 is 2.64 i.e.,

1 2 3 4 5 6 7 8
d-GGAATTCC
d-CCTTAAGG
8 7 6 5 4 3 2 1

if the AH/B form is true in solution, the experimentally observed difference in shift ($\Delta\delta$) between AH2 of base pair 3 and 4 should be 1.31 ppm. The shielding constant data for the AATT double helical region of the dodecamer d-CGCGAATTCGCG for the AH/B-DNA and the Dickerson propeller twist models are summarized in Table II.

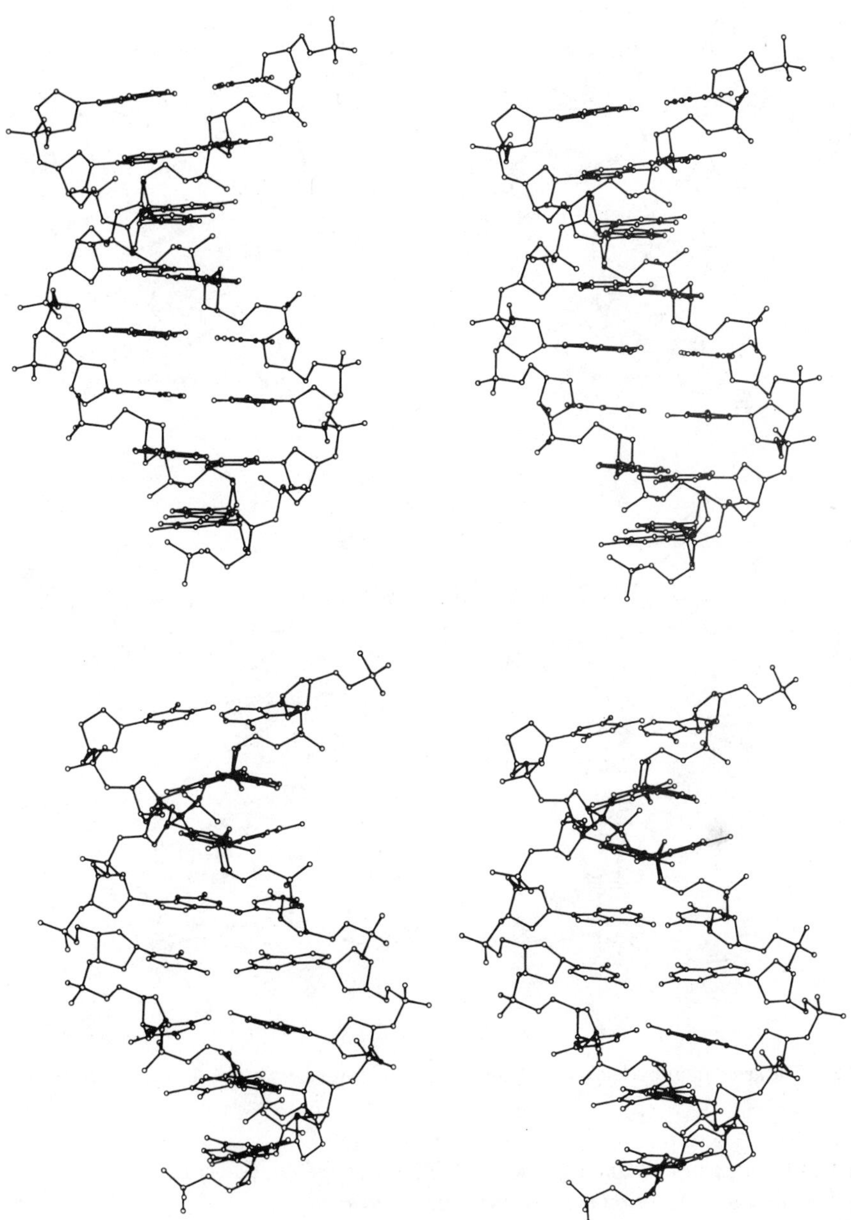

Figure 1. Structure of d-GGAATTCC duplex in stereo in the AH/B-DNA (top) and alternating B-forms (bottom).

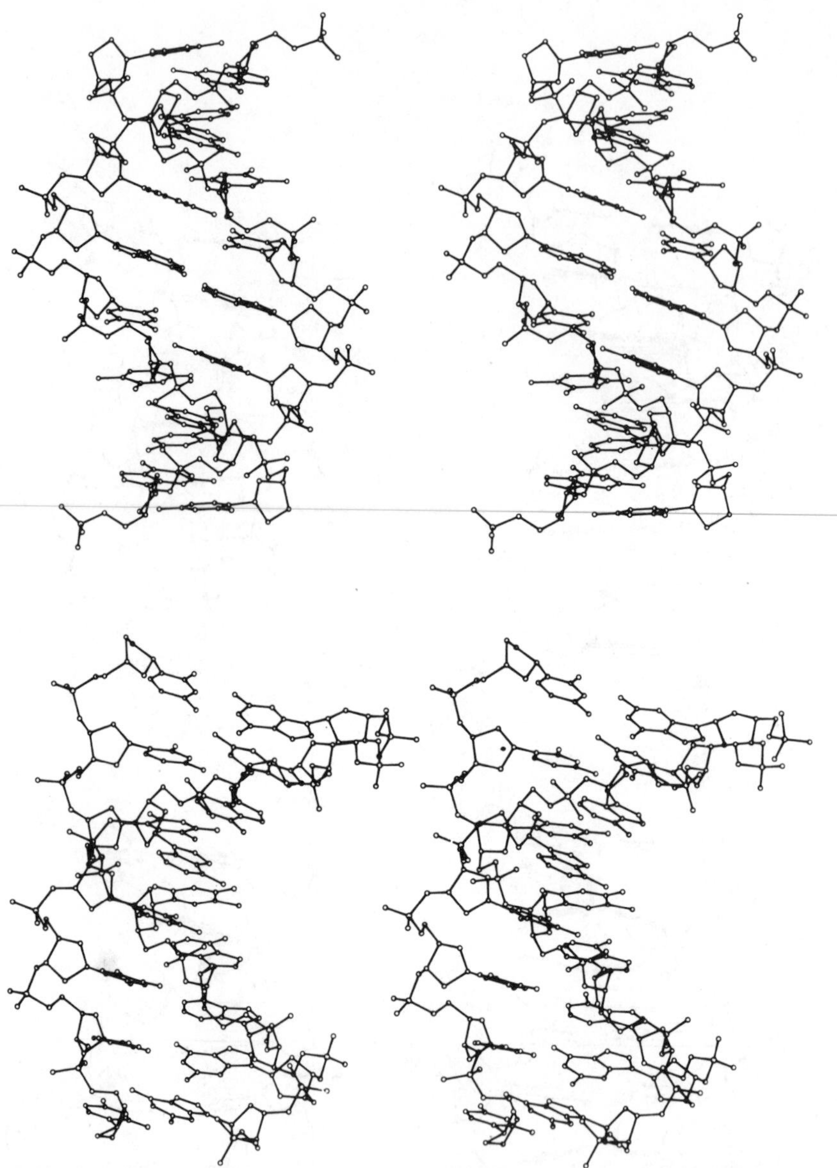

Figure 2. The Levitt (top) and Dickerson (bottom) propeller twisted models for the duplex of d-GGAATTCC in stereo.

The Complementary Strands of d-GGAATTCC and d-CGCGAATTCGCG Double Helices in Solution are Structurally Equivalent

Examination of Dickerson model (Figs. 2 and 3) as well as the various torsion angles in the crystal structure of the duplex d-CGCGAATTCGCG[7] clearly reveals that in the solid state the two complementary strands are structurally inequivalent.

Table I
Computed Magnetic Shielding Constants for the Protons in Base Pairs #3, 4, 5 and 6* in the Duplex of d-GGAATTCC. Data are in ppm

Proton	AH/B-DNA	Alternating B-DNA	Levitt B-DNA	Dickerson B-DNA Strand 1.	Dickerson B-DNA Strand 2.	Average
H2-A3	1.33	0.69	0.91	1.22	1.28	1.25
H8-A3	0.30	0.63	0.57	0.25	0.24	0.25
H2-A4	2.64	0.44	0.34	0.71	0.77	0.74
H8-A4	0.18	0.32	0.48	0.45	0.47	0.46
H6-T5	0.09	0.51	0.61	0.66	0.49	0.57
Me-T5	0.37	1.03	0.79	0.62	0.75	0.69
HN3-T5	0.58	0.29	0.28	0.09	0.34	0.22
H6-T6	0.10	0.14	0.24	0.28	0.27	0.28
Me-T6	0.49	0.16	0.37	0.39	0.44	0.41
HN3-T6	−0.33	0.04	−0.05	−0.10	−0.20	−0.15

*
```
   1 2 3 4 5 6 7 8
d-G G A A T T C C
d-C C T T A A G G
   8 7 6 5 4 3 2 1
```

Table II
The Total Shielding* in ppm for the Various Protons in Arnott-Hukins' B-DNA and the Averaged Dickerson Propeller Twist Model for

```
1 2 3 4 5 6 7 8 9 10 11 12
C G C G A A T T C G C G
G C G C T T A A G C G C
```

Proton	Arnott Hukins B-DNA	Average Dickerson Model
H2-A5	1.82	1.36
H8-A5	0.34	0.33
H2-A6	0.61	0.81
H8-A6	0.37	0.51
HN3-T7	0.43	0.26
H6-T7	0.22	0.60
HMe-T7	0.83	0.71
HN3-T8	−0.10	0.09
H6-T8	0.17	0.34
HMe-T8	0.53	0.54

*The computed magnetic shield constants for the AATT domain of d-GGAATTCC (Table I) and that for d-CGCGAATTCGCG (this table) in the average Dickerson model are different. This is a reflection of (a) the nature of the sequence at sites neighboring the AATT domain and (b) the fact that in one oligomer there are 16 residues and the other one 24 residues.

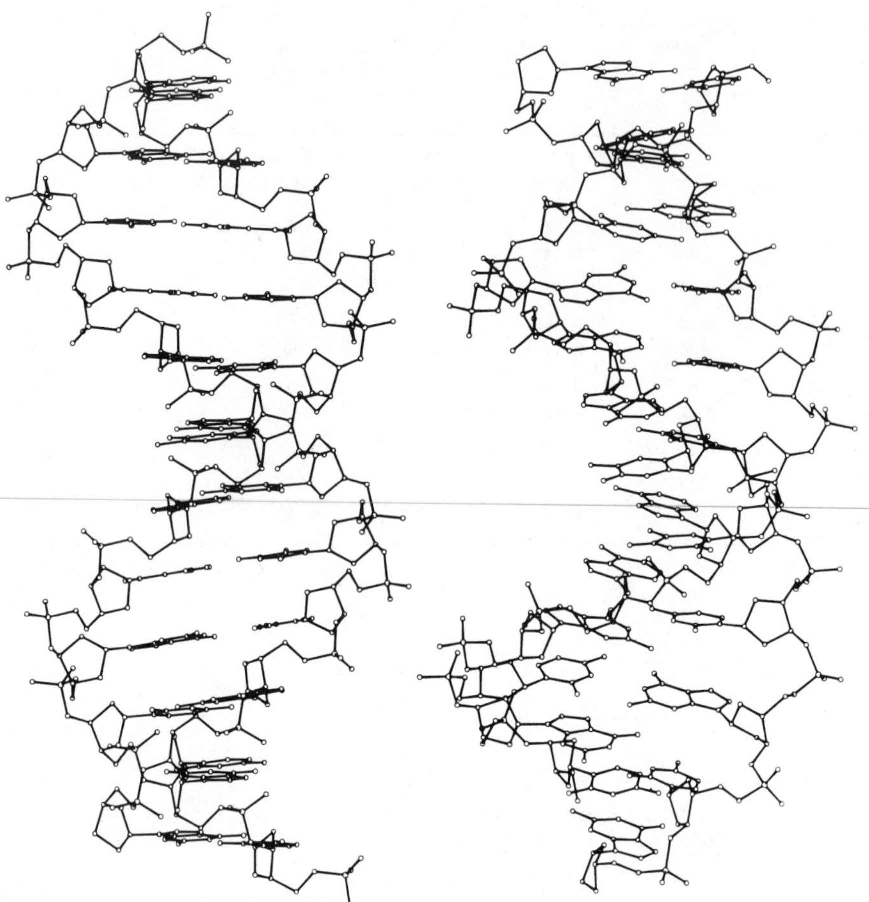

Figure 3. Structure of the dodecamer d-CGCGAATTCGCG in the AH/B- (left) and Dickerson (right) spatial configurations. Note the rigid regularity of the AH/B-form.

In fact this structure has no dyad axis. From the standpoint of ^{1}H NMR spectroscopy these differences between the structures of the two strands are sufficiently different that one should see separate resonances for the same proton in the two strands. In Table I we provide the shielding constants for the AATT residues of each strand of the Dickerson model for d-GGAATTCC. For example, it is seen that the shielding constants of H6-T5, Me-T5 and HN3-T5 of the two strands differ by 0.17, 0.13 and 0.25 ppm. Inspection of data in Table I reveals that the shielding constants of several other protons of identical residues are different for the two strands of the Dickerson model. However the 360 MHz H^1 NMR spectrum of the duplex of d-GGAATTCC[15] clearly shows that the structure is such that in the AATT region there are only two resonances (and not four) for AH8, for AH2, TH6, TMe and TNH3. This clearly indicates that there is symmetry in the molecule and the two strands are structurally equivalent. It is obvious that the specific model of Dickerson as embodied in the xyz coordinates is not precisely true in solution. This is not unexpected because in solution due to internal motions the two strands are expected to mutually negotiate an average structure. Hence in Table I for d-GGAATTCC we have also reported the average magnetic shielding constants for the Dickerson model. In Table II, we report only the *average* computed magnetic shielding constants for the AATT domain of the d-CGCGAATTCGCG double helix. It should also be pointed out that in the energy minimized Levitt model[13] the xyz coordinates indicate very slight differences, within 1° to 2°, in some of the same torsion angles between the two strands. These differences are so small that their effect on the magnetic shielding constants are negligible.

The Adenine/Thymine Base Pairs in the Duplexes of d-GGAATTCC and d-CGCGAATTCGCG in Solution are Propeller Twisted

As was pointed out earlier in the AH/B-DNA the base pairs are flat; in the Klug and Levitt models they are all propeller twisted; in the Dickerson model the A•T pairs are strongly propeller twisted. In Table III we have tabulated the relative shielding ($\Delta\delta$) the protons of the AATT domain will experience if it were to exit as part of a double helical d-GGAATTCC in AH/B, alternating B, the Levitt and average Dickerson propeller twist models. We have also tabulated in Table III the experimentally observed relative shielding ($\Delta\delta$) at 10°C, the latter was obtained from shift data reported by Patel.[15] We have elaborated elsewhere[2-4] the limitations and errors in this approach. We have indicated there that one has to look for agreement between computed $\Delta\delta$ and experimental $\Delta\delta$ for a *collection of protons* and that in each case if the agreement is within 0.1 ppm it is excellent and that in the range of 0.15 to 0.25 ppm is very good to fair. The data in Table III show that the computed $\Delta\delta$ of AH2 between base pairs 3 and 4 for AH/B-form is 1.31 ppm; the experi-

Table III
The Computed $\Delta\delta$ for the various protons of Base-Pairs #3 and 4* in AH/B-DNA, the Alternating B-DNA, the Levitt and Dickerson Propeller Twist Models and the Experimentally Observed $\Delta\delta$.
The data are in ppm

Proton	Computed $\Delta\delta$				Experimental $\Delta\delta$
	AH/ B-DNA	Alternating B-DNA	Levitt B-DNA	Average Dickerson B-DNA	
AH2	1.31	0.25	0.57	0.51	0.36
AH8	0.12	0.31	0.09	0.21	0.0
TH6	0.01	0.37	0.37	0.29	0.23
TMe	0.14	0.87	0.42	0.28	0.28
TNH3	0.92	0.25	0.33	0.36	0.10

* 1 2 3 4
d-G G A A T T C C
d-C C T T A A G G
4 3 2 1

mentally observed $\Delta\delta$ is 0.34 ppm. Also the computed $\Delta\delta$ of TNH3 between base pairs 3 and 4 for the AH/B-DNA is 0.92 ppm; the experimentally observed $\Delta\delta$ in this instance is 0.1 ppm. *The disagreements between the projected values for the AH/B-form and the experimental $\Delta\delta$ are so vast that it is unmistakable that the AH/B-DNA structure in which the base pairs are flat cannot be true for the AATT domain of d-GGAATTCC double helix.* Arguments on similar grounds indicate that the alternating B-DNA structure is also not true for the AATT region. This is not surprising because it was postulated by Klug et al.[12] for alternating adenine thymine copolymers.

Examination of the computed $\Delta\delta$ for AH2, AH8, TH6, TMe and TNH3 for the Levitt and the average Dickerson models with the experimental $\Delta\delta$ (Table III) clearly show that for this collection of protons they show agreement, the quality of the agreement range from excellent in certain cases, good in others to fair in some. There is not a single instance of a vast disagreement as in the case of the AH/B-form or the alternating B-DNA model. *The data thus enable us to conclude that the AATT domains of the double helical d-GGAATTCC in solution exist in a propeller twist mode* a la *Levitt and Dickerson.*[5-8]

The NMR data for the AATT domain of d-CGCGAATTCGCG for the AH-B form and the Dickerson model are summarized in Table IV. Obviously, there is significant deviation of the experimental data from that projected for AH/B-DNA. Thus the experimentally observed $\Delta\delta$ for AH2 is 0.45 ppm, whereas what is projected for the AH/B-form is 1.20 ppm. In the case of the average Dickerson model, there is overall agreement between what is experimentally observed and what is computed for the structure from the theory of NMR. For both d-GGAATTCC (Table III) and d-CGCGAATTCGCG (Table IV), the fit for TNH3 is only barely fair. This may be due to the sensitivity of δTNH3 to the lack of structural stability at the d-AATT regions of the double helix. For example, the temperature profiles for δTNH3 in

Table IV

Experimentally Observed and Computed Differences in Shifts Between Base Pairs 5 and 6* in the AH/B form and the Average Dickerson Model. Data in ppm.

Proton	Exptly Observed $\Delta\delta$	Computed $\Delta\delta$ for AH/B-DNA	Computed for Average Dickerson Model
AH2	0.45	1.20	0.55
AH8	0.0	0.02	0.18
TH6	0.22	0.05	0.26
TCH_3	0.18	0.31	0.19
TNH_3	0.10	0.53	0.35

* 1 2 3 4 5 6 6 5 4 3 2 1
d-C G C G A A T T C G C G
d-G C G C T T A A G C G C

d-GGAATTCC[15] and d-CGCGAATTCGCG[16] display significant sensitivity to temperatures at values considerably below melting. The temperature profiles for[15,16] AH8, AH2 and TH6 again reveal this sensitivity though to a less extent. All these data clearly suggest that the AATT region of the double helix even at a temperature as low as 0°C is undergoing continuous conformational fine tuning and we believe that this is *party* a manifestation of the propeller twisted shapes.

Our findings that the AATT domain of the double helix in solution displays propeller twisted structures *a la* Levitt and Dickerson stand in contrast to our conclusion that the spatial configuration of poly(dG-dC)•poly(dG-dC)[2-5] in low salt is predominantly the AH/B- form. There we conclude that in the AH/B- form the base pairs are either flat or very mildly propeller twisted. Recently Crothers and coworkers[17] conclude that dG-dC stretches are mildly propeller twisted, even though they do not know whether the spatial configuration is AH/B-DNA or the Levitt/ Dickerson DNAs.

The present study along with that of Dickerson[5-8] and that of Crothers on AT rich calf-thymus DNA[9,10] clearly establishes that A•T stretches in double helical DNA, whether in solution or solid state take up a propeller twisted spatial configuration similar to that propounded by Levitt in 1978 by energy minimization studies. It is a triumph of chemical theory over experimental chemistry.

Acknowledgements

The authors are deeply indebted to Professor Bernard Pullman and Dr. Claude Giessner-Prettre for providing them with the tensor elements to compute the contribution to shielding from the paramagnetic and diamagnetic components of the atomic magnetic anisotropy and the ring current constants to compute the contribution to shielding from ring current effects. We thank Professor M. Levitt for making available the coordinates of his propeller twist model. We thank Professor R. E. Dickerson for making available several preprints before publication and the coordinates of his propeller twist model. We thank Professor Donald M. Crothers for communicating his results before publication. We thank Dr. D. J. Patel for directing our attention to his paper on d-GGAATTCC, and consenting to use his experimental data on d-CGCGAATTCGCG from a collaborative program between our and his laboratories.

References and Footnotes

1. Preliminary communication of this work has appeared elsewhere: Wagner, B.J., Mitra, C.K., and Sarma, R.H. *Internatl. J. Quant. Chem.* QBS 8 (in press).
2. Mitra, C.K., Sarma, M.H., and Sarma, R.H., *Biochemistry 20,* 2036 (1981).
3. Mitra, C.K., Sarma, M.H., and Sarma, R.H., *J. Am. Chem. Soc.* (in press).
4. Sarma, R.H., Mitra, C.K., and Sarma, M.H., in *Biomolecular Stereodynamics, Volume I,* Ed., Sarma, R.H., Adenine Press, NY, p. 53 (1981).
5. Wing, R., Drew, H., Takano, T., Broka, C., Tanaka, S., Itakura, K. and Dickerson, R.E. *Nature 287,* 755 (1980).

6. Drew, H.R., Wing, R.M., Takano, T., Broka, C., Tanaka, S., Itakura, K., and Dickerson, R.E., *Proc. Natl. Acad. Sci. USA 78,* 2179 (1981).
7. Dickerson, R.E. and Drew, H.R., *J. Mol. Biol.* (in press).
8. Dickerson, R.E., Drew, H.R., and Conner, B., in *Biomolecular Stereodynamics,* Ed., Sarma, R.H. Adenine Press, NY, p. 1 (1981).
9. Crothers, D.M., Dattagupta, N. and Hogan, M. in *Nucleic Acid Geometry and Dynamics,* Ed., Sarma, R.H., Pergamon Press, p. 397 (1980).
10. Hogan, M., Dattagupta, N. and Crothers, D.M., *Biochemistry 18,* 280 (1980).
11. Arnott, S. and Hukins, D.W.L., *Biochem. Biophys. Res. Commun. 47,* 1504 (1972).
12. Klug, A., Jack, A., Viswamitra, M.A., Kennard, O., Shakked, Z. and Steitz, T.A., *J. Mol. Biol. 131,* 669 (1979).
13. Levitt, M., *Proc. Natl. Acad. Sci. USA 75,* 640 (1978).
14. Prado, F.R., Giessner-Prettre, C., and Pullman, B., *J. Mol. Str.* (in press).
15. Patel, D.J., and Canuel, L.L. *Eur. J. Biochem. 96,* 267 (1979).
16. Wagner, B.J., Mitra, C.K., Sarma, M.H., and Sarma, R.H., and Patel, D.J. *Biochemistry* (submitted).
17. Wu, H.M., Dattagupa, N., and Crothers, D.M., private communcation to RHS from DMC.

Proceedings of the Second SUNYA Conversation in the Discipline Biomolecular Stereodynamics Volume I, ISBN 0-940030-00-4, Ed., Ramaswamy H. Sarma, Adenine Press, New York, ©*Adenine Press*

Fibrous Polynucleotide Duplexes Have Very Polymorphic Secondary Structures

Struther Arnott and Rengaswami Chandrasekaran
Department of Biological Sciences
Purdue University
West Lafayette, Indiana 47907, U.S.A.

Nucleotide Conformations

It is not surprising that the regular secondary structures of polynucleotides are quite polymorphic. Even when the asymmetric unit of structure is a single nucleotide the molecular backbone has the 6 degrees of conformational freedom denoted $\alpha, \beta, \gamma, \delta, \epsilon, \zeta$, in Fig. 1. Moreover each of these rotations about single bonds can assume a wide spectrum of values. In the cases of α(O3′-P-O5′-C5′), γ(O5′-C5′-C4′-C3′), ϵ(C4′-C3′-O3′-P), and ζ(C3′-O3′-P-O5′) these values are trimodally distributed in *gauche minus* (g^-), *trans (t)*, or *gauche plus* (g^+) ranges where $\theta = -60°$ $\pm 60°$, $\theta = 180° \pm 60°$, or $\theta = 60° \pm 60°$ respectively. Of course the full 120° of these ranges is not exploited but a range of half that magnitude is not unusual.[1] Also, because the backbone atoms are usually substituted asymmetrically, the exploited ranges are not always centred on the canonical values of −60°, 180°, +60°. For example, γ in its g^+ range has a mean value[1] of +50° rather than +60°.

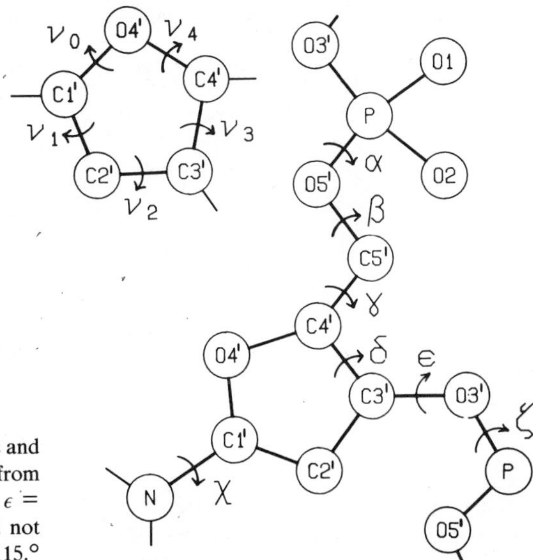

Figure 1. The new convention for atom labels and conformation-angles. In previous publications from our laboratory $\alpha = \psi, \beta = \theta, \gamma = \zeta, \delta = \sigma, \epsilon = \omega, \zeta = \phi$. Now $\chi = \theta$(O4′-C1′-N9-C4) and not θ(C2′-C1′-N9-C4) as previously. $\chi_{new} \simeq \chi_{old} + 115.°$

The rotation δ(C5′-C4′-C3′-O3′) is bimodally distributed in g^+ and t ranges corresponding to C3′-*endo* and C2′-*endo* puckered furanose rings. Furanose rings are quite flexible[1] and consequently the ranges of values accessible to δ is almost as great as that of an exocyclic conformation angle like γ.

The only backbone conformation angle which is monomodally *(t)* distributed is β(P-O5′-C5′-C4′).

An additional source of nucleotide variability is provided by the glycosidic angle χ(O4′-C1′-N9-C4) which is bimodally distributed in *anti (a)* or *syn (s)* ranges with $\theta = -135° \pm 90°$ or $\theta = +45° \pm 90°$ respectively.

Polynucleotide Helices

Helical polynucleotide chains can be described by a variety of parameters. Particularly useful are h and t which describe the intrahelical relationship of successive residues. The axial translation per residue, h, is the distance between equivalent points on successive residues projected onto the helix axis; t, the rotation per

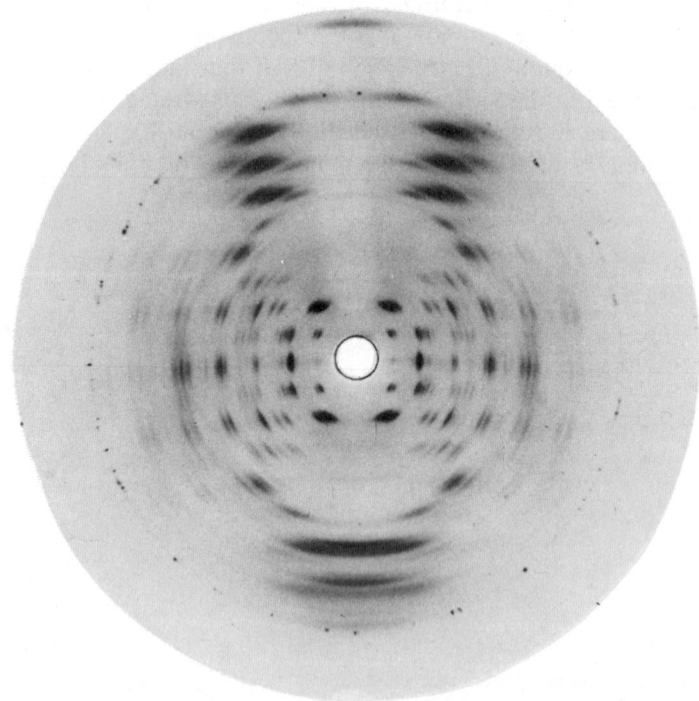

Figure 2 *(a)*. Diffraction from *A*-DNA the member of the $(g^- t\, g^+ a\, g^+ t\, g^-)$ genus with the lowest axial translation and highest rotation per residue. $c = P = 2.82$nm, screw symmetry 11_1, $h = 0.26$nm and $t = +32.7°$.

residue is the amount by which equivalent points on successive residues are rotated about the helix axis.

The axial translation $c = h \times p$ is the smallest length of structure containing an integral number *(p)* of residues which is repeated by simple translation along the helix axis. In the case of "integral" (p_1) helices c is the same as the pitch P, but in

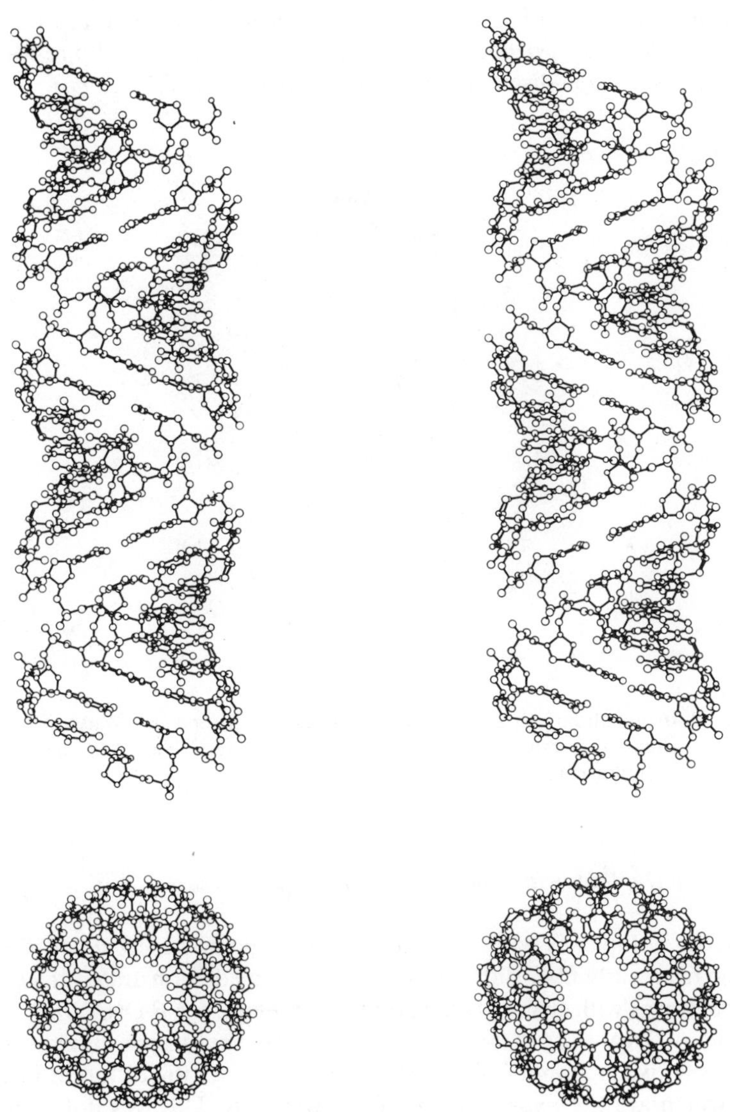

Figure 2 ***(b).*** Pairs of mutually perpendicular stereo projections. The wide and shallow minor groove is in the center of the upper pair. The major groove is deep but clenched shut and inaccessible to anything larger than a water molecule.

the case of "non-integral" (p_q) helices with p residues in $|q|$ turns $c = q \times P$ is larger than P.

Obviously $h = c/p = P/(p/q)$ and $t = 360°/(p/q)$, so that the classical 10_1 *B*-DNA helix[2] with $c = P = 3.38$nm has $h = 0.338$nm and $t = 36.0°$, while the 28_3 *C*-DNA helix[3] has $c = 9.30$nm, $P = 3.10$nm, $h = 0.332$nm, $t = 38.6°$.

For left-handed helices $q<0$ so that Z-DNA[4] is a 6_{-1} helix. Crystallographers would describe the symmetry of Z-DNA[4] as 6_5 which presents no real difficulty since $360°/(p/q) \equiv 360°/(p/(q\text{-}p))$ but requires one to think in terms of p residues in $(p\text{-}q)$ left-handed turns when discussing p_q helices when $q > p/2$.

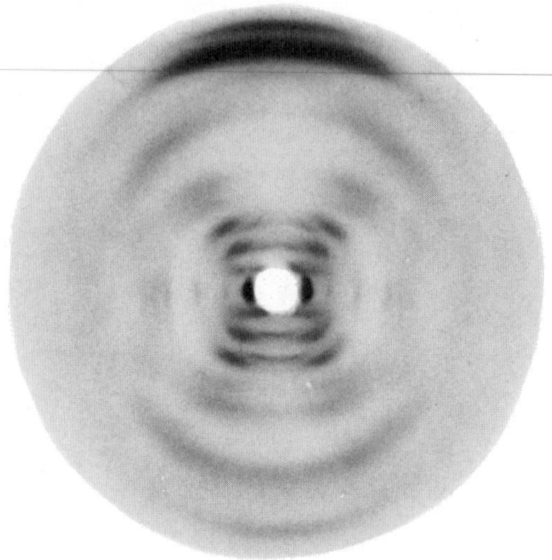

Figure 3 ***(a).*** Diffraction from poly dT • poly dA • poly dT which contains a Watson-Crick base-paired duplex with the largest axial translation and smallest rotation per residue in the $(g^- t\, g^+ a\, g^+ t\, g^-)$ genus. $c = P = 3.92$nm, screw symmetry 12_1, $h = 0.33$nm and $t = +30.0°$.

Polynucleotide Duplexes with Watson-Crick Base-Pairs

Polynucleotide duplexes with Watson-Crick base-pairs were known to be at least dimorphic from early fiber diffraction studies[5] which demonstrated the existence of two helical forms with strikingly different parameters: for *B*-DNA $(h,t) = (0.34$ nm, $36.0°)$ and *A*-DNA $(h,t) = (0.26$ nm, $32.7°)$. We now know that the substantial differences between *A* and *B*-DNA are correlated[6] with the different furanose ring puckers which are C3′-*endo* in *A* and C2′-*endo* in *B*. This was not appreciated at that time and Crick and Watson incorporated into their model[7] for *B*-DNA the wrong (C3′-*endo*) alternative for the entirely creditable reason that it was the only one that had been observed experimentally (in a single crystal study of cytidine[8]).

Qualitatively the conformation of the nucleotides in the Crick and Watson model can be described by a seven letter code *(t t t a* g^+ *t* g^-*)* which details the conformational ranges of individual conformational angles (α β γ χ δ ϵ ζ) in the

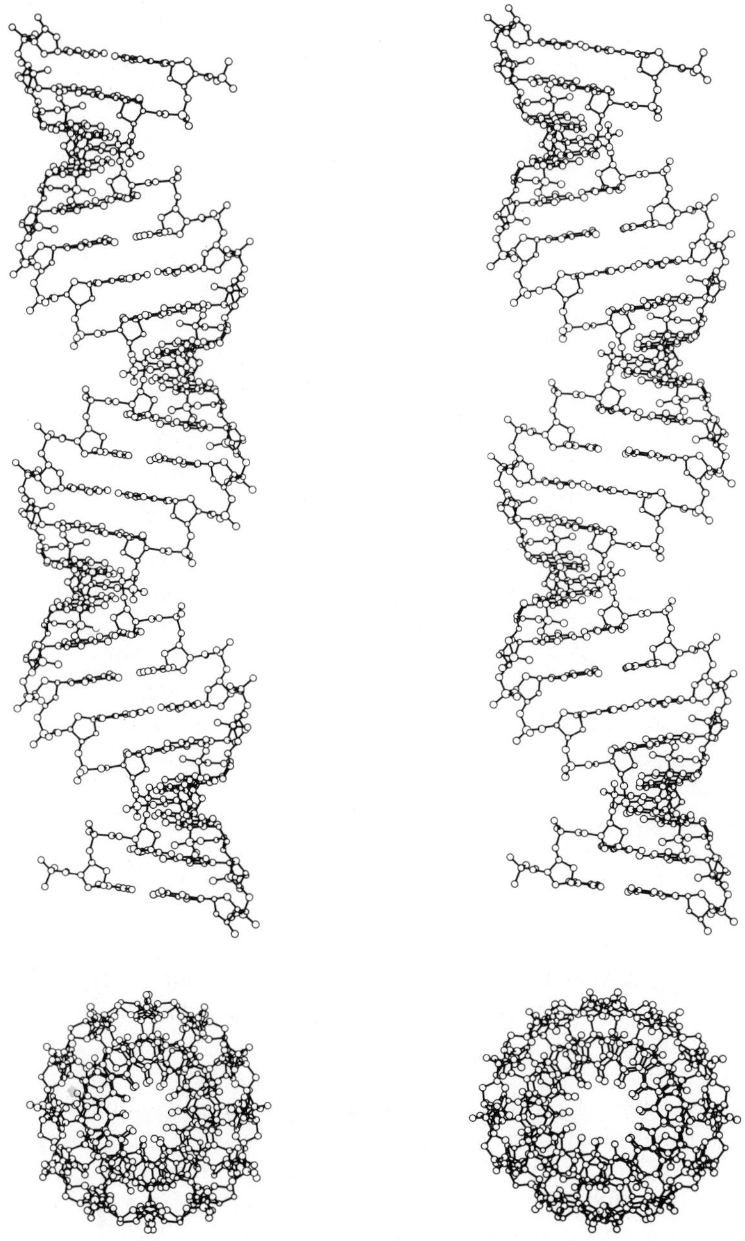

Figure 3 ***(b)***. Pairs of mutually perpendicular stereo projections. The wide and shallow minor groove is retained. The major groove is now wide as well as deep.

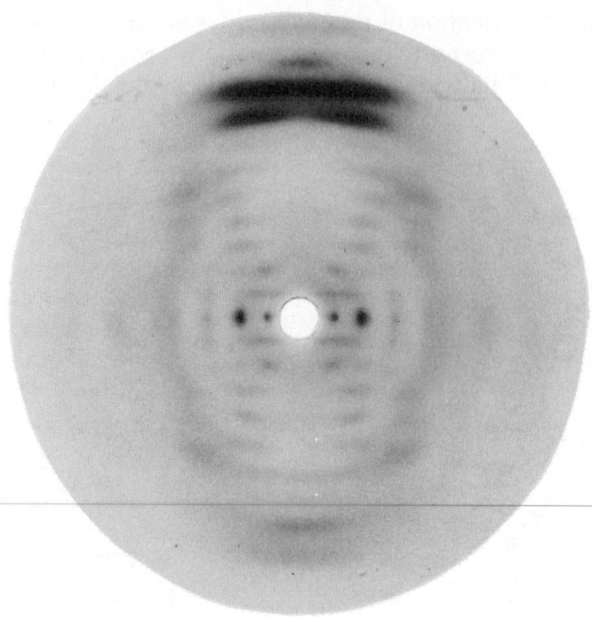

Figure 4 *(a)*. Diffraction from the sole member of the *(t t t a g^+t g^-)* genus. $c = P = 3.13$nm, screw symmetry 10_1, $h = 0.31$nm and $t = +36.0°$.

mononucleotide asymmetric unit. This hypothetical allomorph has a quite different conformation from the one *(g^-t g^+a t t t)* which currently provides the best fit with the *B*-DNA fiber diffraction data. Remedying the original error resulted in a disproportionate amount of attention being devoted to this single allomorph of the DNA duplex. Indeed the relatively monotonous view of this which many have had can properly be described as an artefact of the history of DNA structural studies.

Ironically the conformations ascribed to *A*-DNA, $(\alpha\, \beta\, \gamma\, \chi\, \delta\, \epsilon\, \zeta) = (g^-t\, g^+a\, g^+t\, g^-)$ were qualitatively accurate from the beginning but this allomorph was given little attention until[9] the 1960s when it was discovered that RNA duplexes had analogous structures.[10]

One structure, *C*-DNA, analogous to *B*-DNA was known quite early but it was not until the 1970s that other structures conformationally analogous to *B* were discovered and analysed.[11]

At the end of the 1970s helical polynucleotide duplexes with a dinucleotide as the asymmetric residue of chain structure were postulated by extrapolating from the crystal structure of (TpApTpA).[12] This structure does not form quasi-helical duplex fragments but its structure encouraged the production[13] of a model of a helical duplex in which T and A nucleotides were conformationally different as in the crystal of the tetranucleoside triphosphate.

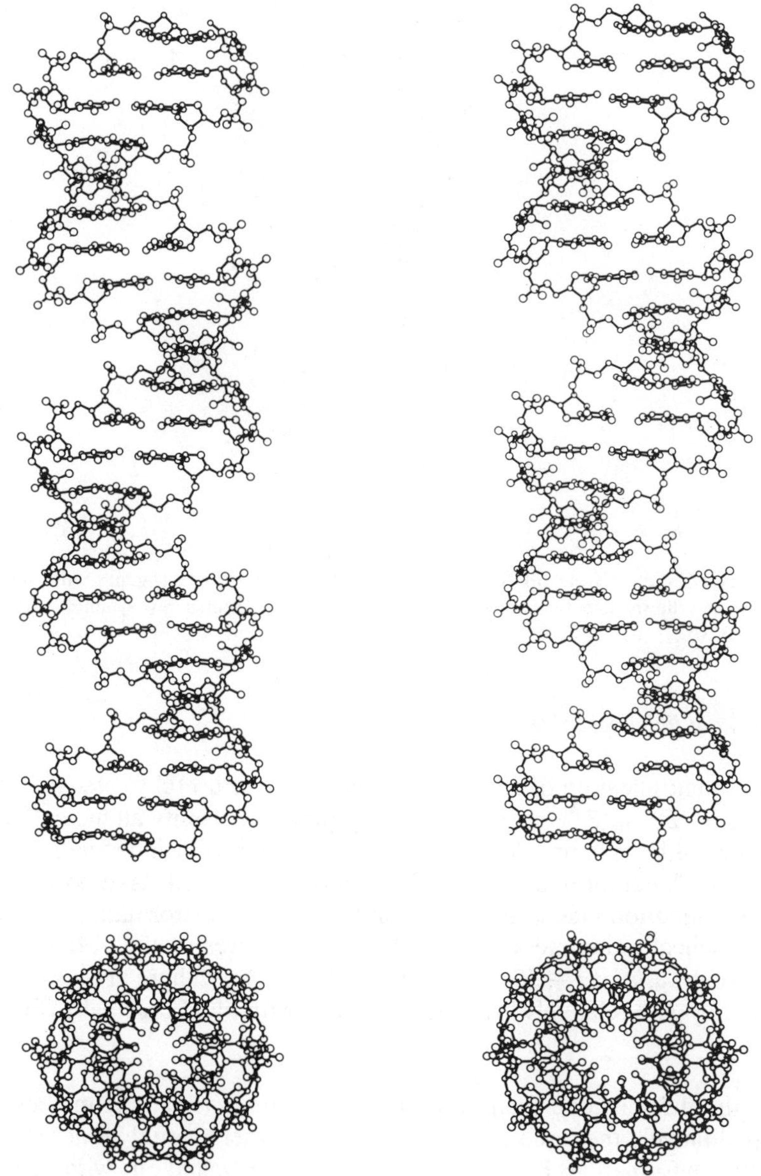

Figure 4 ***(b).*** Pairs of mutually perpendicular stereo projections, The major groove is wide and deep, the minor groove is wide and shallow.

On the other hand the hexanucleoside pentaphosphate (CpGpCpGpCpG) does form quasi-helical duplexes[4] in which alternate nucleosides are C3′-*endo*, *syn* and C2′-*endo, anti.* The extrapolated helix, Z-DNA, is left-handed. Similar helices have been observed in polymer fibers.[14]

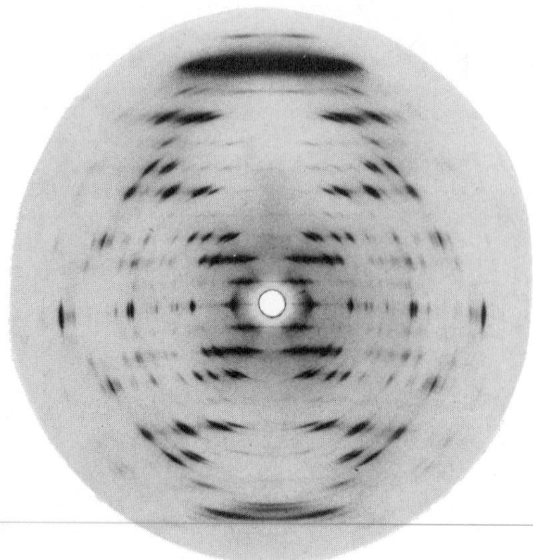

Figure 5 *(a)*. Diffraction from *B*-DNA, the member of the *($g^-t\ g^+a\ t\ t$)* genus with the largest axial translation and the smallest turn angle per residue. $c = P = 3.38$nm, screw symmetry *10_1*, $h = 0.34$nm and $t = +36.0°$.

Categories of Polynucleotide Helices

With the multiplication of allomorphs it is no longer useful to assign trivial names like *A, B, C, D, ..., Z:* it seems more appropriate to classify all helical polynucleotide chains with a mononucleotide asymmetric unit as *A* or *B* depending on whether their furanose conformations are C3′-*endo* or C2′-*endo*. It is important to realize that the distinction indeed should be on the basis of conformation and not on the basis of helical parameters since there are structures of the *A*-family with $h = 0.33$nm (Fig. 3*a,b*) or $t = 36.0°$ (Fig. 4*a,b*) like *B*-DNA. There are also structures in the *B*-family with relatively low values of *h* and with tilted base-pairs (Fig. 6*a,b*) like *A*-DNA.

Within the *A* family one can group together in one *genus* all the molecular *species* with qualitatively the same conformations. Thus there would be at least 2 genera: one with conformations like *A*-DNA *($g^-t\ g^+a\ g^+t\ g^-$)*; and one with conformations like the original Crick and Watson model for *B*-DNA *($t\ t\ t\ a\ g^+t\ g^-$)*. This *A* family structure actually exists as we discovered recently.[15]

The polynucleotide chains in the so-called[13] "alternating *B* structure" and in the *Z*-DNA helix logically should be assigned to a new family (*A*+*B*). Thus would the case also for the polynucleotide strands in the ladder structure[16] with $t = 0.0°$.

Strictly speaking each chain in a duplex should be described. Then *B*-DNA would have a *B* • *B* structure and *Z*-DNA an (*A*+*B*) • (*A*+*B*). A triplex[17] like poly(dT) •

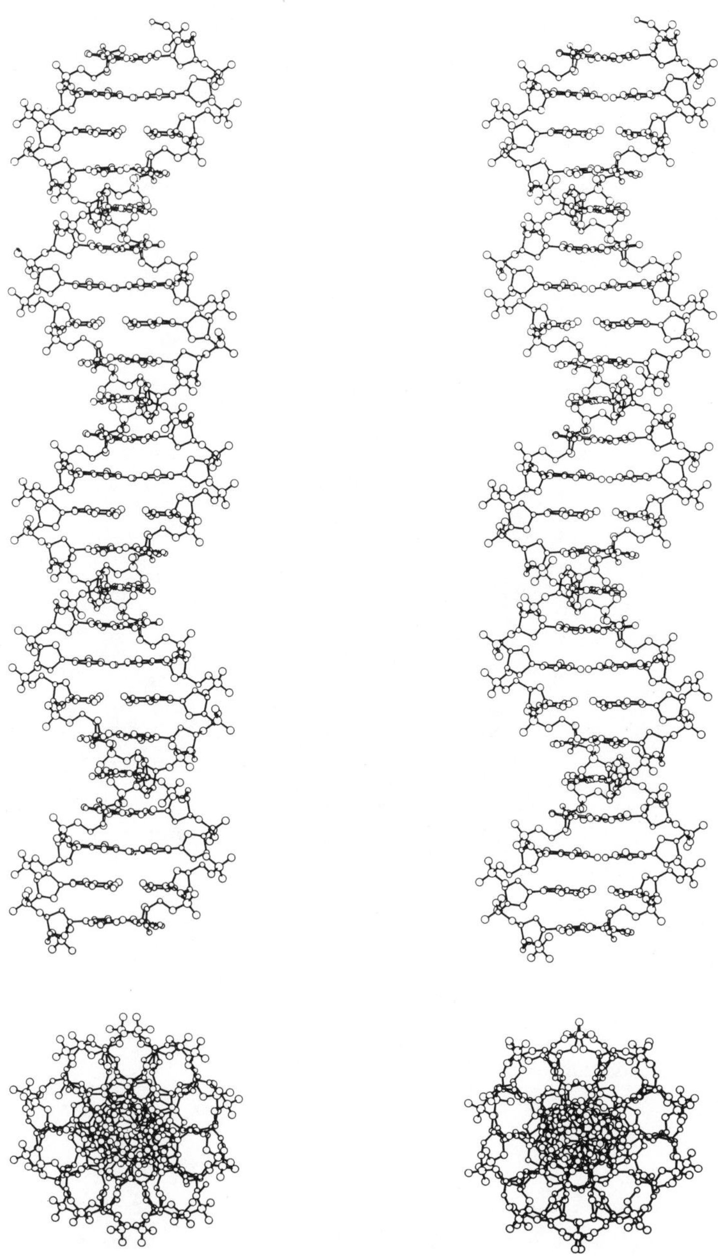

Figure 5 *(b).* Pairs of mutually perpendicular stereo projections. Both the major and minor grooves are wide and about equally deep.

poly(dA) • poly(dT) would be designated $A \bullet A \bullet A$ (or $A_1 \bullet A_2 \bullet A_3$ if one wished to emphasize that the three chains were similiar but not identical). In general, however, such elaboration will not be necessary except where conformational heterogeneity among polynucleotide chains in a multistrand complex is an issue as in the $B \cdot A$ structure being proposed[18] for the DNA-RNA hybrid poly(dT) • poly(A). [We regard such a structure as being quite feasible but not the solution to the structural problem in question.]

The A *Family of Watson-Crick Base-Paired Polynucleotide Duplexes*

The main genus of this family has conformations $(g^-t\ g^+a\ g^+t\ g^-)$ like *A*-DNA (Figs. 2*a,b*). It is of no little interest that *A*-DNA itself lies at one end of a spectrum

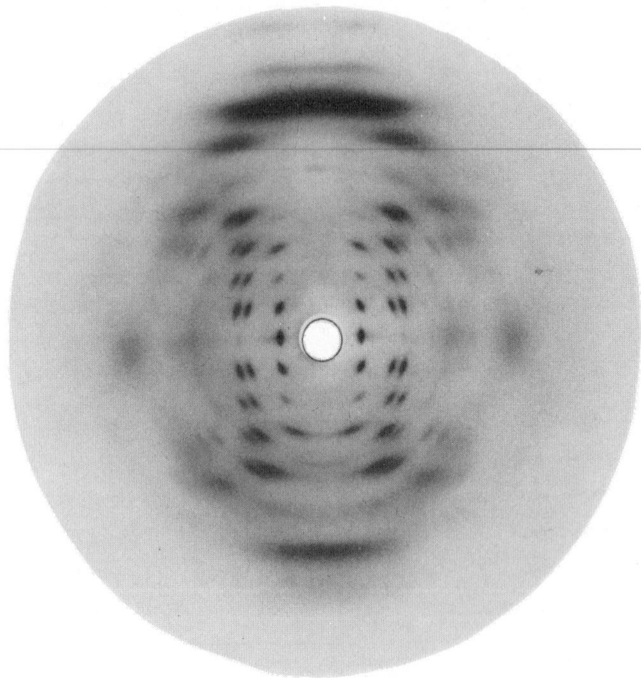

Figure 6 *(a)*. Diffraction from *D*-DNA the member of the $(g^-t\ g^+a\ t\ t)$ genus with the smallest axial translation and the largest turn angle per residue. $c = P = 2.41$nm, screw symmetry 8_1, $h = 0.30$nm and $t = +45.0°$.

of fibrous structures: it has the smallest value of *h* (0.26nm) and the largest value of *t* (+32.7°). In this genus, congeners of *A*-DNA all have similar values of *t* ($+30.0° \leq t \leq +32.7°$) but can have values of *h* up to 0.33nm.

Comparison of the morphology of *A*-DNA ($h = 0.26$nm, $t = +32.7°$) with that of the most distant species observed with the same kind of conformation angles (Figs. 3*a,b*) shows that what changes with *h* is the tilt of the base-pairs and the width of the deep major groove which is clenched shut when *h* is small but wide open when

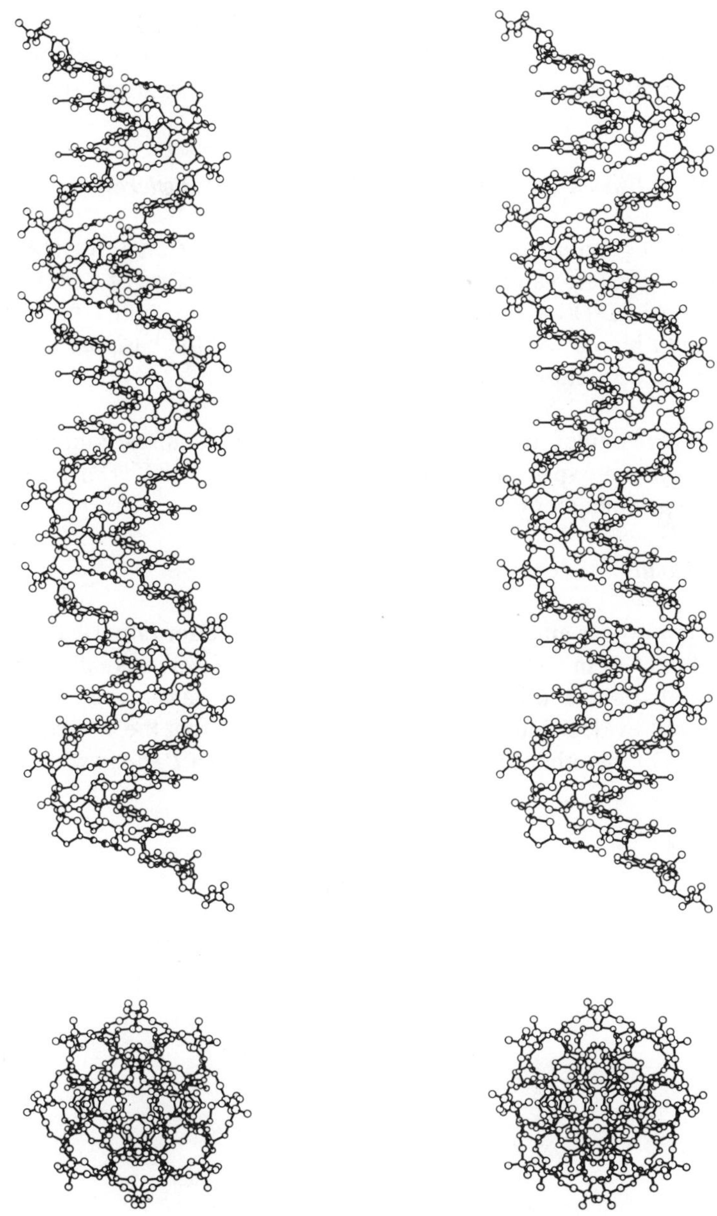

Figure 6 ***(b).*** Pairs of mutually perpendicular stereo projections. The minor groove is extremely narrow.

h is large. An open, shallow minor groove is the morphological feature conserved in this genus.

In the *A* genus *(t t t a* g^+*ṭ* g^-*)* only one fibrous species[15] has been observed (Figs.

4*a,b*) with a structure reminiscent of the original Crick and Watson structure for *B*-DNA. Its gross morphology is typical of the *A*-family but its relatively large rotation per nucleotide apparently requires two $g \rightarrow t$ transitions (in α(O5′-P) and γ(C5′-C4′)).

The B *Family of Watson-Crick Base-Paired Polynucleotide Duplexes*

The commonest genus $(g^- t\, g^+ a\, t\, t\, t)$ contains, at one extreme, *B*-DNA (Figs. 5*a,b*) which has the greatest value of *h* (0.34 nm) and least value of *t* (+36.0°). At the other extreme is *D*-DNA (Figs. 6*a,b*) with h = 0.30 nm and t = +45.0°. As *h* decreases and *t* increases the minor groove closes and the base pairs tilt more (but in the direction opposite to that in *A*-DNA).

To wind up a *B* duplex beyond +45.0° apparently requires conformational $g \rightarrow t$ transitions to give the species with h = 0.32nm and t = +48.0° (Figs. 7*a,b*) which so far is the solitary member of the genus *(t t t a t t t)*.[15,19]

The (A+B) *Family*

An *(A+B)* family was anticipated by the speculative "alternating *B* model" for poly d(AT) • poly d(AT) in solution.[13] No fibrous structure corresponding to this (right-handed) 5_1 polydinucleotide helix has yet been found. Sarma and coworkers have[26] presented NMR evidence in this volume and elsewhere that the "alternating *B*-DNA" is untenable. On the other hand the 6_5 (t = −60.0°) *(A+B)* helix with h = 0.725nm

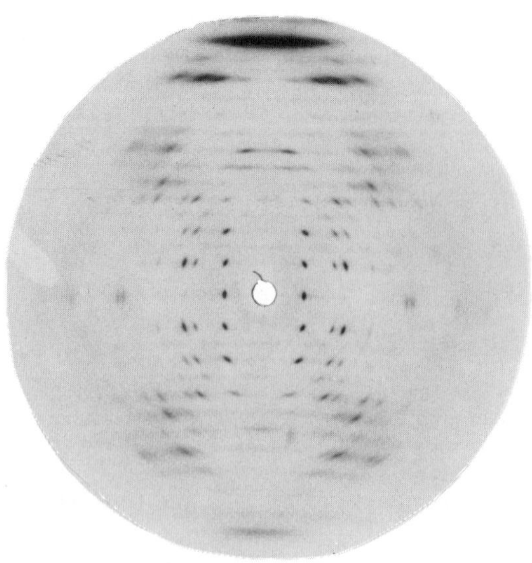

Figure 7 *(a)*. Diffraction from the sole member of the *(t t t a t t t)* genus. c = 4.87nm, P = 2.435nm, screw symmetry 15_2, h = 0.32nm and t = +48.0°.

(Figs. 8*a,b*) which emerged from the crystal structure of $(CG)_3$ has been observed with poly d(GC) • poly d(GC).[14] An isomorphous structure has been observed also with poly d(AC) • poly d(GT) and a similar 7_6 helix ($h = 0.76$nm, $t = -51.4°$) with

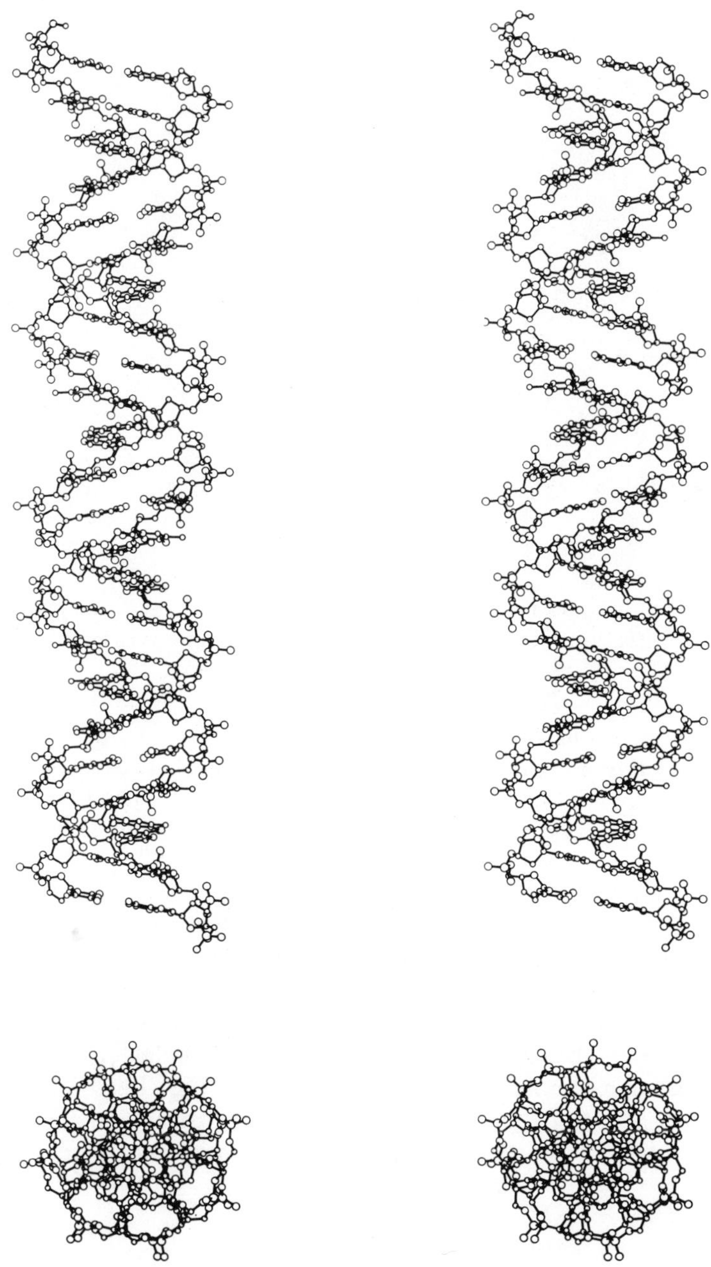

Figure 7 ***(b).*** Pairs of mutually perpendicular stereo projections. The major groove is wide and deep.

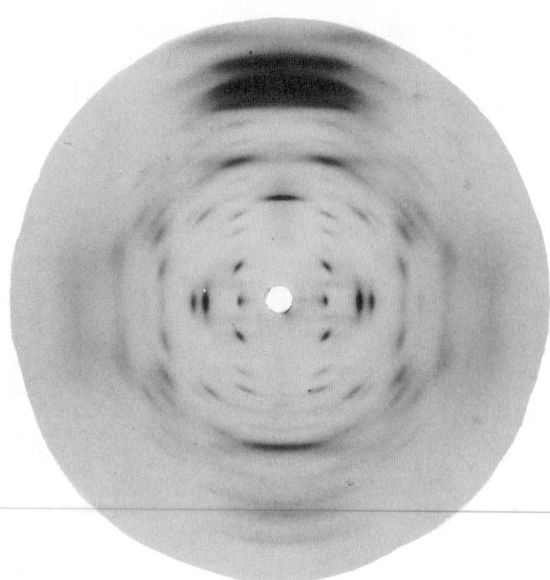

Figure 8 *(a)*. Diffraction from poly dGC • poly dGC, a member of the $(g^+ t\, t\, s\, g^+ t\, g^+) + (t\, t\, g^+ a\, t\, g^- g^+)$ genus. $c = P = 4.35$nm, $h = 0.725$nm, screw symmetry 6_5 and $t = -60.0°$.

poly d(As^4T) • poly d(As^4T). This suggests that this particular family of structures either prefers or is confined to sequences with purine and pyrimidine nucleotides alternating.

These structures belong to a genus $(g^+ t\ t\ s\ g^+ t\ g^+) + (t\ t\ g^+ a\ t\ g^- g^+)$ in which alternate nucleotides are as different as can be. The phosphate ester conformations are g^+g^+ and g^+t. γ(C4′-C5′) is *t* in one but g^+ in the other. The furanose conformations are C3′-*endo* and C2′-*endo* and the glycosidic conformations *s* and *a*.

Another structure (Figs. 9*a,b*) of this family but belonging to another genus $(g^+ t\ t\ s\ g\ t\ g^+) + (g^- t\ g^- a\ t\ g^- g^+)$ has been observed in calf thymus DNA fibers which have been strained. This less underwound duplex[20] has $h = 0.76$nm and $t = -25.7°$.

Finally, the same strategy of using *syn,* C3′-*endo* and *anti*, C2′-*endo* nucleotides alternately is seen in the partially unstacked polydinucleotide duplex observed in a fibrous intercalation complex.[16] This founding member (Figs. 10*a,b*) of a new genus, $(g^+ t\, t\, s\, g^+ t\, t) + (g^- t\, t\, a\, t\, g^- g^+)$, is remarkable for being an achiral duplex, a zipper structure in which the two polynucleotide chains are not intertwined.

DNA and RNA

All the structures discussed are accessible to DNA-DNA duplexes with a restriction only in the case of Z-DNA where certain base sequences may be required.

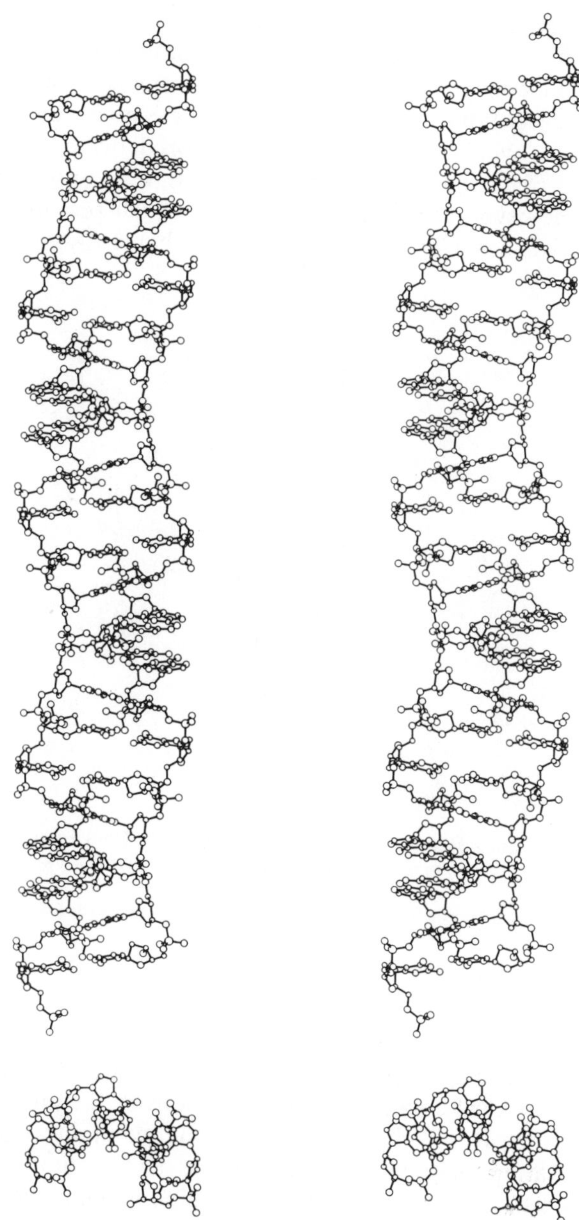

Figure 8 *(b)*. Pairs of mutually perpendicular stereo projections.

Fibrous RNA-RNA duplexes have been observed only in the *A* family. The quasi-helical duplex regions in *t*-RNA are also of this kind. In the known *B* and *(A+B)* duplexes the C2′ hydroxyl would be somewhat sterically compressed as well as shielded from the hydrated exterior.

DNA-RNA hybrid duplexes have now been observed over the whole range of *A* family structure. In general these duplexes are isomorphous with *A*-DNA in less hydrophilic environments and assume allomorphic forms with larger values of *h* in more hydrophilic conditions. Apparently the restrictions on the RNA chain are usually sufficient to over-ride any alternative favored by the DNA chain. The alternative strategy of having a *B* • *A* duplex is an interesting possibiility which has been canvassed. However, the diffraction pattern which prompted this speculation[18] probably derives from a high-*h* *A* structure probably *A*•*A*•*A*.

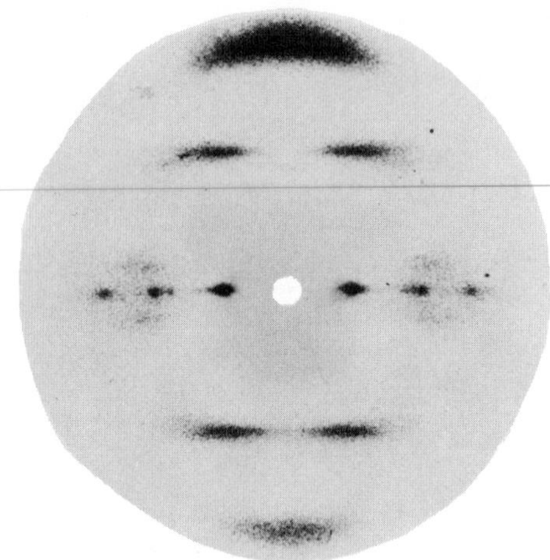

Figure 9 *(a)*. Diffraction from a strained fiber of calf thymus DNA. The structure is a member of the $(g^+ t\, t\, s\, g^+ t\, g^+) + (g^- t\, g^- a\, t\, g^- g^+)$ genus. $c = P = 10.64$ nm. $h = 0.76$nm, screw symmetry 14_{13} and $t = -25.7°$.

Junctions

Clearly one should envisage virtually a continuum of structures within each genus of duplexes since no local conformational transitions are required to change one congener into another. It is likely also that intergenetic but intrafamilial transitions require little activation energy. Relatively smooth deformation of polynucleotide duplex is therefore easy when structures only within one family are involved.

A quite different situation arises when successive blocks of duplex have secondary structures belonging to different conformational families. Then, clearly, there has to be a zone of disruption or transition between the blocks. We have been considering conservative and parsimonious junctions between blocks of *A* and *B*-DNA and between blocks of *B* and *Z*-DNA. By conservative we mean that we have tried to retain Watson-Crick hydrogen-bonding and some base-stacking while ensuring the necessary condition that there be no overshort non-bonded contacts. By par-

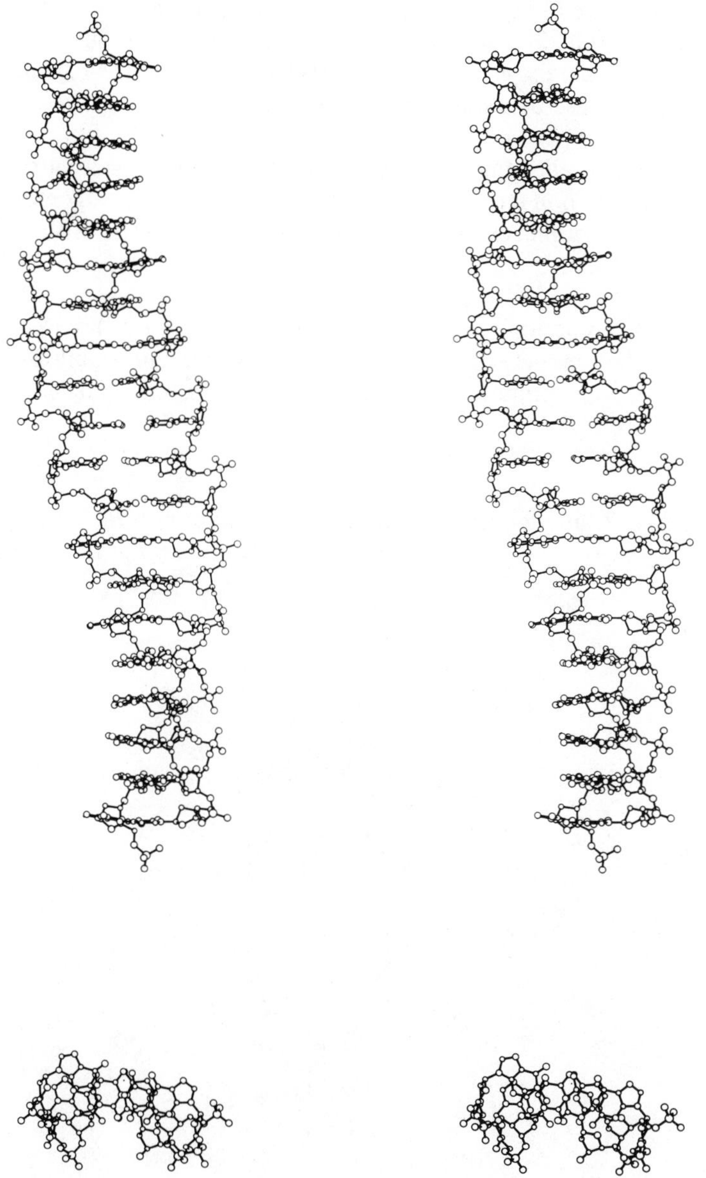

Figure 9 *(b).* Pairs of mutually perpendicular stereo projections showing an almost unwound, "slightly twisted ribbon" structure.

simonious we mean that we have kept the size of the transition zone as small as possible.

We find that with these boundary conditions it is not possible in either case to connect two disparate blocks unless one nucleoside pair and the 4 phosphate ester linkages attached to its 3′ and 5′ ends form a transition zone in which conformations are different from those in the two regular blocks (Figs. 11*a,b;* 12*a,b*). In other words, at least two base-pair stacks have to be different.

It is conceivable that allowing the next set of stacks on each side of the transition zone to be somewhat non-standard would produce an energetically more favorable zone. It is certainly unnecessary to postulate a junction zone with more than 4 non-standard stacks.

The minimum *A,B* junction[21] has conformations $5'A(.\,.\,.\,.\,.\,t\,g^-) + (g^-t\,g^+a\,t\,t\,g^-) + (g^-t\,g^+\,.\,.\,.\,.)\,B3'$ on one strand. (The blanks indicate conformations identical to *A* or *B*-DNA as appropriate). On the (non-identical) complementary strand the conformations are $5'B(.\,.\,.\,.\,.\,t\,g^-) + (g^-t\,g^+a\,g^+t\,g^-) + (g^-t\,g^+\,.\,.\,.\,.)A3'$ (Fig. 11*a,b*).

The minimum *B, Z* junction (Fig. 12*a,b*) has conformations $5'B(.\,.\,.\,.\,.\,t\,t) + (g^+t\,t\,a\,g^+t\,g^-) + (g^+t\,g^+(a)\,.\,.\,.)Z3'$ on one strand, and $5'Z(.\,.\,.\,(s).t\,g^+) + (t\,t\,t\,a\,t\,t\,g^-) + (g^-t\,g^+\,.\,.\,.\,.)B3'$.

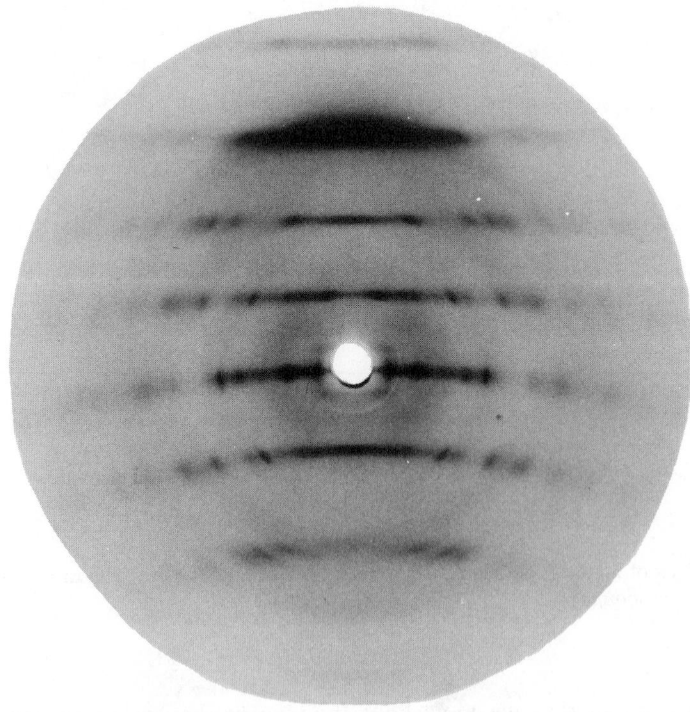

Figure 10 *(a).* Diffraction from a fibre of salmon testes DNA saturated with one intercalating $[(\mathrm{bipy})\mathrm{Pt}(\mathrm{en})]^{2+}$ cation for every two base pairs. The structure belongs to the $(g^+t\,t\,s\,g^+t\,t) + (g^-t\,t\,a\,t\,g^-g^+)$ genus. $c = P = h = 1.02\mathrm{nm}$ and $t = 0°$.

Summary

A nucleotide residue has 6 degrees of conformational freedom in its backbone. Most of these variables have wide ranges of values available to them. Therefore one is not surprised to find a correspondingly wide range of helical secondary structures in polynucleotide duplexes even where the asymmetric unit is only one nucleotide and when restraints for forming hydrogen-bonded duplexes with stacked

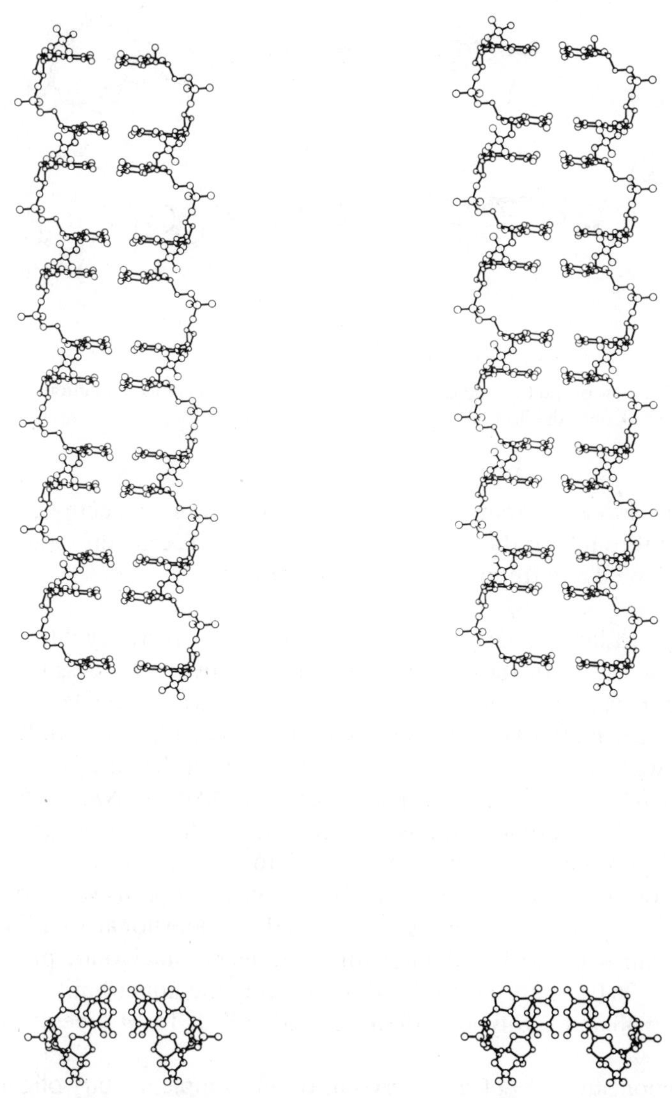

Figure 10 ***(b).*** Pairs of mutually perpendicular stereo projections of the completely unwound DNA duplex.

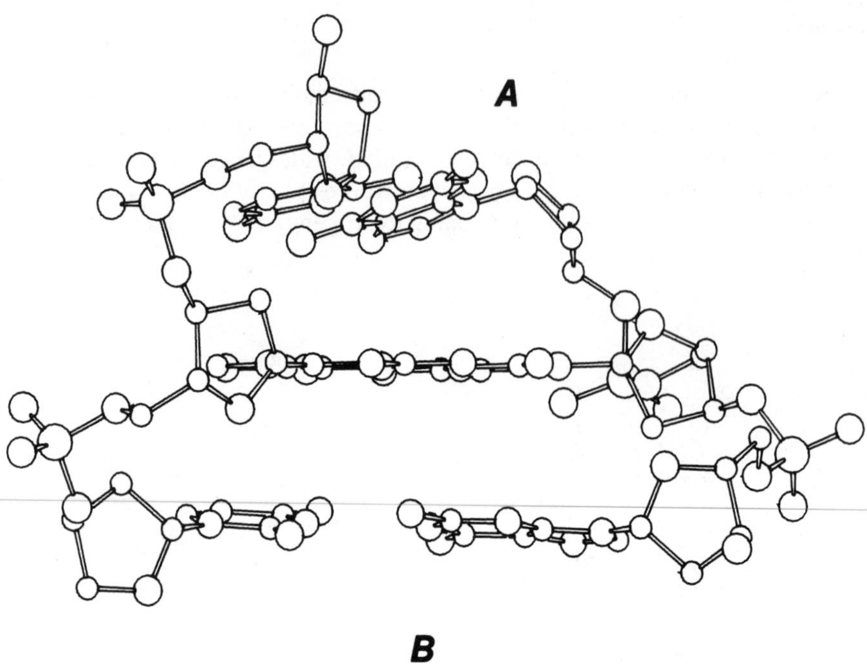

Figure 11 *(a)*. Side view of the trinucleoside diphosphate duplex in an *A* to *B* junction zone. *A*-DNA fits onto the top, *B*-DNA onto the bottom.

base-pairs are imposed. Accumulation of examples of the expected polymorphism has occurred throughout the last 28 years as a result of X-ray diffraction studies of uniaxially oriented and (often) polycrystalline fibers.

Franklin and Gosling[5] as early as 1953 had trapped the two very different *A* and *B* allomorphs of DNA. In these, the rotations *(t)* between successive nucleotides (+32.7° and +36.0°) are not the same, and the lengths *(h)* of nucleotides projected on the helix axes are markedly (31%) different (0.26nm and 0.34nm). Between 1963 and 1974 it was shown that there were at least half a dozen DNA-DNA,[17] DNA-RNA[22,23,24] or RNA-RNA[6,25] allomorphs which were like A-DNA conformationally but had distinctive morphologies associated with modest changes in t (+30.0° to +32.7°) and considerable changes in h (0.26 to 0.33nm). In the same decade it became obvious that there were just as distinctive allomorphs with conformations like *B*-DNA[3,11] but with masses per unit length (proportional to $1/h$) somewhat greater ($0.30\text{nm} \leq h \leq 0.34\text{nm}$) and with rotations per nucleotide proportionately even larger ($+36.0° \leq t \leq +45.0°$). More recently the upper limits of t have been extended to +48.0° for *B*-type[15,19] duplexes and +36.0° for *A*-type duplexes.[15]

X-ray diffraction studies[4,12] of single crystals of self-complementary oligonucleotides with alternating pyrimidine/purine base sequences anticipated polymeric helices in which a dinucleotide would be the asymmetric unit. With more degrees of freedom

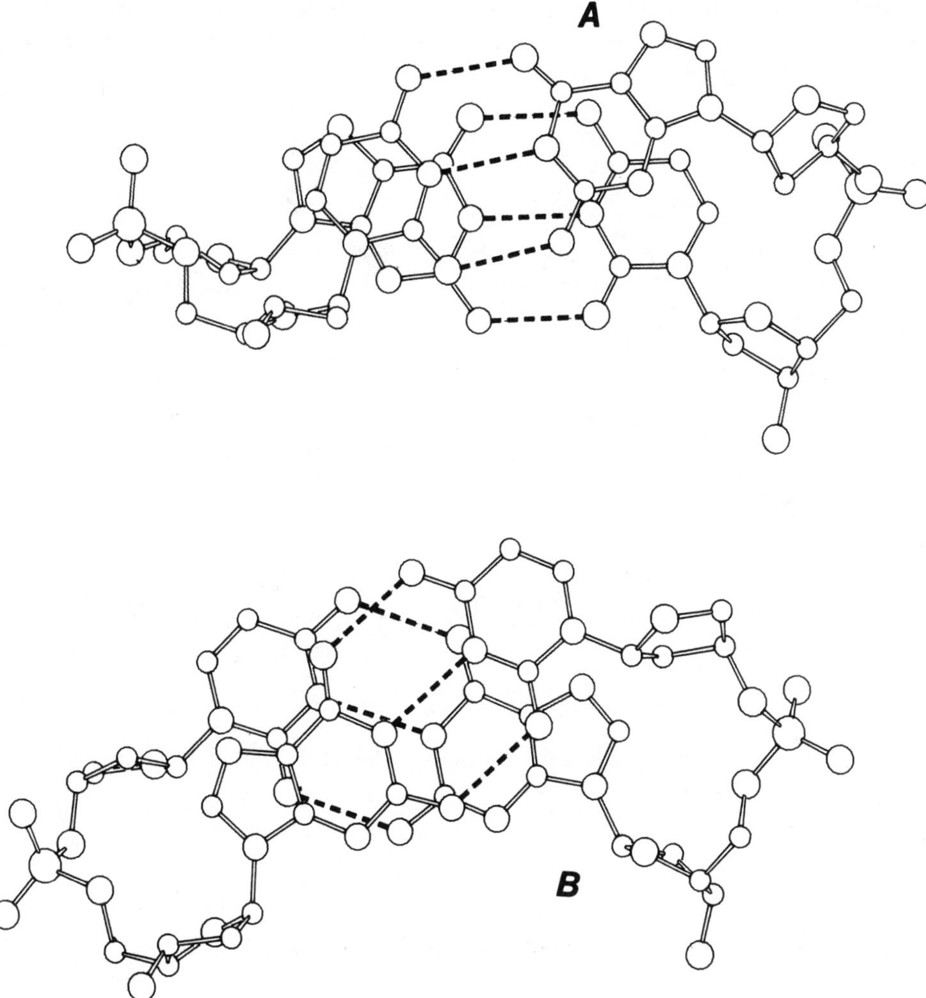

Figure 11 ***(b).*** The two base pair stacks in the junction. The hydrogen bonds in the base pairs are shown by dashed lines.

a wider range of polymeric secondary structures could be conceived including ones with negative[4] values of t. Three left-handed helical polymers have now been characterized[14,20] with values of the average rotation per nucleotide of −30.0°, −25.7°, −12.9°.

A completely unwound duplex can be formed at the expense of unstacking every other base-pair. When stabilised by an intercalator it can be trapped in fibrous form and analysed.[16] Presumably this partially unstacked allomorph presages new groups of polymer secondary structures which can accommodate intercalators.

Junctions between duplex blocks with very different conformations such as *A* and *B*[21] or *B* and *Z* require at least 2 irregular, but quite plausible, stacks and certainly need not involve more than 4.

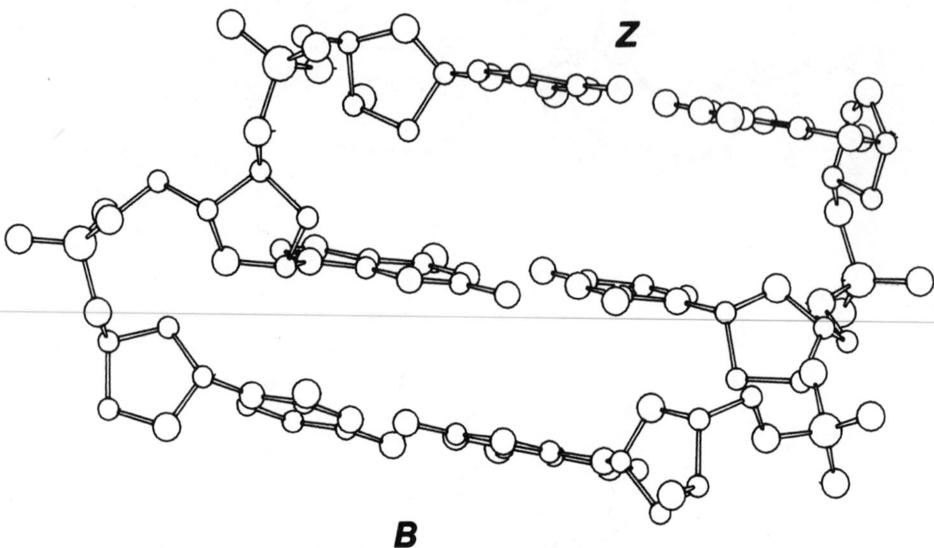

Figure 12 ***(a).*** Side view of the trinucleodide diphosphate duplex in a mimimun *B* to *Z* junction. *Z* and *B*-DNA fit onto the top and bottom respectively.

Acknowledgements

Support from the National Institutes of Health (no. GM 17371) is gratefully acknowledged.

References and Footnotes

1. Arnott, S. and Hukins, D.W.L., *Biochem. J. 130,* 453-465 (1972).
2. Langridge, R., Wilson, H.R., Hooper, C.W., Wilkins, M.H.F., and Hamilton, L.D. *J. Mol. Biol. 2,* 19-37 (1960).
3. Marvin, D.A., Spencer, M., Wilkins, M.H.F. and Hamilton, L.D. *J. Mol. Biol. 3,* 547-565 (1961).
4. Wang, A.H..-J., Quigley, G.J., Kolpak, F.J., Crawford, J. L., van Boom, J.H., van der Marel, G. and Rich, A., *Nature 282,* 680-686 (1979).
5. Franklin, R.E. and Gosling, R.G., *Acta Cryst. 6,* 673-677 (1953).
6. Arnott, S., Fuller, W., Hodgson, A., and Prutton, I., *Nature 220,* 561-564 (1968).
7. Crick, F.H.C., and Watson, J.D., *Proc. Roy. Soc. A. 223,* 80-96 (1954).
8. Furberg, S., *Acta Cryst. 3,* 325-333 (1950).
9. Fuller W., Wilkins, M.H.F., Wilson, H.R., Hamilton, L.D., and Arnott, S., *J. Mol. Biol. 12,* 60-80 (1965).
10. Spencer, M., Fuller W., Wilkins, M.H.F., and Brown, G.L., *Nature 194,* 1014-1020 (1962).

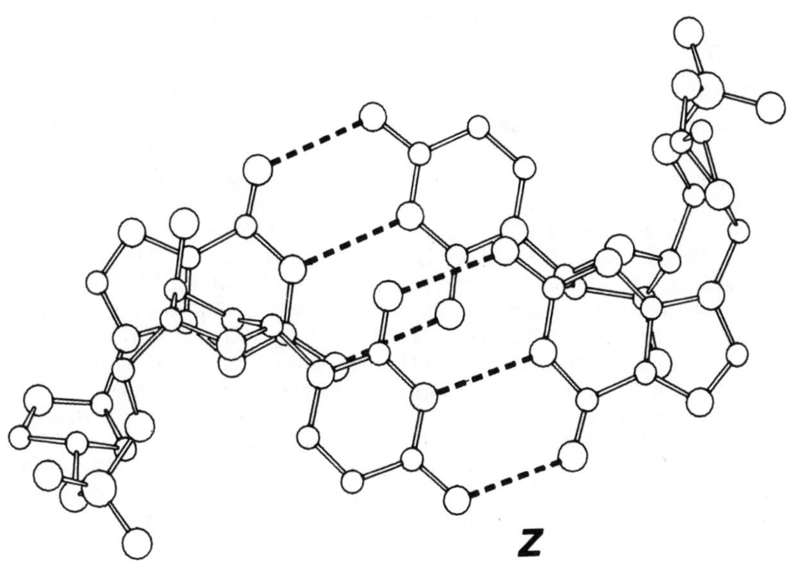

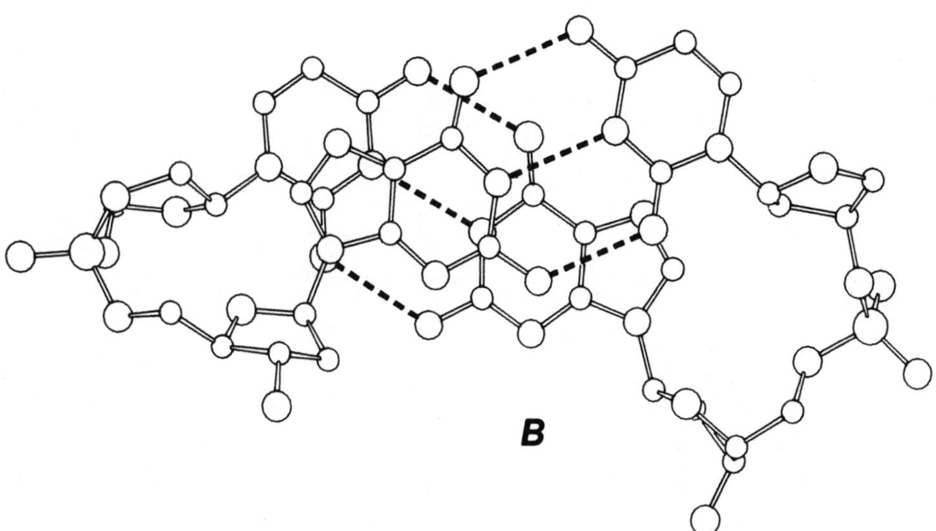

Figure 12 ***(b).*** The two base pair stacks in the junction. The dashed lines show the hydrogen bonds.

11. Arnott, S., Chandrasekaran, R., Hukins, D.W.L., Smith, P.J.C., and Watts, L., *J. Mol. Biol. 88,* 523-533 (1974).
12. Viswamitra, M.A., Kennard, O., Jones, P.G., Sheldrick, G.M., Salisbury, S., Falvello, L., and Shakked, Z., *Nature 273,* 687-688 (1978).

13. Klug, A., Jack, A., Viswamitra M.A., Kennard, O., Shakked, Z., and Steitz, T. A., *J. Mol. Biol. 131,* 669-680 (1979).
14. Arnott, S., Chandrasekaran, R., Birdsall, D.L., Leslie, A.G.W., and Ratliff, R.L., *Nature 283,* 743-745 (1980).
15. Chandrasekaran, R., Arnott, S., Banerjee, A., Campbell-Smith, S., Leslie, A.G.W., and Puigjaner, L., *Fiber Diffraction Methods* (American Chemical Society, Washington, DC, French, A.D. and Gardner, K.H. editors) ACS Symposium Series 141, 483-502 (1980).
16. Arnott, S., Bond, P.J., and Chandrasekaran, R., *Nature 287,* 561-563 (1980).
17. Arnott, S. and Selsing, E., *J. Mol. Biol. 88,* 509-521 (1974).
18. Zimmerman, S.B., and Pheiffer, B.H., *Proc. Natl. Acad. Sci., 78,* 78-82 (1981).
19. Leslie, A.G.W., Arnott, S., Chandrasekaran, R., and Ratliff, R.L., *J. Mol. Biol. 143,* 49-72 (1980).
20. Arnott, S., Chandrasekaran, R., Bond, P.J., Birdsall, D.L., Leslie, A.G.W., and Puigjaner, L.C., in *Structural Aspects of Recognition and Assembly in Biological Macromolecules,* (Seventh Aharon Katzir-Katchalsky Conference, Nof Ginossar, Israel, Balaban, M., Sussman, J., Traub, W., Yonath, A., editors, Balaban ISS, Rehovot and Philadelphia), 2, 487-500 (1981).
21. Selsing, E., Wells, R.D., Alden, C.J., and Arnott, S., *J. Biol. Chem. 254,* 5417-5422 (1979).
22. O'Brien, E.J., and McEwan, A.W., *J. Mol. Biol. 48,* 243-261 (1970).
23. Milman, G., Langridge, R., and Chamberlain, M.J., *Proc. Natl. Aca. Sci. 57,* 1804-1810 (1967).
24. Arnott, S., and Bond, P.J., *Nature New Biology 244,* 99-101 (1973).
25. Arnott, S., Hutchinson, F., Spencer, M., Wilkins, M.H.F., Fuller, W., and Langridge, R., *Nature 211,* 227-232 (1966).
26. Mitra, C.K., Sarma, M.H., and Sarma, R.H. *J. Am. Chem. Soc.* (in press).

Proceedings of the Second SUNYA Conversation in the Discipline Biomolecular Stereodynamics
Volume I, ISBN 0-940030-00-4, Ed., Ramaswamy H. Sarma,
Adenine Press, New York, ©Adenine Press

Conformational Flexibility of DNA: A Theoretical Formalism

V. Sasisekharan, Manju Bansal, Samir K. Bramachari and Goutam Gupta
Molecular Biophysics Unit
Indian Institute of Science
Bangalore—560 012, India

Introduction

The right helical conformation of DNA was suggested[1,2] from the semicrystalline X-ray diffraction pattern of wet fibres of calf thymus DNA (hereafter, B-pattern). Later on, crystalline patterns were obtained for various polymorphous forms (A, B, C, etc.) of DNA.[3-5] It was restated that only the right-handed double-helices could account for the fibre data of various natural DNAs.[6] The same conclusion was also extended to the synthetic polynucleotide fibres.[7] However, our understanding of the scope of these fibre diffraction studies *vis a vis* the structure of DNA was quite different. We pointed out that the fibre diffraction diagram (resolved up to 3Å) gives only the image of the gross structure of the molecule.[8,9] We also recognized that there is enormous conformational flexibility inherent in the polynucleotide backbone which, following a stereochemical guideline, leads to both right and left-handed duplexes for various polymorphous forms of DNA.[10,11] It was also shown that the right and left handed models of DNA in a given form retain the same gross features and thus give similar agreement with the observed data.[11,12] In the present article, we summarize the results of our studies on various DNA duplexes and conclude that DNA in either handedness is a natural consequence of the conformational flexibility inherent in the molecule. We also demonstrate that due to the same flexibility, a stable link could be made between a right and a left-helical segment such that the two handedness can co-exist in the same structure.

For the present purpose, we have chosen two types of DNA duplexes: (i) one with random base sequences e.g. calf thymus DNA, sperm head DNA etc. showing the well known A(n* = 11; p* =28Å) and B(n = 10; p = 34.0Å) forms, (ii) and the other with well defined base sequences e.g. poly d(AT), poly d(GC), poly d(IC) etc. showing predominantly the B- and D(n = 8; p = 24.3 − 25.1Å)-forms and occasionally the novel Z-form (n = 12; p = 45.0Å). It has been shown that for DNA-duplexes with random base sequences, two types of helical structures are possible, viz. the right-handed uniform (RU) and the left-handed uniform (LU) helices.

*abbreviations: n = number of nucleotides per turn of the helix;
p = pitch of the helix

And for the other type of DNA duplexes i.e. with alternating purine-pyrimidine sequences (hereafter, PAPP) which include only poly d(AT), poly d(GC) etc., in addition to RU and LU helices, a novel kind of helical structure called the left-handed zig-zag (LZ) helix has been shown to be possible.

A Stereochemical Guideline for Exploiting the Conformational Flexibility Inherent in the Mononucleotide Repeat Unit of DNA Double Helix

For natural DNAs with random base sequences, it is reasonable to assume that all the nucleotide units have identical conformations irrespective of the nature of the bases attached to them. In such a case, a mononucleotide becomes the repeating unit for the helical duplexes. It is the mononucleotide, the smallest repeating unit

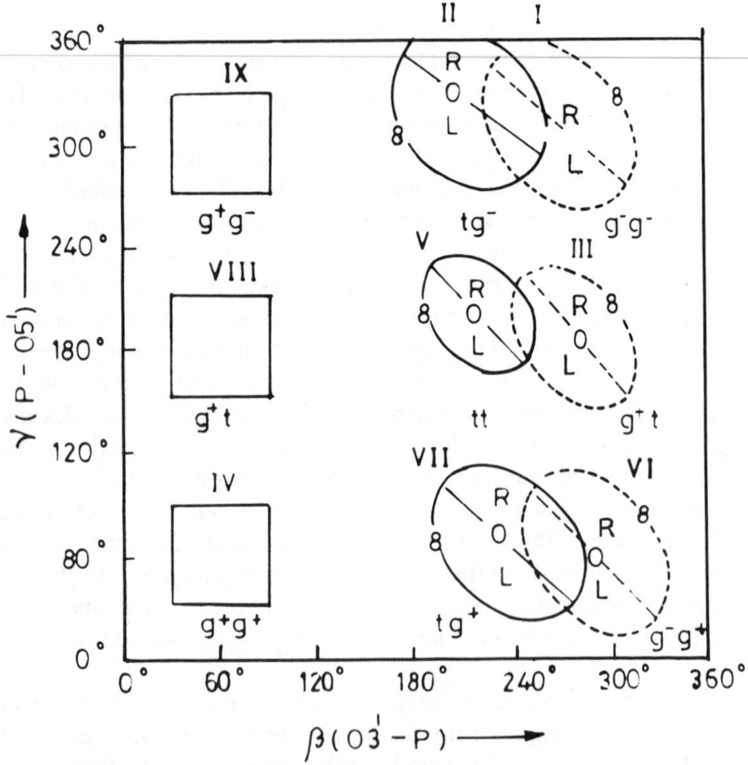

Figure 1. Helical domains in the (β-γ) space obtained using a nucleotide repeat. For the most frequently observed conformation gg around the C4′-C5′ bond, two helical domains (I & II) are obtained: one corresponds to the (C3′-*endo*, g^-g^-) conformation and the other (C2′-*endo*, tg^-) conformation. In each domain both right and left-handed duplexes are possible (see the text). $h = 0$ divides each helical domain into right and left-handed sections. For gt conformation around the C4′-C5′ bond, two helical domains are possible viz. (C3′-*endo*, g^-t) and (C2′-*endo*, tt) (III and V); and similarly for tg conformation around the C4′-C5′ bond the helical domains are respectively (C3′-*endo*, g^-g^+) (VI) and (C2′-*endo*, tg^+) (VII). In the text, only the helical domains I, II and III are discussed. Among the non-helical domains (IV, VIII and IX), only the domain IV i.e. corresponding to (C3′-*endo*, g^+g^+) or (C2′-*endo*, g^+g^+) conformation, is discussed.

Table I
The Conformational Parameters of the B-DNA
(n = 10; p = 34.0Å) Models with a Mononucleotide Repeat

		RU Helices		LU Helix
		($C3'$-*endo*, g^-g^-) conformation	($C2'$-*endo*, tg^-) conformation	($C2'$-*endo*, tg^-) conformation
Backbone torsion angles (degrees)	α (C3′-O3′)	184	225	241
	β (O3′-P)	292	208	204
	γ (P-O5′)	269	313	270
	δ (O5′-C5′)	179	141	135
	ϵ (C5′-C4′)	75	40	36
	ζ (C4′-C3′)	97	141	137
Glycosyl torsion angle (degrees)	χ (C1′-N)	34	70	−3
Furnose conformational angles (degrees)	σ_0 (C2′-C3′)	24.4	−36.7	−38.6
	σ_1 (C3′-C4′)	−19.0	18.6	18.1
	σ_2 (C4′-O1′)	5.0	8.0	11.2
	σ_3 (O1′-C1′)	12.1	−31.4	−36.1
	σ_4 (C1′-C2′)	−23.4	41.0	45.2
Phosphate radius	r_p(Å)	9.2	9.2	9.0
Groove size	s_p(Å)	14.2	14.0	11.2
Base parmeters (degrees)	tilt, θ_x	−7.00	−4.0	−3.0
	twist, θ_y	−7.0	−5.0	0.0
	dis-placement D(Å)	0.25	−0.60	−1.22

of a helical duplex, wherein lies the three major degrees of freedom of a polynucleotide structure. They are as follows: (i) flexibility in the furanose ring puckering; (ii) torsion around the glycosidic bond which links the sugar residue with the base; and (iii) the torsions around the O3′-P-O5′ bonds which join two successive nucleotide units.

The presence of the flexibility in the nucleotide unit implies that there is more than one sterically allowed conformation which could be used to generate helical structures. However, it turns out that not all sterically allowed conformations of the nucleotide unit lead to helical structures. The conformations of the nucleotide units which lead to helical structures are said to fall in the helical domains in the (β-γ) space and the conformations of the nucleotide unit which do not give rise to helical structures are said to fall in the non-helical domains of the (β-γ) space. (See Table I for the nomenclature). Figure 1 shows the helical and the non-helical domains in the (β-γ) space when a mononucleotide is used as a helical repeat.

Following are the systematic features which can be noted from Figure 1. (i) The nucleotide units in the helical domains maintain a preferred correlation between the sugar pucker and the P-O torsions. For example, when the conformation around C4′-C5′ is gg, the helical domains correspond to g^-g^- conformation around the P-O bonds for the C3′-*endo* sugar (called the C3′-*endo*, g^-g^- conformation) and tg^- conformation for the C2′-*endo* sugar (called the C2′-endo, tg^- conformation). Prior to model building, crystal data analyses of thc nucleic acid components and theoretical calculations were performed. These indicated such correlations between the sugar pucker and P-O torsions. These in fact formed the stereochemical basis of molecular model building.[10,11] (ii) In each helical domain, both right and left-helical duplexes can be obtained. It is interesting to note that the backbone torsion angles of the right and left-handed duplexes in a given domain for a given form of DNA are only marginally different. The only difference is in the value of the glycosyl torsion χ; for the right-handed duplex in a given domain it is about 60°-90° different from that of the left-handed duplex in thc same domain. The difference in the value of χ for the right and left-helical duplexes in a given domain, imposes certain restrictions on the feasibility of some structures. Take for example, the (C3′-*endo,* g^-t) conformation (domain III). The right-handed duplexes have χ in the low *anti* region (0°-30°) while the left-handed duplexes either have χ in the high *anti* region (90°-120°) or near *syn* region (300°-330°). Both the high *anti* and near *syn* conformations lead to steric compression for pyrimidine bases and hence the left-handed duplexes in the domain III are rather unlikely. However, in domain II, the right as well as left-handed duplexes are possible because both of them have χ in the *anti* conformation allowed for purines and pyrimidines. (iii) The helical duplexes with mononucleotide as a repeat possess monotonic backbone conformations and sugar-base orientations and hence they are called the uniform helices. And the uniform helices can be either right-handed (RU) or left-handed (LU). (iv) In the non-helical domain, a duplex structure is possible even though the successive nucleotide units do not trace a helical path. This special property of the nucleotide units in the non-helical domains is utilized to generate the LZ structures *(vide infra).*

Structural Features and Scattering Profiles of the RU and LU Helices With Mononucleotide Repeats

The RU and LU helices for a given polymorphic form of DNA are obtained subject to the following four criteria. (i) allowed stereochemistry of the backbone and sugar-base orientation; (ii) favourable stacking arrangement of the bases; (iii) favourable packing of the duplexes in the unit cell and (iv) agreement with the observed diffraction pattern. In what follows, we describe the gross and the fine structures of the RU and LU helices obtained for the, A, B and D-forms of DNA. The gross structure of a duplex refers to the radius (r_p) of the phosphate group, the chain separation (S_p) of the neighbouring phosphate chain across the minor groove and base parameters which include the tilt θ_x twist θ_y and the displacement D of the bases w.r.t. the helix axis. The fine structure corresponds to the backbone torsion angles, glycosyl torsion and the furanoso ring conformation.

The gross and fine structures of the RU and LU helices for the A-, B- and D-forms of DNA are given in Tables I-III.

B-DNA. Table I lists the parameters describing the gross and fine structures of the three B-DNA models. Two of the models are RU helices; one with the (C3′-*endo*, g^-g^-) conformation (domain I in Fig. 1) and the other with the (C2′-*endo*, tg^-) conformation (domain II). The third model is an LU helix with the (C2′-*endo*, tg^-) conformation. Since all the models maintain the preferred correlation, they are stereochemically accepιable. Recently, energy calculations also supported this view showing all the three models are almost equally stable.[13] Figure 2 shows the molecular transforms of the three models. The similarity of the transforms and the crystallographic residual R of the three models suggest that all of them are equally consistent with the X-ray data. The line diagrams of the models, described in Table I, are shown in Figure 3. It may be recalled that the best right-handed B-DNA model by Arnott and Hukins[14] was stereochemically unacceptable because it had

Table II

The Conformational Parameters of the D-DNA Model with a Mononuclectide Repeat as Obtained for poly d(AT) (n = 8; p = 24.3Å) and poly d(IC) (n = 8; p = 25.1Å). All the models have (C2′-*endo*, tg^-) conformation.

		poly d(AT)		poly d(IC)	
		RU Helix	LU Helix	RU Helix	LU Helix
Backbone torsion angles (degrees)	α	208	225	203	211
	β	209	215	218	213
	γ	302	263	308	267
	δ	147	156	148	163
	ϵ	61	32	58	34
	ζ	154	152	147	153
Glycosyl torsion angle (degrees)	χ	72	−4	58	9
Furanose conformational angle (degrees)	σ_0	−29.7	−32.6	−34.4	−34.6
	σ_1	27.1	28.6	22.6	26.1
	σ_2	−13.6	−13.0	−1.7	−6.4
	σ_2	−6.0	−8.0	−20.3	−16.5
	σ_4	22.7	25.0	34.2	32.8
Phosphate radius	r_p(Å)	8.2	7.7	8.2	7.6
Groove size	s_p(Å)	11.2	6.2	6.8	6.5
Base parmeters (degrees)	θ_x	−6.0	−6.0	−25.0	2.0
	θ_y	−8.0	−4.0	0.0	0.0
	D(Å)	−3.8	−2.9	−2.5	−3.0

Table III
The Conformational Parameters of the LU Helix for the A-DNA
(n = 11; p = 28.3Å) with a Mononucleotide Repeat

		LU Helix (C2′-*endo*, tg^-) conformation
Backbone torsion angles (degrees)	α	203
	β	214
	γ	312
	δ	131
	ϵ	32
	ζ	118
Glycosyl torsion angle (degrees)	χ	−20
Furanose conformational angles (degrees)	σ_0	−29.0
	σ_1	−0.8
	σ_2	31.1
	σ_3	51.2
	σ_4	50.6
Phosphate radius	r_p(Å)	10.0
Groove size	s_p(Å)	9.0
Base parameters (degrees)	θ_x(°)	−24.0
	θ_y(°)	−1.7
	D(Å)	1.65

unorthodox (C2′-*endo*, g^-g^-) conformation. However, it is gratifying to note that recently Arnott *et al*[15] chose a (C2′-*endo,* tg^-) conformation for the right-handed B-DNA model with torsion angles very close to the value suggested by us (Table I).

D-DNA. The parameters of the two D-DNA's are given in Table II. Both the models belong to the domain II i.e. (C2′-*endo,* tg^-) conformation; one is an RU helix and other is an LU helix. We for the first time could obtain a stereochemically satisfactory right-handed model for the D-DNA which also gives a good account of the fibre-data. The right handed model of thc D-DNA by Arnott *et al*[16] was stereochemically unacceptable because it had unorthodox (C2′-*endo*, g^-g^-) conformation.[10,11,17] On the contrary, both the RU and LU helices of the D-DNA proposed by us maintain the preferred correlation (i.e. (C2′-*endo*, tg^-) conformation) and hence the stereochemical feasibility of such models can hardly be questioned. The molecular transforms of the two D-DNA models are given in Figure 4 and the line diagrams in Figure 5.

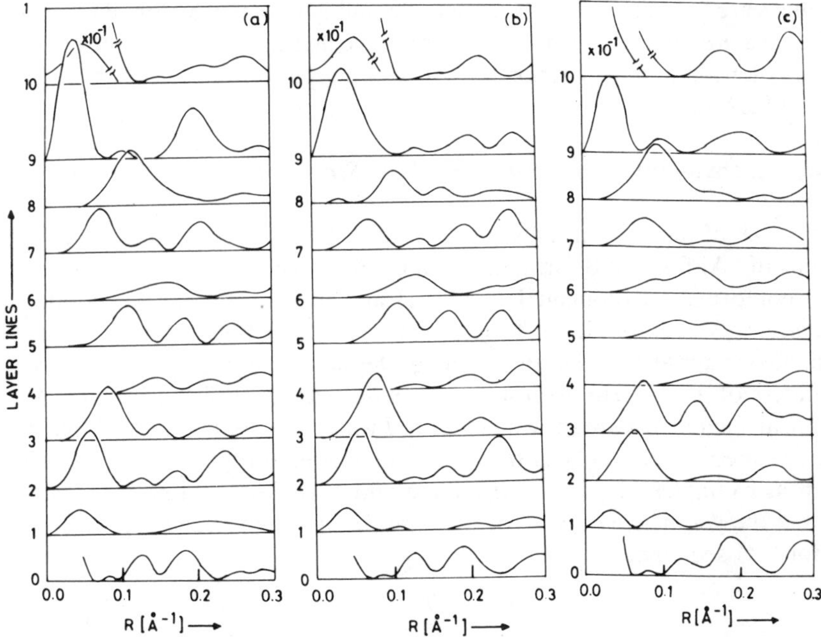

Figure 2. The molecular tranfsorms of the three B-DNA models (Table I) : (a) The RU helix with (C3′-*endo*, g⁻g⁻) conformation which gave an R-factor of 0.36 (b) The RU helix with (C2′-*endo*, tg⁻) conformation (R = 0.30) and (c) the LU helix with (C2′-*end*, tg⁻) conformation (R = 0.37).

A-DNA. The parameters for the LU helix of the A-DNA are given in Table III. The model belongs to the domain II i.e. (C2′-*endo*, tg⁻) conformation. We were the first to report the stereochemically acceptable LU model for the A-DNA which also shows agreement with the fibre pattern (see the molecular transforms in Figure 6).

The possibility of RU and LU helices for the A-, B-and D-forms of DNA is consistent with the smooth A ⇔ B and B ⇔ D transitions observed in the solid phase. It also confirms our hypothesis that DNA in either handedness is a natural outcome of the conformational flexibility inherent in the molecule.

Sequence Specific Molecular Conformations With a Dinucleotide Repeat

The RU and LU helices described in the previous section are poly (mononucleotides). Such models may be expected for DNA polymers with random base sequences but need not hold for PAPP (e.g. poly d(AT), poly d(GC) etc.) which has a well defined base sequence. PAPP is chemically a poly (dinucleotide). Hence, it is reasonable to expect that the structural repeat should also be a dinucleotide. Topologically two distinct types of helical duplexes are obtained for PAPP using a dinucleotide repeat: RU and LZ helices. A RU helix is formed when two neighbouring nucleotides in the dinucleotide repeat lie in two different helical domains.

LZ helices are formed (i) when a nucleotide unit in one helical domain is jointed to another in a non-helical domain and (ii) both the nucleotide units in thc repeat have conformations in non-helical domains. The two types of LZ-helices are referred as LZ1 and LZ2 helices.

RU-Helix in the B-form: Structural Features and Scattering Profile

For the sake of illustration, only one example of a RU helix is discussed. It is the RU helix of PAPP in the B-form, in which the purine nucleotide is in the (C3′-*endo*, g^-g^-) conformation (domain I in Fig. 1) and the pyrimidine nucleotide in the (C2′-*endo*, tg^-) conformation (domain II). Here, the P-O torsions refer to the phosphate groups attached at the 3′-end of the sugar residue.[18] The conformational parameters of a RU helix with a dinucleotide repeat are listed in Table IV. The stacking arrangemcnts are shown in Figure 7 which are not markedly different frem those obtained for RU helices with a mononucleotide repeat.[13] G-C and C-G are chosen as examples of purine-pyrimidine and pyrimidine-purine sequences. It is seen that the geometric overlap of the bases in the same strand is greater for G-C than for C-G sequence.

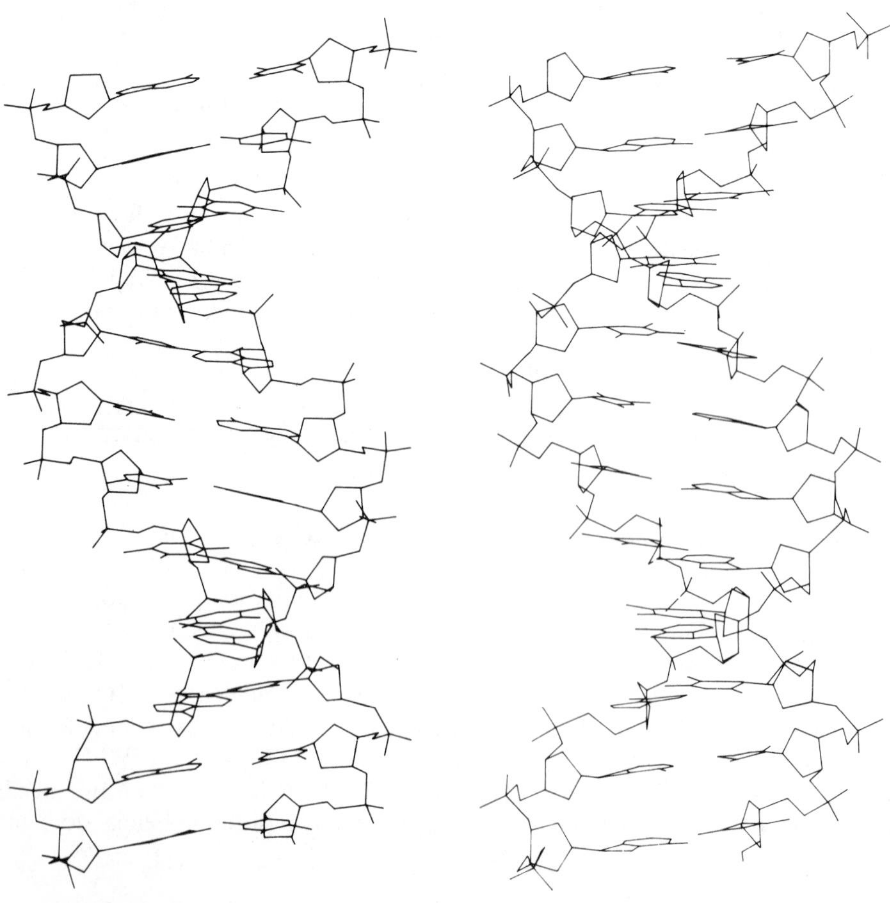

In a RU helix, with a dinucleotide repeat, the glycosyl torsion χ for purines is 24° and χ for pyrimidines is 58°, both the values of χ falling in the *anti* region. Hence, such a structure is also possible with heterogenous base sequence (e.g. calf thymus B-DNA). It, therefore, appears that quantitive analysis of the RU helix (with dinucleotide repeat) is adequately justified with the fibre data of the calf thymus B-DNA. Topologically a RU helix with a dinucleotide repeat is indistingishable from the two with mononucleotide repeats (compare Figs. 3 and 8). Therefore, the molecular transforms of the three RU helices show overall similarity (compare Fig. 2a, 2b and 9). However, for the sequence specific RU helix (with a dinucleotide repeat) a meridional reflection on the 5th layer line is predicted because the former is a 5-fold helix. It has been found that the intensity value at 005 is very small (see Fig. 9) and could be made zero by a small departure from an exact 5-fold helix.

LZ Helices: Two Types

As stated above, the two types of LZ helices are designated as LZ1 and LZ2. Here for the sake of illustration, only one example in each type is considered.The example of LZ1 is the one in which dinucleotide repeat has the purine nucleotides in

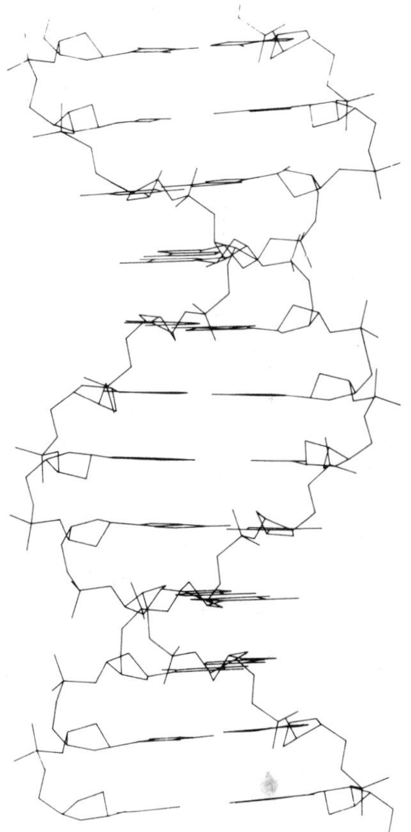

Figure 3. The line diagrams of the three B-DNA models in the same order as in Figure 2.

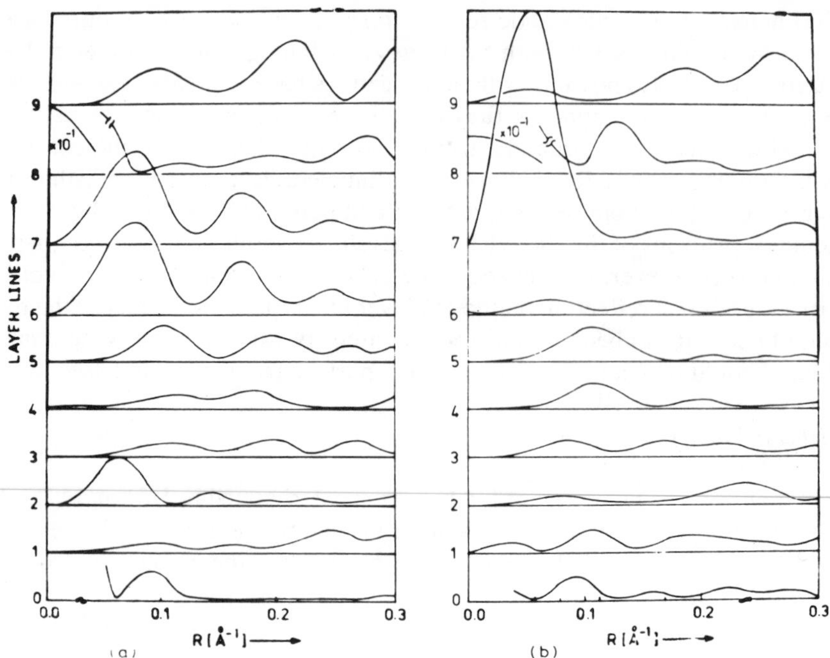

Figure 4. The molecular transforms of the two D-DNA models with (C2′-*endo*, tg−) conformation (Table II): (a) RU helix which gave an R-factor of 0.37 (b) LU helix (R = 0.36).

(C3′-*endo*, g^-t) conformation (a helical domain in Fig.1) and the pyrimidine nucleotides in (C2′-endo, g^+g^+) conformation (a non-helical domain). The example of LZ2 involves the dinucleotide repeat with the purine nucleotides in (C3′-*endo*, g^+g^+) conformation and the pyrimidine nucleotides in the (C2′-*endo*, g^+g^+) conformation (both being in the non-helical domains). In the two types of the LZ-helices, the dinucleotides possess the following common conformational features: (i) all the C3′-*endo* sugars have gt conformation around the C4′-C5′ bond while the C2′-*endo* sugars have gg conformation around the same bond; (ii) the purines are all attached to the C3′-*endo* sugars and have *syn* conformation while the pyrimidines are connected to the C2′-*endo* sugars and have *anti* conformations.[18] Using the two conformations, described above, poly (dinucleotide) models have been generated for PAPP in B(n = 5; p = 34.0Å) and Z(n = 6; p = 45.0Å) forms. The possibility of LZ helices in B-form suggests that the conformational flexibility inherent in dinucleotide unit allows the compression of the helix from a pitch of 45Å (in Z-form) to that of 34Å (in the B-form).

Structural Features and Scattering Profiles of the LZ1 and LZ2 Helices

The B-Form. The conformational parameters of LZ1 and LZ2 helices in the B-form are given in Table V. Although these two types of helices have completely different backbone torsion angles, the stacking patterns of the base in the two structures

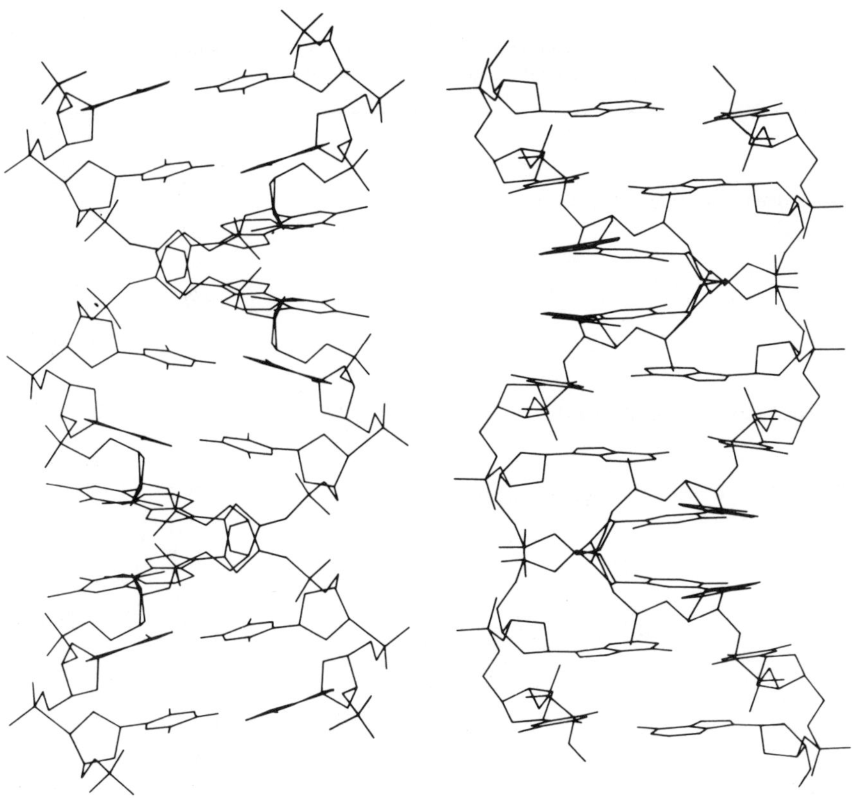

Figure 5. The line diagrams of the two D-DNA models in the same order as in Figure 4.

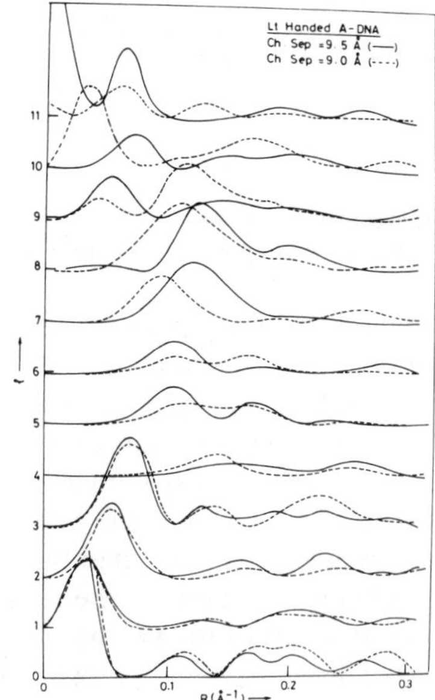

Figure 6. The molecular transforms of the LU helices for the A-DNA. The chain seperation (S_p) are as indicated. Increase of the chain seperation from 9.0 to 9.5Å is obtained by the slight increase in the value of the tilt θ_x (Table III). Note that the differences in the molecular transforms of the two A-DNA models appear only on the higher layer lines.

Table IV*
The Conformational Parameters of the RU Helix with a Dinucleotide Repeat for the B-form of PAPP (n = 5; p = 34Å)*

		purine nucleotide	pyrimidine nucleotide
Backbone torsion angles (degrees)	α	183	205
	β	292	210
	γ	295	286
	δ	147	181
	ϵ	49	66
	ζ	89	156
Glycosyl torsion angle (degrees)	χ (C1′-N9)	24	---
	χ (C1′-N1)	---	58
Furanose conformational angles (degrees)	σ_0	29.6	−32.4
	σ_1	−27.2	22.4
	σ_2	12.7	−3.2
	σ_3	7.4	−17.6
	σ_4	−23.8	30.9
Phosphate radius	r_p(Å)	9.2	9.0
Groove Size	s_p(Å)	14.8	13.8
Twist	t	33	39
Axial rise	h(Å)	3.6	3.2
Base parameters (degrees)	θ_x	15	7
	θ_y	14	

*In Tables IV-VI, the notations for the torsion angles used for the 3′-nucleotides are as follows:

$$\mathrm{O5'} \underset{\delta}{\ddagger} \mathrm{C5'} \underset{\epsilon}{\ddagger} \mathrm{C4'} \underset{\zeta}{\ddagger} \mathrm{C3'} \underset{\alpha}{\ddagger} \mathrm{O3'} \underset{\beta}{\ddagger} \mathrm{P} \underset{\gamma}{\ddagger} \mathrm{O5'}$$

only the absolute magnitude of θ_x and θ_y are indicated. (t,h) values correspond to the helical relation of the two phosphate groups at the 5′- and 3′-ends of the nucleotide unit, e.g. the two phosphates in 5′pGp3′.

show general similarity. Figures 10a and 10b show the stacking arrangements in the purine-pyrimidine and the pyrimidine-purine sequences taking G-C and C-G as examples. In both the LZ1 and LZ2 helices, the G-C sequence shows intrastrand stacking overlap whereas stacking overlap occurs for two Cs in the C-G sequence. There are certain other structural features which the LZ1 and LZ2 helices have in common. (i) In both of them, the phosphate radius r_p in the purine-pyrimidine sequence is always greater than that in the pyrimidine-purine sequence (see Table IV). (ii) C8 and N7 atoms of the purines fall on the outer surface of the helix. (iii) The oxygen atoms of the neighbouring sugar residues along the same strand point in opposite directions while those of the sugar residues across a base-pair point in the same direction.

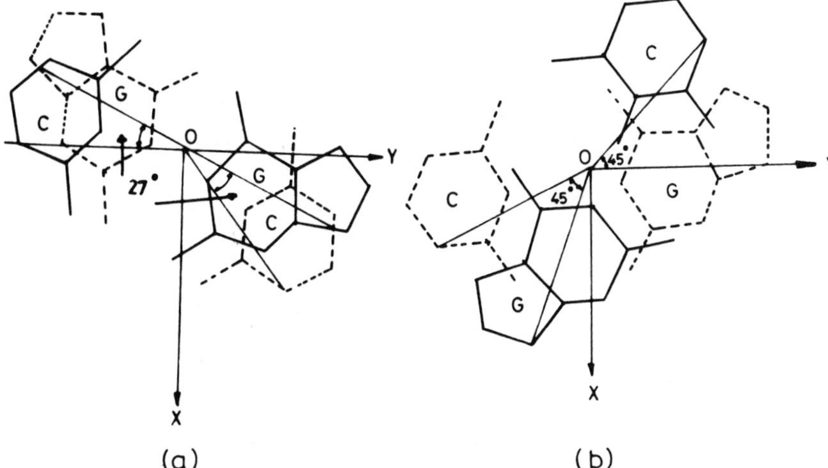

Figure 7. Stacking arrangements in an RU helix (B-form). G-C and CG sequence are taken as examples of pur-pyr and pyr-pur stacking arrangements. (a) G-C stacking with geometric overlap of the intrastrand bases; the twists angle between G and C is 27°. (b) CG stacking with almost no geometric overlap of the bases; the twist angle between C and G is 45°. Thus, even though all the phosphate groups have similar helical twist two neighbouring bases do have different twist angles for different sequences.

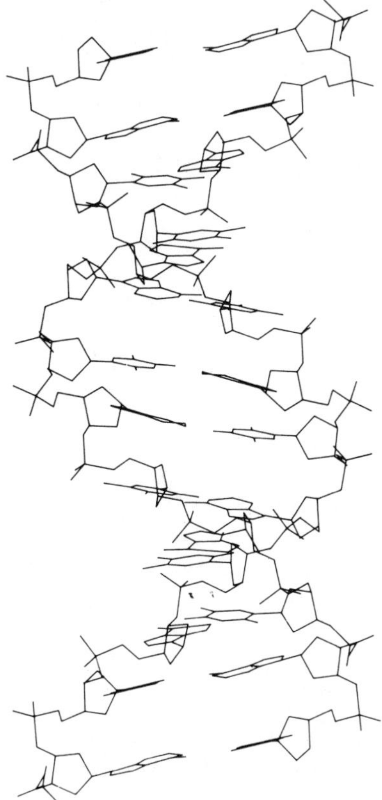

Figure 8. The line diagram of the RU helix in the B-form with a dinucleotide repeat (Table IV).

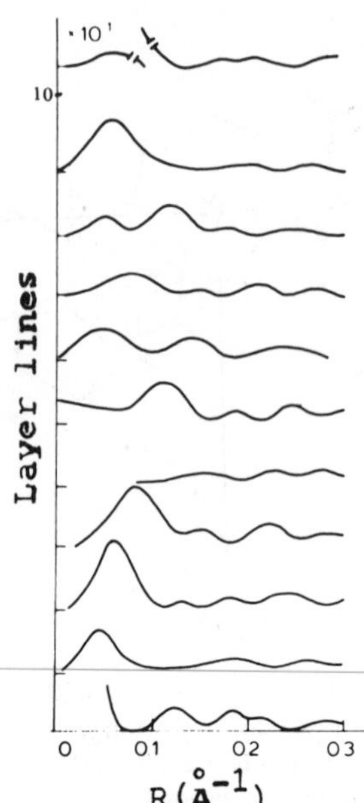

Figure 9. The molecular transform of the RU helix in the B-form with a dinucleotide repeat (Table IV).

For PAPP in B-form, no qantitative X-ray data are available. However, the calf thymus B-DNA X-ray diffraction patterns shows a general similarity with that of PAPP in the B-form.[19] Therefore, the scattering profiles of the LZ1 and LZ2 helices in the B-form are compared with the calf thymus B-DNA pattern. Figures 11a and

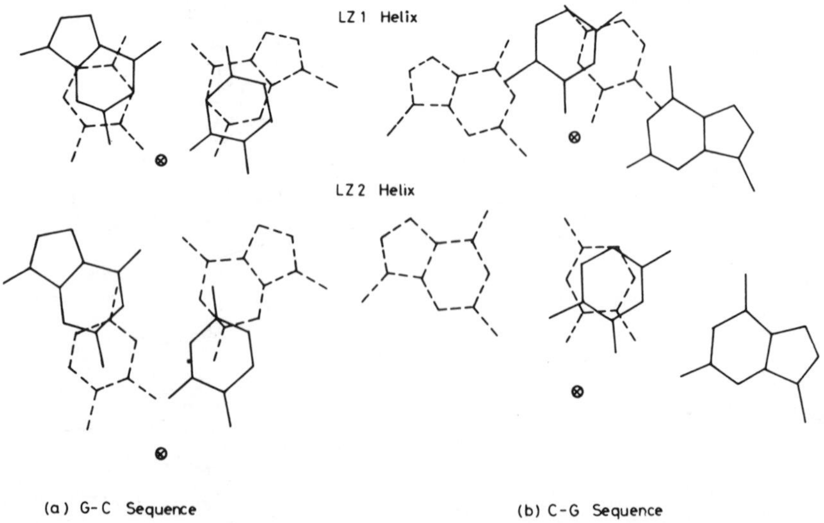

LZ 1 Helix

LZ 2 Helix

(a) G-C Sequence

(b) C-G Sequence

Figure 10. Stacking arrangements of the pur-pyr and pyr-pur sequences in the LZ1 and LZ2 helices in the B-form taking G-C and C-G as examples. (a) G-C stacking arrangements in the LZ1 and LZ2 helices. (b) C-G stacking arrangements in the LZ1 and LZ2 helices.

Table V
The Conformational Parameters of the LZ Helices with a Dinucleotide Repeat for the B-form of PAPP (n = 5; p = 34ÅA)

		LZ1 Helix		LZ2 Helix	
		purine nucleotide	pyrimidine nucleotide	purine nucleotide	pyrimidine nucleotide
Backbone torsion angles (degrees)	α	238	262	189	270
	β	322	103	102	88
	υ	210	85	101	65
	δ	191	179	164	148
	ϵ	159	69	160	72
	ς	95	126	85	158
Glycosyl torsion angle (degrees)	χ (C1′-N9)	225	---	230	---
	χ (C1′-N1)	---	12	---	15
Furnose conformational angles (degrees)	σ_0	30.6	−27.9	39.5	−33.5
	σ_1	−28.0	4.2	−39.0	29.5
	σ_2	13.8	21.3	22.5	−13.2
	σ_3	7.8	−40.3	3.6	−8.6
	σ_4	−24.9	44.0	−27.1	26.0
Phosphate radius	r_p(Å)	8.0	6.0	7.0	5.5
Groove size	s_p(Å)	13.1	7.1		
Twist	t	−23	−49	0	−72
Axial rise	h(Å)	5.3	1.5	6.8	0
Base parameters (degrees)	θ_x	5	3	7	3
	θ_y	7		5	

11b show the molecular transforms of the LZ1 and LZ2 helices in the B-form. The following observations can be made from Figure 11. (i) Both LZ1 and LZ2 being poly (dinucleotide) models the 005 reflection is predicted for both of them. However, only the calculated molecular transform of the LZ1 helix shows a strong meridional reflection at 005 (see Fig. 10a). For the LZ2 helix the calculated transform has a very small intensity value at 005. It may be noted from Table V that the two phosphate groups in the LZ1 helix are either related by a twist, t = 49° and an axial rise, h = 1.5Å or t = 23° and h = 5.3Å while in the LZ2 helix the two phosphate groups are either related by t = 72° and h = 0Å or t = 0° and h = 6.8Å. Thus only for the LZ2 helix (and not for the LZ1 helix) t and h values are always integral multiples of 36° and 3.4Å, corresponding values of a uniform 10-fold B-DNA. The 005 reflection being missing in the molecular transform of a uniform 10-fold helix (i.e. J_5 (0) = 0), it is only logical that the same would hold true for the LZ2 helix with the gross structure as in Table V. (ii) In the calculated transforms of

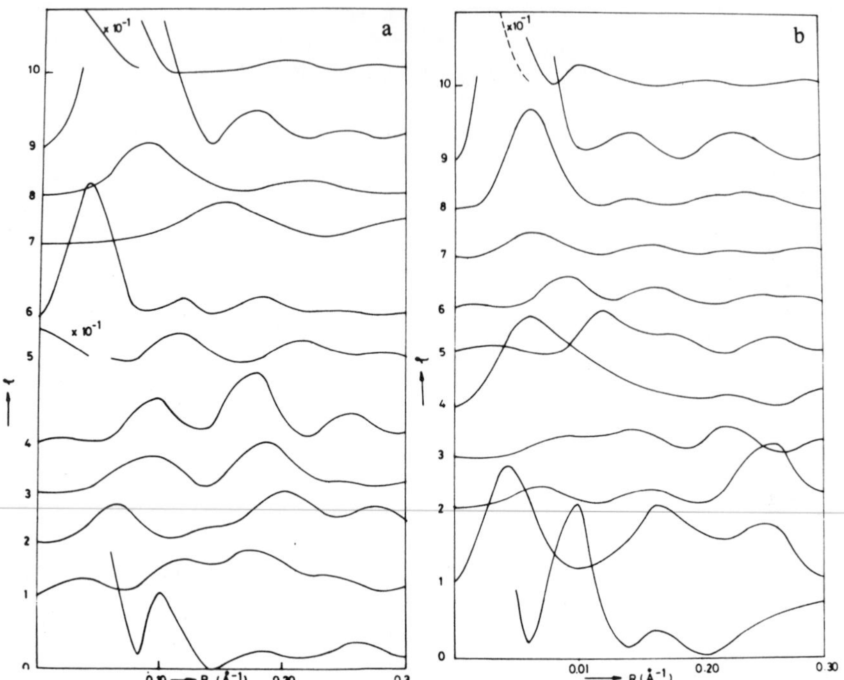

Figure 11. The molecular transforms of the LZ helices in B-form. (a) LZ1 helix in B-form, note the presence of the 005 reflection in the transform. (b) LZ2 helix in B-form, note that the 005 reflection is negligibly small.

both the helices, the 4th layer line shows intensity values much stronger than those observed in the calf thymus B-DNA pattern. (iii) For the LZ1 helix, the calculated transform gives a strong peak at R = 0.04Å^{-1} on the 6th layer line, (largely due to J_{-1} contribution) which may, however, be reduced to almost zero (as observed in the calf thymus B-DNA pattern) because the packing parameter is zero at R = 0.04Å^{-1} on the 6th layer line. However, the calculated transform for the LZ2-helix only gives a weak intensity at R = 0.04Å^{-1} on the 6th layer line. (iv) The molecular transform of the LZ2 helix shows a very strong maximum at R = 0.04Å^{-1} on the lst layer line which is only medium strong in the observed pattern. But the transform of the LZ1 helix shows only a medium strong peak at R = 0.04Å^{-1} on the first layer line. Due to the discrepancies mentioned above, the calculations of the structure factor amplitudes of the LZ1 and LZ2 helices and comparision of the calf thymus B-DNA[14] lead to R-factors with value on the higher side. On omitting about 30 reflections in the far off region the R-factor drops[18] to 0.40. In view of thc fact that the LZ1 and LZ2 helices represent really two extremes it may not be difficult to fit one model or the other with the observed data of PAPP in the B-form. However, in the absence of any quantitative data, it is not possible to refine the models.

The Z-Form. The conformational parameters of the LZ helices in the Z-form (n = 6; p = 45.0Å) are listed in Table VI. They are reminiscent of the single crystal

Table VI
The Conformational Parameters of the LZ Helices with a Dinucleotide Repeat for the Z-form of PAPP (n = 5; p = 34ÅA)

		LZ1 Helix		LZ2 Helix	
		purine nucleotide	pyrimidine nucleotide	purine nucleotide	pyrimidine nucleotide
Backbone torsion angles (degrees)	α	242	263	190	274
	β	318	95	104	83
	γ	211	89	99	65
	δ	194	183	177	142
	ϵ	161	67	177	70
	ζ	90	124	76	156
Glycosyl torsion angle (degrees)	χ (C1′-N9)	230	---	230	---
	χ (C1′-N1)	---	15	---	9
Furnose conformational angles (degrees)	σ_0	35.9	−27.8	43.0	−34.9
	σ_1	−35.3	5.2	−49.0	27.0
	σ_2	18.4	21.0	35.0	−7.4
	σ_3	6.5	39.9	−8.6	−15.6
	σ_4	−27.8	42.4	−22.1	31.2
Phosphate radius	r_p(Å)	7.8	5.8	6.9	5.7
Groove size	s_p(Å)	18.5	11.6	16.0	14.0
Twist	t	−15	−45	2.0	−62.0
Axial rise	h(Å)	5.5	2.0	7.3	0.2
Base parameters (degrees)	θ_x	15	1	7.0	+1.0
	θ_y	15		3	

structures[20,21] of $d(CG)_3$ and $d(CG)_2$. When the structure of $d(CG)_3$ was reported it was said that the P-O torsions for the phosphate groups at the 3′-end of the G bases were the same as those found in the B-DNA i.e. tg^- (see Table I). However, when the structure of $d(G)_3$ was refined[22] it was found to possess predominantly LZ1 conformation as proposed by us[23] except at one G residue which had (C3′-*endo*, g^+t) conformation (a non-helical domain VIII in Fig. 1) a minor variant of (C3′-*endo* g^+g^+) conformation of the LZ2 helix.[23] It is then, only natural that the zig-zag progression of the phosphate groups round the helix axis in the LZ1 helix and in $d(CG)_3$ show close similarity (compare Figures 12a and 12b). Subsequent to our proposal of the LZ2 helix, thc crystal structure of $d(CG)_2$ was solved in which C nucleotides had (C2′-*endo,* g^+g^+) conformation and the G nucleotides presumably had (C1′-*exo,* g^-t) conformation.[21] Thus, in the crystal, both the nucleotides were in the non-helical domains: a characteristic feature of thc LZ2 helices. Therefore, a striking similarity was seen between the structure of $d(CG)_2$ and the LZ2 helix as

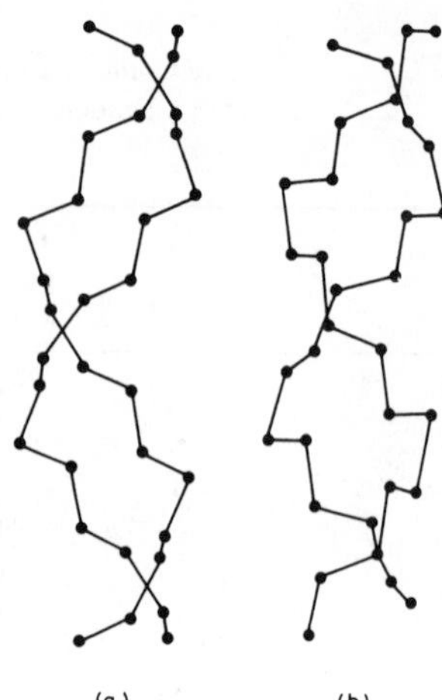

Figure 12. Comparison of the LZ1 helix as described in Table VI with the gross structure of $d(CG)_3$ in the single crystal.[20] (a) The progression of the phosphate groups (dark circles) in thc LZ1 helix (Table VI). (b) progression of thc phosphate groups in the $d(CG)_3$ crystal.[20]

described in Table VI when the zig-zag progression of the phosphate groups are compared (see Fig. 13a and 13b). The stacking arrangements and other structural features of the LZ1 and LZ2 helices in B- and Z-forms are quite similar and hence they are not given here.

The molecular transorms of the LZ1 and LZ2 helices in the Z-form are shown in Figures 14a and 14b. It may be recalled that the so called Z-pattern as reported by Arnott *et al*[24] for the poly d(GC) fibres at high salt has the following features. (i) Presence of meridional reflections on the 6th and 12th layer lines; (ii) Bragg reflections on the layer lines 2, 5, 7, 9, 14 and (iii) streaks on the layer lines 1, 6, 8, 11, 13. The transform of LZ helices (Fig. 14a) shows qualitative agreement with the observed pattern in that it gives (i) meridional reflections on the 6th and 12th layer

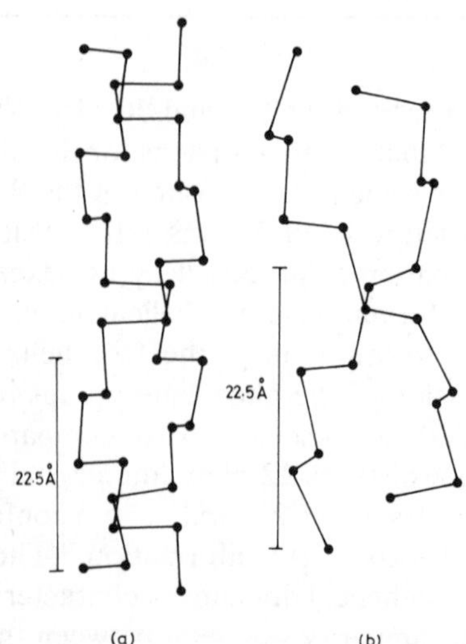

Figure 13. Comparison of the LZ2 helix as dcscribed in Table VI with thc gross structure of $d(CG)_2$ in the single crystal.[21]. (a) The progression of the phosphate groups (dark circles) in the LZ2 helix (Table VI). (b) The progression of the phosphate groups in the $d(CG)_2$ crystal.[21]

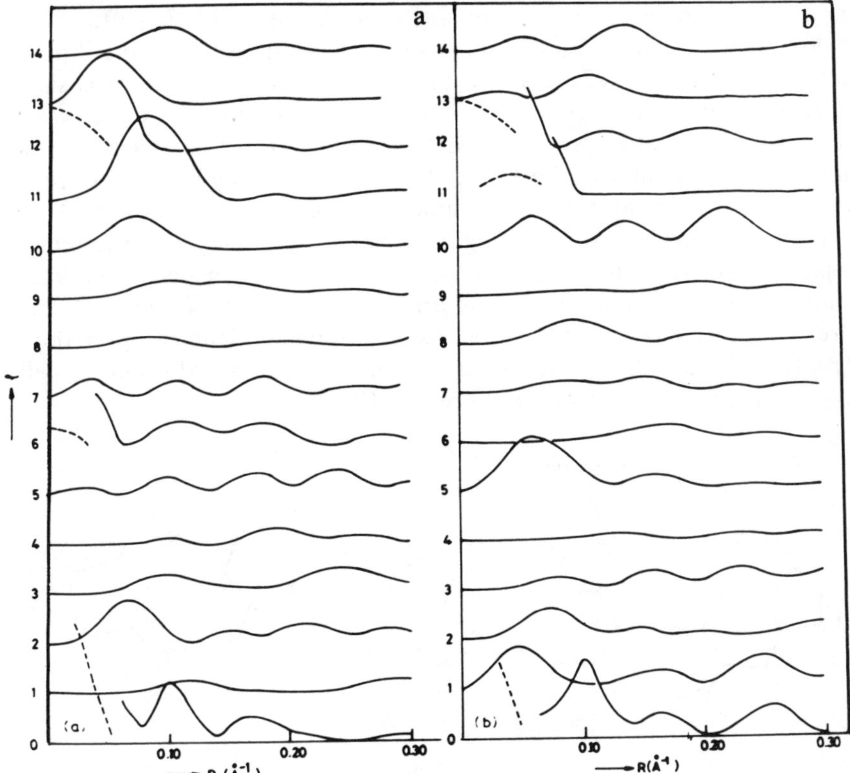

Figure 14. The molecular transforms of the (a) LZ1 and (b) LZ2 helices in the Z-form (n = 6; p = 45Å) Note the presence of a strong 006 reflection in the transform of the LZ1 helix while 006 reflection in the transform of the LZ2 helix is negligibily weak.

lines (ii) meridional defined peaks on the layer lines 2, 5, 7, 12 as observed (except on 14) and (iii) well spread intensity maximum on the layer lines 1, 6, 8, 11, 13. For the LZ2 helix, the calculated transform shows intensity values on layer lines 7 and 9 smaller than those observed. It is to be noted the 006 reflection is vanishingly small (like the 005 reflection in the transform of LZ2 helix in the B- form), though the 0012 is quite strong. However, the agreement of the calculated transform with the observed pattern on the layer lines 1, 2, 5, 8, 11, 13 and 14 is quite good.

Inspired by the observation of the Z-DNA fragments of $d(CG)_2$ and $d(CG)_3$ Arnott *et al* proposed a left-handed model for poly d(GC) at high salt.[24] It was stated that the novel Z-pattern was obtained after prolonged annealing of the poly d(GC) fibres at high salt which otherwise at low salt gave the classical B-pattern. However, we could demonstrate that under subtle changes of environmental conditions and without any drastic measures like prolonged annealing, a smooth B $\rightarrow$ Z transition could be achieved at the solid phase.[25] What we did was as follows. Instead of using the poly d(GC) fibres at high salt, they were drawn from 1:1 (V/V) water-ethanol. The fibre, so obtained, gave a characteristic B-pattern at rh 40% immediately after it was dried. The same fibre after humidity cycle, and equilibration after a few days at rh 40% (at room temperature) undergoes a smooth transition to the Z-pattern as reported.[24] Arnott *et al* also claimed that the proposed structure was similar to Z-DNA fragments in the crystals. However, we showed that the claim was rather weakly founded.[26] Although, the model of Arnott *et el*[24] has backbone torsion angles somewhat similar to the LZ1 helix in Table VI, the base-pairs are moved away from the helix centre (D = 1Å) rather than being brought inward (D = −4Å) as found in the single crystal (see Figures 15a and 15b). Such a positioning of the base-pair neither leads to a zig-zag progression of the sugar phosphate backbone nor does it give the required stacking arrangement as seen in the single crystals. Figures 16a and 16b compare the progression of the phosphate groups in the model of Arnott *et al* and in $d(CG)_3$. It is seen that the former is more like an LU helix and hardly compares with the LZ1 structure of $d(CG)_3$.

Stacking in the single crystals

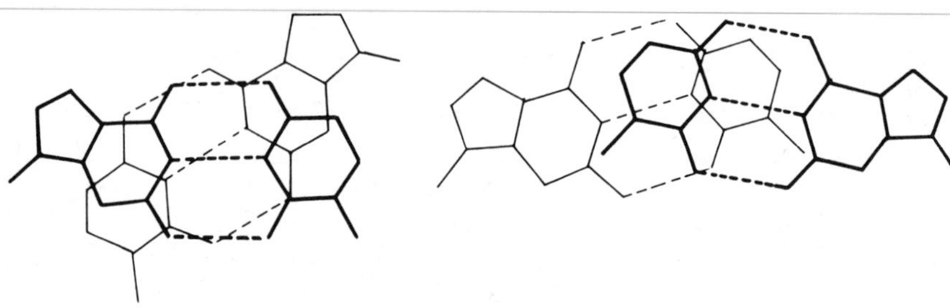

Stacking as suggested by Arnott et.al

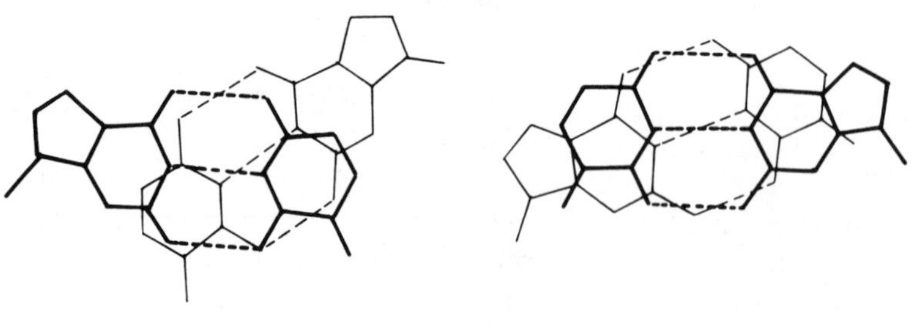

(a) G-C Sequence (b) C-G Sequence

Figure 15. Stacking arrangements of pur-pyr and pyr-pur sequences in the single crystals of $d(CG)_3$ and the model of poly d(GC) at high salt due to Arnott *et al*,[24] taking G-C and C-G sequences as examples. (a) G-C stacking in $d(CG)_3$ and S-DNA. (b) C-G stacking in $d(CG)_3$ and S-DNA.

The D-form. It was earlier pointed out that PAPP in the D-form is the most stable one.[12] Thus, it is of great interest to examine if sequence specific molecular models are also possible for PAPP in the D-form (n = 4; p = 24.3 − 25.1Å). It has been found that LZ2 helices are stereochemically satisfactory for the PAPP in the

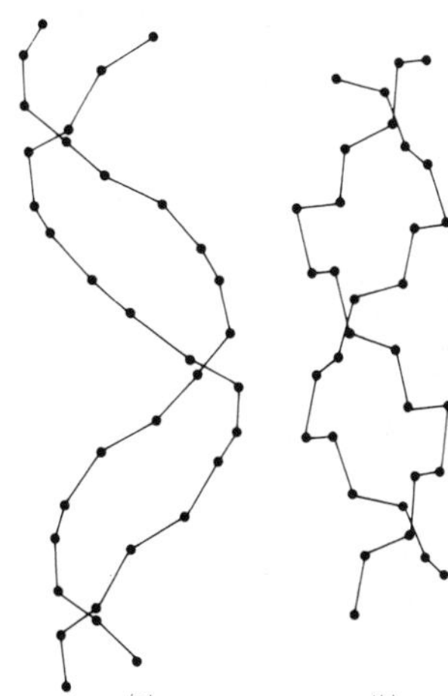

Figure 16. Progression of the phosphate groups (marked by dark circles) (a) in the model of poly d(GC) at high salt due to Arnott *et al* (b) in the single crystal of d(CG)$_3$.

D-form.[18] Thus it turns out that the LZ2 helices can be compressed from a pitch of 45Å (Z-form) to that of 24.3Å (D-form). However, stereochemically inoffensive models appear to be unlikely for the RU helix in the D-form with conformations as given in Table IV and for the LZ1 helix in the D-form with conformations as given in Tables V and VI. The conformational parameters of the LZ2 helix in the D-form are given in Table VII. Note that, the torsion angles for the LZ2 helices in Z-, B- and D-forms are very similar. Figure 17a and 17b show the line diagrams of LZ2 helices in B- and D-forms and bring out the similarity in the topology of the two forms. The molecular transform of the LZ2 helix in the D-form (not shown) gives only a small intensity value at 004 even though the former is a true 4-fold helix. Therefore, it may be generalized that for an n(even)-fold LZ2 helix, the 00n(even) reflection will have a small intensity value. The R-factor as computed for the LZ2 helix in D-form is 0.41 using the data of poly d(AT) fibres[16] and is 0.44 for the poly d(IC) fibres.[17]

Reversal of Handedness in DNA: A Stable Link Between RU and LZ Helices

It may perhaps be emphasized that the possibility of DNA expressing in either handedness, viz, RU, LU and LZ helices, emerges only when the conformational flexibility inherent in the DNA molecule is visualizcd in atomic detail and duly exploited in the molecular model building following a stereochemical guideline. Thus, it is of interest to examine whether the same conformational flexibility allows a ‘stable’ link between a right and a left double-helical segment. By ‘stable’, it is implied that the link should involve (i) allowed stereochemisty of the backbone, (ii) favourable stacking arrangement of the bases and (iii) Watson-Crick base pairing. It was earlier shown that such a link is possible between RU and LU fragments which results in a RL-model of DNA with inverted stacking arrangement at the link.[27] Here we demonstrate the possibility of the stable links between RU and LZ fragments and show that the link involves an inverted stacking arrangement of the bases—a characteristic feature of all RL models. It has been found that an LZ helix

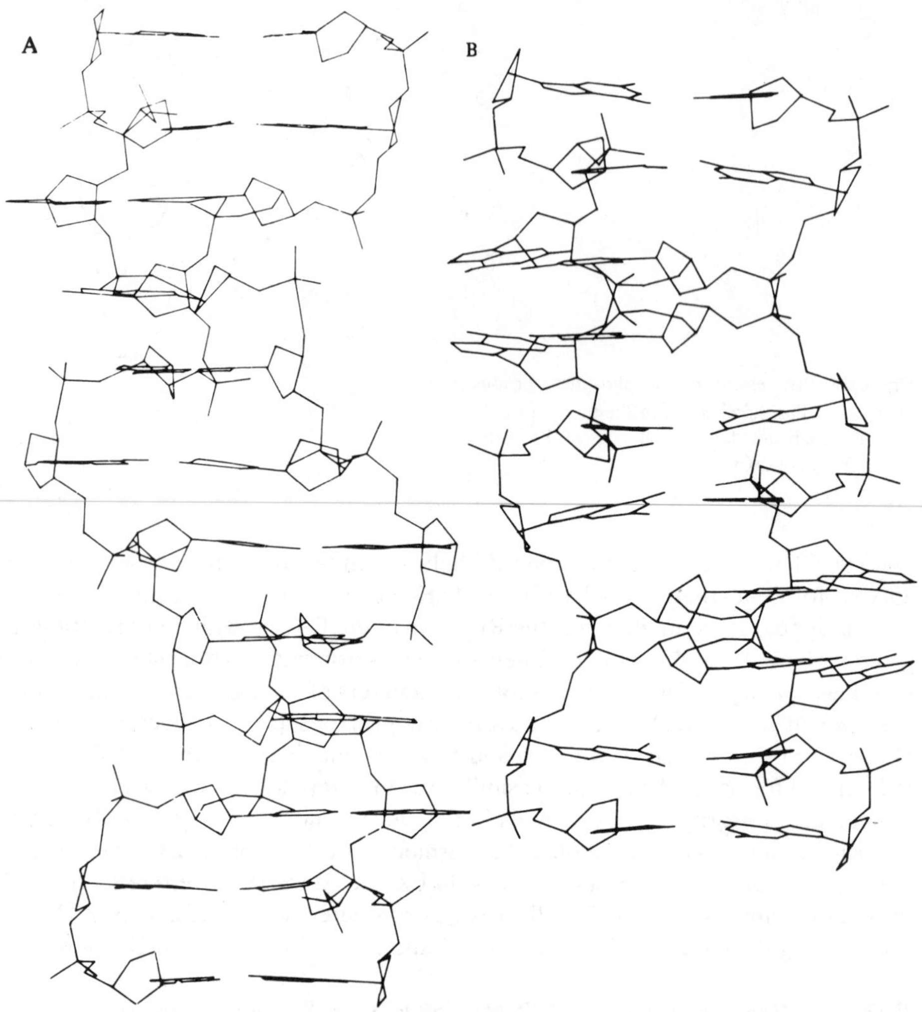

Figure 17. The line diagrams of the LZ2 helices (a) in B- and (b) in the D-form.

can be joined to an RU helix only in a specified manner.[23] The link should be made at the 3′-end of the G residues in the LZ fragments (see Figures 18 and 20). Thus, a segment of an LZ helix in the middle of two RU segments should necessarily have an even number of nucleotides.

The Link Between RU and LZ1 Fragments

Figure 18 describes the conformational features of the link betwecn RU and LZ1 helical fragments. The link consists of a base-paired dinucleoside monophosphate. The lower base-pair marks the end of the LZ1-fragment and hence it has the purine in *syn* conformation and the pyrimidine in *anti* conformation. At the link region, the backbone conformation of the up-chain and the down-chain are different. The

Table VII
The Conformational Parameters of the LZ2 Helix in the D-form
(n = 4; p = 25.1Å)

		purine nucleotide	pyrimidine nucleotide
Backbone torsion angles (degrees)	α	183	261
	β	97	95
	γ	102	66
	δ	146	168
	ϵ	166	74
	ζ	93	159
Glycosyl torsion angle (degrees)	χ (C1′-N9)	225	---
	χ (C1′-N1)	---	10
Furanose conformational angles (degrees)	σ_0	34.0	−35.4
	σ_1	−27.3	34.8
	σ_2	8.7	−19.7
	σ_3	14.8	−4.9
	σ_4	−31.0	25.2
Phosphate radius	r_p(Å)	7.1	5.5
Groove Size	s_p(Å)	11.2	5.5
Twist	t	−18	−72
Axial rise	h(Å)	6.3	0.0
Base parameters (degrees)	θ_x	18	5
	θ_y	15	

up-chain phosphodiester conformations are g^-g^- while in the down-chain they are g^-t. The sugar residues across the lower base-pair have different sugar puckers: the purine is attached to a C3′-*endo* sugar and the pyrimidine to a C2′-*endo* sugar. However, the sugar residues across the upper base-pair have the same C3′-*endo* pucker. Thus, the upper-base pair can form the beginning of an RU-fragment with (C3′-*endo*, g^-g^-) conformation (see Table I). The conformational parameters of the stable link between RU and LZ1 fragments are given in Table VIII. The inverted stacking arrangements in the (pyr--pur) and (pur--pyr) sequence taking (C--G) and (G--C) as examples are shown in Fig. 19. The physical overlap of the bases ensures a stable stacking arrangement. It may be recalled that Wang *et al*[20] suggest a link between the left-handed Z-DNA (LZ1 helix) and the right-handed B-DNA. They also arrived at the inverted arrangement of the bases at the link, but the bases were largely destacked (see Figure 19b).

This was perhaps the reason they felt that the link they suggested was different from the one present in the RL-models proposed by us.[27] However, stacking inter-

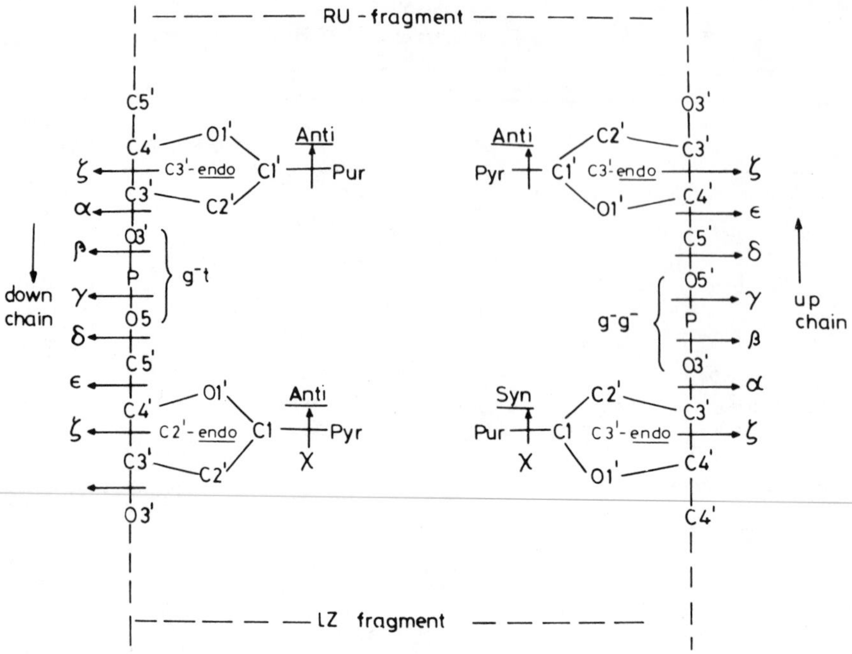

Figure 18. Conformational feature of the link between RU and LZ1 segments. Note that the conformation of the up-chain is different from that of the down-chain.

action being an important stabilizing factor in DNA, a completely destacked, arrangement of the bases would be a rather weak link between two DNA frag-

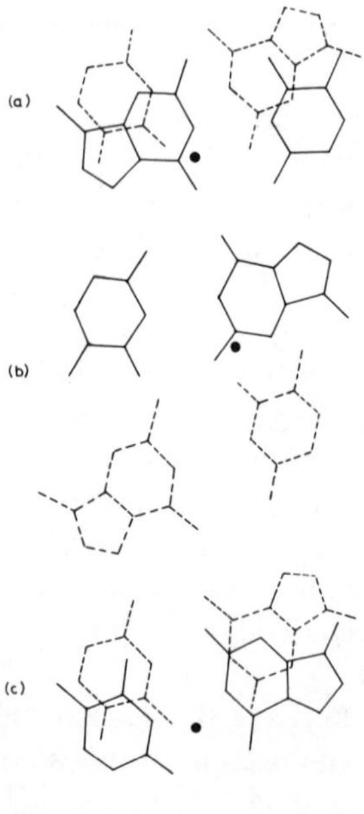

Figure 19. Inverted stacking arrangement at the link between RU and LZ fragments. (a) Inverted stacking arrangement for pur-pyr sequence; (b) from Wang et al;[20] (c)Inverted stacking arrangement for (pur-pur)•(pyr-pyr) sequence.

Table VIII
The Conformational Parameters of the Stable Link Between RU and LZ1 Fragments.*

Torsion Angles (degrees)	Up-chain	Down-chain
(5′-*end*) χ	240	80
(5′-*end*) ζ	85	85
α	182	217
β	315	327
γ	315	202
δ	165	164
ϵ	52	71
(3′-*end*) ζ	84	120
(3′-*end*) χ	50	20

*The torsion angles are as indicated in Figure 18.

ments. It would, therefore, be surprising if such a link occurs at all in the middle of a structure.

The Link Between RU and LZ2 Fragments

The conformational features of the link between a RU and LZ2 helix are shown in

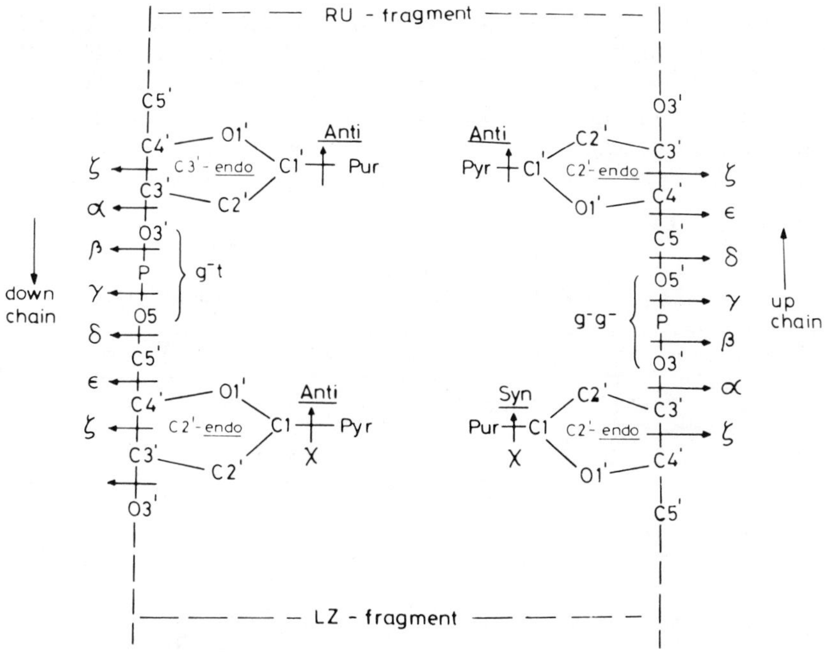

Figure 20. Conformational feature of the link between RU and LZ2 segments. As in Figure 18, the conformation of the up-chain is different from that of the down-chain.

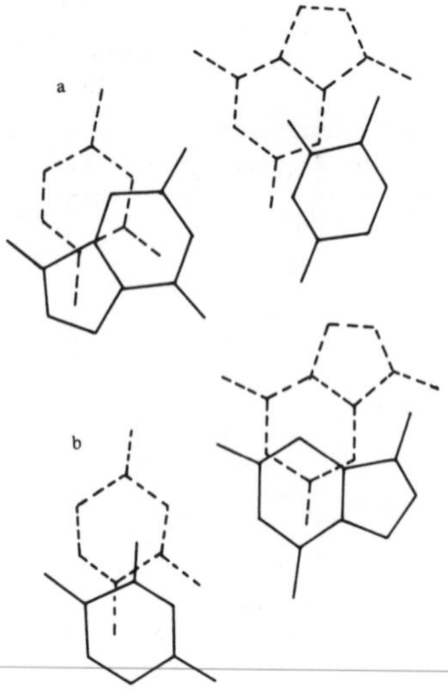

Figure 21. Inverted stacking arrangement at the link between RU and LZ fragments. (a) Inverted stacking arrangement for pur-pyr sequence; (b) Inverted stacking arrangment for (pur-pur)•(pyr-pyr) sequence.

Figure 20. The P-O torsions at the link are the same as described in Figure 17, i.e. the up chain P-O conformations are g^-g^- while they are g^-t^- in the down chain. However, the sugar residues across the upper base-pair have different puckers: the purine is attached to a C3′-*endo* sugar while the pyrimidine is connected to a C2′-*endo* sugar. Thus, even at the link the alternation of sugar pucker is maintained and this would suggest that the upper base-pair at the link may be beginning of an RU-fragment with alternate C3′-*endo* - C2′-*endo* pucker (refer Table IV). The torsion angles of the link are given in Table IX. The inverted stacking arrangement is described in Figure 21. The physical overlap of the bases at the link between RU and LZ2 fragments is similar to that found at the junction of RU and LZ1 fragments.

Table IX
The Conformational Parameters of the Links Between RU and LZ2-Helices.*

Torsion Angles (degrees)	Up-chain	Down-chain
(5′-*end*) χ	240	50
(5′-*end*) ζ	85	86
α	189	203
β	305	325
γ	320	154
δ	140	218
ϵ	68	43
(3′-*end*) ζ	156	153
(3′-*end*) χ	75	65

*The torsion angles are as indicated in Figure 20.

Conclusion

In summary, the present article presents a conceptual development in the stuctural studies on DNA. The conceptual development lies in the recognition of the conformational flexibility inherent in thc DNA molecule. This flexibility when exploited following a stereochemical guideline not only results in both right and left-handed DNA duplexes but also enables one to construct a stable link between two helical segments of opposite handedness.

References and Footnotes

1. Watson, J.D. and Crick, F.H.C., *Nature, 171,* 737 (1953).
2. Crick, F.H.C. and Watson, J.D., *Proc. Roy. Soc. A, 223,* 80 (1954).
3. Fuller, W., Wilkins, M.H.F., Wilson, H.R. and Hamilton, L.D., *J. Mol. Biol., 12,* 60 (1965).
4. Langridge, R., Marvin, D.A., Seeds, W.E., Wilson, H.R., Hooper, C.W., Wilkins, M.H.F. and Hamilton, L.D., *J. Mol. Biol., 2,* 19 (1960).
5. Marvin, D.A., Spencer, M., Wilkins, M.H.F. and Hamilton, L.D., *J. Mol. Biol., 3,* 547 (1961).
6. Arnott, S., *Progress In Biophysics and Molecular Biology, 21,* 265 (1970).
7. Arnott, S., Chandrasekaran and Selsing, E., *Structure and Conformation of Nucleic Acids and Protein-Nucleic Acid Interactions* (Sundarlingman, M. and Rao, S.T., ed.), Univ. Park Press, Baltimore (1975).
8. Gupta, G., Ph.D. thesis, Indian Institute of Science, India (1981).
9. Bansal, M. and Gupta, *Int. Jr. Quant. Chem.* July issue (1981).
10. Sasisekaran, V. and Cupta, G., *Curr. Sci. 49,* 43 (1980).
11. Gupta, G., Bansal, M. and Sasisekharan, V., *Proc. Natl. Acad. Sci. USA, 77,* 6486 (1980).
12. Gupta, G., Bansal, M. and Sasisekharan, V., *Int. Jr. Biol. Macromol., 2,* 368 (1980).
13. Ramaswamy, N., Gupta, G. and Sasisekharan, V., unpublished data.
14. Arnott, S. and Hukins, D.W.L., *J. Mol. Biol., 81,* 93, (1973).
15. Arnott, S., *Diffraction Methods for Structural Determination of Fibrous Polymers* (French, A.D. ed.) ACS Syposium Series, American Chemical Society, Washington, D.C. (1980), in press.
16. Arnott, S., Chandrasekaran, R., Hukins, D.W.L., Smith, P.J.C. and Watts, L., *J. Mol. Biol., 88,* 523 (1974).
17. Ramaswamy, N., Basal, M., Gupta, G. and Sasisekharan, V., manuscript in preparation (1981).
18. Sasisekharan, V., Gupta, G. and Bansal, M., *Int. J. Biol. Macromol., 3,* 3 (1981).
19. Davies, D.R. and Baldwin, R.L., *J. Mol. Biol., 6* 251 (1963).
20. Wang, A.H.J., Quingley, G.J., Kolpak, F.J., Crawford, J.L., van Boom, J.H., van der Marcel, G. and Rich, A. *Nature, 282,* 680 (1979).
21. Drew, H.R., Tanako, T., Tanaka, S., Itakura, K. and Dickerson, R.E., *Nature, 286,* 567 (1980).
22. Wang, A.H.J, Quigley, G.J., Kolpak, F.J., van der Marcel, G., van Boom, J.H. and Rich, A., *Science, 211,* 171 (1981).
23. Gupta, G., Bansal, M. and Sasisekharan, V., *Biochem. Biophys. Res. Commun., 97,* 1258 (1980).
24. Arnott, S., Chandrasekaran, R., Birdsall, D.L., Leslie, A.G.W. and Ratliff, R.L., *Nature, 283,* 743 (1980).
25. Sasisekharan, V., and Brahmachari, S.K., *Curr. Sci., 50,* 10 (1981).
26. Gupta, G., Bansal, M. and Sasisekharan, V., *Biochem. Biophys. Res. Commun., 95,* 728 (1980).
27. Sasisekharan, V., Pattabiraman, N. and Gupta, G., *Proc. Natl. Acad. Sci., USA, 75,* 4092 (1978).

Proceedings of the Second SUNYA Conversation in the Discipline Biomolecular Stereodynamics
Volume I, ISBN 0-940030-00-4, Ed., Ramaswamy H. Sarma,
Adenine Press, New York, ©Adenine Press

Successful Force-Fitting of Nucleic Acid Base Dimerization Energies and Distances With an Empirical-Potential Function That Uses an Adjustable Dielectric and Explicitly Accounts For Lone-Pair Electrons with an Adjustable Set of Charges

Rick L. Ornstein and Jacques R. Fresco
Department of Biochemical Sciences
Princeton University
Princeton, New Jersey 08544

Introduction

Hagler, Lifson and coworkers[1-2] have over the last decade force-fit a simple empirical-potential function (composed only of a pairwise atom-atom summation of Coulombic and Lennard-Jones-like terms) to produce a large variety of crystal energies and structures within reasonable agreement with experimental values. These authors suggest, however, that significant improvements to the fit between calculated and experimental values should result if the function accounted for (1) the anisotropic nature of the electron distribution, (2) polarization, and (3) charge transfer. In the present paper, we describe how we have systematically improved the form of the Hagler, Lifson et al. function along the lines of (1) and (2) above for the calculation of nucleic acid base H-bond interactions. The resultant force-fit function makes it possible to simulate a large body of experimental results on base-base interactions observed in chloroform (for over 20 dimers) and in *in vacuo* (for 4 dimers). These results suggest that electronic factors as well as geometric factors are important in the 'recognition' of nucleic acid bases via H-bonded interactions.

Theoretical and Practical Rationale for Proposed Functional Improvements

Anisotropic nature of electron distribution. In theory the electron distribution of a molecule can be represented exactly in terms of a multipole expansion,[3] while the electrostatic interaction energy between two molecules can be represented by a series of multipole-multipole interactions. When considering molecules of biological interest, which are often relatively large and contain an enormous number of degrees of freedom, one is forced to truncate the expansion at a reasonable point. A satisfactory truncation point in one instance need not apply to another. Consider the following two examples: (1) It is found unnecessary to have an electrostatic term in the total energy function when considering the interactions of simple hydrocarbons. (2) On the other hand, a simple monopole-monopole approximation (MMA) to the electrostatic interaction energy for nucleic acid bases works satisfactorily when the molecules are separated by distances significantly larger than their molecular dimensions, but as the molecules approach each other at separations near the sum of van der Waals radii the MMA is totally unable to lead to meaningful results.[4-6] Thus, when considering the interaction of nucleic acid constituents (or any molecules with high density of electron rich atoms such as oxygen and nitrogen) it is inappropriate to employ the MMA at short separation distances, which is generally the case for systems of biological interest. On the other hand, routine employment of a theoretically more rigorous segmental multipole-multipole approximation up to octopole-octopole terms (at each atomic center)[4-6] for nucleic acid constituent interactions is very cumbersome and computationally expensive.

A multipole expansion of a molecular electron distribution results in atomic anisotropies, while a simple MMA does not. Atomic anisotropies in the electron distribution of nucleic acid constituents which possess a high concentration of atoms with lone pair electrons (LPE) (all oxygen atoms and certain nitrogen atoms) should be reasonably well approximated by superpositioning of atomic dipoles resulting from the lone pair electrons onto the net atomic charge monopoles. The method of handling LPE dipoles in an easy and computationally simple manner is probably best accomplished by treating each LPE as though it were two equal and opposite monopole charges separated by an appropriate distance and positioned such that the resulting dipole vector points along a hybridization orbital corresponding to the so-called 'rabbit ear' projection. Details are discussed in Methods.

Polarization and dielectric attenuation of electrostatic interaction. A variety of ways have been suggested in the literature to account for polarization effects, while some workers prefer to omit this component altogether because of possible errors due to nonadditivity effects.

It is chemically plausible and computationally efficient to attempt to incorporate the effects due to atomic polarization directly in the electrostatic interaction term. Consider two charges separated by a distance where many solvent molecules intervene. The electrostatic interaction will then be attenuated (due to the presence of induced and permanent dipoles that act to oppose the applied field) by a factor approximately equal to the 'bulk' dielectric value of the solvent. What happens to the attenuation factor as the charges move relatively close to each other is at best qualitatively understood.[7-8] Essentially the attenuation drops sharply at short separation distances and is almost independent of the solvent. Thus, to a first approximation, the dielectric value (or attenuation factor) operative in the electrostatic interaction term should be a function of charge-charge separation distance. The distance dependence will be a function not only of the polarizability of the intervening charges, but also of the polarizability of nonintervening charges that are immediately adjacent. To be sure, it is unrealistic to try to precisely account for these effects in an empirical potential function framework. However, we have attempted to see if a chemically-logical distance-dependent effective dielectric (D_e) model could be consistently force-fit with experimental results. The model employed is described below.

Methods

A large body of experimental data exists for nucleic acid base dimerization in various solvents. Our intent is to employ much of this data to force-fit and test the ability of a refined empirical-potential function (EPF) of the Hagler, Lifson et al-type, but which also approximates the effects due to the anisotropic nature of electron distributions resulting from lone-pair electrons and accounts for the attenuation effect of local polarizability on the electrostatic interaction term.

We employed the same Lennard-Jones type 9-6 (repulsive-attractive) potential according to the procedure adopted by Lifson et al.[9] The form of these components is

$$V_{rep_{ab}} = 2\,E_{ab}\,(r^*_{ab}/r_{ab})^9 \tag{1}$$

$$V_{att_{ab}} = -3\,E_{ab}\,(r^*_{ab}/r_{ab})^6 \tag{2}$$

where $V_{rep_{ab}}$ and $V_{att_{ab}}$ are the repulsive and attractive components for the atom pair 'a' and 'b', E_{ab} is the depth of the potential at the minimum r^*_{ab} for atom types 'a' and 'b', while r_{ab} is the actual interatomic separation. Values for E_{ab} and r^*_{ab} are taken from the force-fitting studies of Lifson et al.[9]

Lone-pair electron charges (LPC) were represented by two monopole charges of opposite sign separated by a distance of 0.35Å and 0.39Å for oxygen and nitrogen atoms, respectively, as suggested by Tvaroska and Bleha.[10] The negative charge was positioned along either a tetrahedrally or trigonally oriented hybridized atomic

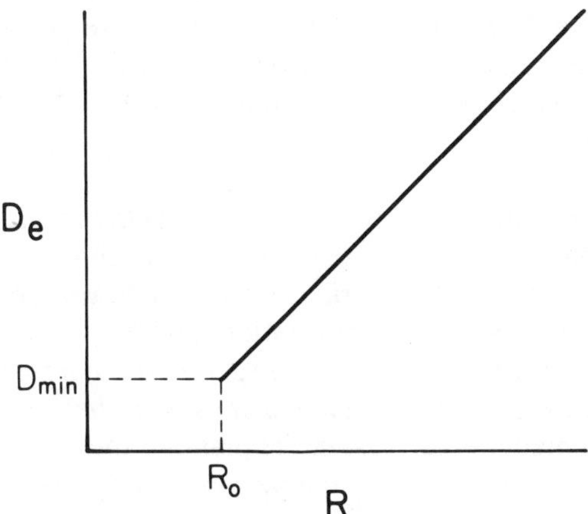

Figure 1. Effective dielectric value (D_e) versus charge-charge separation (R). D_e = SL (R-R_o) + D_{min}, where SL = slope, D_{min} is the minimum D_e value which occurs at a separation of R_o, and R_o = 1.5Å A (see text).

orbital, while the positive charge of each LPE was superimposed on the net atomic charge at the atomic center (its nucleus). The most suitable magnitude for the LPC is to be determined by the force-fitting procedure. Values ranged from 0.0 to 2.0 atomic charge units.

The distance dependent D_e model employed is shown in Figure 1 and is represented by the equation

$$D_e = SL(R-R_o) + D_{min} \tag{3}$$

where SL is the slope of the linear distance-dependence of D_e, R is the actual charge-charge separation, D_{min} is the minimum value of D_e at the separation R_o. R_o was taken to be 1.5Å whereas this corresponds approximately to the closest approach of two interacting charges in an H-bond as shown by the separation distance labelled 'Y'

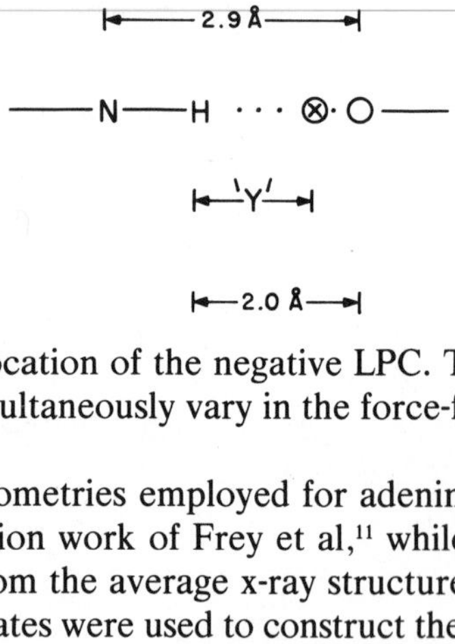

|←— 2.9 Å —→|

—N—H · · · ⊗· O—

|←'Y'→|

|←2.0 Å→|

where ⊗ represents the location of the negative LPC. The variables LPC, SL, and D_{min} are permitted to simultaneously vary in the force-fitting protocol.

The nucleic acid base geometries employed for adenine and thymine were taken from the neutron diffraction work of Frey et al,[11] while the cytosine and guanine geometries were taken from the average x-ray structure given by Voet and Rich.[12] Standard internal coordinates were used to construct the appropriate base analogue geometries. Net atomic charges were computed by the CNDO/2 method (and will be reported elsewhere).

All reported energies, unless otherwise stated, were the result of geometric energy minimization. H-bonding dimers were energy minimized according to the systematic variation of three geometric variables which allowed complete freedom in a plane. The increment sizes were 2° for rotations and 0.25Å for translations. Stacking interactions were energy minimized by four geometric parameters that allowed for complete freedom, except that the base planes were held parallel (no 'propeller-twisting'). The increments were 0.5Å in the plane of the base and 30° for rotations while the translational increment for the perpendicular separation was 0.1Å.

Results and Discussion

Infrared studies indicate that nucleic acid bases dimerize in chloroform via H-bonded interactions without partaking in stacking interactions (reviewed in 13-14), presumably due to effective 'nonspecific' base solvation. In aqueous solution the reverse is true (reviewed in 14), perhaps because of both poor 'nonspecific' base solvation and unfavorable entropic effects, as well as strong H-bond competition by water. It is not known which interaction mode predominates *in vacuo.*

In chloroform the relative enthalpies of dimerization of the base pairs G • C, C • C, A • U, and U • U are 6.8±2, 2.3±1.4, 2.2±1.4, and 0 kcal/mol, respectively.[13] Using these energies along with acceptable intermolecular H-bond distances between the heteroatoms (acceptor-donor distances), it is possible to begin a systematic force-fitting of the three parameters (LPC, SL, and D_{min}) by aiming for a reasonable level of convergence between experimental and calculated (energy and distance) quantities.

Nucleic acid base H-bonded dimers can potentially occur in a variety of different structural forms with two or three H-bonds of the type N-H...N or N-H...O (Figure 2). In principle, all of these should be considered. However, it is necessary for practical reasons to reduce to a minimum the number of different structural forms to be monitored during the force-fitting procedure. Preliminary calculations with a wide range of values for the three variable parameters of the EPF indicate that the most stable H-bonded form for each of the four different base pair dimers remains essentially invarient. This permits us to consider the relative energies of only the single most stable structural form for each different base pair dimer throughout the force-fitting procedure.

In the first stage of force-fitting, we tested 216 different combinations of the three variables (LPC, SL and D_{min}). These values are listed in Table I—top. With each

Table I

Variable Parameter Values Employed in the 2-Stages[1] of Force-Fitting

Stage	Parameters	Values
I	LPC	0.0, 0.25, 0.5, 1.0, 1.5, 2.0
	SL	0.0, 0.5, 1.0, 2.0, 4.0, 8.0
	D_{min}	1.0, 1.5, 2.0, 4.0, 8.0, 16.0
II	LPC	0.5, 0.75, 1.0, 1.25, 1.5, 1.75, 2.0
	SL	0.0, 0.25, 0.5, 0.75, 1.0, 1.25, 1.5
	D_{min}	1.0, 1.25, 1.5, 1.75, 2.0, 2.25, 2.5

[1]All combinations were permitted. Thus 216 and 343 combinations of the three variable parameters were tested in Stage I and II, respectively.

different combination of variables the H-bonded structure was permitted to seek a new energy minimum configuration. The resulting energies and intermolecular H-bond distances were compared to the experimental values. Of the 216 combinations, three gave significantly better results than the others, as monitored by the average relative deviation between experimental and calculated enthalpies ($\Delta\bar{H}$)

G C

a

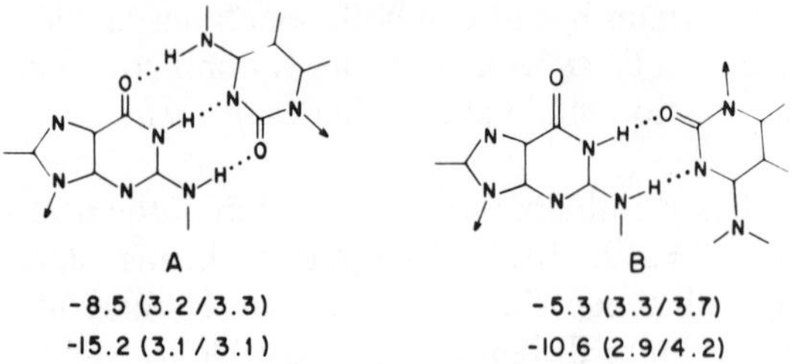

A
-8.5 (3.2 / 3.3)
-15.2 (3.1 / 3.1)

B
-5.3 (3.3/3.7)
-10.6 (2.9/4.2)

C · C

b

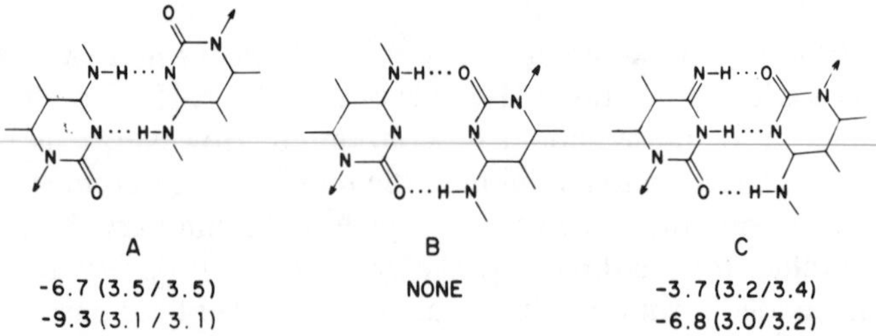

A
-6.7 (3.5 / 3.5)
-9.3 (3.1 / 3.1)

B
NONE

C
-3.7 (3.2/3.4)
-6.8 (3.0/3.2)

A · U

c

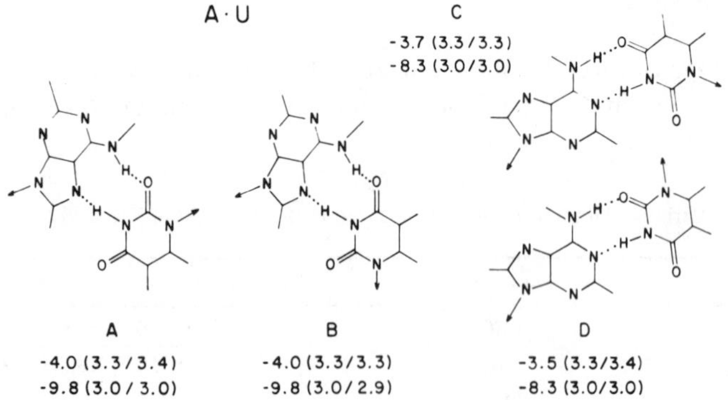

A
-4.0 (3.3 / 3.4)
-9.8 (3.0 / 3.0)

B
-4.0 (3.3/3.3)
-9.8 (3.0/ 2.9)

C
-3.7 (3.3 / 3.3)
-8.3 (3.0/3.0)

D
-3.5 (3.3/3.4)
-8.3 (3.0/3.0)

U · U

d

A
-3.5 (3.2 / 3.2)
-8.0 (3.0 / 3.0)

B
-3.4 (3.2/3.3)
-7.9 (3.0/2.9)

C
-3.4 (3.2/3.1)
-7.8 (2.9/2.9)

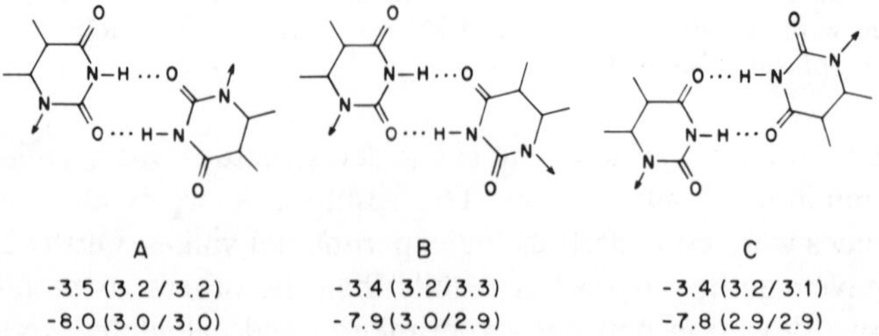

and H-bonded distances ($\Delta\bar{D}$). As shown in Table II, $\Delta\bar{H}$ and $\Delta\bar{D}$ are significantly lower for the three combinations with LPC equal to 1.0 or 1.5 than the case equivalent to the Hagler, Lifson et al. EPF (LPC=0.0, SL=0.0, and D_{min}=1.0). The latter combination corresponds to a constant dielectric value of 1.0, and no account taken of LPC. Also shown in Table II are the corresponding values of $\Delta\bar{H}$ and $\Delta\bar{D}$ when a constant dielectric value of 2.0 or 4.0 is employed without taking account of LPC (The dielectric value of chloroform is about 4.8 at 20°C).

Table II

Comparison of Best-Fit Combinations[1] of Variable Parameters with Combinations of Parameters Equivalent or Similar to Those of the Hagler, Lifson et al. EPF for Nucleic Acid Base Dimerization in Chloroform.

Monitors[2] of the Level of Fit		Variable Parameters			Remark
$\Delta\bar{H}$	$\Delta\bar{D}$	LPC	SL	D_{min}	
.4	.1	1.0	.5	1.5	Best-fit combination
.5	.1	1.0	1.0	1.5	Best-fit combination
.4	.1	1.5	1.0	2.0	Best-fit combination
1.5	.4	0	0	1.0	Constant D_e = 1.0[3]
1.8	.5	0	0	2.0	Constant D_e = 2.0
2.4	.6	0	0	4.0	Constant D_e = 4.0

[1]Taken from Stage I where 216 combinations were tested on the G • C, C • C, A • U, and U • U dimers.
[2]$\Delta\bar{H}$ and $\Delta\bar{D}$ are the average relative deviations between experimental and calculated enthalpies and H-bond distances, respectively. The experimental H-bond distances are taken from the review of Voet and Rich.[12] See text for experimental enthalpies.
[3]Exact conditions of Hagler, Lifson et al. EPF.

We then proceeded to try combinations of the three variables at smaller intervals centered about the successful values shown in Table II. In particular, we explored another 343 combinations as shown in Table I-bottom. The most successful of these were employed to minimize $\Delta\bar{H}$ and $\Delta\bar{D}$ for the dimerization of bases under *in vacuo* conditions (within the chemical criteria discussed below).

Dimerization enthalpies obtained *in vacuo* are available for the same four base pair combinations. (For those measurements, bases methylated at the glycosyl bond position were used, while larger substituents at those positions were present in the chloroform studies. Our calculations were made using the methylated bases.) Rela-

Figure 2. H-bonding alternative structural forms for G • C, C • C, A • U, and U • U with sample enthalpies and H-bond hetero-atom intermolecular distances. Energy minimized enthalpies are shown in kcal/mol for each structural form as calculated by the Hagler, Lifson et al. EPF (upper row) and for the refined version (bottom row) where LPC = 2.0, SL = 0.0, and D_{min} = 2.0. The numbers in parenthesis are the H-bond lengths in the energy minimized configuration (for upper two bonds if 3 are present) such that the number to the right is for the upper H bond. The energies for C_i have been reduced by 6 kcal/mol to account for the tautomeric shift.

tive enthalpies determined by mass spectral analysis for the dimers G • C, C • C, A • U, and U • U are −11.5±2, −6.5±2, −5.0±2, and 0 kcal/mol, respectively.[15] Whereas it is not known if dimerization under *in vacuo* conditions occurs by H-bonding or stacking modes, we monitored both possibilities and assumed that the most preferred of the two alternatives is the one of interest for that particular situation.

The chloroform and *in vacuo* dimerization enthalpies and distances were simultaneously monitored using the best-fit values of the second stage of chloroform force-fit successful combinations of LPC, SL, and D_{min}. In order for a combination of the three variable parameters to be considered chemically plausible, the following chemical criteria were set: First the LPC value must be the same for both conditions (dimerization in chloroform and *in vacuo*) since it is hard to imagine that this parameter would be significantly altered by variations in the solvent. Second, D_{min} for the *in vacuo* case must be less than or equal to D_{min} in the chloroform case, since on average there can not be more reduction of the electrostatic interaction due to polarization effects *in vacuo* than in chloroform, although the reverse is certainly possible.

Table III presents the results of the best combination of the three variables for chloroform dimerization along with a consistent (based on chemical criteria set above) but different combination of variable parameters for *in vacuo* dimerization. The values for LPC, SL, and D_{min} that best fit all these criteria are (2.0, 0.0, and 2.0) and (2.0, 0.5, and 1.5) for chloroform and *in vacuo* dimerization, respectively. It is interesting to note that under *in vacuo* conditions G • C, A • U and U • U dimers prefer H-bonded complexes while C • C prefers a stacked complex.

Table III

Results of Stage-II Best-Fit Combination of Nucleic Acid Base Dimerization in Chloroform and *in vacuo* with Variable Parameters Set at (2.0, 0.0, 2.0) and (2.0, 0.5, 1.5), Respectively.

Medium	Quantities Monitored	G • C	C • C	A • U	U • U	
Chloroform	ΔH (kcal/mol)	−6.8±2	−2.3±1.4	−2.2±1.4	0	Exp.[1]
		−7.2	−1.3	−1.8	0	Cal.
	H-bond distances (Å)	2.82/2.91	3.04/3.04	2.88/2.91	2.82/2.82	Exp.[2]
		3.1/3.1	3.1/3.1	2.9/3.0	3.0/3.0	Cal.
in vacuo	ΔH (kcal/mol)	−11.5±2	−6.5±2	−5.0±2	0	Exp.[3]
		−11.5	−5.6	−3.2	0	Cal.
	H-bond distances (Å)	2.82/2.91	3.04/3.04	2.88/2.91	2.82/2.82	Exp.[2]
		2.8/2.7	2.7/2.9	2.7/2.8	2.7/2.7	Cal.
	inter-planar stacking separation (Å)	3.25	3.4	3.3	3.3	Exp.[4]
		3.2	3.0	3.0	3.2	Cal.

[1]Taken from reference 13. [2]Taken from reference 12. [3]Taken from reference 15.
[4]Taken from reference 16.

Using the best-fit chloroform parameters of Table III, we now proceeded to monitor how well these parameters obtained with the EPF methodology could reproduce the relative energies of H-bonded dimers measured for over 20 different base pairs in chloroform. Table IV lists the appropriate experimental and calculated

Table IV

Experimental and Calculated Quantities for 21 Dimers Observed in Chloroform

Exptl Rank-Order	Dimer[1]	ΔG^{2}_{exp}	ΔH^{3}_{cal}	ΔG^{4}_{cal}	Rank-Order (based on ΔG_{cal})	Rank-Order Deviation
1	A • A	−0.7	−5.5	−6.2	1	0
2	U • U	−1.1	−8.0	−8.7	6	4
3	2AP • 5, 6DU	−1.4	−7.2	−7.9	3	0
4	8BA • 8BA	−1.5	−6.8	−7.2	2	2
5	C • C	−2.0	−9.3	−9.3	10	5
6	A • 5, 6DU	−2.0	−8.8	−9.2	9	3
7	2AP • U	−2.3	−8.2	−8.6	5	2
8	8BA • U	−2.3	−8.6	−9.0	7	1
9	2 AP • T	−2.4	−8.1	−8.5	4	5
10	2AP • 5BU	−2.6	−8.8	−9.2	8	2
11	A • U	−2.7	−9.8	−10.2	13	2
12	2, 6DAP • 5,6DU	−2.7	−8.7	−9.6	11	1
13	A • T	−2.9	−9.7	−10.1	12	1
14	2, 6DAP • U	−3.0	−9.8	−10.6	14	0
15	2, 6DAP • T	−3.2	−9.8	−10.6	15	0
16	A • 5BU	−3.2	−10.5	−10.9	16	0
17	I • I	−3.6	−13.2	−13.2	19	2
18	2, 6DAP • 5BU	−3.7	−10.5	−11.4	17	1
19	I • C	−4.5	−12.8	−12.8	18	1
20	G • G	−4.8	−15.9	−15.9	21	1
21	G • C	−6.1	−15.2	−15.2	20	1

[1]The bases are substituted in the glycosyl position as described in text. A = adenine; 2AP = 2 aminopurine; 8BA = 8-Br adenine; 2, 6DAP = 2,6 diamino purine; U = uracil; 5, 6DU = 5,6 dihydro uracil; T = thymine; 5BU = 5-Br uracil; C = cytosine; I = inosine; G = guanine.

[2]ΔG_{exp} = -RT InK where K's are taken from reference 13. Values for dimers 20 and 21 are the average of the range reported.

[3]Values calculated with LPC = 2.0, SL = 0.0, and D_{min} = 2.0 for the most stable structural form.

[4]$\Delta G_{exp} = \Delta H_{cal}$ + EI, where EI is the entropic increment due to degeneracy (see text).

quantities for the various dimer pairs. Column three contains the enthalpy, ΔH_{cal}, of dimerization for the most stable structural form determined by calculation, while the calculated relative free energies (ΔG_{cal}) is in column four. ΔG_{cal} is obtained by adding to ΔH_{cal} an entropic increment to the free energy due to 'degeneracy' of structural forms that lie within 0.5 kcal/mol of the most stable form.

This increment is determined by the Boltzmann equation for the entropy (S) of the system

$$S = K_B \ln \Omega$$

where Ω is the 'degeneracy'. The entropic energy contribution at 25°C is 0.0, 0.41, 0.65, and 0.82 kcal/mol when Ω is 1, 2, 3 and 4, respectively.

The rank order of the calculated entropy-corrected energies in column five is in reasonable agreement with the order of the experimental association constants; thus, 17 out of 21 dimers have a calculated rank-order within 2 units of the experimental order. The worst case differs in rank by 5, while the average deviation is 1.6. This level of discrepancy is not unreasonable in view of the fact that many of the experimental values are very close and that different groups were attached to the glycosyl position in the actual experiments, i.e. , ethyl for adenine derivatives, cyclohexyl for uracil derivatives, and 2′, 3′-benzylidine-5′-tritylribose for cytosine, guanine, and inosine. Another way to observe the extent of agreement between the calculated and observed energies is shown graphically in Figure 3. It is evident from the correlations exhibited between the experimental and calculated values of the free energies (Figure 3a) and the calculated enthalpies (Figure 3b) that the level of agreement is quite satisfactory, indeed more so than the rank order comparison may suggest.

The base dimers so far considered are either self-associates or analogues of the normal Watson-Crick pairing combinations. It is note-worthy that when two-component mixtures were made the following dimers did not occur in chloroform in detectable amounts: A • C, A • G, G • U and C • U; only the self-associates were observed. On the other hand, A • I was observed to form at a significant but modest level. (It is interesting that in aqueous solution, pairs of homopolymers show comparable tendencies; that is, only poly A and poly I interact to form a helical

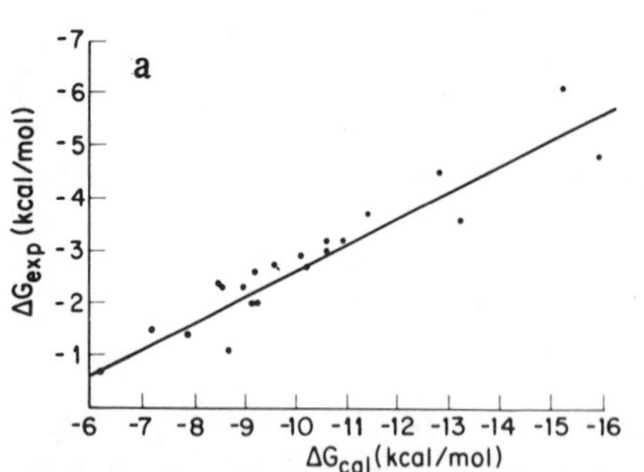

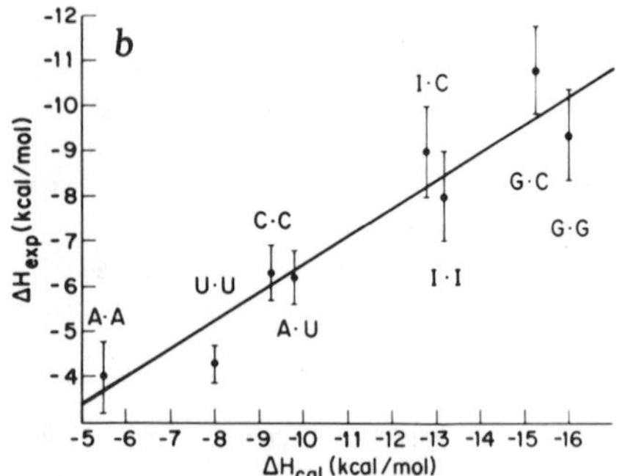

Figure 3. Depiction of linear correlation between experimental (in chloroform) and calculated (LPC = 2.0, SL = 0.0, D_{min} = 2.0) (a) ΔG values for 21 dimers (see Table IV for details), and (b) ΔH values for all cases where experimental quantities are available.

structure; the others remain as pairs of single strands H-bonded only to water (J. Fresco-unpublished)). These situations are computationally testable by the procedures described above and follow the relation:

$$X \bullet X + Y \bullet Y \Leftrightarrow 2\,X \bullet Y \tag{4}$$

Table V depicts the data for these five cases. In the four cases where the self-associates are preferred the calculated difference in enthalpy between the reactants and products of equation (4) is at least 2.04 kcal/mol in favor of the reactants, while for A • I the reactants are preferred by only 0.50 kcal/mole which corresponds to a 30% probability for A • I but a probability of ≤ 3% for the other four non-self-associates in Table V.

Table V
Comparison of Experimental (in Chloroform) and Calculated (LPC = 2.0, SL = 0.0, D_{min} = 2.0) Preferences for Self-Associates Versus Hetero-Associates for Non-Watson-Crick Dimer

Hetero-Associates	Experimentally Observed[1]	Calculated Quantities	
		ΔH_i (kcal/mol)	P_i^2 (percent)
A • C	–	2.06	3
A • G	–	2.04	3
G • U	–	3.60	<1
C • U	–	7.64	≪1
A • I	+	0.50	30

[1]Taken from reference 13.

$$2P_i = \frac{\exp(-\Delta H_i/kT)}{\sum_i \exp(-\Delta H_i/kT)}$$

Conclusion

In summary, we have successfully refined an EPF of the Hagler Lifson et al-type by force-fitting nucleic acid base dimerization (H-bonding or stacking) energies and distances in *in vacuo* and in chloroform (for over 20 cases) with a chemically plausible set of adjustable parameters. The adjustable parameters were (1) the explicit incorporation of lone-pair electrons, and (2) a distance-dependent dielectric. The former parameter is introduced to account for the atomic anisotropic nature of the electron distributions, while the latter parameter provides a logical and computationally simple means of estimating the attenuation effect of local polarizabilities on the electrostatic interaction. The adjustable parameters found to lead to the best-fit were (LPC = 2.0, SL = 0.0, D_{min} = 2.0) for chloroform and (LPC = 2.0, SL = 0.5, D_{min} = 1.5) for *in vacuo* nucleic acid base dimerization.

In view of the convincing fit achieved between observed and calculated energies and distances for the 26 different base dimers in chloroform, it seems appropriate to exploit the current results to better understand the determinants of nucleic acid base H-bonded interactions. The present data indicate that in addition to geometric specificity, an important determinant of H-bond stability is the anisotropic nature of the electron distribution within the interacting bases, i.e. their electronic specificities. Hence, the number of H-bonds (2 *vs* 3) of the N-H....N or N-H....O type cannot be the sole determinant of the strength of the composite H-bonded interaction. A suggestion that this might be the case was made earlier, based merely upon the base dimerization affinities measured in chloroform.[cf. 13] A more in-depth analysis of these first principles will be presented elsewhere. Meanwhile, the current methods and results are being extended to the case of nucleic acid helices in order to properly account for H-bond and stacking interactions in their interior. Such an analysis should provide useful insights into the role of H-bonded base mispairing in the template-dependent macromolecular synthesis processes of replication, transcription and translation.

Acknowledgement

This work was supported by a National Research Service Award to R.L.O. from NIH Grant GM-07625, and by NSF Grant PCM-8023706. Generous access to the facilities of the Princeton University Computer Center is also gratefully acknowledged.

References and Footnotes

1. Hagler, A.T., Huler, E., and Lifson, S. *J. Am. Chem. Soc. 96,* 5319-5327 (1974).
2. Dauber, P. and Hagler, A.T. *Acc. Chem. Res. 13,* 105-112 (1980).
3. London, F. *J. Phys. Chem. 46,* 305-316 (1942).
4. Rein, R. Adv. *Quantum Chem. 7,* 335-396 (1973).
5. Stamatiadou, M.N., Swissler, T.J., Rabinowitz, J.R., and Rein, R. *Biopolymers 11,* 1217-1234 (1972).
6. Egan, J.T., Swissler, T.J., and Rein, R. *Int. J. Quantum Chem. QBS 1,* 71-79 (1974).
7. Ramachandran, G.N., and Srinivasan, R. *Indian J. Biochem. 7,* 95-97 (1970).
8. Kirkwood, J.G. and Westheimer, F.H. *J. Chem. Phys. 6,* 506-512 (1938).
9. Lifson S., Hagler, A.T., and Dauber, P. *J. Am. Chem. Soc. 101,* 5111-5121 (1979).
10. Tvaroska, I. and Bleha, T. Biopolymers 18, 2537-2547 (1979).
11. Frey, M.N., Koetzle, T.F., Lehmann, M.S., and Hamilton, W.C. *J. Chem. Phy. 59,* 915-924 (1973).
12. Voet, D., and Rich, A. *Prog. Nucl. Acid. Res. Mol. Biol. 10,* 183-265 (1970).
13. Hartman K.A., Lord, R.C., and Thomas, Jr., G.J. in *Physico-Chemical Properties of Nucleic Acids,* Ed by J. Duchesne, Academic Press, N.Y. pp 1-89 (1973).
14. Ts'o, P.O.P. in *Basic Principles in Nucleic Acid Chemistry,* Ed by P.O.P. Ts'o, Academic Press, N.Y. Vol. 1, pp 453-584 (1974).
15. Yanson I.K., Teplitsky, A.B., and Sukhodub, L.F. *Biopolymers 18,* 1149-1170 (1979).
16. Bugg, C.E., Thomas, J.M., Sundaralingam M., and Rao, S.T. *Biopolymers 10,* 175-219 (1971).

Proceedings of the Second SUNYA Conversation in the Discipline Biomolecular Stereodynamics
Volume I, ISBN 0-940030-00-4, Ed., Ramaswamy H. Sarma,
Adenine Press, New York, ©Adenine Press

Electrostatic Molecular Potentials and Accessibilities in Z-DNA Versus B-DNA

Krystyna Zakrzewska, Richard Lavery and Bernard Pullman
Institut de Biologie Physico-Chimique,
Laboratoire de Biochimie Théorique
associé au C.N.R.S.,
13, rue P. et M. Curie - 75005 Paris, France

Introduction

A number of recent crystallographic investigations[1-5] have brought to light the existence of a family of structurally related DNA's which are distinctive in having left-handed helical conformations. This family of nucleic acids termed Z-DNA's, have currently only been proved to exist in oligonucleotide crystals. However, Sarma *et al.*[6] have produced evidence for a left-handed conformation of poly(dG-dC)•poly(dG-dC) in solution and Wang *et al.*[1,2] have suggested that it is geometrically feasible that sections of Z-DNA may be interspersed within longer B-DNA double helices. These latter authors have proposed idealized geometries for two related Z-DNA double helices, termed Z_I and Z_{II}, which we shall adopt in our present theoretical study. For details see Rich *et al.* in this volume.

In this publication, we wish to compare two theoretically computed fundamental properties of Z-DNA's, namely, the molecular electrostatic potential and the atomic steric accessibilities with the same properties in the more familiar B-DNA. These properties have been employed in a series of publications from this laboratory for the exploration of the chemical and biochemical reactivity of nucleic acids: B-DNA,[7-11] $tRNA^{Phe}$[12-16] and in preliminary way for one Z-DNA conformer.[17]

In view of the encouraging results obtained, in particular for the interpretation of some important features of the interaction of nucleic acids with carcinogens, it seemed useful to carry out a deeper comparison of B- and Z-DNA's, the more so as the suggestion was made that Z-DNA fragments interspersed within a B-DNA helix could present sites of enhanced reactivity toward such agents.[1]

Method

The technique employed for calculating the molecular electrostatic potential of nucleic acids has been described in detail in our previous publications.[7-8-10-11] In essence, it consists of subdividing the macromolecule into a number of smaller units, namely, phosphates, sugars and bases, for each of which an *ab initio* SCF wave function is calculated. From these wave functions, multicenter multipole expansions are developed. These expansions have been shown to reproduce very satisfactorily the electrostatic potentials of the molecules concerned down to a distance of 2Å from their constituent atoms.[18] Below this distance, exact electrostatic potentials must be calculated. The electrostatic potential of a model nucleic acid is subsequently constructed by the superposition of the potentials of the subunits, appropriately oriented to form the nucleic acid helix.

We shall use here three representations for describing these potentials. The first one relates to the *surface envelope* surrounding the macromolecule.[10] This envelope is formed by the intersection of spheres centered on each atom of the molecule, with radii equal to the van der Waals radius of the atom concerned, multiplied, if so desired, by a factor F. This factor enables us to investigate envelopes situated closer or further from the macromolecule and thus to monitor changes in the distribution of the potential as a function of the distance. In the present study, we shall, however, limit ourselves to a single envelope constructed with F = 1.7 which is the closest envelope to the macromolecule for which the potentials may be calculated from multipolar expansions.[18] These surface potentials have been calculated for one full turn of each double helix.

In order to view the surface envelope two "windows" were employed. These windows are simply planes containing uniform grids of points, placed at some distance from the face of the molecule that one wishes to view. Subsequently, perpendiculars are dropped from each grid point until they touch the surface envelope. At each of these points, the electrostatic potential of the molecule is calculated. The resulting potentials are then represented by shaded zones projected onto the plane of the window. In the present study, two windows were used for each helix. These windows being parallel to the helical axis, one facing the minor groove (or the sole groove in Z-DNA) and the second, diametrically opposed, facing the major groove (or the convex face in Z-DNA). These windows were centered vertically on the midpoint of the full turn of the helices studied and view the middle half of each helix. By helical symmetry, the resulting views can be joined together vertically yielding an image of the full turn. This approach minimizes the introduction of end-effects due to the finite length of the helices.

The second representation of the potentials consists of drawing isopotential maps in planes which bisect the model double helices perpendicularly to the helical axis.[7] These maps allow a precise view of the distribution of the potential at varying distances from the helical axis and complement the surface envelope representation which indicates potentials at fixed distances from the nucleic acid atoms. For both these representations, the model employed consisted of a full helical turn of the appropriate acid.

The third representation of the potentials indicates their values at sites associated with reactive atoms on the bases or the phosphates.[8] In order to be able to compare these values directly for different DNA's a standard model helix of 11 phosphates in each strand was used in this case. This corresponds to the length of the helix

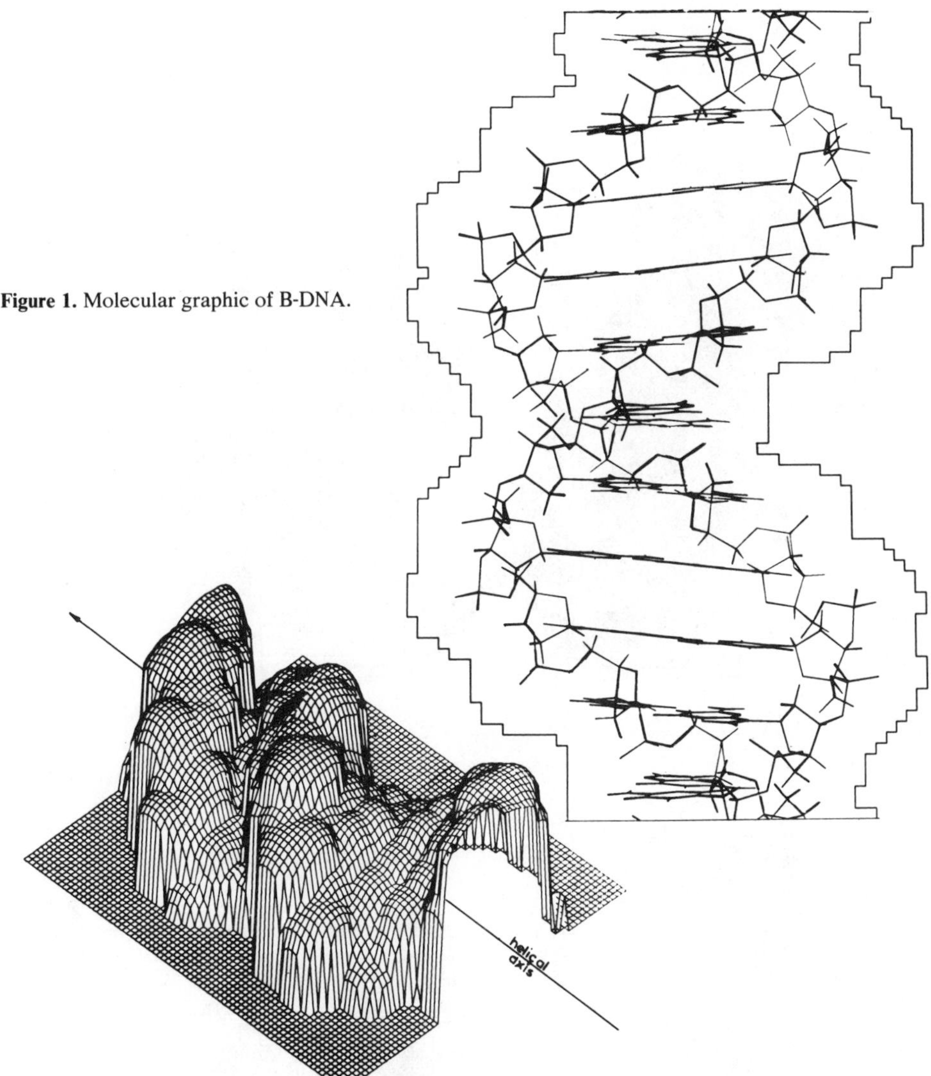

Figure 1. Molecular graphic of B-DNA.

Figure 2. Surface envelope graphic of B-DNA.

studied in our previous publications on B-DNA,[8] but involves truncating the Z-DNA model to slightly less than one full turn. Although it is clear that longer model helices yield more negative site potentials, due essentially to the increased number of negatively charged phosphates, preliminary studies have shown that this does not involve significant changes in the *ordering* of the site potentials.

In each of these three representations we have, in addition, studied one case of counterion screening of the double helices. This example consists of binding one

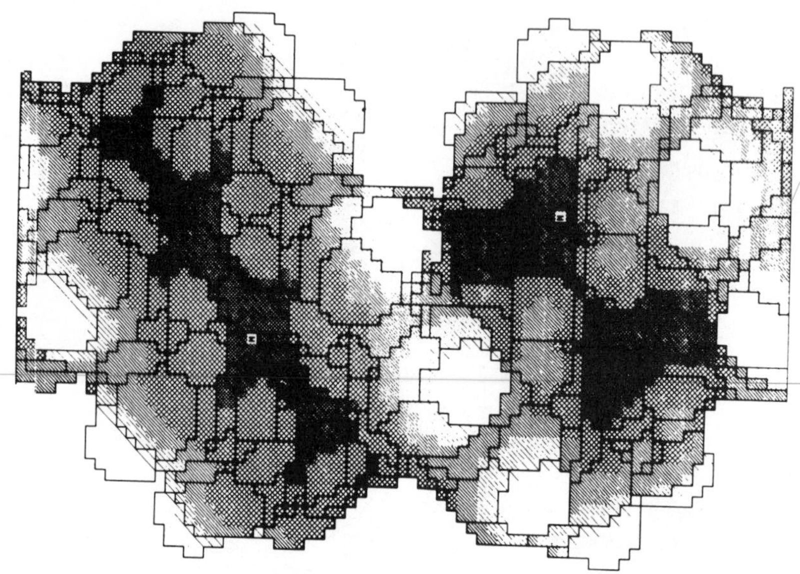

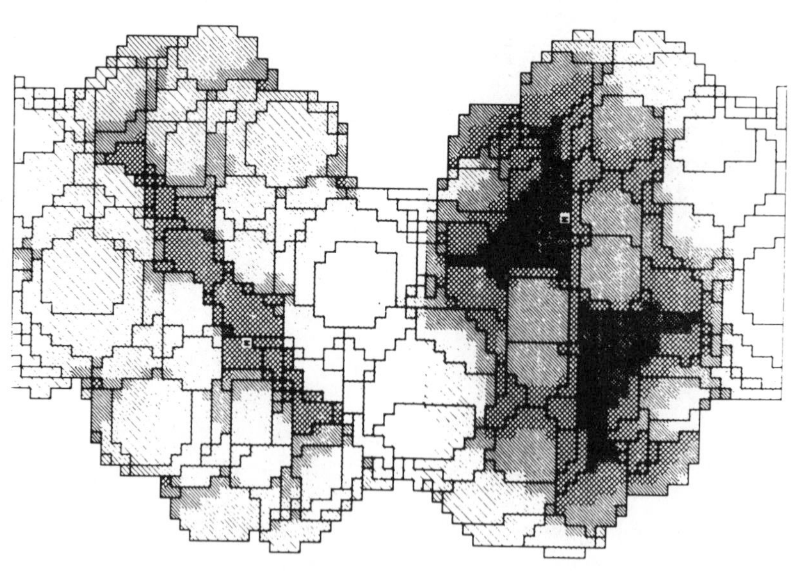

Figure 3. Surface potential of B-DNA unscreened (left) and screened (right) by Na^+ cations (kcal/mole). Also see text and the display diagram.

sodium ion symmetrically in a bridge position between the two anionic oxygens of each phosphate of the acids and represents a case of maximal screening.[7]

The second property studied for the model helices of the nucleic acids is the static steric accessibility associated with the reactive atoms of the bases or the phosphate groups. The method of calculating these accessibilities, which are expressed as accessible areas on the van der Waals spheres surrounding the atoms concerned, is described in detail in our previous publication.[11] The accessible areas calculated in the present study correspond to a simple spherical attacking species with a radius of 1.2Å.

The double helices studied are the following:

(1) B-DNA with the geometry due to Arnott and Hukins,[19] composed of the alternating (CG)•(CG) base sequence, for comparability with Z-DNA.

(2) Z_I-DNA, in the idealized geometry proposed by Wang *et al.*[1] for a continuous left-handed double helix with base sequence (CG)•(CG), based on observed oligonucleotide crystal structures.

(3) Z_{II}-DNA, a second idealized geometry for the left-handed double helix also due to Wang et al.[2] The principal difference between Z_I and Z_{II} resides in the GpC phosphate whose conformation changes from g^-t in the form to g^+t in the latter.

SHADING	B-DNA	B-DNA SCREENED	Z_I- or Z_{II}-DNA	Z_I- or Z_{II}-DNA SCREENED
□ (blank)	-488 → -508	79 → 51	-469 → -513	106 → 76
◿ (single diagonal)	-508 → -529	51 → 23	-513 → -557	76 → 45
(double diagonal)	-529 → -550	23 → -4	-557 → -601	45 → 15
(triple diagonal)	-550 → -571	-4 → -32	-601 → -646	15 → -15
(cross-hatched)	-571 → -592	-32 → -60	-646 → -690	-15 → -45
(dense cross-hatched)	-592 → -614	-60 → -89	-690 → -735	-45 → -77

Details of Shading of the Surface Envelope Potentials (all values in kcals/mole)

Potentials at the Surface Envelopes

The results for B-DNA are shown in Figures 1-3. These surface potentials have been calculated for one full turn of the double helix, viewed from the direction indicated by the molecular diagram in Figure 1. In this diagram, the minor groove is located in the upper half of the figure and the major groove in the lower half. A computer generated drawing of the surface envelope is reproduced in Figure 2 in which the two grooves are particularly well recognizable. The surface potentials for the unscreened helix are presented in Figure 3, left, where various grades of shading are used to represent the results, darker shadings implying more negative potentials. (For details of shading in this figure and in subsequent figures, see the display diagram. It should be noticed that the same shadings correspond to different values of potentials in screened and unscreened B- or Z-DNA's).

It is usually expected that the most negative potentials will be associated with the negatively charged phosphates in the backbones of the double helix. It may, however, be seen, in Figure 3, left, (see also ref. 11) that, in fact, the deepest potentials are located in the grooves of the acid. Local surface potential minima in Figure 3 are indicated by the letters M. Their values are −594 kcal/mole and −613 kcal/mole in the minor and major grooves, respectively. The latter value is a global minimum for the surface potential. It is associated with the atom N3 of a guanine atomic sphere (Notation N3(G); similar notations will be used throughout this paper for related situations concerning the other reactive centers).

The results for a Na^+ screened B-DNA are also given in Figure 3. Although the magnitude of the surface potential has now diminished considerably, the local minima in the minor and major grooves being −84 kcal/mole and −88 kcal/mole, respectively, the overall distribution of the potential has not changed much. Thus, more negative potentials are still seen to be in the grooves of the double helix and the global surface minimum is still located on a N3(G) sphere in the major groove.

The results for Z_I-DNA are presented in Figures 4-6. As for B-DNA, one full turn of the helix has been studied, the orientation of the helix being as shown in the molecular diagram in Figure 4. It is to be noted that this conformer has only a single groove, visible in the upper half of Figures 4-6, related to the minor groove of B-DNA. The major groove is replaced (lower half of Figures 4-6) by a convex face formed of protruding base pairs. A computer generated perspective drawing of this helix is shown in Figure 5 in which the left-handedness of the helix, the presence of a deep although narrow minor groove and the absence of the major groove are all clearly discernible.

The surface potentials of the uncreened helix are given in Figure 6 (left) and the most negative zone of potential is clearly visible in the groove. On the convex face, the potentials are weaker by roughly 50 kcal/mole and no concentration of negative potential can be seen there. The global surface minimum of −735 kcal/mole lies on the edge of the groove of the helix on the sphere of a phosphate anionic

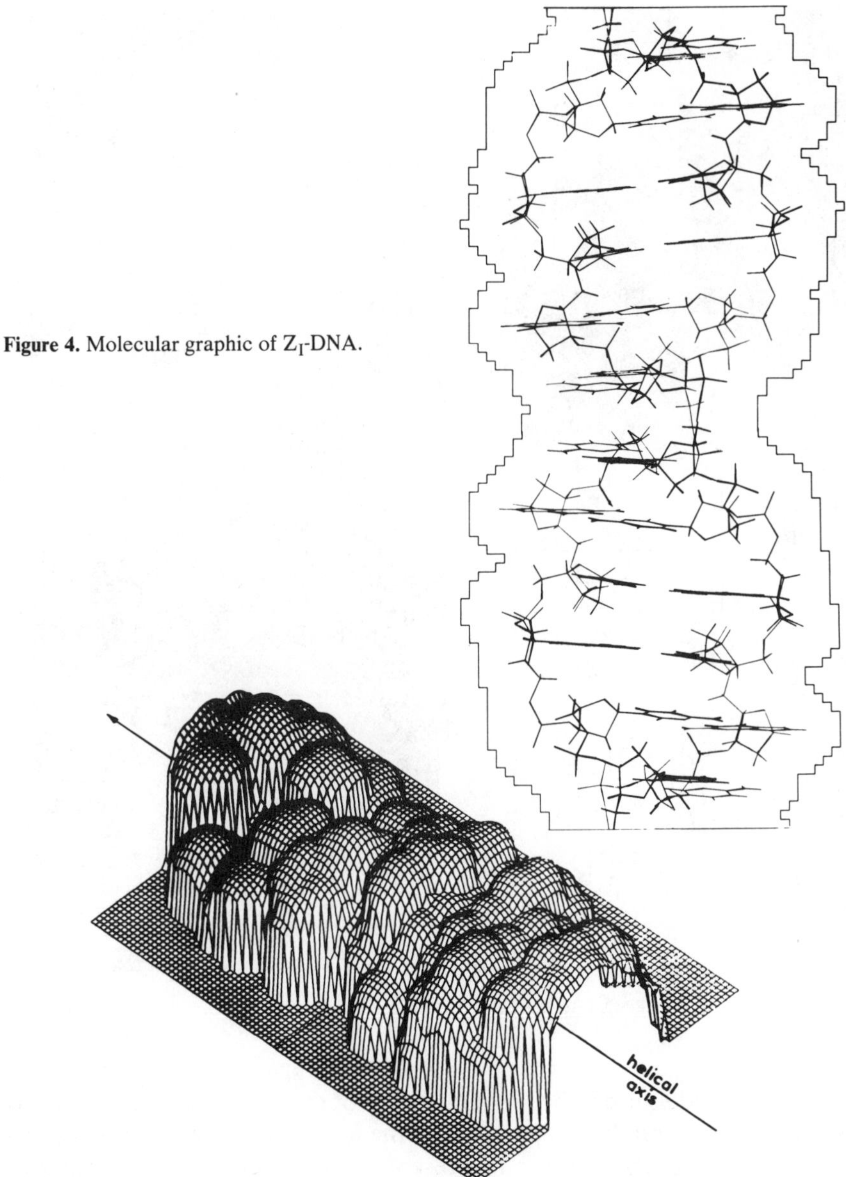

Figure 4. Molecular graphic of Z_I-DNA.

Figure 5. Surface envelope graphic of Z_I-DNA.

oxygen. The fact that this "visible" minimum is located on a phosphate oxygen rather than on an electronegative atom of a base, as in B-DNA, is probably due to the narrowness of the Z_I groove compared to the minor groove of B-DNA, which leads to an occlusion of the base atom spheres from the surface envelope by the sugar-phosphate backbones.

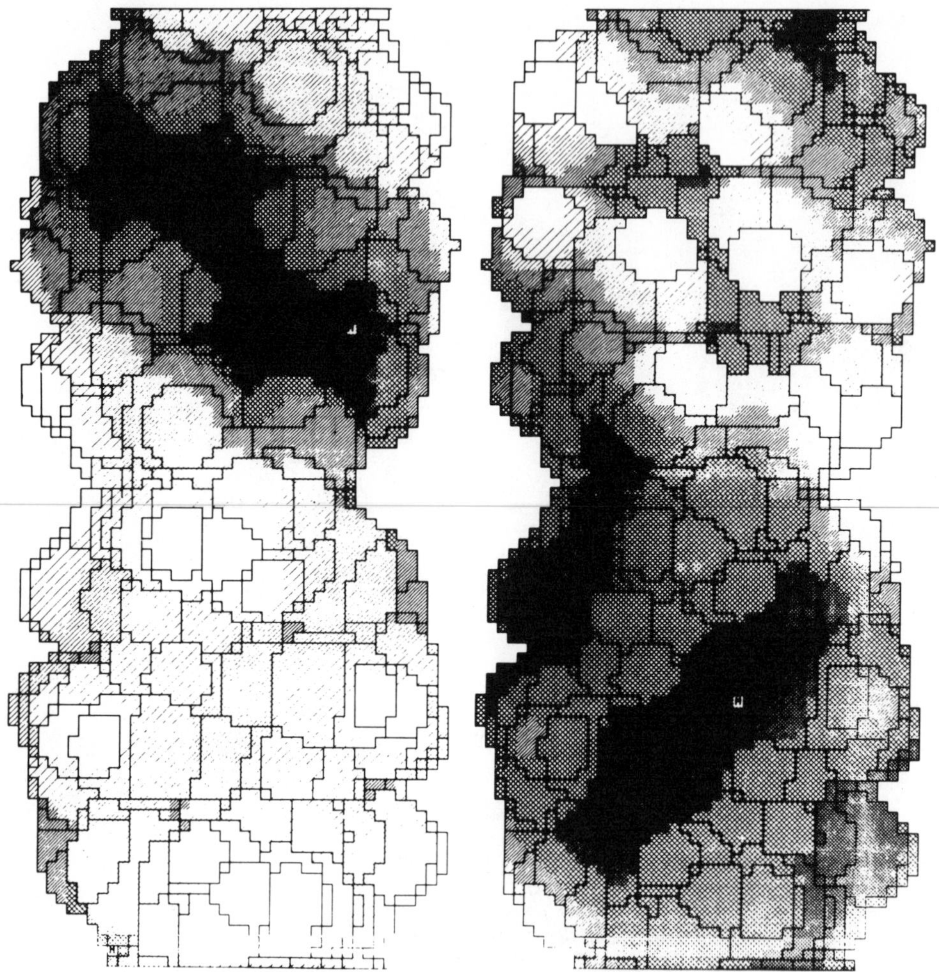

Figure 6. Surface potential of Z_I-DNA unscreened (left) and screened (right) by Na^+ cations (kcal/mole).

Note that with the models used, the absolute values of the potential for B-DNA and Z-DNA should not be directly compared since the latter has more phosphates in a full turn and consequently is liable to present for that reasons more negative surface potentials.

For the Na^+ screened Z_I-DNA, the surface potentials are shown in Figure 6, right. These results indicate a marked change from the situation in the unscreened acid of Figure 6, left: not only have the absolute magnitudes of the potentials decreased, but also the most negative potential zone has moved from the groove of the double helix to its convex face, in the lower portion of the figure. The global surface minimum, calculated to be −76 kcal/mole, now lies on the sphere of an O6 atom of guanine. This change in the distribution of the potential upon screening is quite unlike that in B-DNA and is probably a result of the closeness of the phosphate groups, and, consequently, of the strength of the screening, across that groove.

The results for Z_{II}-DNA are given in Figures 7-9. The orientation of the helix, shown in Figure 7, is as for Z_I with the sole groove of the helix visible in the upper half of the figure. The computer generated perspective drawing for this helix is

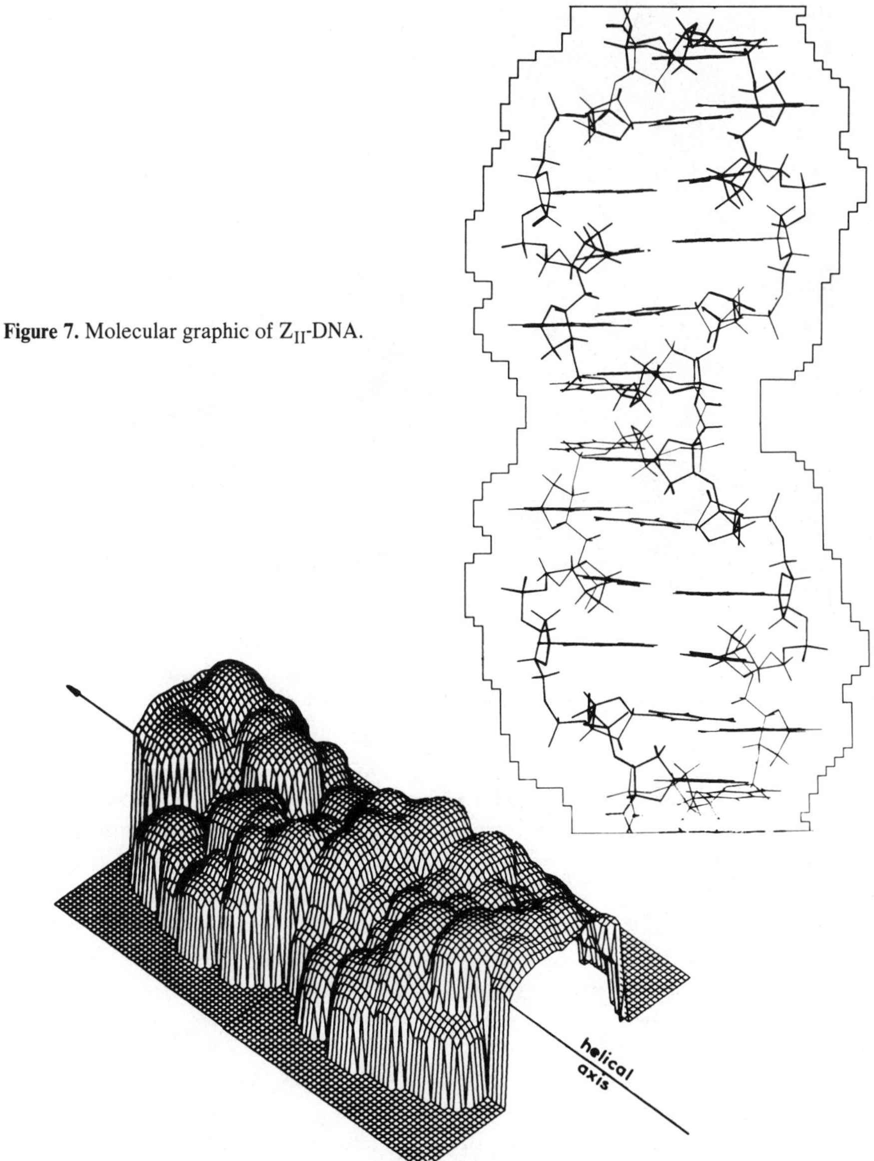

Figure 7. Molecular graphic of Z_{II}-DNA.

Figure 8. Surface envelope graphic of Z_{II}-DNA.

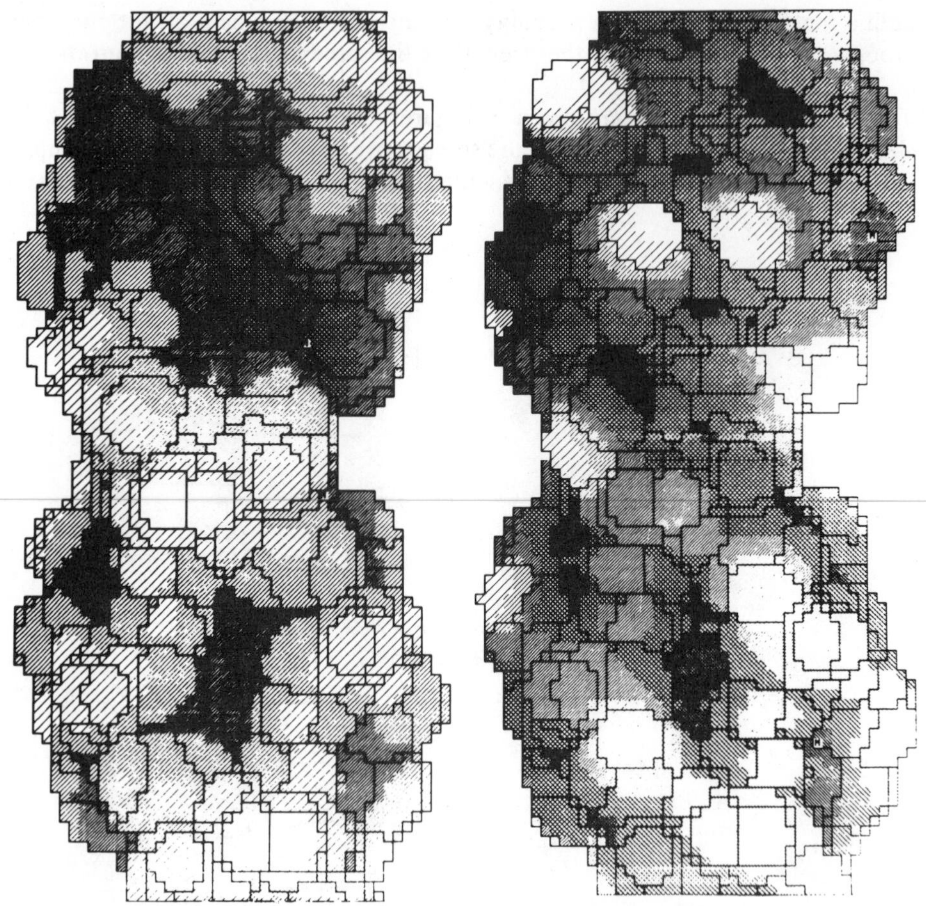

Figure 9. Surface potential of Z_{II}-DNA unscreened (left) and screened (right) by Na^+ cations (kcal/mole).

shown in Figure 8, altogether similar to Figure 5. The surface potentials for the unscreened acid are presented in Figure 9, left. Although a certain concentration of negative potential in the groove of this double helix can again be seen, the difference in potential between the groove and the convex face (30 kcal/mole) is less than for Z_I-DNA (50 kcal/mole). In fact, the surface potentials for Z_{II} are everywhere less negative than for Z_I and the global minimum, although still on the surface of a phosphate anionic oxygen, has decreased by 50 kcal/mole, to −680 kcal/mole.

The surface potentials for the Na^+ screened helix are given in Figure 9, right. The effect of adding counterions to the Z_{II} helix can be seen to have destroyed the concentration of negative potential in the groove, without creating any substantial concentration of potential on the convex face of the helix, as was the case for Z_I. The global surface minimum of −61 kcal/mole is situated on the sphere of an anionic phosphate oxygen in the lower half of the figure. A local minimum found in the region of the groove is, however, only 8 kcal/mole less attractive. It may be noted that, as for the unscreened helices, the surface potentials for screened Z_{II}-DNA are generally less negative than those for Z_I-DNA.

It is seen, thus, that, quite generally, screening of either Z_I- or Z_{II}-DNA reduces the magnitude of the negative surface potentials considerably more than in the case of B-DNA. Also, the effect of screening is quite different for the three helices; while the distribution of the potentials in B-DNA is little affected, the Z_I negative zone situated in the groove of Z_I shifts to the convex face of the helix, and the negative zone situated in the groove of the unscreened helix of Z_{II} is destroyed to yield a much more uniform surface potential distribution in the screened system.

Mid-Plane Isopotential Maps

In Figures 10-15, we present isopotential maps for unscreened and screened B-, Z_I- and Z_{II}-DNA's. These maps are drawn in planes perpendicular to the helical axis of each conformer and are placed between the two central base pairs of the helices. These base pairs are indicated on each map by solid lines for the pair above the plane and dotted lines for the pair below. The phosphorus atoms of the phosphates, between these two base pairs, are also indicated by the letters P. The sign X on each map indicates the helical axis and a vector from this point indicates radial distances in Å. The positions of the grooves of the double helices are indicated in each figure. All isopotential values are kcal/mole.

For unscreened B-DNA, the results are presented in Figure 10. The different isopotential lines are elliptical near the helical axis and almost circular, centered on this axis, when more distant. The potentials are somewhat more negative on the side of the major groove, which correlates with the results on the surface potentials. When B-DNA is screened by Na^+ cations the situation is more complex (Figure 11). The magnitudes of the potentials have strongly diminished and at long distances from the helical axis the potential is positive everywhere. Closer to the base pairs the potential remains negative, slightly more so on the side of the major groove, than in the minor groove.

The results for unscreened Z_I-DNA are presented in Figure 12 and, as for B-DNA, the isopotential curves are roughly circular about the helical axis. The potentials are more negative on the side of the groove than on the convex side. (In this figure, the displacement of the base pairs to one side of the helical axis to form the convex face of the double helix is particularly clear).

The effect of screening on these isopotentials is indicated in Figure 13. The reversal of the distribution of the potentials with respect to the situation in the unscreened acid, similar to that noted in studies on the surface envelopes, is clear. The isopotential curves are now positive in the groove while the convex face of the double helix remains negative, although weakly so.

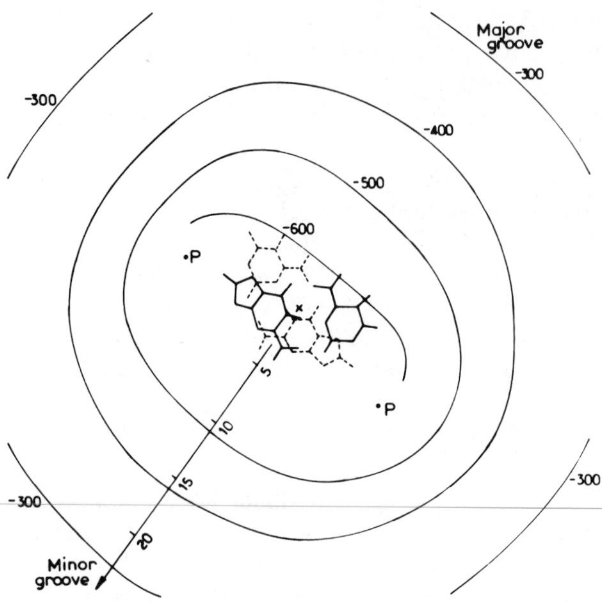

Figure 10. Mid-plane isopotentials for B-DNA (kcal/mole).

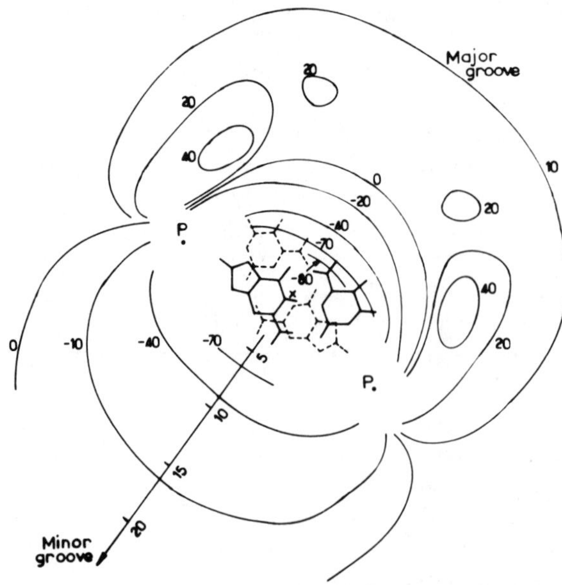

Figure 11. Mid-plane isopotentials for B-DNA screened by Na^+ cations (kcal/mole).

The results for unscreened Z_{II}-DNA are shown in Figure 14. The isopotential lines, as for the other unscreened helices, are roughly circular about the helical axis. The potentials are slightly more negative on the side of the groove, but the difference is

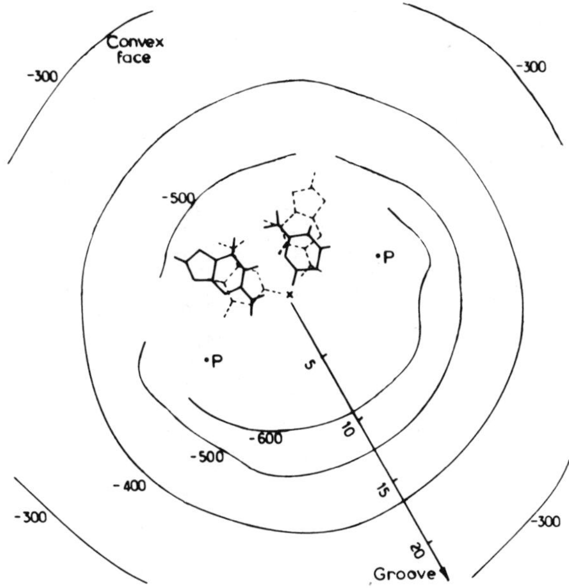

Figure 12. Mid-plane isopotentials for Z_I-DNA (kcal/mole).

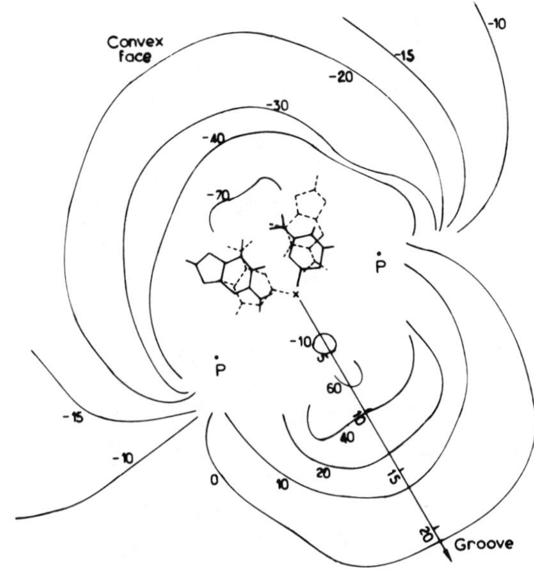

Figure 13. Mid-plane isopotentials for Z_I-DNA screened by Na^+ cations (kcal/mole).

not as marked as for Z_I-DNA. In Figure 15 are the isopotentials for the screened helix. At long distance, the potentials are positive everywhere. Closer to the center a zero isopotential line runs almost entirely around the helical axis. Within its limit,

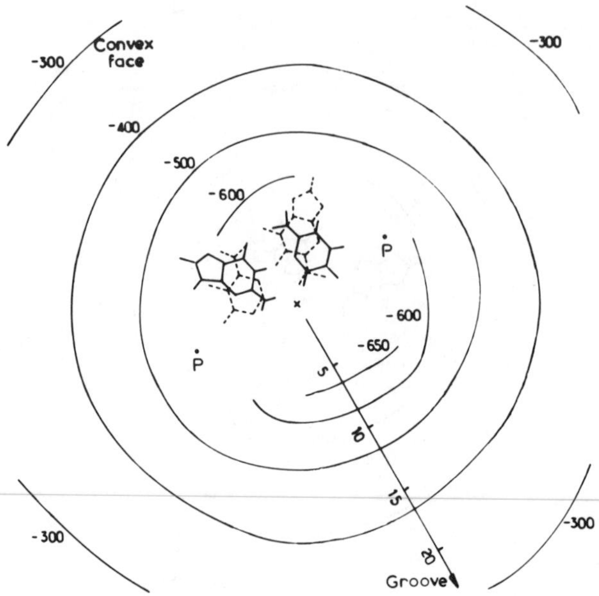

Figure 14. Mid-plane isopotentials for Z_{II}-DNa (kcal/mole).

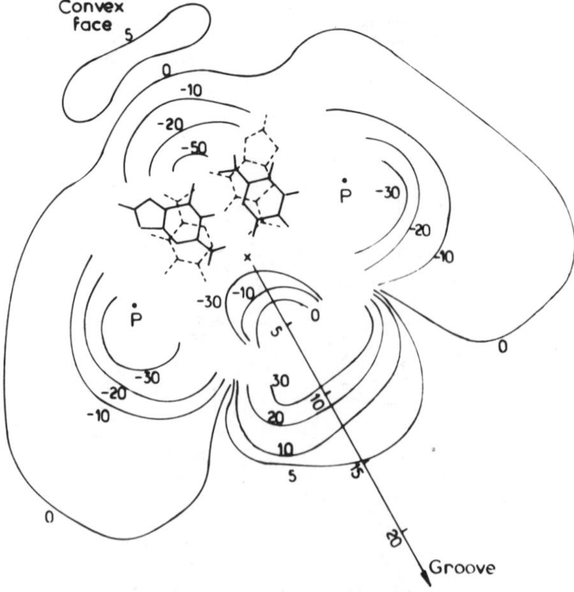

Figure 15. Mid-plane isopotentials for Z_{II}-DNA screened by Na^+ cations (kcal/mole).

close to the center there are four zones of small negative potential: in the groove of the helix, on the convex face and surrounding the two phosphate backbones. This situation corresponds to the observation made from the study of the surface envelope potentials that there is no specially favored zone of negative potential in screened Z_{II}-DNA.

Base and Phosphate Site Potentials

The electrostatic potentials at the usual sites associated with the atoms susceptible to electrophilic attack are given in Table I for B-, Z_I- and Z_{II}-DNA. Results are

Table I
Base and Phosphate Site Potentials (kcal/mole)

Unit	Site	Unscreened			Screened		
		B-DNA	Z_I-DNA	Z_{II}-DNA	B-DNA	Z_I-DNA	Z_{II}-DNA
Guanine	N2(3′)	−624.9	−620.9	−602.5	−110.2	− 39.5	−30.1
	N2(5′)	−617.4	−643.0	−627.4	−105.8	− 44.5	−49.9
	N3	−668.7	−657.8	−662.4	−153.4	− 80.6	−97.4
	N7	−664.7	−601.2	−645.7	−138.0	−120.5	−82.9
	C8(3′)	−605.3	−511.6	−541.9	− 79.6	− 27.1	−12.6
	C8(5′)	−579.8	−546.5	−563.2	− 52.1	− 31.9	− 4.3
	O6	−645.2	−599.5	−641.9	−124.7	−104.7	−91.1
Cytosine	N4(3′)	−615.7	−545.0	−577.6	− 94.1	− 36.5	−26.3
	N4(5′)	−607.9	−550.1	−568.3	− 89.9	− 35.4	−27.3
	C5(3′)	−592.5	−575.6	−595.7	− 64.3	− 55.9	−42.1
	C5(5′)	−585.1	−580.8	−596.2	− 63.1	− 43.3	−50.9
	O2	−652.0	−695.4	−607.4	−134.8	− 81.3	−83.2
Phosphate	Bridge GpC		−665.1	−602.1			
		−582.8					
	Bridge CpG		−696.4	−644.2			

presented for both unscreened and Na^+ screened double helices. For an easier comparison of the site potentials at the bases, we also produce a graphical representation in Figure 16, left (unscreened helices) and Figure 16, right (screened helices) where the potentials are arranged on vertical axis, more negative potentials being towards the bottom of these figures. We recall that the potentials calculated for the three forms of DNA in this section are directly comparable as the model B, Z_I and Z_{II} helices employed in this study contain identical numbers of phosphates namely, 22.

We begin by considering the potential for the unscreened acids at the base reactive sites. The first point to be noted is that the range of potentials for the sites is rather similar in the three helices, but that their ordering changes significantly. For B-DNA N7(G) and N3(G) are associated with the most negative potentials, followed by the two carbonyl oxygens O2(C) and O6(G). These most negative sites all lie in the plane of the bases. The four remaining sites which occur out of the base planes, N2(G), C8(G), N4(C) and C5(C) are all somewhat less negative.

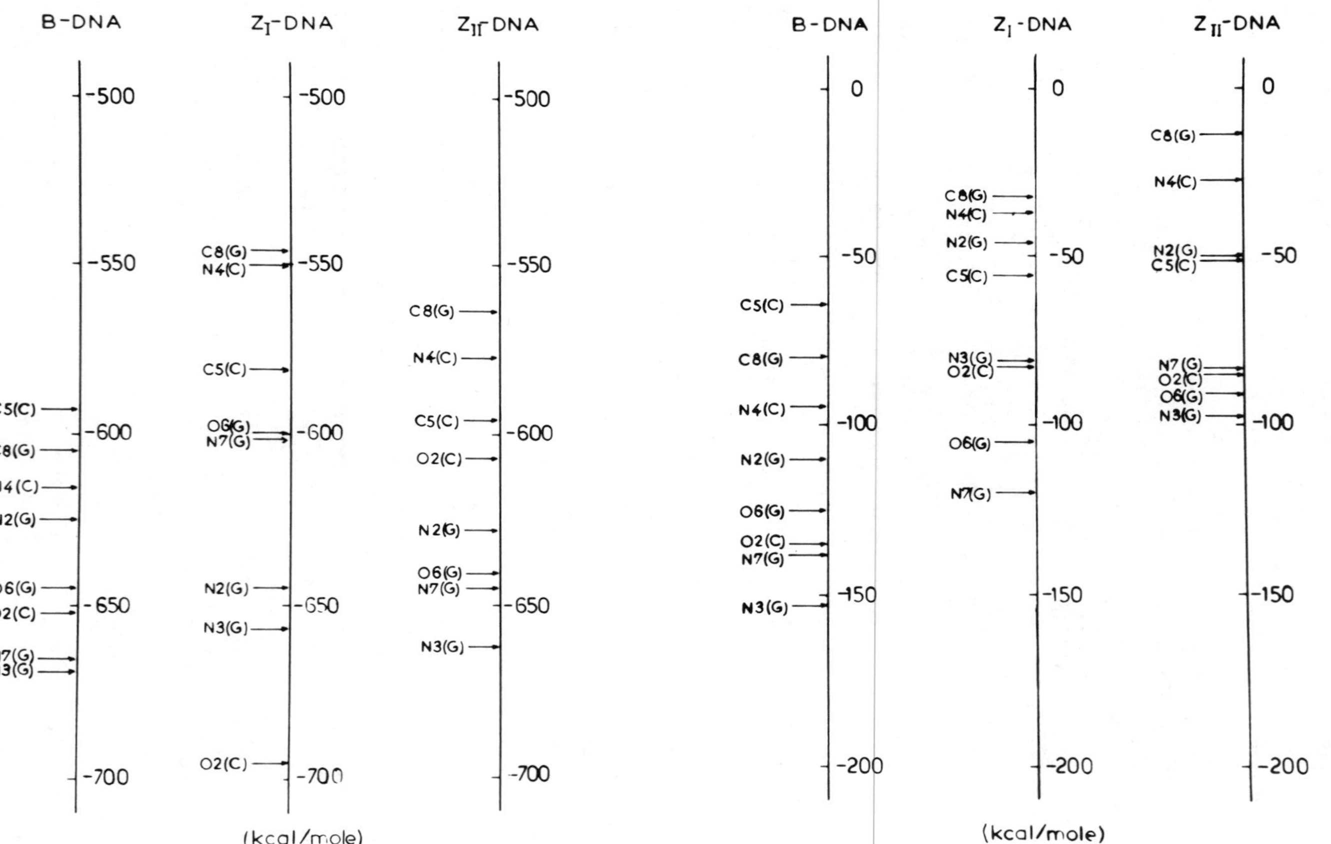

Figure 16. Comparison of the base site potentials for the unscreened helices (left) and screened helices (right) in kcal/mole.

For Z_I-DNA, a clear distinction may be seen between the sites in the groove and the sites occurring on the convex face. Thus O2(C), N3(G) and N2(G), the most negative sites, all lie in the groove. N7(G) and O6(G) which were associated with deep negative potentials in B-DNA are much less so in Z_I and they lie on the convex face of the helix. The least negative sites in B-DNA C5(C), N4(C) and C8(G) remain so in Z_I but the latter two sites are still less negative for Z_I-DNA. The range of potentials has, in fact, increased from 77 kcal/mole to 149 kcal/mole, in passing from B− to Z_I-DNA.

For Z_{II}-DNA, the range of base site potentials is closer to that of B-DNA (100 kcal/mole) and the distinction between the groove sites and convex face sites, seen for Z_I, no longer exists. Now N3(G) (in the groove) and N7(G) (on the convex face) are the most negative sites. Among the remaining in plane sites, O6(G) has a similar potential to that in B-DNA while O2(C) is much less negative. This is the reverse of the situation in Z_I, where the potential at O2(C) is almost 100 kcal/mole deeper than at O6(G).

The results obtained for the base site potentials in the screened helices are given graphically in Figure 16, right. For all three helices studied, the magnitude of the potentials has decreased by roughly 500 kcal/mole upon Na^+ screening. For B-DNA, no change in ordering of the site potentials has occurred. This is strongly contrasted by the situation for Z_I-DNA where a profound rearrangement has taken place. Whereas for unscreened Z_I, it was the sites in the groove that had the most negative potentials, in the screened helix, it is the sites at the convex face which dominate, namely, N7(G) and O6(G). This reversal on screening, for Z_I-DNA, has already been noted in the preceding sections where surface potentials and mid-plane potentials were studied. For Z_{II}-DNA, there is only a slight reordering which brings O2(C) into a relatively more negative position and N2(G) into a relatively less negative position in the screened helix with the respect to the unscreened one. Finally, a trend in the deepest site potentials may be noted for the screened helices: from B-DNA, with the most negative values, to Z_I and then Z_{II}-DNA. Such a trend is not observed for the unscreened case.

Finally, we may describe the potentials at the bridge position between the two anionic oxygens of the phosphate groups. This position corresponds to the preferential site of binding to phosphates of metal cations. The resuls are given at the bottom of Table I. It may be noted that the most negative potentials are associated with the phosphates of Z_I-DNA, the phosphates of B-DNA having considerably less negative potentials and the values found for the phosphates of Z_{II}-DNA being intermediate. Further, for both Z-DNA's, the site potential of the CpG phosphate is roughly 40 kcal/mole more negative than that of the GpC phosphate.

Base and Phosphate Atom Accessibilities

The accessible areas for atoms of B-, Z_I-, and Z_{II}-DNA towards a spherical attacking species (radius = 1.2Å) are given in Table II.

Table II
Base and Phosphate Atom Accessibilities (Å²)

Unit	Site	B-DNA	Z_I-DNA		Z_{II}-DNA	
Guanine	N2	0.00	0.32		0.56	
	N3	0.19	0.00		0.00	
	N7	3.01	5.61		3.99	
	C8	0.47	2.00		1.42	
	O6	2.46	3.43		2.62	
Cytosine	N4	0.37	0.00		0.00	
	C5	0.97	0.00		0.00	
	O2	0.24	1.05		1.90	
			GpC	CpG	GpC	CpG
Phosphate	O1	9.08	9.04	8.76	4.24	9.02
	O2	6.99	4.16	9.57	8.92	8.80
	O3′	1.94	0.57	1.13	0.24	0.85
	O5′	0.00	0.12	2.06	0.16	1.74

The results for the bases show that all but one (N3) of the atoms studied for guanine are more accessible in Z-DNA than in B-DNA and moreover, that three atoms N7, C8 and O6 are considerably more accessible in Z_I than in Z_{II}. For cytosine, only O2 becomes more accessible in Z-DNA with respect to B-DNA and this more so in Z_{II} than in Z_I. The remaining atoms of this base, N4 and C5, are both inaccessible in the Z-DNA's.

We have also studied the accessibilities of the phosphate oxygens in the three helices and note that the accessibilities of the anionic oxygens are always considerably larger than those of the esteric oxygens. However, for the GpC phosphates of the Z-DNA's, there is an imbalance between the anionic oxygens, O1 being twice as accessible as O2 in Z_I, while the reverse is true in Z_{II}.

Conclusion

It has been shown in our earlier publications that although the interaction energy between two molecular species is composed of at least four essential components,[20] namely, electrostatic, polarization, charge transfer and exchange repulsion, the knowledge of the electrostatic molecular potential of the nucleic acids provides a very useful guide for the elucidation of important aspects of the reactivity of these polyanionic species towards attacking electrophiles, in particular, when these are in the form of cations, as is the case for the majority if not the totality of the ultimate forms of chemical carcinogens.[21] It seems, therefore, useful to consider the results obtained in this paper on the potentials, in particular the base site potentials, of the different forms of the nucleic acids, in conjunction with the results on their steric accessibilities, for the detection of differences which may exist from that point of view between the B- and Z-DNA's.

Summarizing our findings, we may observe that, within the restrictions indicated above, our deduction would be that for B-DNA, whether screened or unscreened, N7(G) and O6(G) should be altogether the most favored sites for electrophilic attack. The remaining sites, inasmuch as they are accessible and are associated also with significant potentials, are, of course, also possible targets for such attacks, in particular, when account is taken of the "specificity" of the attacking species with respect to the nature of the attacked site, a problem which is, in fact, related to the role of the components of the interaction energy not included in this treatment.[20] It seems, nevertheless, that, in similar reactions, the low accessibility of e.g. N3(G) should disfavor this site, in spite of its relatively deep potential, with respect to N7(G). This conclusion seems to be very satisfactorily verified by the available experimental data.[22]

In Z-DNA's, one must distinguish more carefully between the screened and unscreened cases.

For unscreened Z_I-DNA, it is O2(C) which has the most negative potential and a high accessibility. N2(G) also has a relatively strong negative potential, stronger than in B-DNA, and in contrast to B-DNA, is accessible even in the static approximation. N7(G) and O6(G) have relatively great accessibilities but moderate potentials. The situation changes upon screening the phosphates and, as for B-DNA, N7(G) and O6(G) become the most favored sites for electrophilic attack. In contrast, the potential of O2(C) and N2(G) are reduced.

In unscreened Z_{II}-DNA N7(G) and O6(G) have, once again, the best combination of negative potentials and high accessibilities. It is worth mentioning that in this double helix, N2(G) and O2(C) have the highest accessibilities found for these atoms in any of the conformers studied. They are both associated with intermediate potentials that of N2(G) being slightly deeper than in B-DNA. In screened Z_{II}-DNA, the only significant change concerns the increase of the negative potential associated with O2(C).

It may also be interesting to add that C8(G) is significantly more accessible in the Z-helices than in B-DNA, but that its potential is deeper in the latter than in the former.

Comparing the magnitudes of the potentials associated with the three helices, it is seen that they span a comparable zone in the unscreened case, but that for the screened case, B-DNA is associated with the most negative potentials, followed, in order, by Z_I-DNA and then Z_{II}-DNA. We may remark that this comparison is made for model DNA's with equal numbers of phosphate groups. An alternative approach would have been to take equal lengths of the helix for the three conformers. This would have resulted in more phosphates being considered for the B-DNA model (for example, 26 phosphates in 44.6Å, the pitch of Z-DNA, against only 24 phosphates for the latter conformation) and thus still greater negative potentials than for Z-DNA in either the screened or unscreened helices. Thus, when similar states are considered from the point of view of the potential, the Z-DNA's are not expected to manifest a higher intrinsic affinity towards electrophiles than B-DNA, with the possible exception, for the unscreened acids, of N2(G) in Z_I and Z_{II}-DNA's and of O2(C) in Z_I-DNA.

On the other hand, the accessibilities of a number of sites, N7(G), O6(G), C8(G), N2(G), O2(C) are substantially increased in the Z-DNA's with respect to B-DNA. Inasmuch as the potentials associated with these sites remain appreciable, this increased accessibility could well favor their attack by an appropriate carcinogenic species. If observed such an increase of reactivity could then be essentially due to this factor. A decrease of affinities would, on the contrary, point essentially to the primary importance of the magnitude of the potential.

A very recent publication by R.M. Santella, D. Grunberger, I.B. Weinstein and A. Rich (Proc. Natl. Acad. Sci. U.S.A., (78), 1451-1455, 1981) demonstrates that the reactivity of poly(dG-dC)•poly(dG-dC) for binding N-2-acetylaminofluorene at the C8 position of guanine residues is substantially smaller in condition in which the polymer exists in the Z-form (in 55% ethanol or 1 M $MgCl_2$) than when it exists in the B-form (in 25% ethanol or Low Mg^{++} concentration). This result may be considered as an indication of the preponderance of the effect of the decrease of the electrostatic potential at C8 of guanines in Z-DNA with respect to B-DNA (as evaluated in our paper) which seems thus to dominate over the greater accessibility of that position in Z-DNA in comparison to B-DNA.

Acknowledgement

This work was supported by the National Foundation for Cancer Research to which the authors wish to express their thanks.

References and Footnotes

1. Wang, A. H.-J., Quigley, G.J., Kolpak, F.J., Crawford, J.L., van Boom, J.H., van der Marel, G. and Rich, A., *Nature 282,* 680-686 (1979).
2. Wang, A. H-J., Quigley, G.J., Kolpak, F.J., van der Marel, G., van Boom, J.H. and Rich, A., *Science, 211,* 171-176 (1980).
3. Drew, H.R., Tanako, T., Tanaka, S., Itakura, K., and Dickerson, R.E., *Nature 286,* 567-573 (1980).
4. Crawford, J.L., Kolpak, F.J., Wang, A.H-J., Quigley, G.J., van Boom, J.H., van der Marel, G. and Rich, A., *Proc. Natl. Acad. Sci. (U.S.A.) 77,* 4016-4020 (1980).
5. Arnott, S., Chardrasekaran, R., Birdsall, D.C., Leslie, A.G.W. and Ratliff, R.L., *Nature 283,* 743-745 (1980).
6. Mitra, C.K., Sarma, M.H. and Sarma, R.H., *Biochemistry, 20,* 2036-2041 (1981).
7. Perahia, D., Pullman, A. and Pullman, B., *Int. J. Quant. Chem., Quant. Biol. Symp. 6,* 353-363 (1979).
8. Pullman, B., Perahia, D. and Cauchy, D., *Nucl. Acids Res. 6,* 3821-3829 (1979).
9. Cauchy, D., Lavery, R. and Pullman, B., *Theoret. Chim. Acta 57,* 323-327 (1980).
10. Lavery, R. and Pullman, B., *Int. J. Quant. Chem., Quant. Biol. Symp. 8,* in press.
11. Lavery, R., Pullman, A. and Pullman, B., *Int. J. Quant. Chem., Quant. Biol. Symp. 8,* in press.
12. Lavery, R., Pullman, A. and Pullman, B., *Nucl. Acids Res. 8,* 1061-1079 (1980).
13. Lavery, R., Pullman, A. and Pullman, B., *Theoret. Chim. Acta 57,* 233-243 (1980).

14. Lavery, R., Pullman, A., Pullman, B., and de Oliveira, M., *Nucl. Acids Res. 8,* 5095-5110 (1980).
15. Lavery, R., Pullman, A. and Corbin, S., in the present volume.
16. Lavery, R., Corbin, S. and Pullman, B., *Int. J. Quant. Chem., Quant. Biol. Symp. 8,* in press.
17. Zakrzewska, K., Lavery, R., Pullman, A. and Pullman, B., *Nucl. Acids Res. 8,* 3917-3932 (1980).
18. Goldblum, A., Perahia, D. and Pullman, A., *Int. J. Quant. Chem. 15,* 121-129 (1979).
19. Arnott, S. and Hukins, D.W.L., *Biochem. Biophys. Res. Comm. 47,* 1504-1509 (1972).
20. Pullman, A. and Armbruster, A.M., *Theoret. Chim. Acta 45,* 249-256 (1977).
21. Pullman, A. and Pullman, B., *Int. J. Quant. Chem., Q.B.S., 7,* 245-260 (1980).
22. Singer, B., *Prog. Nucl. Acid. Res. Mol. Biol. 15,* 219-283 (1975).

Proceedings of the Second SUNYA Conversation in the Discipline Biomolecular Stereodynamics
Volume I, ISBN 0-940030-00-4, Ed., Ramaswamy H. Sarma,
Adenine Press, New York, ©Adenine Press

The Molecular Electrostatic Potential on the Surface Envelopes of Macromolecules: Yeast tRNA^{Phe}

Richard Lavery, Alberte Pullman and Sylvie Corbin
Institut de Biologie Physico-Chimique
Laboratoire de Biochimie Théorique
associé au C.N.R.S.
13, rue P. et M. Curie—75005 Paris—France.

Introduction

The various functions of transfer ribonucleic acids in the cellular machinery make a detailed understanding of their reactivity an important objective. Since the resolution of the structure of one such macromolecule, yeast tRNA^{Phe}, the elucidation of this reactivity in terms of structural characteristics of the polymer seems to become feasible. So far, attention has been essentially centered on correlations between the reactivity of specific sites within tRNA^{Phe} towards various chemical reagents and their *steric accessibility,* judged at first qualitatively from the tertiary folding of the macromolecule[1,2] and evaluated more recently by more quantitative procedures by Alden and Kim,[3] Thiyagarajan and Ponnuswamy[4] and in the authors' laboratory.[5]

Recently, a series of publications from our laboratory[5-8] has drawn attention also to the significance for the reactivity of tRNA of its *electronic structure* in particular of the *electrostatic molecular potential* associated with this macromolecular skeleton. These publications have shown the utility of *combining the data on the potential with the data on the accessibilities* for the interpretation of the reactivity of different sites of tRNA^{Phe} (at the bases and on the phosphates) toward a variety of chemical reagents.

The computations of the electrostatic molecular potential associated with tRNA^{Phe} have so far been carried out either for different planes cutting through the macromolecule or for specific localized sites. They suffer from the drawback of not permitting an easy view of the potential surrounding interesting *regions* of the macromolecule. While obviously useful for exploring some aspects of the interaction of relatively small reactants with this substrate, they are therefore insufficient for the treatment of interactions with larger reagents, such as the polyamines which are bound to tRNA^{Phe} in the crystal, or with other macromolecules, such as aminoacyl tRNA synthetase or the components of ribosomes.

In the present publication, we attempt to provide this overall view by calculating

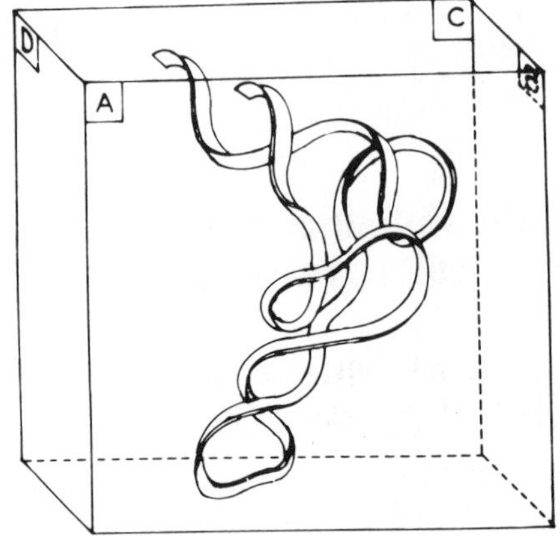

Figure 1. The orientation of tRNA^{Phe} and the four windows used to view its surface envelope.

SHADING	POTENTIAL DIVISION (kcal/mole)
	-547 to -707
	-708 to -867
	-868 to -1027
	-1028 to -1187
	-1188 to -1347
	-1348 to -1507

Definition of Shading Used for Surface Potential of Graphics

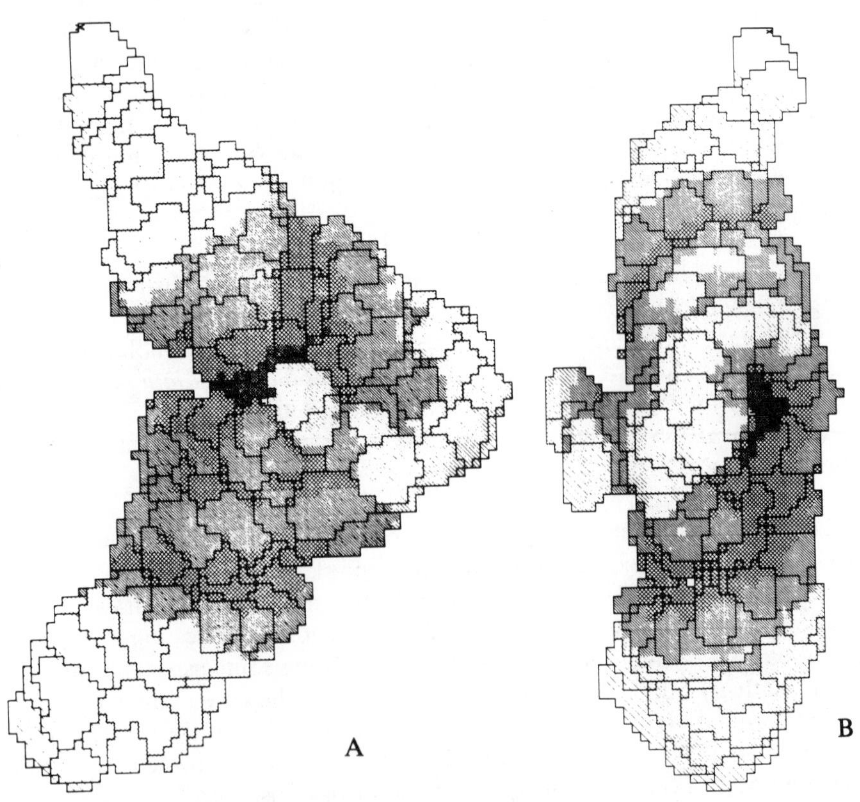

Figure 2. The surface potentials of tRNA^{Phe} viewed through the windows A, B, C, and D, as indicated (for details of shading, see table at top of page).

the electrostatic potential on a *surface envelope* surrounding the macromolecule. This technique, recently applied to B-DNA,[9] yields a simple, but informative representation of the distribution of the potential around biopolymers and is capable of visualizing features of this distribution which are difficult to deduce from the study of site potentials alone.

Method

The general procedure for calculating the electrostatic potential of nucleic acids has been described in our preceding publications.[8] In essence, the potential of the macromolecule is determined by a superposition of the potentials of a number of subunits: the bases, sugars and phosphates in case of the nucleic acid. Each of these subunit potentials is calculated from an overlap multipole expansion[10,11] developed from the electron density of the unit, itself derived from the calculation of its *ab initio* wave function.

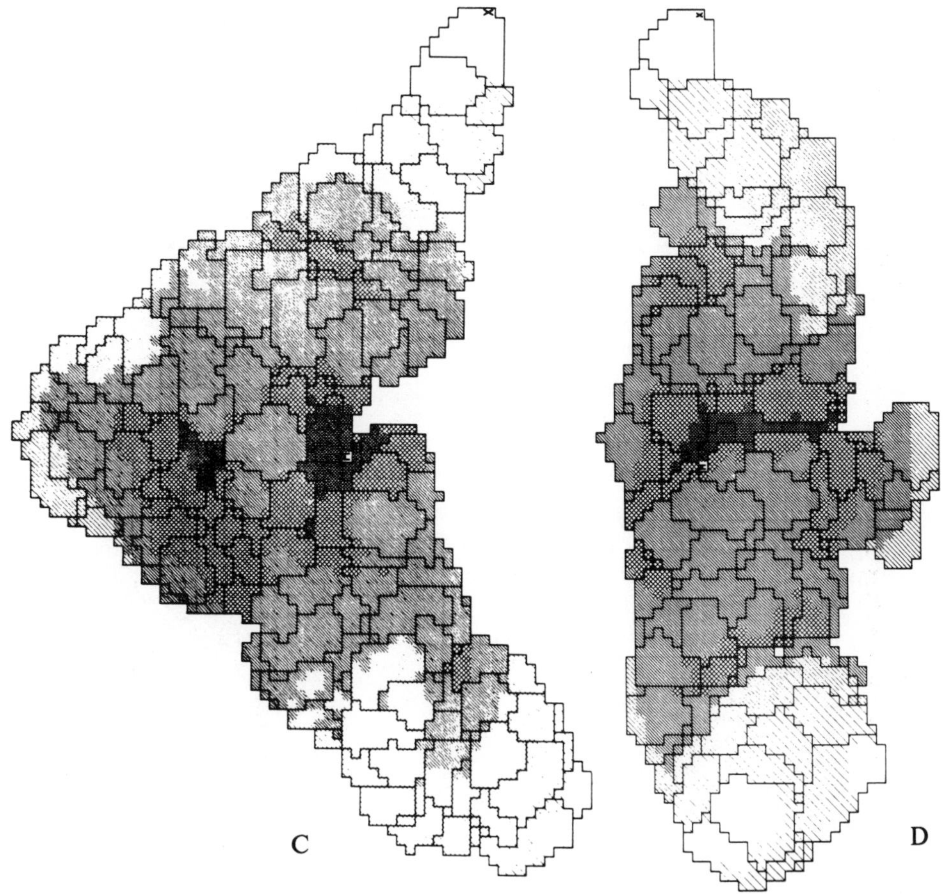

Figure 2 continued from previous page

The use of such multipole expansions in the calculation of electrostatic potentials has been shown to yield accurate results provided no atoms of the molecule concerned are approached to less than 2Å.[12]

The construction of surface envelopes around macromolecules and the subsequent calculation of the potential on these envelopes is described in our recent treatment of B-DNA.[9] The envelope itself is formed as the outer boundary of spheres centered on each of the atoms of the molecule. The radii of these spheres are equal to the appropriate van der Waal's atomic radii multiplied by a common factor 'F'. This factor may be adjusted to yield envelopes more or less distant from the macromolecule, but must be at least 1.7 if multipole expansions are employed in the potential calculations because of the aforementioned limit of approach, this value was used in the present study. The geometry of $tRNA^{Phe}$ employed is that due to Sussman *et al.*[13]

Electrostatic Potentials on the Surface Envelope of tRNA

In presenting the electrostatic potentials on the surface envelope of $tRNA^{Phe}$ we make use of four views of the macromolecule. These views are obtained by looking

Figure 3. The $tRNA^{Phe}$ molecular structure viewed through the windows A, B, C, and D, as indicated.

through four "windows" A, B, C, D, positioned to form a rectangular box around the molecule as illustrated in Figure 1. The molecule has been oriented with its largest dimension vertical, the anticodon loop being placed as the lowest region in the windows and the acceptor end as the highest. Using the conventional "L" description of $tRNA^{Phe}$ windows A and C look at the two large faces of the "L" while B looks at its external angle and D at its internal angle.

The surface potentials calculated for these various views are represented, as described in Ref. 8, by the projection of the surface onto the plane of each of the windows. Shading is used to indicate the magnitude of the potentials, the most negative zones being the most darkly shaded (see display of shading for details of the convention).

Figure 2 presents the surface potentials seen through windows A, B, C, and D. In Figure 3 are given the corresponding views of the tRNA molecule (hydrogen atoms excluded), which may be used to assist in associating zones of surface potential with the underlying molecular structure. Further, in Figure 4 we present perspective views of the surface as seen through the two largest windows A and C which show more clearly the 3-dimensional form of this surface with its various grooves and

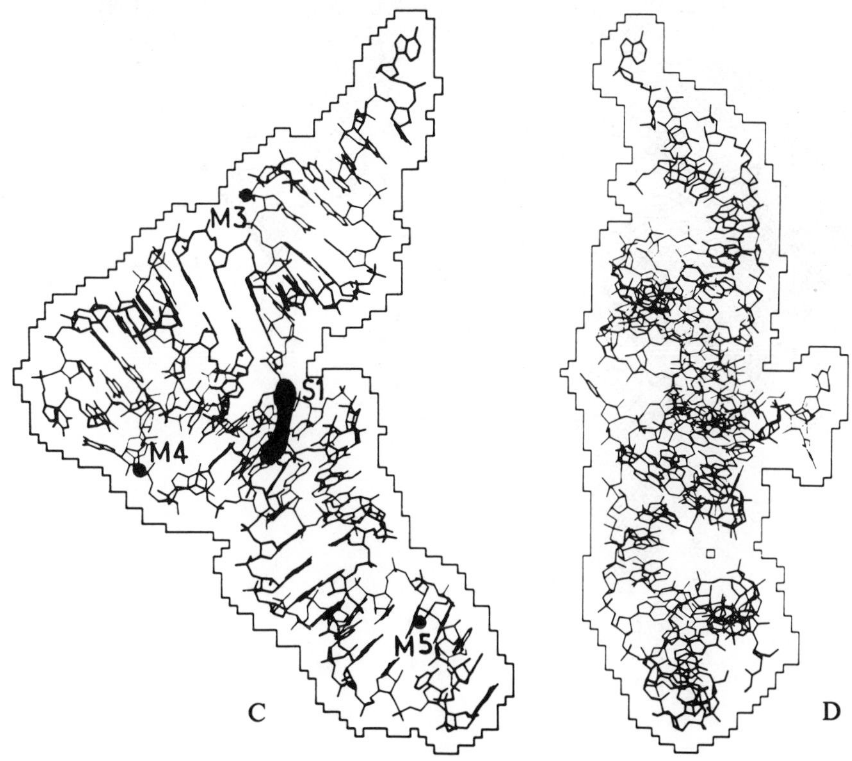

Figure 3 continued from previous page

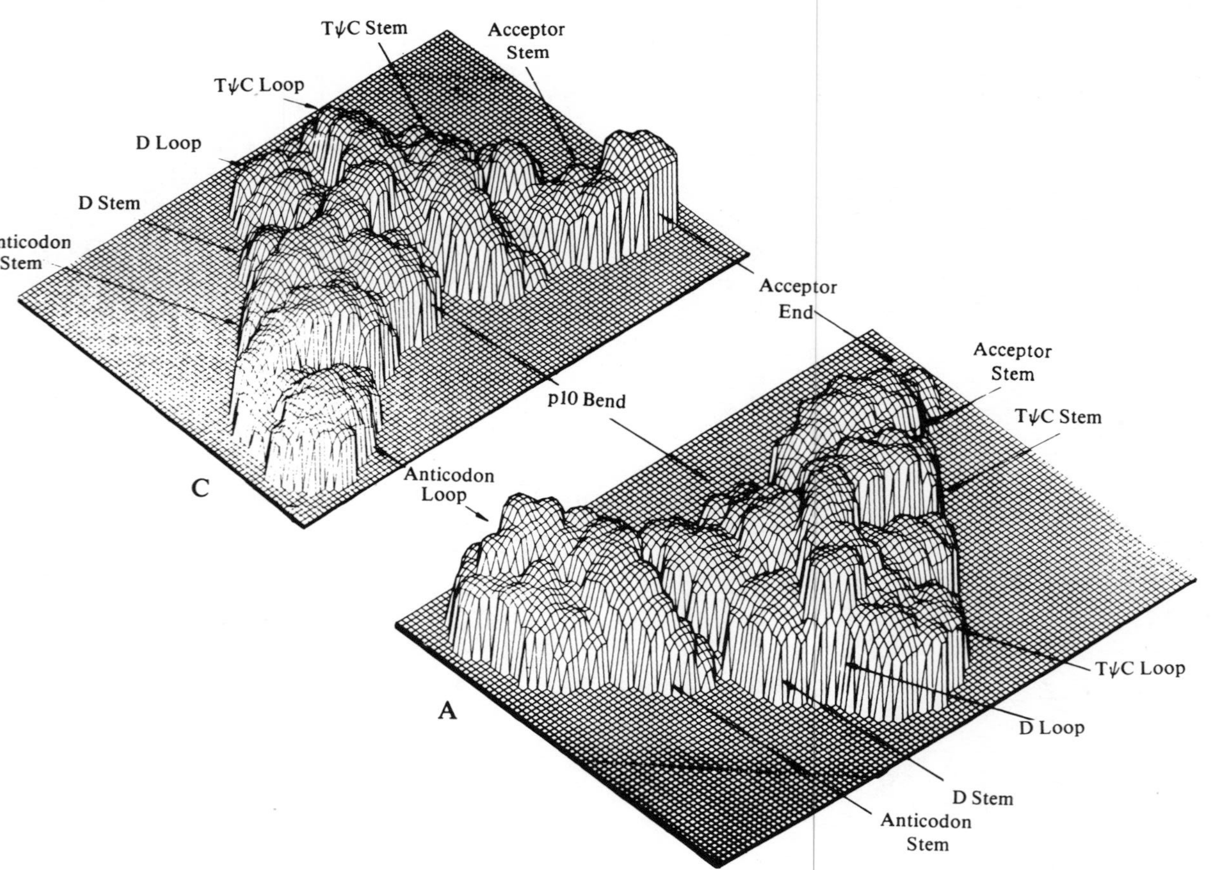

Figure 4. Perspective representations of the surface envelope viewed through windows A and C.

hills. On each view in Figure 2, the local maximum and minimum are marked by the letters X and M, respectively. Altogether, the variation of the potential on the surface is very large, extending from a maximum of −547 kcal/mole to a minimum of −1507 kcal/mole. These overall extrema correspond to the maximum in view D and the minimum in view A.

The most negative potentials are concentrated in the internal angle of the "L," close to the P10 bend. This zone, visible in the center of views A and C and crossing the center of view D, is strongly negative because of the tight P10 bend. Moving out from this central zone towards the extremities of the molecule, namely, the anticodon loop, the acceptor end and the TΨC loop, the magnitude of the surface potential diminishes rapidly. This situation was observed already in the study of planes of potential cutting through $tRNA^{Phe}$ reported in ref. 6. Now, with the aid of the surface envelope technique, we are able to observe more detailed correlations between the potentials and the molecular structure of $tRNA^{Phe}$.

Thus, in view A of Figure 2, a zone of deep negative potential may be seen in the lower arm of the "L." It corresponds to the deep groove at the junction of the D stem and the anticodon stem, visible in the perspective view of A (Figure 4). This zone of strong negative potential spreads around this arm and can be seen in views B and D. Further, in view A we observe a strong negative zone in the upper arm of the "L," visible also in views B and C, which correlates with the groove at the junction of the PΨC stem and the acceptor stem (see perspective representation in Figure 4).

Several other features may also be noted. Firstly, the rather weak negative potentials associated with the dihydrouracil residues 16 and 17, which protrude from the main body of the macromolecule (see center left of view B). Secondly, the strong negative potentials between the D loop and the TΨC loop in view C and also in the inward curved pocket of the anticodon loop, (see view C, Figure 2).

This distribution of the negative potential on the surface of $tRNA^{Phe}$ may be considered in relation with the relative affinity of various regions of the macromolecule for electrophilic reagents. Although, as we have stressed continuously in our publications related to the use of electrostatic molecular potentials (see e.g., references 5-10), the total interaction energy between two molecular species is composed of a number of contributions, among which the electrostatic component is only one, it may nevertheless be expected and is, in fact, observed, that for interactions involving ionic species, as in the case of interactions between the anionic nucleic acids and cations, the electrostatic component plays an important if not a dominant role. In such cases, the potential is expected to give a good indication of the reactive zones. The precise selection of a given zone for a given type of interaction being, of course, determined by the precise structure of the attacking agent and its possibilities for specific associations with the different constituents of the receptor site. Rephrasing, while the distribution of the surface potentials is expected to provide useful indications on the main zones of affinity between the interacting species, it is not expected to account for the specificity of the interactions.

That this is indeed the case may be seen by comparing the binding positions of the five Mg^{2+} ions (M1 – M5) and the two spermine polyions (S1, S2) located in the orthorhombic crystal of $tRNA^{Phe}$[14,15] as indicated in Figure 3, with the surface potentials in Figure 2.

Magnesium M1 and one extremity of spermine S1 are seen to be in the central and most negative potential zone of the macromolecule. Spermine S2 is bound in the strong negative potential zone at the junction of the anticodon stem and the D stem. Magnesium M2 is in a relatively strong potential zone in the D loop. Magnesium M3 is within the strong negative potential zone associated with the junction of the acceptor and TΨC stems. Magnesium M4 is, like M3, in a relatively strong potential zone in the D loop and, finally, magnesium M5 is within the strong negative potential zone in the pocket of the anticodon loop.

It is therefore evident that all these counterions are located in zones of highly negative potential. The exploration of their exact positioning at the various binding sites necessitates going beyond the present analysis. In particular, the various degrees of hydration of the Mg^{2+} ions and the numerous hydrogen bonds which the bound water molecules establish with various atoms at the receptor sites play, of course, an important role in the definite selection and architecture of binding sites. It may further be remarked that magnesium ions M2 and M4 are involved in indirect binding to a second tRNA molecule in the crystal lattice[15] and thus their positioning will not depend solely on the single tRNA molecule we study here. Work is continuing in our laboratory with a view to account for the specificity of all these binding sites. The interest of the present approach for the nucleic acids is nevertheless underlined by the recent demonstration, using ^{25}Mg NMR spectroscopy in DNA solution, of the predominance of "site" binding in this type of interaction[16] contrary to earlier predictions[17] that such associations should be essentially "delocalized."

A further confirmation of the significance for cation binding to $tRNA^{Phe}$ of the zones of deep molecular electrostatic potential is provided by the elegant study of Sundaralingam *et al.*[18] on the crystal structure of the ethidium bromide-$tRNA^{Phe}$ molecular complex. This work demonstrates that the drug binds specifically and is lodged in the cavity at the mouth of the "P10 loop," it means in the zone of the deepest potential. The binding is stabilized by hydrogen bonds between the ethidium amino groups and the anionic oxygens of phosphates P8 and P15 and by stacking of the ethidium ring over U8 and, partially, G15. Nevertheless, the significance of the deep potential may be deduced from Sundaralingam's own observation that although similar conformations for the polynucleotide fold are observed in the anticodon and pseudouridine hairpin loops, no ethidium binding is observed in these regions. It may also be interesting to add that a reinterpretation by Sundaralingam, *et al.*[19] of the NMR data relating to the same complex in solution, which have originally been interpreted by some authors[19] as involving intercalation between the base pairs U6 - A67 and U7 - A66 of the amino acid stem, indicates that the interactions in solution are, in fact, similar to those in the crystal.

References and Footnotes

1. Rich, A. and Rajbhandary, U.L., *Ann. Rev. Biochem. 45,* 805-860 (1976).
2. Goddard, J.P.,Progress in *Biophys. and Mol. Biol. 32,* 233-308 (1977).
3. Alden, C.J. and Kim, S.H., in *Stereodynamics of Molecular Systems,* Ed. R.H. Sarma, Pergamon Press, New York, 331-350 (1979).
4. Thiyagarajan, P. and Ponnuswamy, P.K. *Biopolymers 18,* 2233-2247 (1979).
5. Lavery, R., Pullman, A. and Pullman, B., *Theoret. Chim. Acta,* 233-243 (1980).
6. Lavery, R., Pullman, A. and Pullman, B., *Nucl. Acids Res. 8,* 1061-1078 (1980).
7. Lavery, R., De Oliveira, M. and Pullman, B., *J. Comp. Chem. 1,* 301-306 (1980).
8. Lavery, R., Pullman, A., Pullman, B., and De Oliveira, M., *Nucl. Acids Res. 8,* 5095-5111 (1980).
9. Lavery, R., Pullman, A., and Pullman, B., *Int. J. Quant. Chem. Quant. Biol. Symp.* 1981 in press.
10. Dreyfus, M., Thèse de 3ème Cycle, Paris, 1970.
11. Port. J.M.G., and Pullman, A., *FEBS Letters 31,* 70-73 (1973).
12. Goldblum, A., Perahia, D. and Pullman, A., *Int. J. Quant. Chem. 15,* 121-129 (1979).
13. Sussman, J.L., Holbrook, J.B., Warrant, R.W. and Kim, S.-H., *J. Mol. Biol. 123,* 607-630 (1978). We have used in this work a further refined coordinate system obtained as a personal communication from Dr. S.-H. Kim.
14. Holbrook, S.R., Sussman, J.L., Warrant, R.W., Church, G.M. and Kim. S.-H., *Nucl. Acids Res. 4,* 2811-2820 (1977).
15. Teeter, M.M., Quigley, G.J. and Rich, A. in *Nucleic Acid-Metal Ion Interactions,* Ed. T.G. Spiro, Wiley, P., 145 (1980).
16. Rose, D.M., Bleam, M.L., Record, J.R., M.T. and Bryant, R.C., *Proc. Natl. Acad. Sci. 77,* 6289-6296 (1980).
17. Manning, G.S., *Acct. Chem. Res. 12,* 443-449 (1979).
18. Liebman, M., Rubin, J. and Sundaralingam, M., *Proc. Natl. Acad. Sci. 74,* 4821-4825 (1977).
19. Jones, C.R. and Kearns, D.R., *Biochemistry 14,* 2660-2665 (1975).

Proceedings of the Second SUNYA Conversation in the Discipline Biomolecular Stereodynamics
Volume I, ISBN 0-940030-00-4, Ed., Ramaswamy H. Sarma,
Adenine Press, New York, ©Adenine Press

Nuclear Overhauser Effect Studies of tRNA: A Progress Report

Alfred G. Redfield, Siddhartha Roy, Valentina Sánchez
James Tropp, and Nara Figueroa
Department of Biochemistry
Brandeis University
Waltham, Mass. 02254

Introduction

The transfer RNA's are of interest in their own right and because of their interactions with diverse proteins. They also provide a possible model for understanding structure and dynamics of other nucleic acids. Our laboratory has specialized in NMR of tRNA, with the overall motivation of probing its dynamics through measurement of exchange rates of certain labile protons with solvent protons.[1-8] We hope that this exchange reflects local flexibility of the molecule. The protons we study in this way are ring nitrogen protons formally belonging to uracil and guanosine which are involved in internal hydrogen bonds, either in standard Watson-Crick base pairs or in tertiary interaction base pairs and base triples. Selected examples of these are shown in Figure 1, and the NMR spectrum and structure of yeast $tRNA^{Phe}$ are shown in Figures 2-4. Both the ring NH and the methyl region of the spectrum are relatively simple because there are relatively few NMR lines in both regions, compared to the confused central spectral region. Our NMR studies of tRNA have rested on earlier investigations (reviewed in reference 9), especially the work of Kearns, Shulman, and Reid and coworkers on the ring NH region, and Kan and Sprinzl and coworkers on the methyl region.

A description of exchange-rate studies of, mainly, yeast $tRNA^{Phe}$, by Dr. Paul Johnston in our laboratory has recently been published.[8] Exchange rate measurements are performed by studying line broadening at high temperature, by following disappearance of resonances after a change from H_2O to D_2O solvent at low temperatures,[3] and by a saturation-recovery method for moderately high temperatures.[1] NMR technology can cover rates from $years^{-1}$ to around 200 sec^{-1}. There is an unfortunate inaccessible gap, for tRNA, between 10^{-1} and $10^{3} sec^{-1}$, where interesting phenomena probably occur, but where exchange is masked by magnetic relaxation.

Our work on exchange-rate measurement has been temporarily sidetracked because we have been working on identification of NMR lines, primarily by nuclear Overhauser effect. In the present article we will summarize our methods and progress in

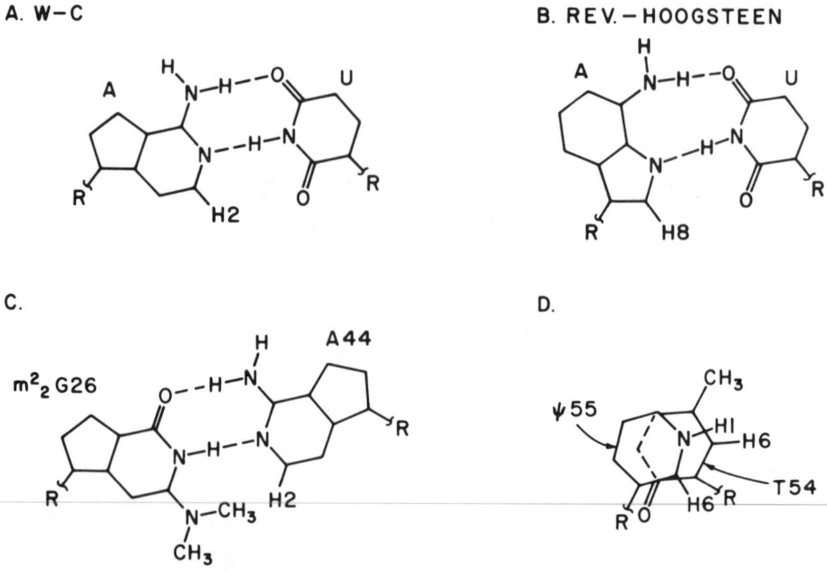

Figure 1. Base pairs occurring in the yeast tRNA^{Phe} x-ray structure. Only those protons involved in various NOE's are shown. (a) Watson-Crick AU pairs. (b) Reverse-Hoogsteen AU base pair. (c) The m^2_2G26-A44 tertiary pair. (d) The stacking of Ψ55 and T54.

this area, rather than our rate measurements which were described in previous reviews.

NMR Methods

All our experiments use the simple sequence shown in Figure 5. The 214 pulse flips over all protons in a large section of the NMR spectrum; these subsequently precess at their own rate and induce a signal in the NMR coil which is eventually digitized, accumulated, and Fourier transformed to yield the spectrum. This pulse is designed to not excite the H_2O proton resonance which would be so strong (55 molar) as to overwhelm the ~1 mM tRNA signals. The 214 pulse provides time-resolution as well as greater flexibility and sensitivity than other alternatives including the brute-force engineering approach to increased dynamic range and linearity that is widely advertised by Varian Associates. (Of course, such engineering improvements are generally desirable even if they do not solve the problem of working in H_2O buffer with high sensitivity for macromolecules. The advantage of an antiselective pulse such as the 214 pulse over a short pulse is that water is never saturated, avoiding the resulting loss of signal due to magnetic energy transfer ("spin diffusion") to H_2O and the necessity to let the entire water plus macromolecule system recover between sequences. We have discussed the 214 and related sequences elsewhere.[10])

The presaturation pulse in Figure 5 is designed to saturate a particular line in the

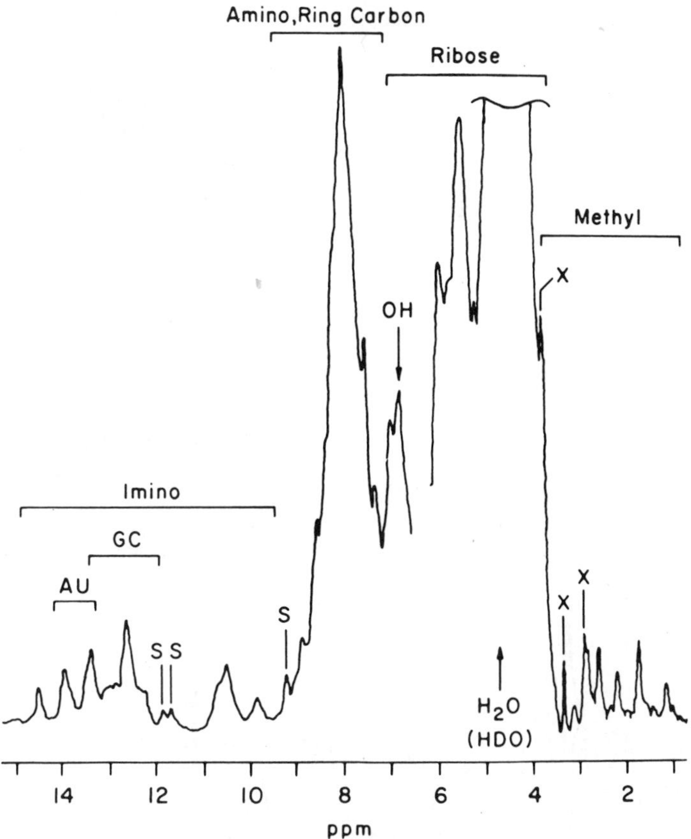

Figure 2. The entire NMR spectrum of yeast tRNA^Phe. This is a composite of a D_2O spectrum (right) and an H_2O spectrum (left). The x's are EDTA and the resonances marked S are single proton resonances. From reference 9. Reproduced with permission, from the *Annual Review of Biophysics and Bioengineering,* Vol. 9. Copyright 1980 by Annual Reviews Inc.

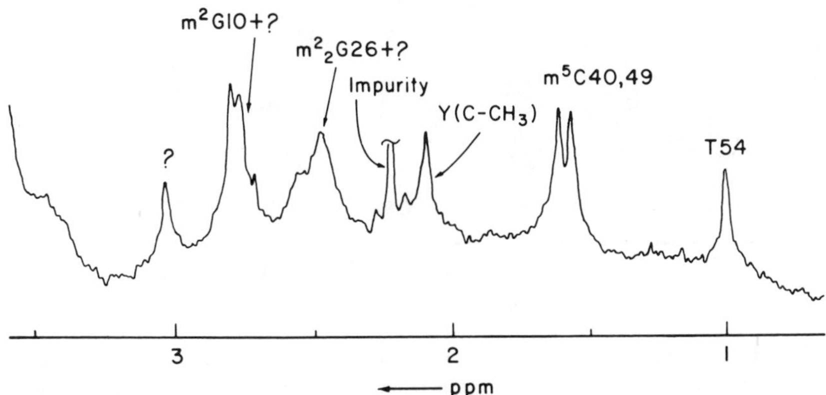

Figure 3. The methyl region of yeast tRNA^Phe in the presence of Mg^{++}, obtained on the 600 MHz instrument at Carnegie Mellon University, Pittsburgh, PA. Rapid scan correlation NMR was used. Obtained through the courtesy of Dr. J. Dadok.

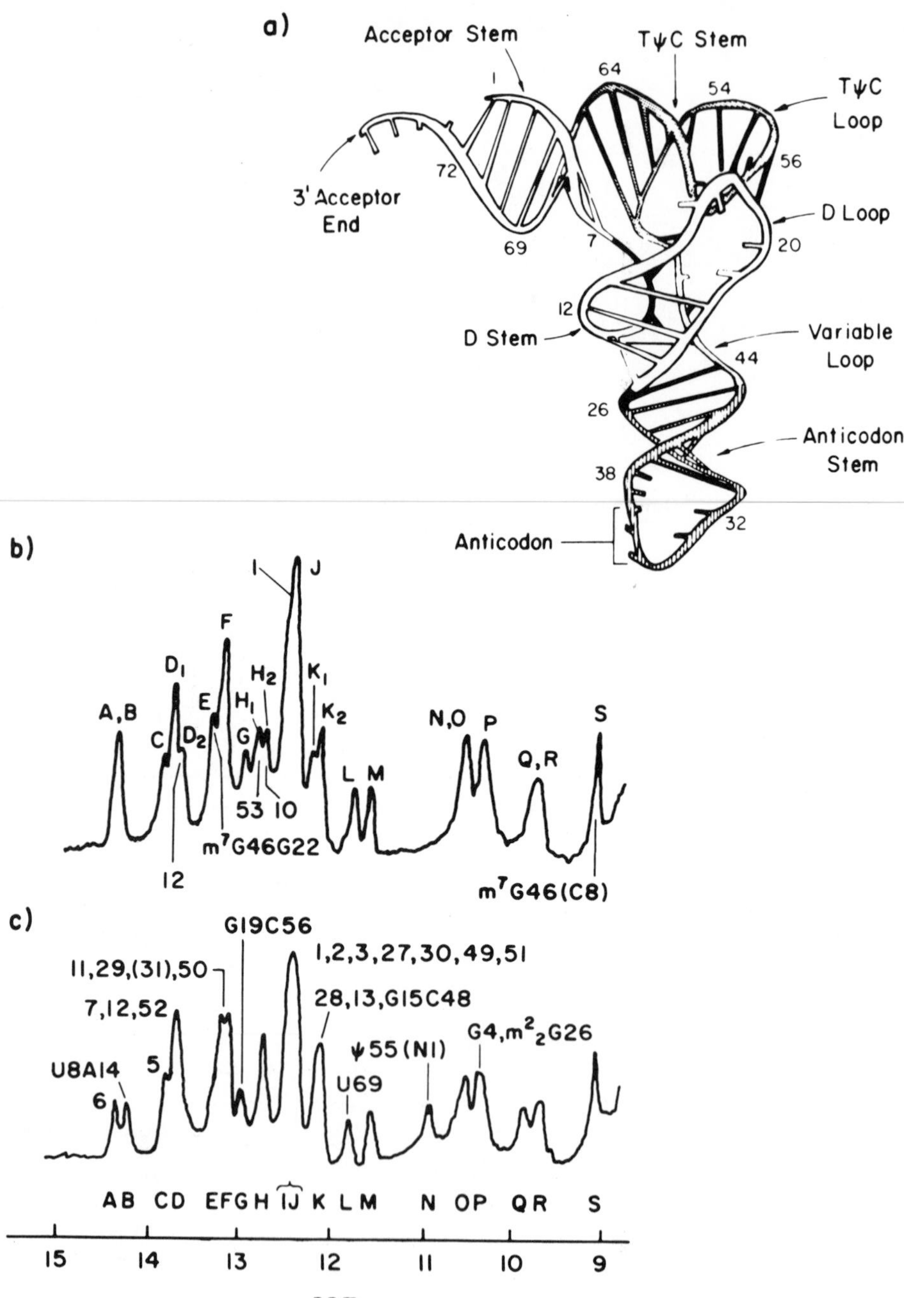

Figure 4. (a) The three dimensional structure of yeast tRNA^{Phe}. From Quigley, *et al., Proc. Nat. Acad. Sci. 72*, 4866 (1975). (b) Downfield spectrum in high magnesium. (c) Same in zero magnesium. Letters A,B,C are labels only. Numbers 1, 2, 3, 5, etc. refer to secondary pairs; tertiary pairs are fully specified. About half of the identifications are highly speculative. From reference 5. Reproduced with permission from *Biochemistry 20,* 1147-1156 (1981). Copyright 1981 American Chemical Society.

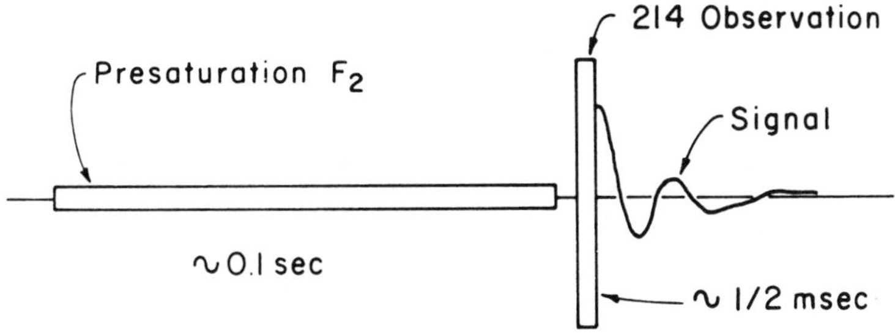

Figure 5. Radio frequency sequence used in our experiments.

spectrum at a frequency f_2 (saturation is similar to bleaching and occurs when a set of spins in the spectrum are externally irradiated, equalizing their populations and eliminating their magnetization and subsequent signal). It is possible to adjust the saturating power so that a single line or even a partly resolved shoulder in the spectrum is predominantly decreased in intensity. For relaxation or exchange measurements we program a delay between the presaturation and the 214 pulse, to observe recovery of the line from saturation, as described elsewhere.[1]

The nuclear Overhauser effect, for larger macromolecules, is similar in many ways to optical resonance transfer: Energy from one species of protons that has been pumped in by saturation can be transferred to near neighbors through the magnetic field that one spin exerts on its neighbors. In smaller molecules this process is complicated but in slowly tumbling macromolecules, it is rather simple. As in optical resonance, energy is transferred at a rate proportional to the inverse sixth power of the distance. However, the requirement that absorption lines overlap is replaced by a weaker condition; and the uncertainty in optical transfer rates due to orientation of dipolar transition moments does not exist in NMR. Thus the result of saturating one line is a decrease in intensity of resonances of nearest neighbors. An example is shown in Figure 6.

By various obvious strategies it is possible to quantify and interpret these effects and obtain the rate of transfer of energy and thus the distance between spins. The extraction of distances involves two uncertainties, namely the tumbling rate of the molecule, and the effects of rapid internal motion; but because of the r^{-6} dependence of the transfer rate, the distance estimates thus obtained remain far more precise than those obtainable by x-ray diffraction.

Unfortunately, the impression has been received by many NMR spectroscopists that the power of NOE in macromolecules is limited because energy pumped in could be passed around among many spins resulting in hopeless confusion; and also that highly selective excitation is very difficult. Neither problem has been serious in our work. We use a saturating pulse length of around 0.1 sec which is the time it typically takes for energy to transfer once between a pair of spins. Pulse sensitivity is a function of power and length, adjusted with the help of experience. Effects of

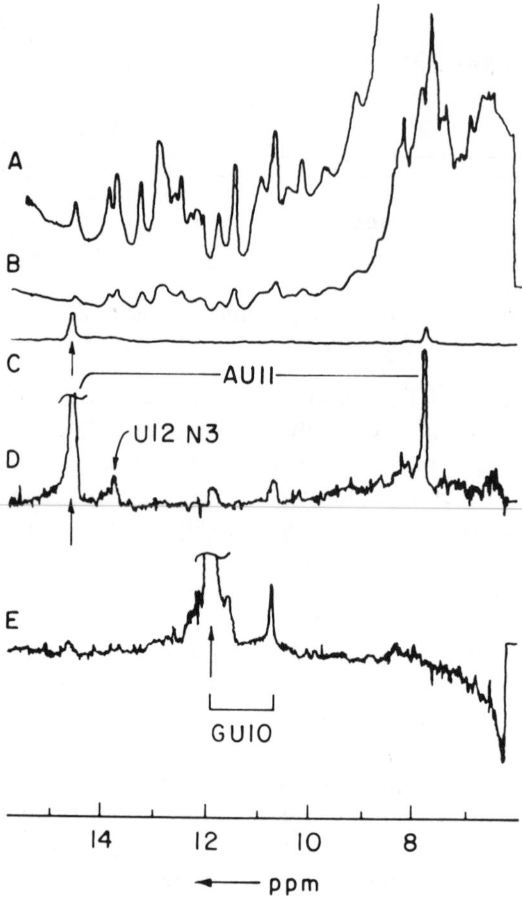

Figure 6. (a), (b) Downfield portion of the spectrum of yeast $tRNA^{Asp}$ at two gain settings. Buffer pH 7, 0.1 M NaCl, no $MgCl_2$. (c), (d) Difference NOE spectrum obtained by subtracting, from the spectrum of (b) a spectrum with presaturation at the most downfield peak. (c) is at the same gain as (b) while (d) is the same data at 16 x higher gain. (e) Similar to (d), irradiating at 12 ppm to show the GU10 NOE. Arrows indicate irradiation points in (c)-(e).

multiple flips (domino effects) are rarely seen in tRNA, nor do they interfere with studies in proteins except perhaps on highly proton-rich side chains. In many cases the existence of multiple flips can be tested by studying magnitudes of several NOE's (see below), leading to interesting results. Their existence can also be inferred in principle by studying the buildup of NOE's as a function of irradiation pulse length, or by similar methods, but so far we find this impractical in tRNA because of the noise limitations on reasonable-length NMR runs.

Distances can be inferred by combining a relaxation-time (T_1) measurement with NOE, as we recently described[7] for tRNA. This is only feasible if multiple spin flip effects are small (which is often true); if both of the resonances involved are in unpopulated regions of the spectrum; and if one of them is resolved well enough for a T_1 measurement. Otherwise, it is necessary to measure the size of the NOE as a function of irradiation time.

We have not yet had occasion to want to know any distance in tRNA with great accuracy so that we have not used this latter method much. However, we have used a primitive version of it with three or four irradiation times such as 50, 100, 200 and 400 msec, to get some idea of transfer rates and thus a fairly good idea of distances, to aid in assignment.

Our discussion would be incomplete without mention of the two-pulse sequence developed by Kumar, Ernst, and Wuthrich[11] to replace the presaturation pulse of Figure 5. Their technique permits a survey of all the NOE's in a molecule, in a single run taking about one tenth the time needed for the equivalent survey using the sequence of Figure 5. They have demonstrated their method in $H_2$0 solvent for a small protein.[12] Extension of their methods to tRNA will most likely lead to technical problems having to do with transfer of saturation from solvent to macromolecule via spin diffusion, but these problems can probably be overcome by use of antiselective pulses similar to the 214 pulse.

Assignment of the tRNA Spectrum: Early Work

It was relatively easy to assign spectral regions to classes of protons: methyl, sugar, aromatic carbon, and amino protons resonate in their normal places.[9] Ring NH protons of Watson Crick pairs were relatively easily assigned to 12-13.5 ppm for GC pairs and 13.2 to 14.5 ppm for AU pairs, on the basis of studies of fragments and models. There are relatively few methyl groups in tRNA's-generally from three to ten-and many methyl lines can be assigned through the similarity of their shifts in monomer and tRNA. The 6-carbon proton of 5-methyl cytidine was assigned on the basis of its unusually rapid exchange (for a carbon proton) with solvent. We have reviewed this early work elsewhere.[9]

However, further progress was difficult. Serious efforts were made to develop semiempirical predictive schemes for NMR shifts in tRNA, and use these together with comparisons between tRNA species to make assignments. These schemes are nearly untested and perhaps untestable for ring N protons in secondary GC pairs, and all appear to fail for such protons in AU pairs.[5] We assume the problem is that minute changes in relative position of A versus U in such a pair produces relatively large changes in electron density around the NH proton.

Ring NH protons of tertiary base pairs, and of naturally modified bases are especially interesting. These resonances have been assigned by looking for specific changes in well-characterized chemically modified tRNA's, or tRNA's to which are bound magnetic ions in specific sites, or by looking for resonances sensitive to Mg^{++}. One resonance, that of the sU8 ring N proton in *E. coli* $tRNA^{Val}$, has been assigned by chemical modification and this assignment tested by NOE's. Otherwise most of these assignments, though reasonable and plausible in most cases, are untested by independent means and should be accepted cautiously.

Solvent Exchange Rates Used For Assignment

Early in the history of tRNA NMR it was suggested, quite plausibly, that protons involved in tertiary interactions might exchange more rapidly, as temperature is increased, then secondary protons, and that such early exchange might be used to help identify this important class of protons. Since it is rather easy to study this early exchange simply by raising the temperature, this remains a useful idea in attempts to understand the spectrum of a newly isolated tRNA. However, there are problems; for one thing it then becomes partly circular to say that the experiment shows that tertiary structure unfolds first. Fortunately, several assignments made in this way have withstood later tests by NOE. However, mistakes have been made: the proton exchange rate for the resonance marked *A* in Figure 4 has very similar temperature dependence to that of the resonance marked *B*. Resonance *A* was once thought to be a tertiary proton resonance because, in part, its exchange rate vs. temperature paralleled that of a group of other early exchanging protons. However, NOE experiments on purine C-8 deuterated samples (below) have shown that it is a secondary proton, and our current view is that it is one of three AU base pairs in the acceptor stem. The acceptor stem most likely melts early under the conditions of Figure 4C (zero Mg^{++}) just after the tertiary structure, as monitored by the GU4 resonance which we identified by NOE.

Two interesting resonances, namely those at 13.65 ppm in yeast $tRNA^{Phe}$ and at 13.6 ppm in *E. coli* $tRNA^{Val}$, are distinguished[3,6] by a markedly slower exchange rate than any other ring NH proton, namely roughly one $(day)^{-1}$ in high Mg^{++} at 15°C. The next slowest-exchanging resonance rate found by us is about one $(hour)^{-1}$. There is no such one-day proton in yeast $tRNA^{Ala}$. We assign this resonance to UA12 in the D-stem because yeast $tRNA^{Phe}$ and *E. coli* $tRNA^{Val}$ have similar D-stem sequences, and this AU base pair is more or less in the core of the molecule, stabilized by interaction with A9 as well as by stacking interactions with surrounding stable pairs. The position of these slowly exchanging resonances is roughly that expected for a secondary AU pair. Yeast $tRNA^{Ala}$ does not have an AU pair in the D-stem. Obviously, this assignment is not rigorous; unfortunately, we have seen no NOE from these resonances, but this non-observation is interesting since it suggests a distorted AU pair compared to most other AU pairs.

Nuclear Overhauser Effect

The first NOE in a macromolecule was observed in 1970 by Gupta,[13] in cytochrome-C, but it took several years before observation of NOE's reached its present explosion. To some extent this lag was due to the requirement of much instrument time. It is not interesting to merely observe an NOE without having any idea what it is due to; and it is of limited interest to find an NOE where it is already expected although such an observation does often increase by one the inventory of identified resonances, for later use. Situations where NOE may be of use are those where unusual types of protons are close (within about 3.5Å) to each other and where a crystal structure exists. Examples which we have used in tRNA (also likely to be applicable to proteins) are: 1. Ring NH protons close to each other; 2. Ring NH protons close to methyls; 3. Aromatic protons very close to methyl protons; 4. Aromatic protons close to ring NH protons; 5. Almost any NOE involving a previously identified resonance; 6. Almost any NOE involving a carbon proton that can be biosynthetically deuterated. We should add that in the case of smaller proteins, decoupling experiments can be used in much the same way as NOE, or combined with it, but that spin-spin coupling effects are masked (or at least have not been identified) in tRNA because of line broadening. It is likely that decoupling will be useful in ^{13}C or ^{15}N labeled tRNA's.

The GU Wobble Base Pair

Early in our NOE surveys Paul Johnston found a unique NOE between two downfield exchangeable resonances in yeast $tRNA^{Phe}$. He concluded that it could only be due to the two ring NH protons of GU4 which is in the acceptor stem. Brian Reid kindly gave us samples of *E. coli* $tRNA^{Val}$ and $tRNA^{fMet}$, both of which have a single GU pair. We found single unique NOE between ring NH resonances for these species also, as described elsewhere.[4] This is the first evidence that the GU pair exists in the form predicted by Crick's wobble hypothesis. The GU pair is not necessarily relatively unstable, as is often assumed: In yeast $tRNA^{Phe}$ it shows early solvent exchange, but in *E. coli* $tRNA^{Val}$, where it is in the T-stem, it does not exchange earlier than the majority of secondary protons.

NOE of Purine C-8 Deuterated tRNA's

Reverse Hoogsteen AU base pairs (Figure 1B) exist at two places in tRNA's, according to the yeast $tRNA^{Phe}$ crystal structure, namely at U8-A14 and T54-A58. There was no way to distinguish these either by chemical shift or by NOE; the Watson Crick AU pairs often have a U ring NH to adenine C2H NOE, but reverse Hoogsteen pairs could have a similar NOE to adenine C8H (see Figure 1B). It was relatively easy to grow both *E. coli* and yeast adenine auxotrophes on C8 deuterated adenine (made inexpensively by heating adenine in D_2O) and see which NOE's from the AU to the aromatic region disappear. A full account appears elsewhere[5] and we claimed that the results disproved the usefulness of all existing ring current shift theories for the prediction of ring NH resonance positions, at least for AU pairs. Of course, such theories are still useful for interpreting possible strain-induced shifts of AU resonances, and for carbon protons. For a discussion of the inadequacy of ring current theory to predict magnetic shielding constants in nucleic acids, see Sarma and coworkers in this volume.

In addition to our published work, we have used comparison of C8-deuterated and native NOE's to help in identification of several other classes of protons, and have verified the previous identification of the sU8-A14 resonance in *E. coli* $tRNA^{Val}$ and $tRNA^{fMet}$.

The general method of using deuterium as a negative label for NOE's should be extendable to other classes of protons in tRNA, and to many problems in larger proteins. Large scale deuteration of amino acid C_α protons can be catalyzed inexpensively with aminotransferases, and aromatic sites deuterated by heating in D_2O. Essential amino acids, at least, could be incorporated into mammals, as well as into tissue culture and microorganisms.

The TΨC Loop

The methyl of thymidine 54 had been identified by Kan and coworkers, aided by the fact that it is the most upfield methyl resonance and is present in all tRNA's except eucaryotic initiator. In *E. coli* $tRNA^{Val}$ there is a large expected NOE from this proton to its neighboring T54 C6 proton and also weak NOE's to protons we identified as Ψ55 N1 and C6 protons, and the G53 C8 proton.[7] The latter was identified by disappearance in tRNA that was purine C8 labeled (guanosine is biosynthesized from adenine in the auxotrophes we use, so all purines are C8 deuterated). Since both the T54 methyl and the Ψ55 NH protons are resolved their T_1's could be easily measured and thus precise distance estimates could be made simply from the sizes of NOE's for a long preirradiation time. The theory for methyl-single spin NOE had to be worked out in order to get methyl to single-spin distances. The resulting distances confirmed the general stacking of T54 and Ψ55 as determined by x-ray crystalography. Distances obtained are more accurate than x-ray distances.

The Ψ55N1 proton resonance is of particular interest. It resonates at 10.65 to 10.95 ppm, depending on Mg^{++} content in the buffer, and is the most Mg^{++} sensitive resonance. It is one of a growing number of resonances (including GU resonances) identified in the relatively unpopulated region from 9 to 12 ppm. We believe that many of the resonances in this region come from ring NH protons that are not internally base-paired but that are somehow stabilized sufficiently by the tRNA structure so that they exchange slowly enough to be observed (a ring NH proton in G or U monomer exchanges with H_2O so rapidly at pH 7 that it would be unobservable). The Ψ55 N1H proton is also unusual in that its solvent exchange rate behaves oppositely from all other ring-NH protons we have studied: the rate increases with increasing Mg^{++}, which stabilizes all other ring NH protons.

Another unexpected result of this research concerns the C8 proton of G53, which we found by NOE at 8.4 ppm in yeast $tRNA^{Phe}$, 6.76 ppm in yeast $tRNA^{Ala}$, and 6.73 ppm in both *E. coli* $tRNA^{Val}$ and $tRNA^{fMet}$. All four tRNA's have similar sequences near G53, judging from the yeast $tRNA^{Phe}$ x-ray structure and the sequences, except that the residue next to G53 is U52 in the first species and G52 in the last three. Conventional ring current theory would not predict nearly such a shift difference resulting from this base substitution.

The Uses of Second Order NOE's

Rather than being a nuisance, we have found second order NOE's to be useful in disentangling spectra. In bovine superoxide dismutase James Stoesz[14] found NOE's from both of two histidine NH proton resonances, previously studied by Lippard and coworkers, to a single histidine C2 proton that had previously been identified by Bannister and coworkers as that of His 41. The natural inference that these were the N1 and N3 proton resonances of the same histidine was confirmed by observation of a second-order NOE between them via the C2 proton. Final confirmation was made by deuterating the C2 position which eliminated these NOE's.

In yeast $tRNA^{Phe}$, we have observed a strong NOE from the previously mentioned N1H resonance of Ψ55, to 7.34 ppm and a weak one to the T54 methyl resonance.[7] A strong NOE was also observed from 7.34 ppm to and from the T methyl resonance. We suspected that the 7.34 ppm NOE's were separate NOE's from the T54 methyl to its C6 proton and the Ψ55 N1 to C6 proton. An alternative could have been that they involved a single proton resonating at 7.34 ppm, and that the weak NOE from the Ψ55 N1 proton to the T54 methyl proton was a second order NOE via this single proton. This explanation was eliminated because the putative second order NOE was too small, i.e., less than the product of the first-order NOE's from ΨN1H to the 7.34 ppm resonance and from the latter to the T54 methyl resonances. This doesn't rule out a second order contribution to the Ψ55 N1H-T54 methyl NOE, but shows that more than one 7.34 ppm proton is involved.

A third application of such ideas occurred in disentangling methyl-downfield NOE's in yeast $tRNA^{Phe}$. The resonance marked m^2_2G26 in Figure 3 is broad and difficult to integrate. It shows a complicated NOE pattern to the aromatic region (Figure 7, top) including a strong one to an exchangeable resonance at 10.4 ppm. We were skeptical of the methyl assignment but confirmed it by isolating yeast $tRNA^{Ala}$ which has m^2_2G26 but fewer other methyl groups. In this latter species we also found a methyl resonance at the same place with an NOE to about 10.4 ppm. In both cases, there is a strong NOE to about 7.8 ppm (see Figure 7, top). These NOE's we attribute, therefore, to the G26 N1H and the A44 C2 protons; another NOE due to the G10 C8 proton is not present in Figure 7 because this is a purine C8-deuterated sample, but is present in unlabelled tRNA (see reference 5, Figure 2d and e). Observation of the G26 N1H proton at 10.4 ppm is unexpected because this is roughly the expected position of an imino proton that is not internally hydrogen bonded.

Thus about half the NOE's seen in Figure 7 (top) are accounted for, but we suspected that the "m^2_2G26" methyl resonance contained some other overlapping resonance. To establish this, we looked at the second-order NOE from the 10.4 ppm

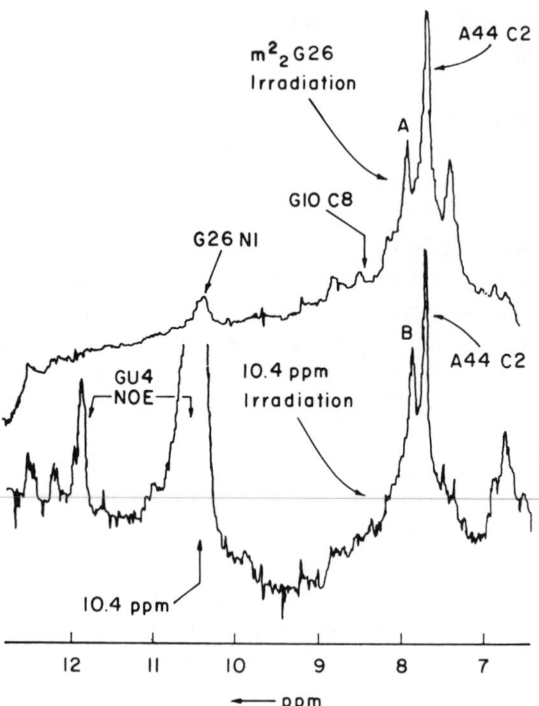

Figure 7. *Top.* Yeast tRNA^{Phe} NOE's observed in the downfield region when the resonance marked m^2_2G26, in Figure 3, is irradiated. The non-NOE marked G10 C8 is missing because the sample is purine C8 deuterated; see reference 5. *Bottom.* NOE pattern observed when the overlapping resonances at 10.4 ppm are irradiated. The aromatic NOE's are discussed in the text. The GU4 NOE was previously established, and overlaps the m^2_2G26 N1H resonance by chance.

resonance via the G26 methyls to the aromatic region (Figure 7, bottom). Two NOE's observed on direct irradiation of the "m^2_2G26" resonance are missing in the second order spectrum. The NOE lines marked A and B in Figure 7 are definitely not in the same place. We do not yet have any definite hypothesis for the identity of the NOE's or resonances not already mentioned above.

Finally, we have seen NOE's from m^5C methyl resonances to the downfield region which are probably third-order NOE's to the N3H imino proton on the same base, via the intervening two amino protons.

The D Stem of Yeast tRNA^{Asp}

This species is unusual in having the sequence of base pairs GΨ AU AU GU in the D stem (Figure 8). This allows a remarkable series of NMR identifications from NOE's between base pairs, rather than within a single base pair as were most of the NOE's previously described. This is possible because these NOE's, which are so weak that we would tend to ignore them as artifacts, are just barely observable and

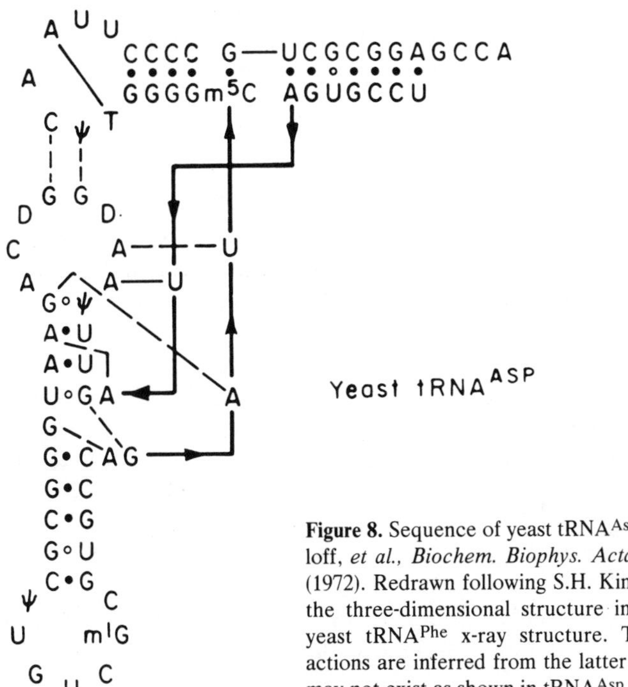

Figure 8. Sequence of yeast $tRNA^{Asp}$, from Gangloff, *et al., Biochem. Biophys. Acta 259*, 210-222 (1972). Redrawn following S.H. Kim to represent the three-dimensional structure implied by the yeast $tRNA^{Phe}$ x-ray structure. Tertiary interactions are inferred from the latter structure and may not exist as shown in $tRNA^{Asp}$.

are interpretable because of the relatively unusual juxtaposition of GU and AU pairs. In addition, the reverse Hoogsteen U8-A14 pair is stacked close enough to the GΨ pair for us to see an NOE between these base pairs. Finally, the m^7G46 tertiary interaction found in yeast $tRNA^{Phe}$ is not present; this eliminates an NOE that could be confused with a GU or GΨ NOE.

This species is also of interest because its crystal structure is being determined by Moras, *et al.*[15] in Strasbourg, France, and our NMR study is being carried out in collaboration with them.

There are four GU pairs (including ΨG13) in this tRNA (Figure 8) and we found three sets of NOE's characteristic of GU pairs (we still have not found the other one). There are two likely reverse Hoogsteen pairs (U8-A14 and T54-A58) and we found resonances of both of them by the criterion that a sharp NOE in the aromatic region, from a resonance far downfield, disappears upon purine C8 deuteration. One of these shows an NOE to one of the GU resonances, which can only be between U8A14 and ΨG13, identifying both of these and, by elimination, T54-A58. A resonance showing the characteristic NOE pattern of a standard secondary AU pair shows an NOE to one of the other GU pairs (see Figure 6C and D), identifying these as UA11 and GU10. Another NOE is also visible in Fig. 6, to AU12. Thus, we have markers extending over the entire D-stem. These NOE's can be used to measure distances in this stem; their intensity implies that the corresponding protons must be more or less facing each other in their respective base-pair planes. Two of these sets of resonance (T54-A58 and the G Ψ resonances) have never been identified previously. T54-A58 resonates at 13.76 ppm, and the GΨ pair resonates at 10.33 and 11.44 ppm.

Conclusion

We have described methods for elucidation of the structure and the NMR spectrum of tRNA. The main strategies are NOE combined with selective deuteration and creative use of second-order effects. Many of these techniques could be used for protein studies.

In view of this progress in the understanding of the tRNA spectrum, it is now more attractive to return to selective studies of proton exchange rates; of conformational change reported by NOE distance measurements; and of tRNA interacting with proteins.

Acknowledgements

This work was initiated in our lab by Dr. Paul Johnston. The *E. coli* mutant used for purine C8 deuteration was kindly supplied by Dr. Barbara Bachmann of Yale University. We thank Drs. J.E. Haber and R. Schleif and their coworkers for help in producing the deuterated tRNA's. A.G. Redfield is also at the Physics Department and the Rosenstiel Center at Brandeis, and N.F. is in the Biophysics Program and was supported by a Gillette Fellowship. Supported by USPHS grant GM20168 and the Research Corporation. This is contribution number 1367 from the Biochemistry Department of Brandeis University.

References and Footnotes

1. Johnston, P.D., and Redfield, A.G., *Nucleic Acids Research 4* 3599-3615 (1977).
2. Johnston, P.D. and Redfield, A.G., *Nucleic Acids Research 5* 3913-3927 (1978).
3. Johnston, P.D., Figueroa, N., and Redfield, A.G., *Proc. Nat. Acad. Sci. USA 76* 3130-3134 (1979).
4. Johnston, P.D. and Redfield, A.G. in *Transfer RNA, Structure, Function and Recognition*, Ed. by P. Schimmel, J. Abelson, and D. Soll, Cold Spring Harbor Laboratory, pp. 191-206 (1979).
5. Sánchez, V., Redfield, A.G., Johnston, P.D., and Tropp, J., *Proc. Nat. Acad. Sci. USA 77* 5659-5662 (1980).
6. Johnston, P.D. and Redfield, A.G., *Biochemistry 20*, 1147-1156 (1981).
7. Tropp, J.S. and Redfield, A.G., *Biochemistry 20* (1981).
8. Johnston, P.D. and Redfield, A.G., *Biochemistry 20* (1981).
9. Schimmel, P., and Redfield, A.G., *Ann. Rev. Biophys. Bioeng.* , 181-222 (1980).
10. Redfield, A.G. and Kunz, S., in *NMR and Biochemistry* Ed. by S. Opella and P. Lu, M. Dekker Co., New York, pp. 225-239 (1979).
11. Kumar, A., Ernst, R.R., and Wuthrich, K., *Biochem. Biophys. Res. Comm. 95*, 1-6 (1980).
12. Kumar, A., Wagner, G., Ernst, R.R. and Wuthrich, K., *Biochem. Biophys. Res. Commun. 96*, 1156-1163 (1980).
13. Redfield, A.G. and Gupta, R.K., *Cold Spring Harbor Symp. Quant. Biol. 36* 405-411 (1971).
14. Stoesz, J.D., Malinowski, D.P. and Redfield, A.G., *Biochemistry 18* 4669-4675 (1979).
15. Moras, D., Comarmond, M.B., Fischer, J., Weiss, R., Thierry, J.C., Ebel, J.P., & Giegé, R., *Nature 288* 669-674 (1980).

Proceedings of the Second SUNYA Conversation in the Discipline Biomolecular Stereodynamics
Volume I, ISBN 0-940030-00-4, Ed. , Ramaswamy H. Sarma,
Adenine Press, New York, ©Adenine Press

Solvation of DNA at 300 K: Counter-ion Structure, Base-pair Sequence Recognition and Conformational Transitions A Computer Experiment

Enrico Clementi and Giorgina Corongiu[61]
International Business Machine Corporation
Poughkeepsie, N. Y. 12602

Introduction

Previously we have reported on the interaction between one water molecule and the bases[1] and base-pairs[2] of the nucleic acids, A-DNA single helix[3], B-DNA single[4] and double helices[4,5]. In addition, *Monte Carlo simulations* (at 300 K) have been presented for a cluster of water molecules enclosing the bases and the base-pairs[2], or a limited region around A-DNA single helix[3] and B-DNA single helix.[4-6] *These studies represent preliminary steps.* We extend our previous effort by considering, via simulations, not only a much larger number of water molecules than previously studied, but also the effect of counter-ions, initially the Na^+ ion. We shall discuss B-DNA interacting with either a few or many water molecules per nucleotide unit, presenting our findings as a *set of three successive approximations.* In this computer experiment the simulated temperature is 300 K. *At first* we considered B-DNA and water molecules; *secondly,* we consider the same system, but to each phosphate unit we add a sodium counter-ion placed at a *pre-determined* position. *Finally* we shall consider a larger B-DNA fragment and the water molecules and the ions will be let *free* to assume the statistically optimal positions and orientations for the temperature analyzed. This third step brings about a newly determined structure for the counter-ions. At each approximation we consider the full range of relative humidities.

This study is applied to *conformational transitions.* On the basis of new findings (counter-ion structure) we present a base-pair sequence *recognition mechanism* and a quantitative simulation.

Graphical Representation of the Probability Density

Some of the analysis reported below is not carried out using probability density maps, as done in the past[2-9] since these are somewhat difficult to read. The probability maps have been replaced by an algorithm described in the following four steps: 1) after computation of the probability density maps, the probability density maxima for the oxygen atoms are located, 2) for the hydrogen atoms, the probability maxima are determined subject to the constraint of being located on a sphere of radius equal to one half of the O-H internuclear separation in H_2O, 3) a sphere of radius 0.5 A is centered at the oxygen and at the two hydrogen atoms probability maxima, and 4) the conformations from the Monte Carlo simulation are scanned to determine how many times a water molecule fell into the volume defined in Ref.3. We note that the assumption of a sphere around the oxygen and hydrogen atoms, implies an isotropic probability distribution; as pointed out previously[2-4] the probability distribution is often anisotropic, especially at room temperature. On the other hand, the advantage of the new method is that it allows a graphical representation of immediate understanding. The technique described here brings 60% to 80% of the full set of the simulated conformations into the three spherical volumes associated to each water molecule. This representation is hereafter referred to as *"the average configuration"*.

The water structure at high humidity is analyzed in greater detail, since it corresponds to a situation expected to be rather near to the one for Na^+-B-DNA in a physiological solution; low humidity is of interest mainly for comparison with physico-chemical (rather than biological) experiments, for example on DNA fibers. The two strands and the constituent chemical groups are differentiated in this paper by presence or absence of an asterisk; thus, for example, P(n) and P*(n), G(n) and G(n)* refer to the n-th phosphate group, or to n-th base (guanine) of either the h or the h* helix, namely the two strands forming the double helix (h and h* are short notation for the 5′-3′ and 3′-5′ strands, respectively). Water molecules are labelled with an index and, in general, a water molecule with an index m has its oxygen atom *above* (projection on the Z axis) any water molecule with an index value larger than m.

We recall that in the first solvation shell one finds water molecules *"bound"* to sites (hydrophilic sites), and water molecules *near* the hydrophobic sites, generally acting as bridges between hydrophilic sites. Thus the *"first solvation shell"* distribution contains the *"bound"* distribution and additional water molecules. The difference between the *"total"* and *"the first solvation shell"* distributions provides the *"grooves"* distribution. It is clearly a matter of taste to speculate how much a "groove distribution" coincides with "second and third solvation shells distributions". We have opted for the "groove" terminology, since it allows further differentiation between major and minor grooves, a distinction of relevance in discussing nucleic acids. Lewin[10] subdivides the major groove's water molecules into three belts (upper, middle and lower); our subdivision is about equivalent, since the water molecules of the upper and lower belts correspond essentially to our *first solvation shell's* water molecules.

We note that since a water molecule experience the entire field of the DNA fragment, there is necessarily an element of ambiguity in deciding if a given water molecule "solvates" only a given site, unless an exact definition is provided. Our selection criteria to assign a water molecule as "bound" to a given atom "a" of the B-DNA fragment are: 1) the oxygen atom of water and the atom "a" internuclear

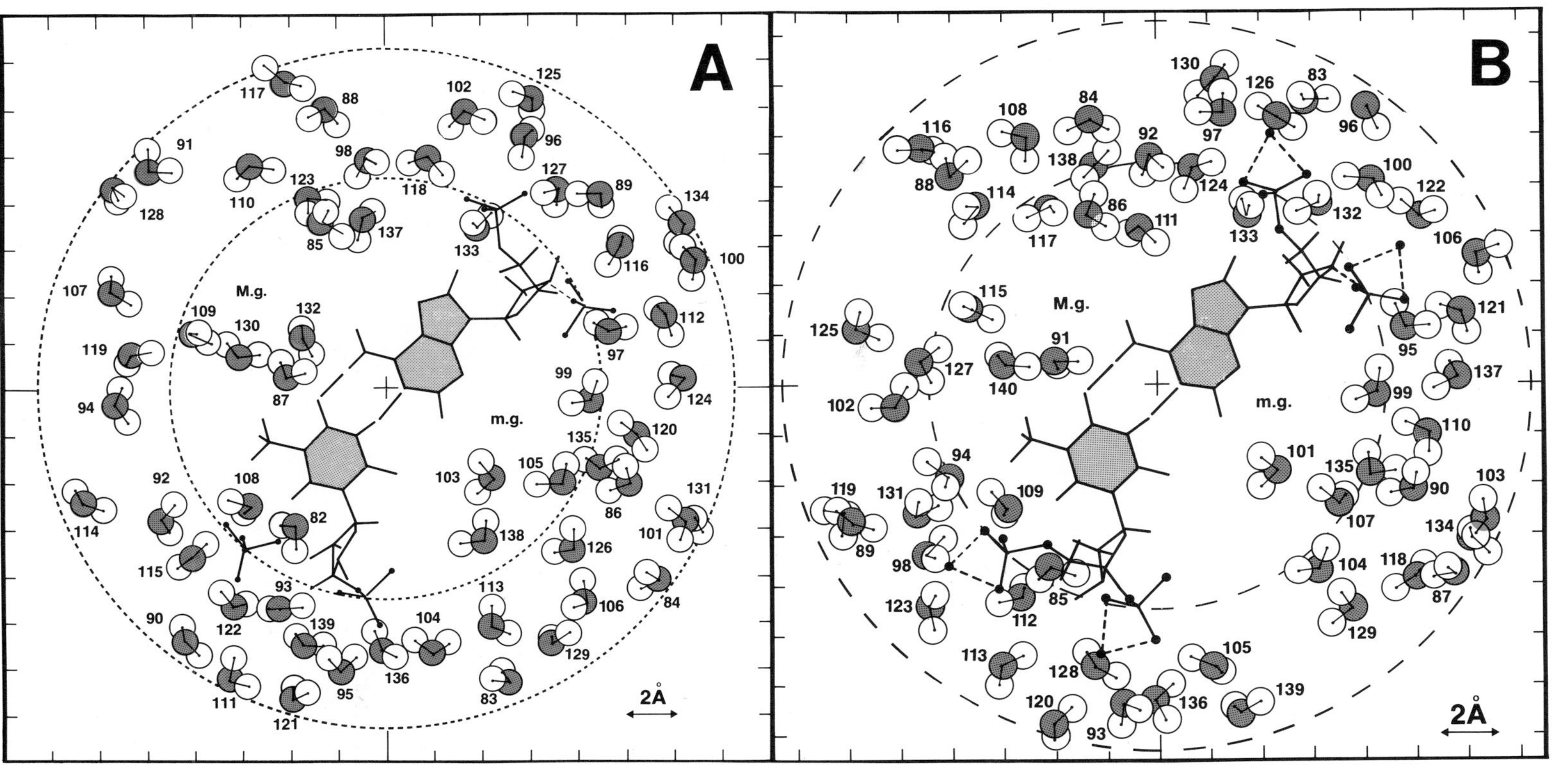

Figure 4. Water molecules contained in a disk of 4 A thickness and solvating B-DNA (insert A) and Na^+-B-DNA (insert B); the radii of the two circumferences are 14 A and 8 A, respectively.

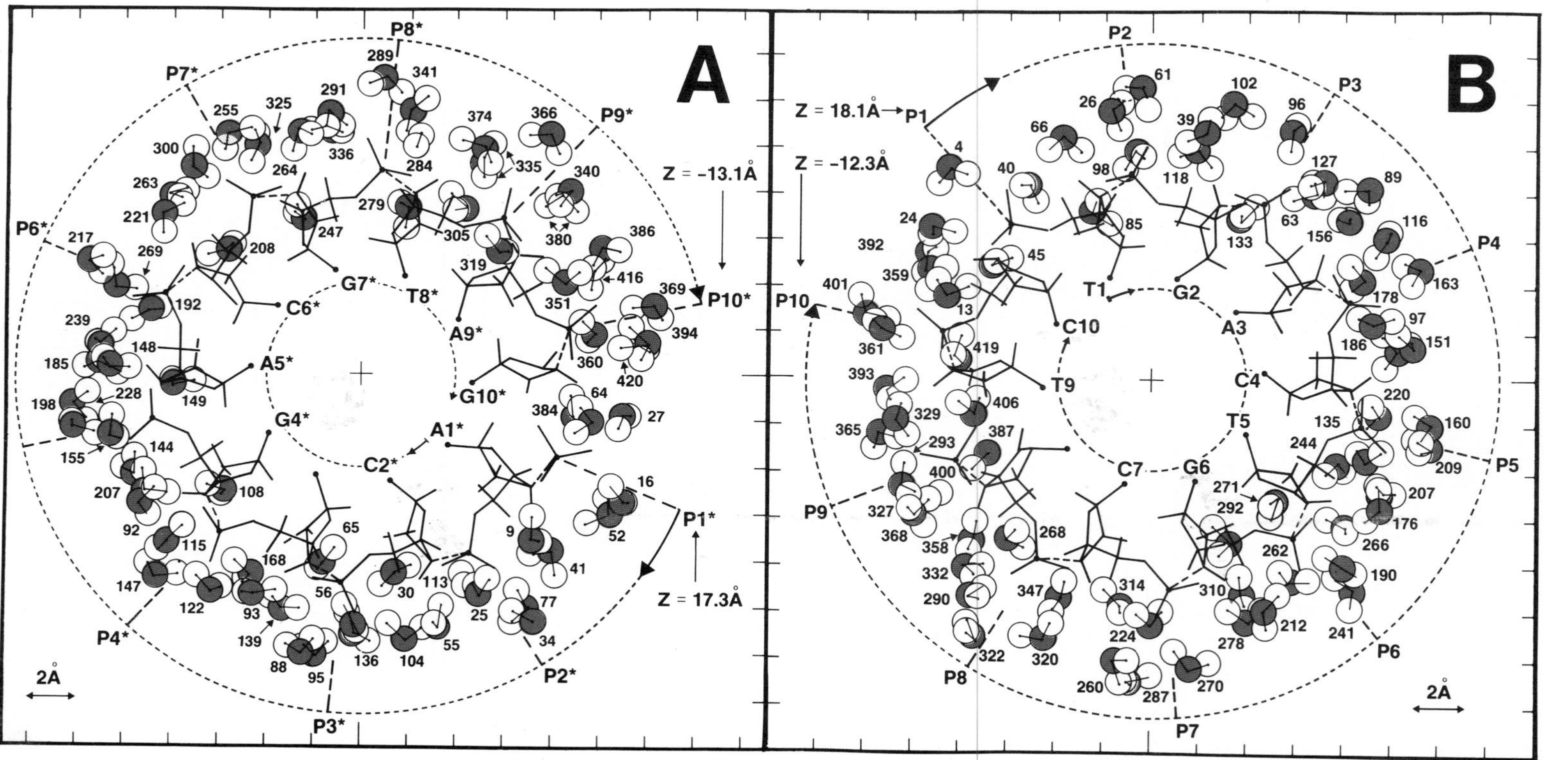

Figure 3. Water molecules solvating the h helix in (insert A) and the h* helix (insert B) in B-DNA.

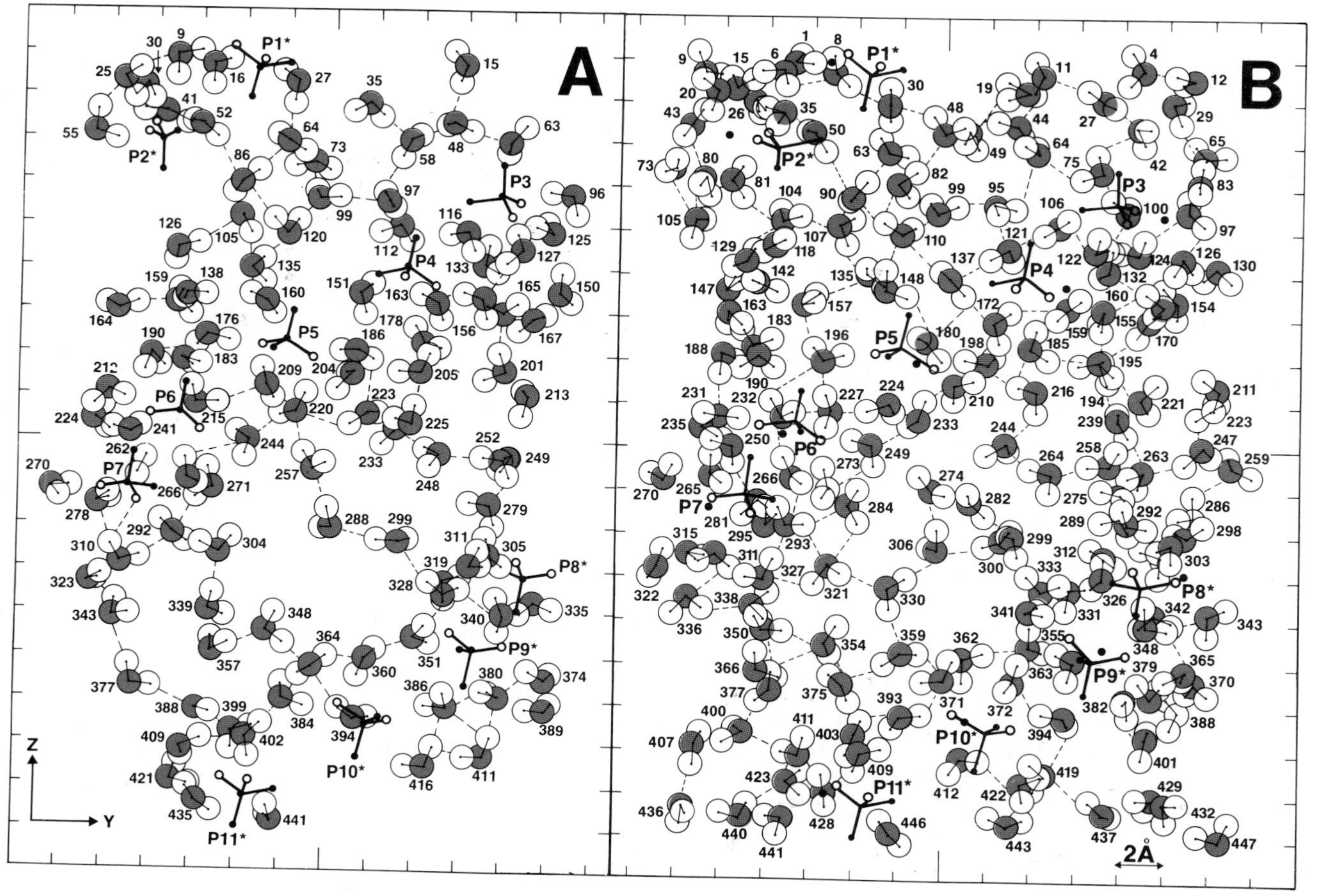

Figure 6. Network of water molecules in B-DNA and Na^{+}-B-DNA (see text).

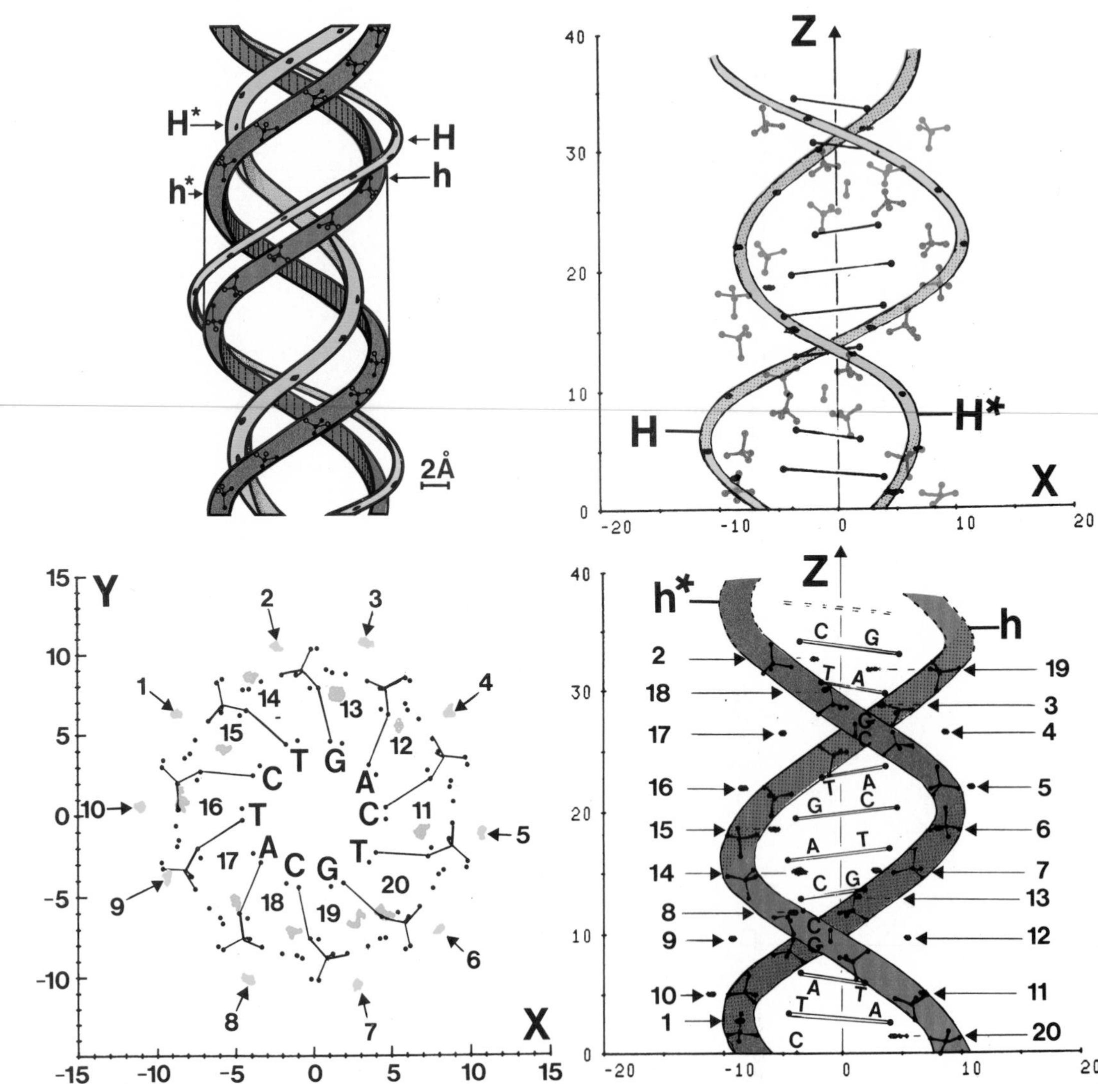

Figure 16. *Bottom-left:* projection of the statistical distribution of the twenty counter-ions in the X-Y plane. *Bottom-right:* same distribution projected in the X-Z plane. *Top-right:* same as bottom-right, but with connecting helices for the ions, rather than the phosphates. *Top-left:* combination of top-right and bottom-right for more than one B-DNA turn.

separation, R(a, O), must be smaller or equal to a threshold value T(a, O) and 2) one hydrogen atom of a given water molecule and the atom "a" internuclear separation, R(a, H) must be compared to a threshold value T(a, H) to ensure the proper orientation of that water molecule relative to the atom "a". In this way we can distinguish the case "a"—H-O from the case "a"—O-H (for example N: (in a base) from H of NH_2).

In Fig.1 we report the projection onto the X-Y plane of the third (A3-T3*) and fourth (C4-G4*) base-pair of our fragment (and the corresponding sugar-phosphate groups) in order to clarify our definitions of "*bound* water", "*first* solvation shell water" and "*groove* water." The thick solid line, *composed* by a family of circumpherences of radius T(a, O) and *enclosing* either A3-T3* or C4-G4*, represents regions of "bound" water molecules. A number of atoms "a" have been identified in the figure with a larger dot to represent the nuclear position and the radius origin. Some of the spheres are inside the thick solid line, like O3′ in A3-T3* and O2P, O5', N3, N5 in C4-G4*. One can notice at a glance the strong overlap of these spheres, and this fact constitutes the main reason for the *"non-additivity"* namely, the total number of water molecules found to be *bound* to the DNA fragment is smaller than the sum of the numbers of water molecules found to be *bound* to the atoms composing the DNA fragment. However the "orientation" criterion imposed on the hydrogen atoms (threshold T(a, H)) reduces considerably the above inconvenience. The shaded areas at the periphery of A3-T3* and C4-G4* refer to water molecules in the *first solvation shell,* but not bound to hydrophilic sites (see for example the area for CH_3 of thymine in A3-T3*). Water molecules in this volume might be counted more than once in our analyses, once as being in the first solvation shell (for example of CH_3 for T3*) and once as being bound to a given nearby site. Finally, in C4-G4* we have presented two PO_3^--Na^+ groups near to guanine, G4*. Notice the overlap of O1P (at one phosphate group) with Na^+ (on the second group), or the overlap of O2P at one group with O5′ at the second group. Thus we stress once more that a water molecule *should not be* physically associated to a *single* site; its location and orientation are due to the *entire* field of the fragment. An assignment is, however, essential in order to interpret experimental data, in particular infrared, Raman and scattering data; and also it is very important to describe and understand the solvent structure.

First Approximation: B-DNA and Water Molecules

The B-DNA double helix fragment we consider has been previously discussed (see Ref.2), and consist of twelve base-pairs (namely, two more base-pairs than needed to reproduce a full B-DNA double helix turn) with the corresponding sugar and PO_4-CH_2 units. The following sequence of base-pairs has been selected: AT*, TA*, GC*, AT*, CG*, TA*, GC*, CG*, AT*, TA*, CG*, GC*. The B-DNA double helix fragment is enclosed into a cylinder with its axis co-axial to the B-DNA long axis (z axis). The cylinder height is 36.0 A with a base radius of 14.5 A. The two base-pairs and the corresponding sugar units have been added in order to improve the interaction field descriptions at the bottom and top ends of the B-DNA frag-

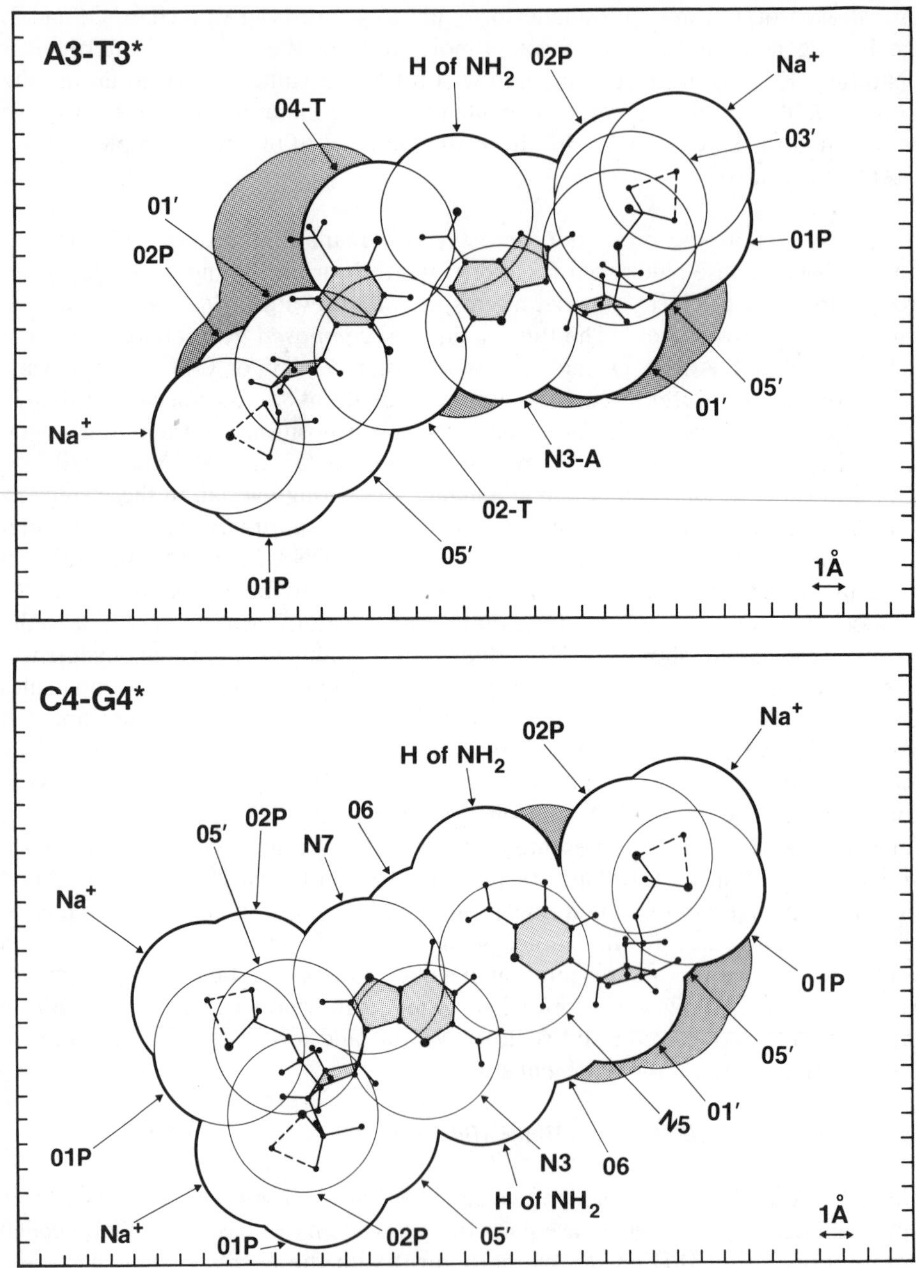

Figure 1. First solvation shell decomposition. Example for two subunits of the Na^+-B-DNA fragment.

ment. In the Monte Carlo simulation below reported 447 water molecules have been placed into the cylinder. The equilibration process was carried out for $2x10^6$ conformations; the statistical data below analyzed are obtained from additional

$2x10^6$ Monte Carlo "moves" (these computations have been carried out on either an IBM-370/3033 or a IBM-370/3081 computer).

Structure of First Solvation Shell

Let us start with a gross analysis of the computed data. In Figure 2 (insert A) we report the distribution of the internuclear distances from an atom (either O or H) of water to the nearest atom of the B-DNA fragment. The internuclear distance is designated as R(i-a), where "i" refers either to a hydrogen or the oxygen atom in water and "a" refers to the atom of B-DNA nearest to "i". We indicate as N (i) the number of atoms of type "i" located (from analyses of the Monte Carlo data) at a R(i-a) distance from "a". The R(O-"a") internuclear distance distribution (see the solid line of Figure 2) shows a very distinct peak with a maximum at 2.6 A (first region), a second peak with a maximum at 3.4 A (second region) and a set of peaks that approximately can be associated to an asymmetrical gaussian distribution with a maximum at about 4.5 A (third region). The first, second and third regions contains approximately 180, 40 and 220 water molecules, respectively.

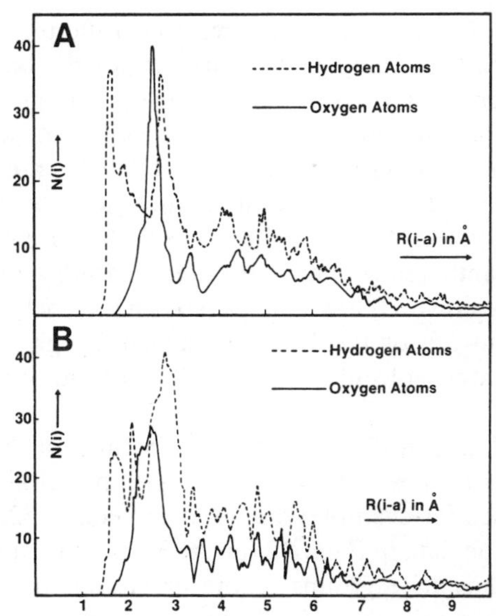

Figure 2. Distribution of oxygen (solid line) and hydrogen atoms (dashed line) internuclear distances from the "a" atom of B-DNA (see text). Insert A for B-DNA, insert B for Na-B-DNA.

The R(H-"a") internuclear distance distribution (see the broken line in Figure 2) is characterized by two very distinct peaks with maxima at 1.8 A and 2.8 A and a distribution of peaks with decreasing intensity for large R(H-"a") values. From these distributions we learn that *most* of the water molecules in the first region are oriented with a hydrogen atom pointing to "a"; since the (H-"a") peak with a maximum at 1.8 A is somewhat broader and a bit more intense than the second one (at 2.8 A), we can also conclude that a small number of water molecules have *both* hydrogen atoms pointing towards B-DNA. The hydrogen and oxygen peaks separations inform us (2.6 A − 1.8 A = 0.8 A and 2.6 A − 2.8 A = −0.2 A) that, in general, for the first region the hydrogen bond forms structures for the type "a"—H-O.

A detailed analysis of the oxygen atom distribution in the first region, indicates that relatively few water molecules have the oxygen atom (rather than a hydrogen atom) in proximity to an "a" atom. In general, we can associate the water molecules of the first region to the first hydration shell. The water molecules of the second R(O-"a") region are less easily defined.

Let us now analyze the distribution of water molecules in more detail, by considering the probability distribution maps. In Figures 3 and 4 we report the water molecules solvating either the phosphate groups in each of the two helices of B-DNA or the water molecules contained in a disk of about 5 A thickness, (slicing the cylinder perpendicular to the z-axis). In the figures, the water molecules are represented with the new algorithm described above; we shall talk of "water molecule number N", as a short expression to indicate "the ensemble of water molecules that falls within the volume number N, consisting of the previously described three spheres of radius 0.5 A."

In Figure 3 (insert A), we report the ten phosphate groups P1 to P10 of h and the corresponding sugar unit, but not the base-pairs, that are, however, indicated by reporting the terminal nitrogen atom and by using the notation, T1, G2, ..., C10. The outmost circumference has 14.5 A radius; the marks on the figure's frame are at 2 A interval. The water molecules are seen from positive values looking down toward negative z values. Only the water molecules very *near* the phosphate P1 (at z = 18.1 A) to P10 (at z = −12.3 A) are reported. Since the "water molecules" are numbered with an index of increasing value along the z direction (from positive z to negative z), low indices (starting from 1) correspond to water molecules solvating the top of the B-DNA fragment, high indices (approaching 447) correspond to water molecules solvating the bottom of the B-DNA fragment.

Notice in Figure 3 the dual features of the water clusters: not only that the water molecules enclose the PO_4^- group, but also they form *hydrogen bonded filaments* (see for example, the water molecules 322, 230, 332, and 358 in the P8-P9 region). The data in Figure 3 (insert B) provides a view of the water molecules solvating the phosphate groups of the h* strand. In Figure 4 (insert A), we report only one base-pair, the A3-T3* base-pair; the water molecules experience the immediate fields of the G2-C2* and C4-G4* base-pairs (not reported in the Figure in order not to further complicate the drawing). Notice, in addition that in the major groove (Mg) there are ten water molecules (82, 108, 87, 132, 130, 109, 123, 85, 137 and 133) whereas only four water molecules are found in the minor groove (mg) (138, 103, 105, and 99); this situation changes when we extend the major groove and the minor groove volume up to a radius of 14.5 A. The physical reason for these findings is that near to the bases there is more "free" space in Mg than in mg, but further out toward R = 13 or 14 A the field generated by the phosphates in the mg

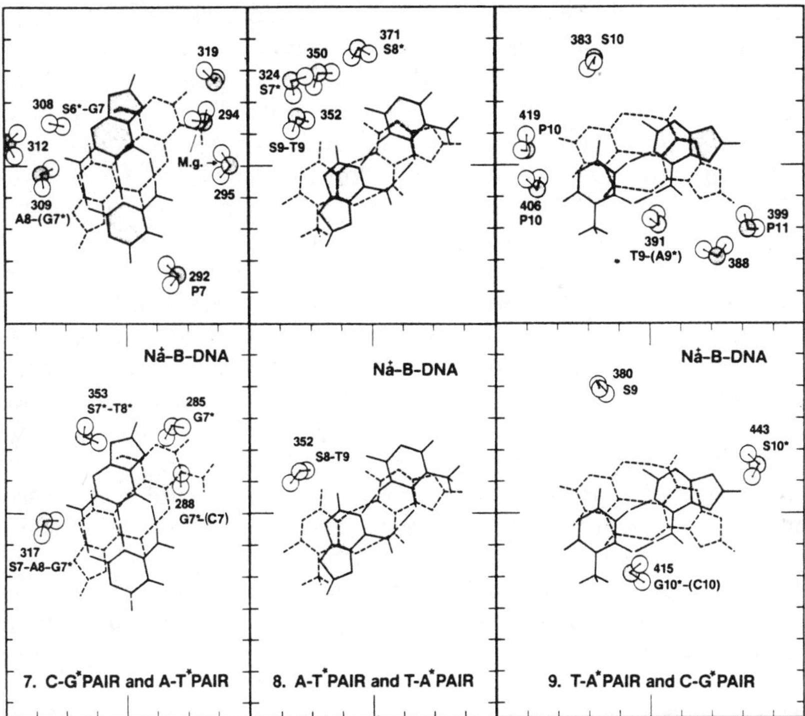

Figure 5. Water molecules in the vicinity of C7-G7*, A8-T8* (insert 7) and A8-T8*, T9-A9* (insert 8) and T9-A9*, C10-G10* (insert 9).

is stronger than the field generated by the phosphates in the Mg. A more quantitative analysis is provided in Tables 1 and 2 of reference 11.

In Figure 5 (top inserts), we report the water molecules solvating three base-pairs. One point should be immediately stressed: we report in these figures not only the water molecules hydrogen bonded to one (or more) bases but also some of those *"nearby"* the bases. For a given water molecule we use the following notation (in addition to the index for the water "volume"): *very strongly* hydrogen bonded water molecules to B-DNA are differentiated from *strongly* hydrogen bonded ones by writing the solvated site either without or within parentheses. Additional figures for the base-pairs are reported in reference 11.

For example, in Figure 5 (insert 7 top) the water molecules reported are in the vicinity of the C(7)-G*(7) and the A(8)-T*(8) pairs; the base-pair (C-G*) is represented with full lines since it is at higher z-value than the lower base-pair (A-T*), represented by dashed lines. The water molecule 292 solvates very strongly the phosphate group in h (P7); water 308 solvates very strongly S6* and also guanine G7 of h. Water 295 is too far from any B-DNA atom to be assigned as solvating a specific group and therefore, it is labeled only as Mg, namely, one of the molecules in the major groove. Finally water 319 is unlabeled, because it is not strongly hydrogen bonded to any atom of B-DNA. Notice that *often one water binds two bases of two successive base-pairs* (for details see reference 11). In conclusion, we have considered three types of water molecules: *very strongly* hydrogen bonded (to one or more atoms of B-DNA) *strongly* hydrogen bonded, and *weakly* hydrogen bonded. More precisely, water reported as *very strongly* hydrogen bonded are those with an O-"a" internuclear distance not larger than 3.2 A. If the hydrogen bond length is larger, then we classify it as a *weak* hydrogen bond. These criteria are rather restrictive and somewhat arbitrary: the PO_4^- field is very intense and a water molecule "strongly" hydrogen bonded to one of the oxygen atoms in PO_4^-, necessarily strongly feels the field of the remaining atoms in the PO_4^- group. In our classification of "hydrogen bond" we have included, as an additional criterion, the requirement that the overall orientation of a water molecule must be one intuitively reasonable; for example, the oxygen (of H_2O) to oxygen (of PO_4^-) internuclear distance must be larger than the hydrogen (of H_2O) to oxygen (of PO_4^-) distance. In Tables 3 and 4 of reference 11 we have reported in detail the water molecules hydrogen bonded to the sugar units and to the bases, respectively. From these data it is clear that a water molecule bridges very often two bases of two nearby base-pairs, less often two bases of the same base-pair and seldom is *bound* only to one base of a base-pair.

Average Interaction Energies

With these restrictive definitions and considering only *strongly* hydrogen bonded water molecules, we summarize as follows:there are 5.9 water molecules solvating each PO_4^- group, 0.3 water molecules solvating each sugar group, 0.5 water molecules solvating both the sugars and bases (namely, hydrogen bonded bridges between a sugar and a base) and 0.9 water molecules solvating a base. These average values refer only to the first solvation shell, in the strict sense above defined; in this way about 160 water molecules out of a total of 447 are considered. The average water-B-DNA interaction energies (in Kj/mol) are -101.9 ± 5.8, -86.9 ± 4.3, -85.9 ± 4.4 and -63.4 ± 4.2 for the PO_4^-, sugar, sugar and base, base, respectively. The average water-water interaction energies are -6.1 ± 3.8, -12.6 ± 3.7, -12.6 ± 3.5 and -16.6 ± 4.3 Kj/mole for the same groups above listed. We recall[12] that the interaction energy in bulk water is -35.6 Kj/mol.

Trans-groove and Interphosphate Water Filaments

The complexity of the structure of water in the two grooves is evident in Figure 6 (left insert), where we consider the water molecules enclosed in the volume between two co-axial cylinders of radii 8 A and 12 A, respectively. The figure reports the water of only one half of the cylinder volume on a y, z projection. To us the striking feature of this figure is *the clear evidence of hydrogen bonded water filaments from a phosphate group of h to a phosphate groups of h*, spanning the major groove, and of hydrogen bonded water filaments connecting two successive phos-*

phate groups, from P(i) and P(i+1) in the h (and/or in h*) helix. We have previously commented on these *filaments*[2-5] which were surmised from the iso-energy maps[5] or determined clearly in the A-DNA and B-DNA single helix Monte Carlo simulations;[3,4] however the limited number of water molecules considered in our previous simulations was limiting the validity of our suggestion. We note that this feature—the filament existence—*has been previously encountered in ion-pairs in solution.*[13] *We feel that this feature is basic in any water solution containing ions, and will have profound consequences to the understanding of dynamical and temperature dependent properties of solutions with ions.*

The hydrogen bonds shown in the figure are reported only if the oxygen-oxygen distance (between two waters) is equal or smaller than 3.5 A, and if the oxygen-hydrogen distance is smaller than the corresponding oxygen-oxygen distance. Typical water filaments (see Figure 6) are formed by the waters 310, 323, 343, 377, 388 and 399 linking P7 of h to P11* of h*; or 271, 266, 292, 304, 339, 357, 348, 364 linking P6 of h to P10* of h*; or 220, 257, 288, 299, 328 and 340 linking P5 of h to P9* of h*. These are *transgroove* filaments. Other structured filaments are present and connect a phosphate to a successive phosphate in the same helix. Notice for example, the water molecules for the *inter-phosphate* filaments 402, 384, 364 and 394 from P11* to P10* and 364, 360, 351 from P10* to P9*. In the above examples waters 310 and 399 are *terminal waters of a transgroove* filament; the structure in the *inter-phosphate* filaments is different since the filament 402, 384 and 369 (from P11* to P10*) *continues* with waters 360, 351 leading to P9*. It is stressed that these structures *do not correspond to data obtained by analyzing one or few conformations, but are statistically "stable" and meaningful structures.* It is very tempting to postulate that proton "are transferred preferentially along these filaments." Hence, these filaments are of importance in reactivity studies. *These structures are "dynamical" in the sense that a given structure can evolve into a different structure, at relatively little expense for the total energy of the system.* We refer to a number of early experimental studies[14-25] and to a review paper[26] for comparison to our data.

The experimental sequence of the hydration at different molecular sites[14] is reported to be first at the two free oxygen atoms of PO_4^-, then at the two bonded oxygen atoms of PO_4^-, then at the oxygen atom of the sugar and finally at the bases. These findings are both corroborated by our simulation and complemented with energetic data on the water-site interaction energy. We find that the most attractive site is PO_4^-, followed by the sugar unit and finally by the bases. From our results (and reference 11) one can distinguish between the two free-oxygen atoms and the two bonded oxygen atoms in PO_4^-. However, one must keep in mind the fact that the PO_4^- field is very attractive for water molecules and therefore, a sharp distinction on the attraction for a water molecule from one of the two types of oxygen atoms *independently from the other* is not too meaningful. On the other hand, we recall that the field near the two free oxygen atoms is reinforced by the field of the two bonded oxygen atoms, whereas the field at one of the two bonded oxgyen atoms is weakened by the field of the -CH_2 group. These qualitative data are analyzed below in a quantitative way.

The number of water molecules solvating the PO_4^- group at room temperature has been estimated[14] to be between 5 and 6 in DNA neutralized by Na^+ counter-ions. Our simulation yields 5.9 water molecules for B-DNA double helix without counter-ions.

Indirect experimental evidences (angular distribution of near elastic scattering by neutron diffraction[27]) suggests the existence of transgroove water molecules (preferential orientation along the main DNA axis).

Second Approximation: B-DNA, Water Molecules and Na^+ Ions at Fixed Position

In the previous sections we have considered aspects of the solvation problem for a B-DNA double helix fragment. In this section the PO_4^- units have been neutralized with Na^+ counter-ions. As known the double helix either in solution or in fibers, is stabilized by the presence of counter-ions. As previously done, we use the Monte Carlo techniques and *ab-initio derived atom-atom pair potentials*[1,12,13,28-30] and a variable number of water molecules (up to 447) to simulate the hydration at several relative humidity values; all computations are performed at a simulated temperature of 300 K.

For the B-DNA fragment's geometrical characterization, we refer to the previous sections and to the original reference[31]. The Na^+ ion is placed near the free oxygens of PO_4^- at its minimum energy position (the Na^+ atom-atom pair-potentials with model compounds containing the PO_4^- group are obtained in our standard way[29]). *The ion is kept at this fixed position during the Monte Carlo simulation;* this is a non-unreasonable approximation, because the very strong attraction from PO_4^- and Na^+ yields a well pronounced, localized and deep minimum. Since, however, *alternative positions for the Na^+ counter-ions could be considered,* as later discussed in this paper, we present this computation as a *model* study rather than as a *realistic simulation,* which would require the Na^+ counter-ions positions not to be fixed as input but to be determined by the Monte Carlo technique.

In Figure 2 (insert B), we present the histogram for the number of oxygen (or hydrogen) atoms, N(i), (i=O or i=H) having a distance R(i-a) from the nearest atom "a" of Na^+-B-DNA. Comparing this histogram, with the equivalent one for B-DNA (see Fig.2) we notice a broad oxygen atom's peak with its maximum at about R(O-a)=2.5 A and extending up to somewhat more than 3.0 A and three (no longer symmetrically placed) peaks at about R(H-a)=1.8 A, 2.1 A and 2.8 A. Taking as first shell all the water molecules up to about R(i-a)=3.0 A, we conclude that there are more water molecules in Na^+-B-DNA's first shell than in B-DNA. The reason for the hydrogen atom orientation is to be found in the different orientation assumed by a water molecule in the field of Na^+ ions[8,9,13] relative to the orientation in the field of the PO_4^- groups[5]. From Figure 2 we note a richer fine structure in Na^+-B-DNA than in B-DNA, especially in the region between R(i-a)=3.0 A and R(i-a)=9.0 A, suggesting *a more structured water pattern;* in the region R(i-a)=0.0 A to R(i-a)= 3.0 A the integrated area of the distribution N(i) versus R(i-a) is larger than for the equivalent histogram in B-DNA, giving evidences of more densely packed water. The *ion-induced compression effect* is fully expected on the basis of the energetic differences for a water molecule in Na^+-B-DNA, relative to a water molecule in B-DNA as discussed in Reference 11. Less qualitatively, we obtain for B-DNA

$$n(O) = \int_{R_1}^{R_2} N(O)\, dR = 176 \qquad \text{for R1} = 0.\ \text{A and R2} = 3.0\ \text{A}$$

$$n(H) = \int_{R_1}^{R_2} N(H)\, dR = 373 \qquad \text{for R1} = 0.\ \text{A and R2} = 3.0\ \text{A}$$

and for Na^+-B-DNA

$$n'(O) = \int_{R_1}^{R_2} N(O)\, dR = 211 \qquad \text{for R1} = 0.0\ \text{A and R2} = 3.0\ \text{A}$$

$$n'(H) = \int_{R_1}^{R_2} N(H)\, dR = 418 \qquad \text{for R1} = 0.0\ \text{A and R2} = 3.0\ \text{A}$$

Clearly, the corresponding integrals from R1 = 3.0 to R2 = 14.5 A are:

for B-DNA

$m(O) = 447 - n(O) = 271$
$m(H) = 894 - n(H) = 521$

for Na^+-B-DNA

$m'(O) = 447 - n'(O) = 236$
$m'(H) = 894 - n'(H) = 476$

In conclusion, considering a distance up to R=3.0 A from the nearest atoms of the solute, we find 176 water molecules (and 21 residual hydrogen atoms, belonging to other water molecules) in B-DNA; this number increases to 209 (and two residual oxygen atoms) in Na^+-B-DNA. Alternatively stated, the 22 Na^+ ions in B-DNA have crowded in about 30 additional water molecules into the first solvation shell (or about 1.5 water molecules per Na^+ ion) relative to B-DNA. It is noted that the distance of 3 A, obtained from Fig.2 appears also in the early literature on DNA solvation, particularly in the notable work by Lewin[10].

An obvious consequence (as we shall later discuss in detail) of the ion-induced *compression* effect is that the double helix conformation is *sterically stabilized relative to a less packed situation,* (for example, B-DNA without counter-ions). A second obvious consequence is that the ion-induced compression effect must be some function of the counter-ions interaction energy with water, *thus it will vary from ion to ion.* Experimentally, it is well known that temperature, relative humidity, ionic concentration and specificity are the variables capable to induce conformational transition in DNA (see below).

A complementary global analysis on the structure of water in Na^+-B-DNA and in B-DNA can be obtained by considering the number of water molecules enclosed in the volume between two coaxial cylinders, with the main axis coincident with the B-DNA fragment's main axis (Z direction). We designate as R(i) and R(o) the radius

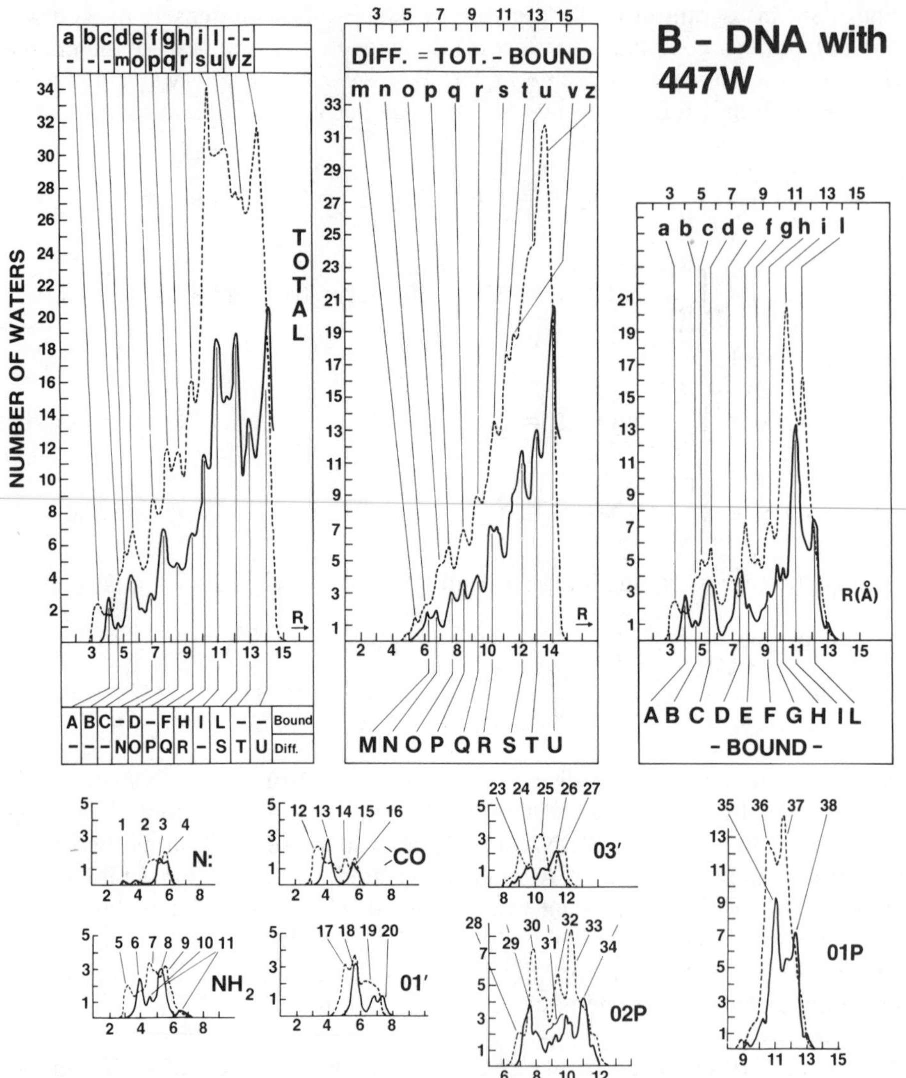

Figure 7. Probabily distributions for water molecule's hydrogen and oxygen atoms as function of R solvating B-DNA and Na^+-B-DNA. The three inserts at the top refer to total, "remainder" and "bound" water distribution (left to right). Figure 7 is continued to right side page.

of the inside and of the outside cylinders, and we select $\Delta R = R(o) - R(i) = 0.2$ A. The resulting diagrams for the hydrogen probability distribution as function of R (dashed line) and for the oxygen atom (full line) are reported in Fig.7 for B-DNA and Na^+-B-DNA. These diagrams can be compared also with the iso-energy contour maps, previously reported for A-DNA, B-DNA double or single helix[2-6,29].

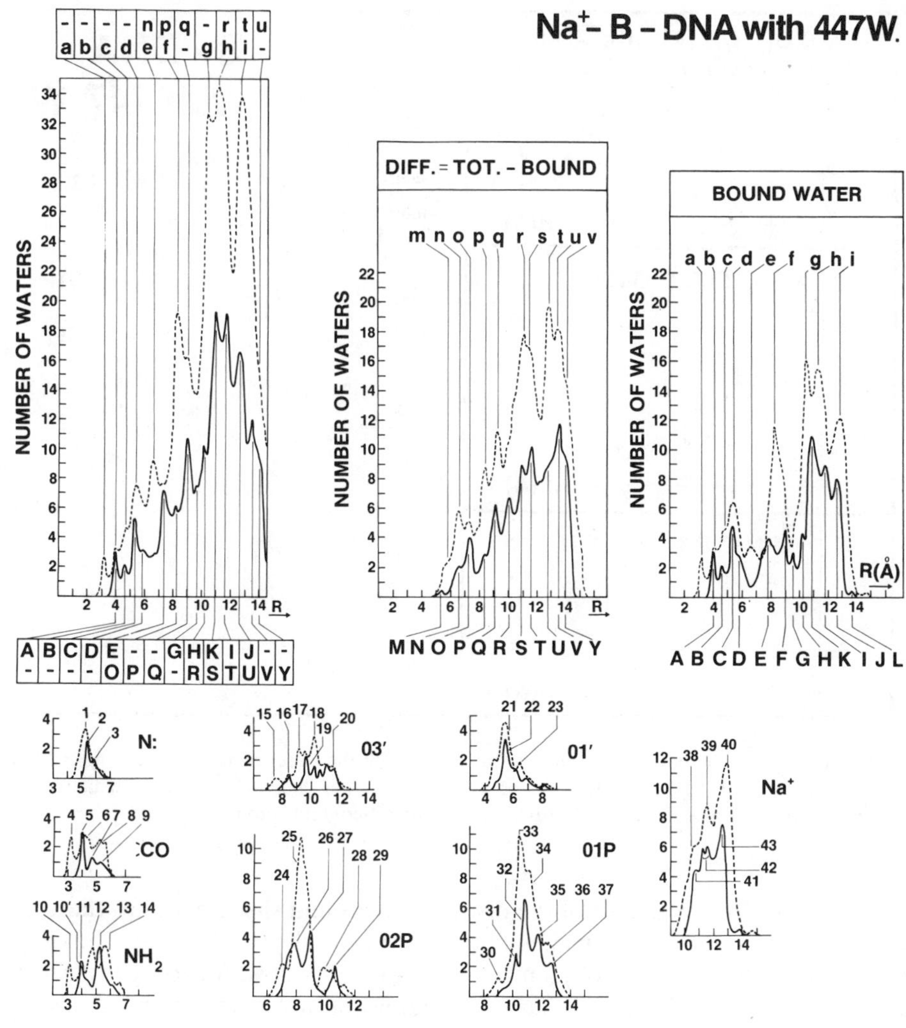

Continuation of Figure 7 from left-hand side page.

In the figures we report the probability distribution for the 447 water molecules (*total"* distribution), a partial distribution related to the water molecules bound to hydrophilic sites ("bound" water molecules distribution) and a second partial distribution *("remainder")* defined as the difference between *"total"* minus *"bound"* distributions.

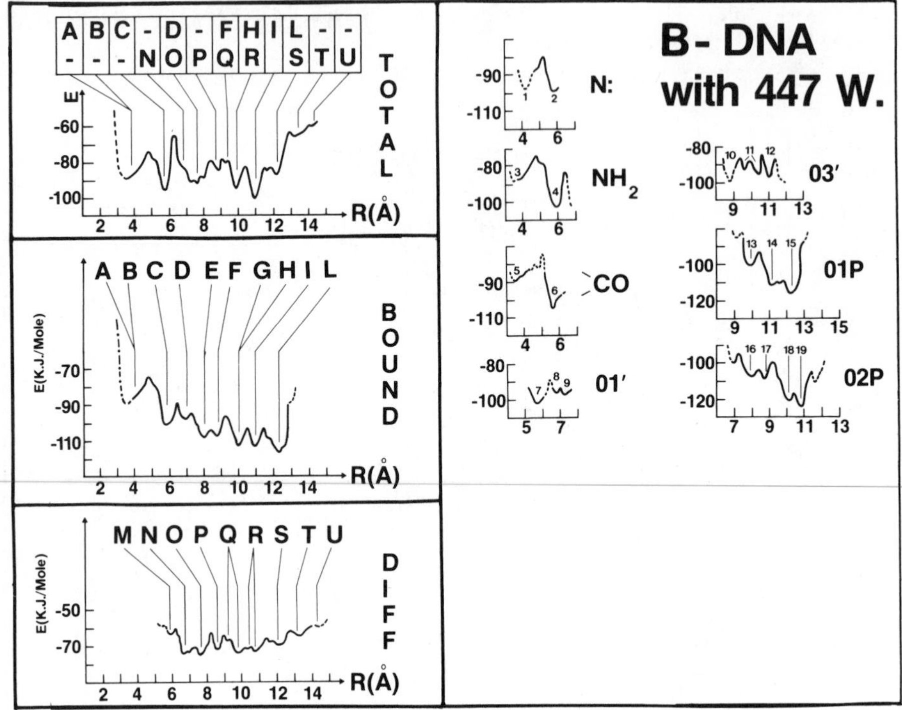

Figure 8. Energy distributions (in Kj/mol) versus R(in A) for the water molecules solvating B-DNA at 300 K. Top left, middle left and bottom left for *"total"* distribution, "bound" water and "remainder"; the other inserts refer to specific "sites" distributions. Figure 8 is continued to right-side page.

In Figure 7 we report details of the distribution around specific groups (hereafter, referred as *"sites analyses"*), namely the N:, NH_2, CO sites of the bases, the oxygen O1′ of the sugar, the bound oxygen O3′ and of PO_4^-, the two free oxygens O2P and O1P of PO_4^-, and Na^+, the O5′ has no bound water molecules. In the abscissa we report R in A and in the ordinate we plot the number of oxygen (or hydrogen) atoms contained in the previously described volume element defined by R(i) and R(i)+0.2 A. Since a given water molecule (as previously pointed out) can belong to the first solvation shell of more than one atom (of the solute) the distributions of the water molecules for the N:, NH_2, CO, O1′, O3′, O2P, O1P and Na^+ groups are *not* additive.

In Figure 7 the oxygen atom peaks are identified with capital letters, A, B, ... U and the hydrogen atoms peaks are identified with lower case letters a, b, ... z. For the site's diagrams the peaks are indicated with numerals.

Before analyzing the distribution data in Figure 7, we explain the main features of Figure 8. For each probability distribution (Figure 7) there is a corresponding energy distribution, reported in Figure 8. The energy units are Kj/mol. The minima

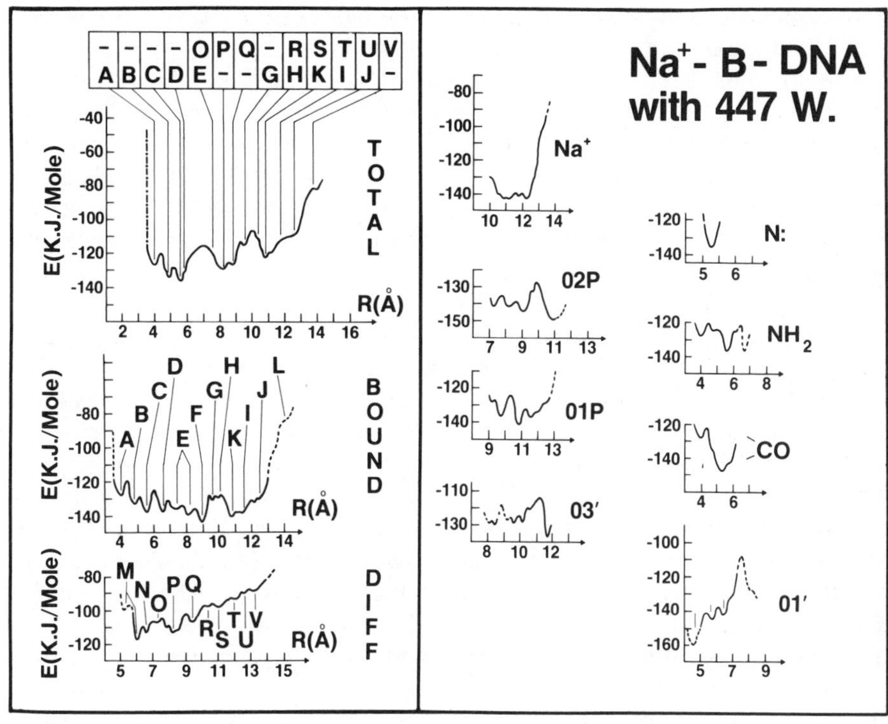

Continuation of Figure 8 from left-hand side page.

in the energy distribution are given by capital letters for the *"total", "bound"* and *"remainder"* waters, by numerical indices in the "sites" analyses. The energy distribution are given as full lines or as dashed lines depending on the statistical significance of the energy. For example, in Figure 8, the energy distribution for the N: atoms (bases) shows two minima, designated as 1 and 2; the corresponding probability density distribution (Fig.7) is low in the interval R=2 A to R=4 A, therefore the corresponding energy distribution is statistically not too meaningful, and is reported as a dotted line rather than a full line.

The analyses of Figures 7 and 8 is facilitated by recalling the R value of few important groups or atoms of the DNA fragments (where R is the distance of a given DNA atom from the long axis of DNA). For *adenine* the two hydrogen atoms of NH_2, are at R=1.8 A and R=2.8 A and the lone pair nitrogen atoms are at R=3.2 A (N3) and R=4.0 A(N7). For *cytosine* the two hydrogen atoms of NH_2 are at R=2.1 A and R=3.7 A, and the oxygen atom at R=3.7 A(O2). For *guanine* the two hydrogen atoms of NH_2 are at R=3.0 A and R=4.0 A, the two lone pair nitrogen are at R=3.3 A (N3) and R=3.9 A(N7) and the oxygen atom at R=1.7 A(06). For *thymine* one oxygen atom is at R=2.8 A(04) and the other is at R=3.6 A(02). These are not the possible bonding sites for the single bases or the base-pairs in solution, but for the base-pair in B-DNA double helix. As known, more sites are known in solution[2] for separated base-pairs, but these are not present in B-DNA mainly due to steric hindrance because of the base-pair stacking and the presence of the sugar groups. For the sugar the oxygen atom O1′ is at R=6.2 A. For the *phosphate groups* the two free oxygen atoms are at R=10.2 A (O1P) and R=9.1 A(O2P), whereas the two bound oxygen atoms are R=6.7 A(O5′) and at R=8.8 A(O3′); we recall that O5′ is near the CH_2 hydrophobic group. The *sodium* ions are placed as previously discussed at R=10.7 A.

With all the above in mind, we can now analyze the data of Figures 7 and 8. Let us start by comparing the *"total"* distributions of the 447 water molecules in B-DNA and in Na^+-B-DNA (top left inserts in Figure 7). We notice a set of peaks for the hydrogen and for the oxygen atoms; the two higher peaks for the oxygens (peaks K and I) at R=10.8 A and R=11.7 A for Na^+-B-DNA can be compared to the highest peak, U at R=14.2 A, in B-DNA. The shift confirms the ion induced *compression effect,* previously mentioned in analyzing Figure 2 (insert B). Equally evident is the main overall feature in the hydrogen distribution (relative to the oxygen distribution); the hydrogens are shifted toward smaller R values in B-DNA than in Na^+-B-DNA; in addition the B-DNA and the Na^+-B-DNA patterns of peaks are different. A peak by peak identification can not be made without reference to the distributions of the *"remainder"* water (Figure 7, top middle inserts) and of the *"bound"* water molecules (Figure 7, top right inserts). By integration of the "bound" distribution we find that in B-DNA 157±2 water molecules out of the 447 are "bound" water. The "remainder" water molecules population increases with R (as fully expected), with a characteristic pattern (peaks M to U in B-DNA, M to Y in Na^+-B-DNA). By comparing Na^+-B-DNA with B-DNA we notice that the "remainder" remarkably shows the ion induced compression effect. This is to be expected; indeed these molecules can be compressed more readily than the bound molecules, the latter being trapped at the sites, that is, at the most intensive and attractive field region, as shown by the energy distributions for the water molecules in the "remainder" relative to those in the "bound" (see also Figs.11 and 12, the three inserts at the left side).

The presence of the counter-ion brings about another global effect: in Na^+-B-DNA the *bound* water molecules have approximately a constant energy attraction from R=4.0 A to R=12.0 A, whereas in B-DNA the bound water molecules are more attracted at R=13.0 than at lower values of R. *Thus, the counter-ion brings about an over-all increase in the attraction for all the water molecules, and this effect is more prominent for the water molecules near the base-pairs than for those at large R values.* Another important feature is that the "remainder" water molecules show the lowest energy at R=6.0 in Na^+-B-DNA, (minima M and N) whereas the lowest energy is between R=6.0 A and R=11.0 A in B-DNA (minima N, P, Q, and R). This observation brings about at least two conclusions. *First, a molecule* (for example, a carcinogenic or an anti-carcinogenic molecule with polar groups) *is expected to be "pressed" toward the base-pair by the field of the counter-ions, displacing groove's water molecules. Second, theoretical computations on carcinogenic or anti-carcinogenic compounds performed on models where only a single base or a base-pair are considered, rather than a full fragment of DNA with its counter-ions might be not too relevant to problems related to the DNA interaction with such compounds,* unless, case by case, it has been quantitatively demonstrated that the solvent and counter-ion effects are small relative to the computed interactions between DNA and the chosen compound (at the temperature in consideration).

Let us continue with the analyses of the data in Figures 7 and 8; in this section we shall focus on the *"bound"* water molecules. To understand the density distribution peak by peak and the energy minima, we have to consider the *"sites* analyses" reported in Figure 7 and 8. Let us start with B-DNA (Figure 7, top right insert and the seven small insert at the bottom). *Peak A* is formed by peak 6 of NH_2 and by peak 13 of CO; *peak B* is formed by peak 11 of NH_2; *peak C* is formed by peak 3 of N:, peak 8 of NH_2, peak 18 of O1′ and peak 16 of CO; *peak D* is formed by peak 29 of O2P, and peak 20 of O1′; *peak E* is formed by the right shoulder of peak 29 of O2P (not identified by a number); *peak F* is formed by peak 24 of O3′, by peak 31 of O2P (the peaks at R=8.8 A, 9.3 A, 9.8 A and 10.2 A have been collectively labelled as peak 31 in O2P) and by the beginning of peak 35 of O1P; *peak G* is formed mainly

Table I

Solvation of B-DNA without and with counter-ions at T=300 K:
water's population bound to sites and its interaction energy (Kj/mol)*

#	Site's Name	B-DNA		Na+-B-DNA		# of Sites
		Waters	Energy	Waters	Energy	
1	O1P	62.29	−108.18	47.47	−133.54	19
2	O2P	46.44	−111.12	35.91	−141.97	19
3	O5′	1.59	−90.40	0.95	−131.25	19
4	O3′	16.49	−91.09	17.56	−124.92	19
5	Na+	------	------	65.87	−134.73	19
6	O1′	15.29	−98.68	16.02	−139.87	19
7	NH_2 (A)	5.53	−83.93	4.50	−126.93	5
8	NH_2 (C)	9.84	−79.96	10.56	−120.87	5
9	NH_2 (G)	4.86	−93.80	5.58	−141.66	5
10	N3 (A)	2.27	−85.10	1.78	−136.07	5
11	N7 (A)	0.95	−84.63	1.34	−121.55	5
12	N3 (G)	2.54	−98.23	2.68	−139.84	5
13	N7 (G)	1.16	−83.46	2.45	−120.66	5
14	02 (T)	1.98	−108.00	2.96	−151.03	5
15	04 (T)	4.58	−80.10	4.87	−125.50	5
16	02 (C)	2.00	−98.51	2.00	−142.77	5
17	06 (G)	5.16	−86.58	5.26	−126.29	5
18	Boundary (type 1)	14.99	−106.94	11.62	−130.72	---
19	Boundary (type 2)	0.00	------	7.78	−130.84	---
20	Total	158.87	−103.87	192.52	−133.75	

*In the last column we report the number of times a given site is present in the fragment (excluding the top and bottom which are considered in the boundaries).

Table II
Solvation of B-DNA without and with counter-ions at T=300 K:
population of water bound to groups of sites and its interaction energy (Kj/mol)*.

#	Groups of Sites	B-DNA		Na+-B-DNA	
		Waters	Energy	Waters	Energy
1	O1P, O2P, 05′, 03′	113.36	−107.35	93.54	−135.84
2	O1P, 02P, 05′, 03′, Na+	------	------	141.70	−134.93
3	NH_2 (A, C, G)	17.56	−84.39	18.72	−127.56
4	N: (A, G)	6.92	−89.57	8.25	−130.36
5	O (C, G, T)	11.32	−90.99	12.96	−134.09
6	Adenine	8.45	−83.90	7.62	−128.12
7	Cytosine	11.83	−83.09	12.56	−124.36
8	Guanine	13.25	−91.00	14.81	−132.78
9	Thymine	6.56	−88.52	7.83	−135.15
10	A-T	12.56	−86.60	13.87	−132.55
11	G-C	21.14	−87.15	22.76	−129.18
12	A-T, G-C	24.56	−87.04	30.04	−130.16
13	A-T, G-C, O1′	31.15	−89.20	35.56	−130.17
14	(ph-S) (base-pair) (S-ph)	15.05 ± 0.10		19.00 ± 0.10	

*See Table I (last column) for number of sites (and, therefore, of groups of sites).

by peak 31 of O2P, peak 24 of O3′ and peak 35 of O1P; *peak H* is formed by peak 35 of O1P, by peak 31 of O2P, and by peak 26 of O3′; *peak I* is formed by peak 35 of O1P (the maximum of this peak), by peak 34 of O2P and by peak 26 of O3′; finally, *peak L* is formed by peak 38 of O1P (the low sub peak of L corresponds to the low sub-peak of peak 38 of O1P (at R=13.0 A)). Thus each feature of the "bound" water distribution is now *fully identified.*

In Tables I to IV we condense the finding discussed up to now. In these tables additional "site" decompositions are presented relative to those given in Figure 7, mainly to further characterize those sites known to be of basic importance in problems related to carcinogenic activity and intercalations. We have *excluded* from the analyses the top and the bottom base-pairs (A0-T0* and G11-G11*) and the connecting three-phosphate groups and five sugar units. The exclusion, was made in order to delete possibly spurious data due to boundary condition artifacts. The water molecules at the top and bottom's boundaries are designated as *"Boundary" type 1 or type 2* depending if the binding is of type "a"-H-O or "a"-O. Table I considers water molecules "bound" to specific sites (like O1P or O2P); Table II considers groups of sites (like all the oxygen atoms in PO_4^-); Table III considers not only "bound" water molecules, but all those forming the first hydration shell; Table IV considers the water molecules around hydrophobic groups containing H atoms (-CH, CH_2 and CH_3) and is given to allow further interpretation on infrared or Raman and nmr studies at the carbon atoms. *The information provided by these self-explaining tables compared to the experimental data accumulated in the last twenty years can be taken as an indication of the evolution in simulation techniques.*

Table III
Solvation of B-DNA without and with counter-ions at T=300 K: population of water and its interaction energy (Kj/mol) in the first Solvation shell.

#	"Regions" within first Solvation Shell*	B-DNA		Na-B-DNA	
		Waters	Energy	Waters	Energy
1	O1P, O2P, O5%, O3′, CH_2	136.78	−104.51	146.10	−133.39
2	O1P, O2P, O5′, O3′, CH_2, Na+	------	-----	170.09	−133.38
3	O1′	15.36	−98.04	16.21	−139.90
4	sugar (C_4OH_5)	83.83	−97.79	100.53	−134.21
5	NH_2 (A, C, G)	19.97	−85.93	20.77	−127.85
6	N: (A, G)	7.08	−89.71	8.66	−129.92
7	O (C, G, T)	11.32	−90.99	12.96	−134.09
8	Adenine	11.81	−86.63	13.93	−133.89
9	Cytosine	19.56	−87.33	18.11	−124.48
10	Guanine	19.54	−95.24	20.47	−135.05
11	Thymine	15.63	−87.14	15.33	−132.18
12	A-T	24.37	−87.30	23.93	−132.47
13	G-C	34.84	−91.73	33.40	−130.46
14	A-T, G-C	47.80	−91.14	50.18	−131.08
15	A-T, G-C, (C_4OH_5)	106.74	−94.32	122.72	−121.94
16	Total for first solvation shell	185.31	−99.74	223.39	−130.63
17	Total for Grooves	261.70	−64.39	223.62	−91.74
18	Total (water-water & water B-DNA)	447.00	−79.36	447.00	−111.30
---	(Water-water + total interaction)	−15.61 ± 0.05		−22.0 ± 0.05	
---	(Water-B-DNA + total interaction)	−63.75 ± 0.10		−89.3 ± 0.1	

*See the last column of Table 1.

Table IV
Water molecules perturbing hydrogen atoms (hydrophobic) in B-DNA without and with counter-ions at T=300 K*.

Group	Atoms of group	B-DNA		Na+-B-DNA	
		Waters	Energy*	Waters	Energy*
CH_2	H5′1, H5′2	43.32	−95.12	46.95	−128.30
C_4OH_5	H3, H4	55.97	−95.80	64.17	−131.06
"	H2′1, H2′2	22.13	−99.89	29.69	−137.11
"	H1′	12.44	−98.98	10.80	−139.79
Bases	H in A**	3.24	−92.57	4.63	−144.97
"	H in C**	9.56	−89.08	8.23	−123.30
"	H in G**	5.00	−104.24	7.09	−134.98
"	H in T**	10.04	−95.12	9.38	−128.49

*Kj/mol.
**The hydrogen atoms either involved in the base-base hydrogen-bonds or in the NH_2 groups are *not* considered in this Table.

The "sharing" of water molecules at different sites has been pointed out previously in the single-helix (A-DNA and B-DNA) solvation studies[3,4] as well by Lewin[10]. In this paper a more definite answer is provided. For example the sites at the bases (lines 6 to 9 in Table II) are cumulatively solvated by 24.56 water molecules, but considering the sites, one by one, 40.09 molecules of water are involved, namely about 40% (15.53) of the water molecules are "shared". The "sharing" percent in the water bound to the phosphate sites (lines 7 to 10 and composite analyses at line 11) is rather small and we interpret this fact as a consequence of our stringent criteria for the thresholds T(a-H) and T(a-O); relaxing these values we would include "second" solvation shells, where the "sharing" is even more prominent (see Ref.4 and 9).

Structure of Water at High Relative Humidity: "Average Configuration" Representation

Let us now analyze the water structure making use of the *"average configuration"* data. In Figures 4 (insert B) and 9, we analyze the water molecules either included in a disk volume 4.4 A thick (from Z=12.1 A to Z=7.7 A, see also Figure 4 (insert A) or the water molecules hydrogen bonded to the h* helix (see also Figure 3 (insert B).

In Figure 4 (insert B), we report the third base-pair A3-T3* and those water molecules *fully inside the disk.* Several water molecules below discussed are only partly within the disk and *are not reported in the figure.* To simplify the figure we have omitted the G2-C2* base-pair immediately above A3-T3* and the C4-G4* base-pair, immediately below. There are four phosphate groups in this disk, P3, P4 (top of figure) and P3*, P4* (bottom of figure). The four Na^+ ions are indicated with a full dot connected to the free oxygen atoms of PO_4^- by dashed lines. The P3 phosphate group is hydrogen bonded to the water molecules 60, 100, 124, 132 and 133; the P4 phosphate group is hydrogen bonded to the water molecules 95, 121, 160, 172 and 185; the P3* phosphate group is hydrogen bonded to the water molecules 73, 85, 105, 128 and 147; the P4* phosphate group is hydrogen bonded to the water molecules 109, 112, 131 and 156. The Na^+ ions are solvated as follows:Na^+(3) is hydrogen bonded to the water molecules 83, 97 and 126; Na^+(4) is hydrogen bonded to the water molecules 122, 159 and 160; Na^+(3)* is hydrogen bonded to the water molecules 74, 93 and 128; Na^+(4)* is hydrogen bonded to the water molecules 123, 131, 167 and 169. The water molecules 128, 131 and 160 establish a hydrogen bond to both the phosphate and the Na^+ ion. Concerning the base-pairs, the water molecule 91 is hydrogen bonded to G2, T3* and C2*; 111 is hydrogen bonded to A3; 140 is hydrogen bonded to T3* and G4*. The remaining water molecules of Figure 13 are either second or third shell solvation and therefore fill the major or the minor groove.

In Figure 9, we consider the water molecules hydrogen bonded to the Na^+-PO_4^- groups in the h* helix; in the figure we have indicated the sugar residues and the N1 or N9 atoms of the bases. Comparing this figure with Figure 3 the denser water packing in Na^+-B-DNA relative to B-DNA is apparent.

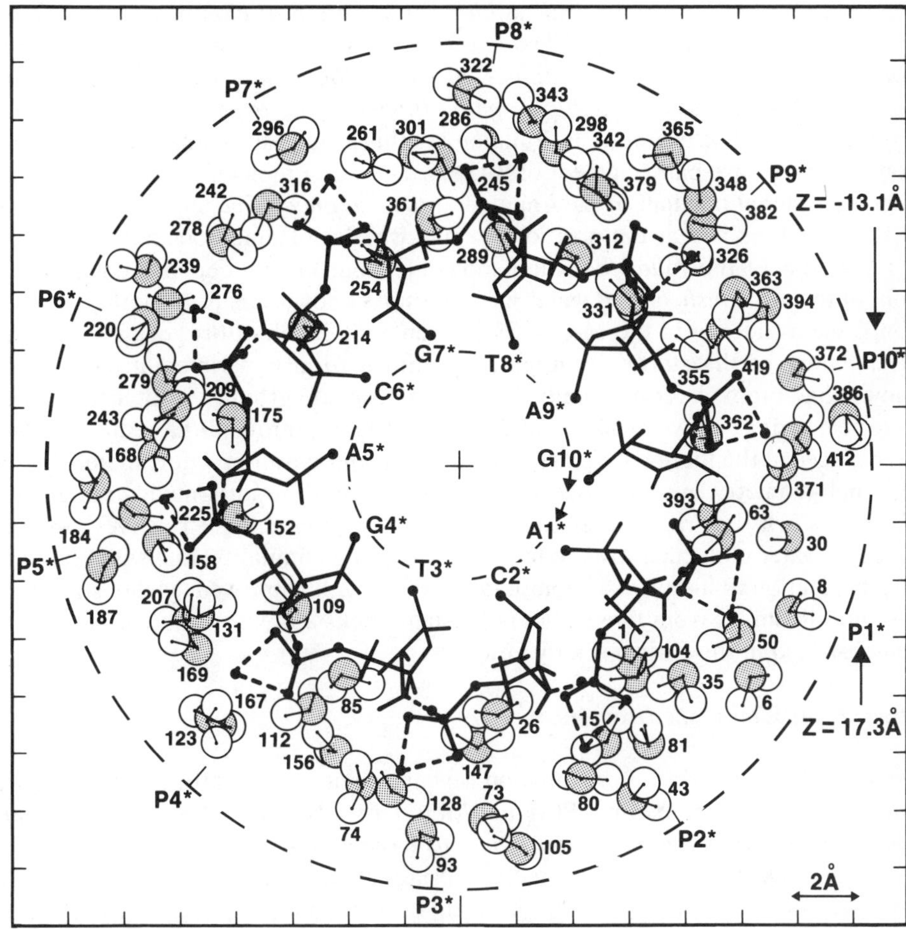

Figure 9. Water molecules solvating the sodium-phosphate groups of the h* strand of Na^{+}-B-DNA; "average" configuration.

In Figure 5 (bottom inserts) we report the water molecules "bound" to three base-pairs or in its immediate neighborhood. Additional figures for the base-pair sequence of the B-DNA fragment are reported in reference 32. One base-pair (the one at higher Z) is presented more markedly than the second, which lies immediately below (looking from the Z axis). From this figure, the average position and orientation of the water molecule solvating the base-pair can be obtained. The structure accepted by the water molecules represents the effect of the totality of the forces acting on each water molecule, namely the effect of the interaction energy between a given water molecule and the rest of the solute-solvent system. By shifting the stress from *forces* to interaction energy we implicitly restrict ourself to a static representation (Monte Carlo) and we neglect the dynamic representation (molecular dynamics).

Today's literature on "theoretical" studies for solvation of biomolecules often neglects temperature averaging and makes use of a "single configuration" despite the availability of Monte Carlo techniques, proposed about twenty years ago. In addition, in today's literature often too small a fragment and too few water molecules are considered. For example, B. Pullman and his group[33] have analyzed the solvation of a three base-pair fragment of B-DNA. The water molecules were placed one after the other at one half of the fragment, *not all at once.* The implicit assumption of these authors is that the position for one water molecule does not depend on the positions of the *following* water molecules, but mainly depends on the positions of the *previously* placed water molecules. As known (see Ref. 34) this assumption is incorrect and in the Monte Carlo technique much effort goes into erasing the memory of the starting configuration. Since from experimental data it was known that 4 to 6 water molecules solvate a phosphate group, 17 water molecules were selected[33]; five were placed at each phosphate group and the remaining two were placed at the three base-pairs. This starting configuration unfortunately biases the final geometry output. Therefore, the simulation[33] reports mainly "boundary effects" (since the fragment is too small), at low-intermediate relative humidity (since 17 water molecules are too few to describe high humidity) and with temperature neglection. This type of computational study, if improved, yields structural data not dissimilar from those obtained using "sticks and balls" models, as clearly demonstrated in the early work by Lewin.[10]

Structure of Water in the Grooves

We recall that in the first approximation we have commented on the *intra-phosphate water filaments* (connecting P(n) to P(n+1) or P*(n) to P*(n+1)) and *inter-phosphate or "trans-groove" water filaments* (connecting P(n) to P*(m)). In Na^+-B-DNA the ion-induced compression effect brings about an added feature relative to B-DNA, namely we observe hydrogen-bond cross-linking between inter-phosphate water filaments.

In Figure 6 (right insert), we examine the structure of water in the minor and in the major grooves selecting waters with $8 \leq R \leq 13$ A. The water molecules are projected onto the Y-Z plane, namely a plane containing the long axis (Z axis) of DNA (see Fig. 10); we have selected those water molecules with positive value for the X coordinate of the oxygen atom. The figure is seen from the +X direction; the five atoms forming a phosphate group are reported; full dots represent the two bound oxygen atoms O3′ and O5′, whereas open dots represent the two free oxygen atoms O1P and O2P. The Na^+ ion is represented by a full dot. The space enclosed by two lines passing through P3, P4, P5, P6 and P7 and through P8*, P9*, P10*, P11* corresponds to the major groove; above and below it are portions of the minor groove. Dotted lines between atoms of water molecules indicates a water-water hydrogen bond.

For Na^+-B-DNA the solvation pattern can be described either in term of *cross-linked* inter-phosphate water filaments or in terms of local *cyclic structures* composed of about 5±2 water molecules. For example, the water molecules 293, 321, 330, 359, 371 can be considered as a inter-phosphate filament; a second filament is formed by the water molecules 293, 327, 338, 350, 354, 375, 403. The cross-linking water molecules for these two inter-phosphate filaments are water 354 and 375. Alternatively, cyclic structure representations are given, for example, by the water molecules 327, 321, 330, 359, 375, 354, 350, 338 or 375, 359, 371, 393, 403. At this stage we do not see any reason to prefer the cross-linked filament "model" over the cyclic structure "model".

The above "low" resolution structure of water molecules in the grooves is now complemented with the "high" resolution obtained not from the "average" configuration, but from a full statistical analysis on the 2x10^6 Monte Carlo configurations.

First we partition the water molecules into those belonging either to the major or to the minor groove. The cylinder containing the solvent-solute system is subdivided into sectors (see Fig. 10, left side insert). For a full B-DNA turn there are N sectors of angle α and height (along Z) equal to 13.2 A for the minor groove and 20.08 A for the major groove; for a full B-DNA turn there are $360/\alpha$=N sectors (in the figure we have represented four sectors in the minor groove). We select a clockwise rotation for α (with α=0, y=0 and x positive, see left side of Fig. 10). Finally, to increase the resolution in locating a water molecule in the grooves, each sector is partitioned into four sub-volumes, as indicated at the bottom left insert of Fig. 10, by subdividing R into four segments, namely 0.0A≤R<4.0. A, 4.0A≤R<8.0A, 8.0A≤R<12.0A and 12.0A≤R<R(max)=14.5 A. In the right side insert of Fig. 10 we report the phosphorous atoms of the h and h* strands; these atoms are on a cylindrical surface and have been projected onto XZ plane. When α increases from 0° to 180° (or 180° to 360°), we consider all those water molecules with negative (or positive) value for Y. We count the water molecules which fell into one of the four sub-volumes of a given sector selecting those left out from the previous analyses performed for the first solvation shell (therefore, only true "groove" molecules are

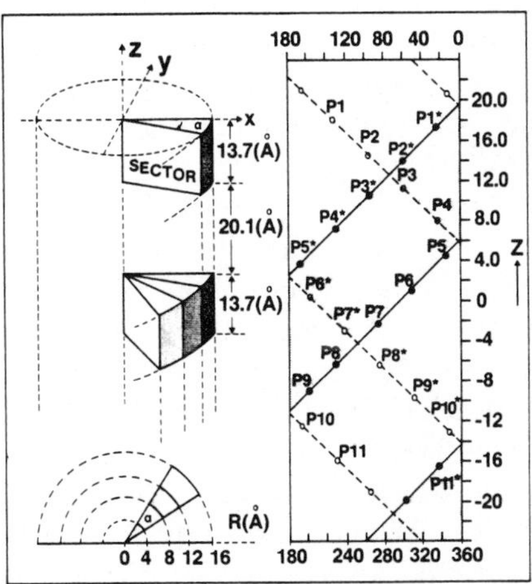

Figure 10. Grooves analyses: sectors and sub-volumes (left), major and minor (dashed) grooves (right); the bottom scale is seen from the -Y axis, the upper scale from the Y axis.

considered). After several trials we have selected $\alpha=4°$ as the value yielding a clear graphical representation which is reported for the major groove in Fig. 11 (top insert) (the values reported on the abscissa follow the definition used in Fig. 10); the dashed vertical lines identify the position for the P* atoms (of the h* strand) and the marks at the top of the figure identify the position for the P atoms (of the h strand). In the figure we compare Na^+-B-DNA with B-DNA. In the region 0. A$\leq$R$<$4.0A there is no groove water, as previously pointed out for example, by considering iso-energy maps[2,6,29].

The number of water molecules (Nw) and the average interaction energy per water molecule, E (in Kj/mol) varies with R as reported below:

Limits	Nw	E	Groove	Solute
$4\leq R<8$	0.38	−125.38	minor	Na^+-B-DNA
$4\leq R<8$	15.82	−104.25	major	Na^+-B-DNA
$4\leq R<8$	0.0	0.0	minor	B-DNA
$4\leq R<8$	9.22	−66.93	major	B-DNA
$8\leq R<12$	19.70	−98.52	minor	Na^+-B-DNA
$8\leq R<12$	74.02	−97.22	major	Na^+-B-DNA
$8\leq R<12$	18.94	−72.47	minor	B-DNA
$8\leq R<12$	64.87	−66.77	major	B-DNA
$12\leq R\leq R(MAX)$	49.77	−88.54	minor	Na^+-B-DNA
$12\leq R\leq R(MAX)$	64.41	−82.58	major	Na^+-B-DNA
$12\leq R\leq R(MAX)$	64.92	−63.75	minor	B-DNA
$12\leq R\leq R(MAX)$	102.95	−61.76	major	B-DNA

From the above data we learn new feature on the packing of the water molecules in the grooves.

The interaction energy in the minor groove is larger than in the major groove (as expected because the attractive sites are nearer in the minor groove). For B-DNA the interaction energy is nearly constant in the major groove, as previously predicted[2,5] but it increases sharply by decreasing R for Na^+-B-DNA. *Thus the grooves can be seen as channels to convey reactants to the base-pairs;* the "flow" in such channels depends both on the counter-ion position and specificity, and on sterical factors of the reactant.

In Fig. 11 a more detailed representation of the "groove" water molecules in the major groove is given. An equivalent diagram for the minor groove shows the same periodicity of peaks but with lower intensity; these peaks represent the "intra" phosphate filaments previously discussed. The gross characterization of the intensity distribution given in Fig. 11 is the nearly periodic existence of peaks, approximately evenly spaced ($\Delta\alpha = 10°$), well developed in the regions $8\leq R<12$ and $21\leq R\leq R(max)$ and weakly as well as infrequently developed in the region 4A$\leq$R$\leq$8. The peak pattern *clearly points out the existence of a structural organization which even if temperature dependent, is not destroyed by thermal motion at room temperature.* The intensity of the peaks (i. e. , the number of water molecules bound to the phosphates) adds up to "filaments" of about 5 to 7 water molecules.

The number of water molecules, Nw, and the average interaction energy per water molecule, E, with consideration given not only to the groove's water but also to the

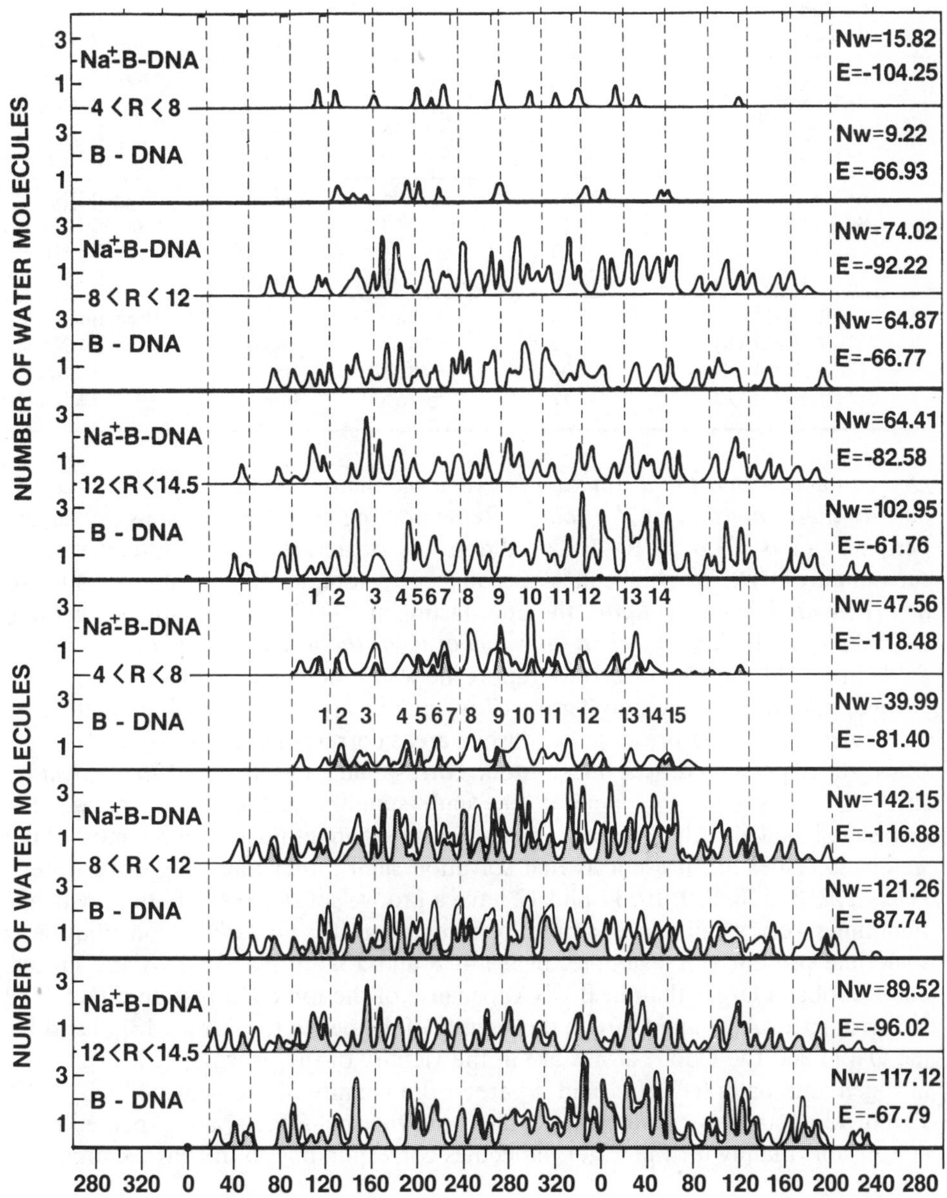

Figure 11. Probability distributions for water molecules in B-DNA and Na+-B-DNA major groove (top) first solvation shell and groove water molecules (bottom).

first solvation shell water molecules, are partitioned into the sector's sub-volumes are as reported below:

Limits	Nw	E	Groove	Solute
$0 \leq R < 4$	0.0	0.0	minor	Na^+-B-DNA
$0 \leq R < 4$	1.15	−125.25	major	Na^+-B-DNA
$0 \leq R < 4$	0.0	0.0	minor	B-DNA
$0 \leq R < 4$	1.92	−87.10	major	B-DNA
$4 \leq R < 8$	20.25	−137.12	minor	Na^+-B-DNA
$4 \leq R < 8$	47.56	−118.48	major	Na^+-B-DNA
$4 \leq R < 8$	19.79	−94.56	minor	B-DNA
$4 \leq R < 8$	39.99	−81.40	major	B-DNA
$8 \leq R < 12$	80.83	−117.46	minor	Na^+-B-DNA
$8 \leq R < 12$	142.15	−116.88	major	Na^+-B-DNA
$8 \leq R < 12$	74.36	−87.54	minor	B-DNA
$8 \leq R < 12$	121.26	−87.74	major	B-DNA
$12 \leq R \leq R$ (MAX)	65.56	−98.90	minor	Na^+-B-DNA
$12 \leq R \leq R$ (MAX)	89.52	−96.02	major	Na^+-B-DNA
$12 \leq R \leq R$ (MAX)	72.56	−68.33	minor	B-DNA
$12 \leq R \leq R$ (MAX)	117.12	−67.96	major	B-DNA

One can notice that the distinction between the major and minor groove, extends outside the dimension of the solute. Remembering that the Na^+ ions are at the periphery of B-DNA (at R=10 A), the groove structure extends into the liquid surrounding DNA. We shall call "classical groove volume" the one up to R=12. A, and *"extended groove volume"* the one starting at R=12. A. The latter ends when E=−36 Kj/mol[12], *namely when the value of R for the water is equal to about 20 to 25 A.* In Fig. 11 (bottom insert) we report the water molecules present in the major groove, considering not only "groove" water but also first solvation shell water molecules. The peaks presented as dashed areas correspond to *groove* water molecules; those presented as a line-contour correspond to *groove* and *first solvation* shell water molecules. To facilitate the understanding of the figure, some of the peaks are labelled with numerals. In the first sub-volumes (O.<R<4.A), most of the water molecules are present as first solvation shell rather than as groove waters. Peaks 1, 2, 3, 5, 6, 7, 9, 10, 11 and 12 have a groove's water contribution, which is sometimes very small (for example, peak number 10); no *groove* contribution is sometime present (for example, peaks 4, 8 and 13). In Na^+-B-DNA the groove contribution is larger than in B-DNA, because of the ion compression effect. The NH_2 groups for A, G and C are in the vicinity of the peaks 1, 2, 3, 8, 9, 13, 14 and 15; the O4(T) and the O6(G) atoms are in the vicinity of the peaks 4, 5, 7, 11 and 12; nitrogen lone pairs (N7 of G and A) are in the vicinity of the peaks 4, 6, 8, 9, 10, 12 and 13. Rotations of water molecules (and thermal effects) are expected to be appreciable mainly for the water molecules corresponding to the dashed area.

In the sub-volumes from R=4 A to R=8 A each first solvation peak is enhanced by groove water contributions (exceptions are at the boundaries). The peaks intensity is higher for Na^+-B-DNA than for B-DNA, as expected. *The emerging picture is that a water filament starts as bound water, continues as groove water to end as bound water, in agreement with the previous findings.*

In the sub-volumes from R=12 A to R=14 A the groove water is responsible for most of the intensity of the peaks, especially in Na^+-B-DNA.

Water Solvation of DNA at Low and Medium Relative Humidity

As known (see Refs. 14, 15, 16 and 26) in adsorption-desorption experiments, the isotherm presents a characteristic growth curve. The experiments are generally performed on DNA fibers, and in these conditions there are several parameters to keep in mind, even at constant temperature, constant relative humidity and constant ionic concentration. By decreasing (or increasing) the number of water molecules, there is a reorganization in the entire water system. In addition the relative position of the DNA units in the fiber varies (a swallowing of the fiber is observed). Moreover, concomitant with the reorganization of the water and of the DNA units within the fiber, conformational variation in the DNA can occur. *Finally,* fiber diffraction data are limited to about 3 A resolution. An advantage in simulating the water structure around a fixed-geometry Na^+-B-DNA fragment is that the energetic and structural variations of the water system can be followed, without the worry of the notable additional complications associated with the interactions of one DNA molecule (and its solvent) with other DNA molecules in the fiber.

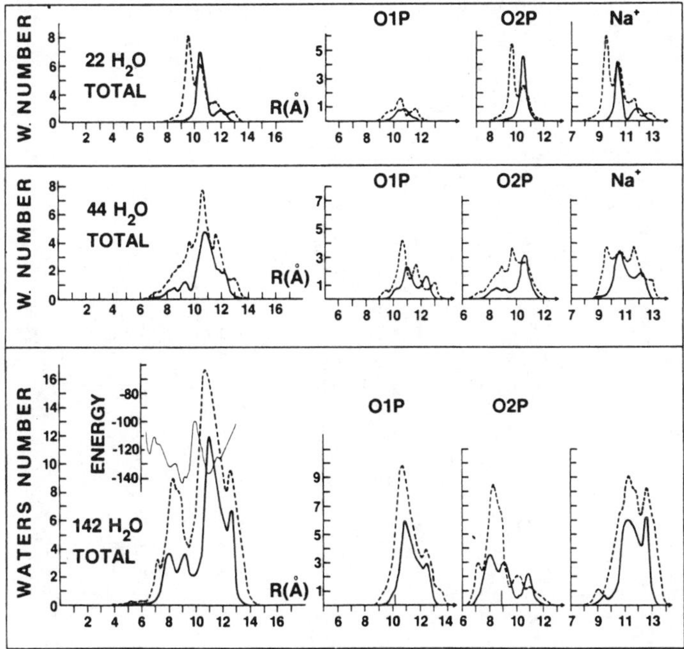

Figure 12. Probability distributions for 22, 44 and 142 water molecules in Na^+-B-DNA.

We precede the following analyses by reporting average distributions computed by adding to the Na^+-B-DNA fragment either 22, 44, 142 or 257 water molecules. The results are reported in full detail in Tables XI to XIV of reference 32 and are summarized here in Figures 12 and 13.

For 22 waters the sites interested are Na^+, O1P and O2P; one water molecule is bound to each PO′-Na^+ unit. A water molecule is bound with about equal probability to Na^+ and O2P, but less to O1P. The O5′ and O3′ site "do not participate" (the analyses on O1P and O2P yields a sum of 17.80 water molecules); by considering also O5′ and O3′ and the CH_2 group, this value does not change. Essentially two distributions are found, one with a water bound both to Na^+ and O2P, and a second, less probable with a water bound both to Na^+ and O1P.

In order to accommodate 22 *additional* water molecules, the original distribution of the 22 water molecules is fully re-arranged. Indeed for the case of 44 water

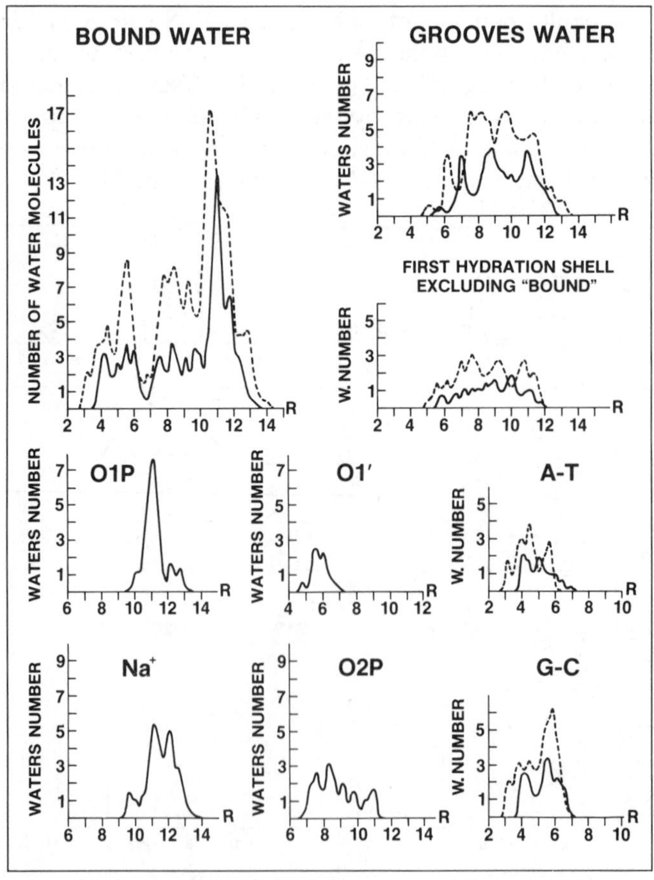

Figure 13. Probability distributions for 257 water molecules in Na+-B-DNA.

molecules, the O3′ site starts to be interested, the O1P and the O2P site's difference, in water population, becomes smaller than for the case of 22 water molecules, and the average energy of a water molecule interacting with PO′-Na^+ drops from −155.75 Kj/mol (for 22 waters) to −148.45 Kj/mol. The rearrangement pushes the water molecules (in average) to smaller R values, and not only the hydrogen atoms of sugar but also those of the A and G bases become interested, even if marginally.

Let us now fill the DNA sample with 142 water molecules; the grooves remain essentially empty (1.43 water molecules out of 142 are found in the grooves) but the finding of some, even if very little water in it, indicates that the "bound" water number has attained its saturation value. This statement is not equivalent to saying that the structure or the maximum number of the *bound* water molecules for 142 water and for 447 water molecules solvating the DNA fragment are the same. Indeed by filling up the grooves, additional water penetrates the "bound" region, and because of the interaction between water in the "grooves" and water "bound", a rearrangement can be induced in the "bound water" structure. In other words, we are gathering preliminary evidences that the adsorption process is not "monotonic"; at low relative humidity the water molecules go mainly to the DNA sites, at higher, mainly to the grooves, at even higher to the sites and the grooves and, finally, to the "extended" grooves. Incidentally, it should be by now evident why the interpretation of experiments at different humidity (desorption-adsorption studies) are very hard (even neglecting the experimental difficulty in obtaining reproducible raw data), especially when experiments are performed on DNA fibers, where additional variables must be considered. For the case of 142 water, about 2.97, 2.18, and 1.87 water molecules are bound to Na^+, O1P and O2P, respectively; the total populations of *bound* water for PO′ and PO′-Na^+ are 4.48 and 6.40 water molecules, respectively. The total populations for the *first hydration shells* for PO′ and PO′-Na^+ are 5.71 and 6.52, respectively. The previously noted drop in the binding energy for the "bound" water molecules around PO′-Na^+ continues and the energy is now −134.48 Kj/mol. The bases are interested at the solvation and the solvation order are $A<G<C<T$ by considering the number of "bound water" molecules, $A<C<G<T$ by considering the number of waters in the first solvation shell and $A<C<T<G$ by considering the binding energy for the water molecules in the first solvation shell. The distribution of the total water's population at different R values shows a well defined structure; by comparing the shift of the most intense peak one can gain a feeling of the intense water structure re-arrangement which follows the repeated addition of water molecules. The AT pair is more solvated than the GC pair for the cases of 44 and 142 water molecules, in agreement with experimental findings at low humidity. In addition we note that at low humidity O2P is more solvated than O1P, a situation that will be reversed by increasing the relative humidity. Finally, in the presence of Na^+ counter-ions, the O1P and the O2P water distributions are nearly additive, contrary to what is found for B-DNA without counter-ions.

Let us now solvate the DNA fragment with 257 water molecules (see Fig. 13). This case deserves to be analyzed more deeply, since it provides an interesting step in the adsorption process. With 257 water molecules the first hydration shell is nearly saturated and in addition there are about 71 water molecules in the grooves. From Fig. 13 we learn that the bound water molecules exhibit the same ten peaks as found in the 447 water molecules case; however the peaks L and J (see Fig. 7) are more developed by adding water to the 257 molecules now present, because the Na^+ and the O1P sites are not fully solvated. With 257 water molecules the hydration at the base-pairs is nearly as much evolved as with 447 water molecules (peak C is, however, higher that in the case of 447 water molecules). The water molecules in the "classical" grooves are well developed at smaller R values but far from saturated at larger R values, as expected. Therefore, the term *"groove"* water could be substituted by the term *"second solvation shell"* for the case of 257 waters. Relative hydrations at the bases are not discussed, since these are given in full detail in reference 32.

In a study in progress, we are analyzing the differences in rotational freedom associated to the water molecules either bound or into the grooves at different relative humidities. This is an aspect of interest to interpret angular distribution data from scattered neutrons. We note that the findings concerning the water filaments are in agreement with the scattering data by Dahlborg and Rupprecht[27].

By considering the above partial humidity simulations, the following model for an adsorption process emerges. *First the water molecules go to the phosphate groups. If the humidity is very low* (one water molecule per phosphate—22 water case—) *then a given water has several nearly equivalent positions to select at each phosphate and the same position will not be selected. By an increase in the humidity,* such that two water molecules are available for each phosphate group (—44 waters case—), *the best solvation positions are not those left unoccupied by the first water molecule, but new ones.* This is to be expected:the Na^+-PO′ system is different from the H_2O-Na^+-PO′ system and the latter has only a vague memory of the former. By increasing the humidity such that about 6 water molecules are available for each phosphate group (142 waters case), on the average all waters solvate the Na^+-PO′ groups, but some molecules solvate even the base-pairs or are present in the groove's region. The presence of water in the groove indicates that a saturation level has been reached. Alternatively, we can state that the initial saturation of the grooves is nearly concomitant to the solvation at the base-pairs. By increasing the humidity, such that about 11 water molecules are available for each phosphate group (—257 waters case—) *more water molecules go to the groove's region and the solvation of the sugar's oxygen and of the base-pairs is nearly completed.* Finally, by making available about 20 water molecules per phosphate group (—447 water case—) additional water is packed into the groove's region and around the solvated sites, but most of it goes to the outside of Na^+-B-DNA, to the "extended grooves" region, still strongly perturbed by the Na^+-B-DNA field, so much so that the bulk water interaction is far from being reached. By adding two additional water layers (that is, by adding 1600 water molecules) one would have nearly reached the bulk water region, at an R value of about 20 A.

Conformational Transitions: A Model and a Preliminary Example

In considering the adsorption process, we should keep in mind some overall energetic aspects, reported in Fig. 14. Repeatedly, we have pointed out that the interaction of water with DNA increases by adding counter-ions to the solute therefore stabilizing Na^+-B-DNA relative to B-DNA. This increase is drastic as shown in Fig. 14, where the total (namely the sum of water-DNA and water-water) average interaction energy per water molecule is reported as a function of the number of water molecules, Nw, solvating the DNA fragment. In the top insert of Fig. 22 the full line refers to the *total* interaction, the dashed line to the *water-DNA* interaction; the *water-water* interaction is reported in the bottom insert of the figure. By a substantial increase of solvent all curves will eventually end at about −36 Kj/mol, the bulk-water value. *The total interaction could induce to think that the adsorption process is monotonic; the partial interactions offer indications for an opposite conclusion.* Physically, the following model clearly emerges *in B-DNA:* the loss of

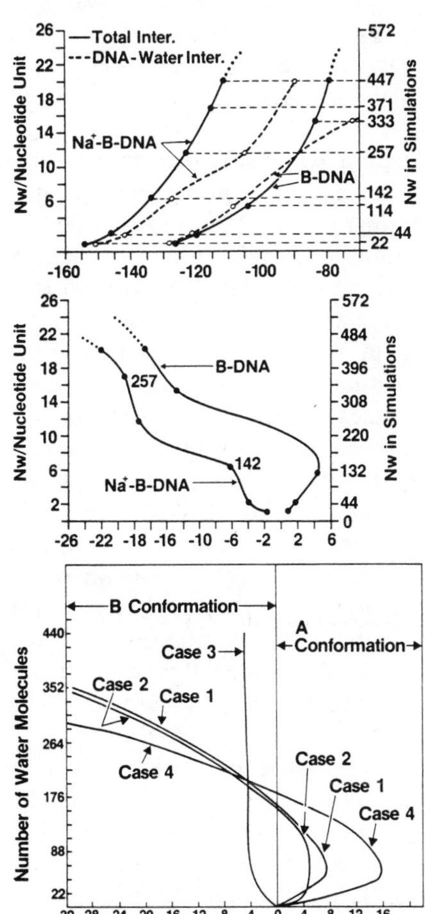

Figure 14. *Top insert:* Total interactions (solid line), water B-DNA interactions (dashed lines). *Middle insert:* water-water interaction per water molecule. *Bottom insert:* Solvent stabilizations at constant temperature (300 K) and different humidities for the A→B transition.

bulk water energy (water-water interaction) is more than compensated by the strong PO′ attraction; in addition, the water-water interaction at low humidity would amount to only a fraction of the bulk-water value. But, when the PO′ groups are screened by few solvent molecules, and when the humidity increases, then the system attempts to regain the bulk-water interaction energy since the water-water interaction becomes more and more important. *In Na^+-B-DNA* (and in any B-DNA counter-ions system) the initial adsorption situation is more complex. At low humidity, a water molecule does not experience only the PO′ field, but also the counter-ion field. The latter exerts a request on a water molecule that would demand a very different orientation from the one demanded by the PO′ field. Thus the water molecule compromises its orientation between the two conflicting requests, and by so doing an orientation is achieved, which is not as repulsive to another water molecule as the one accepted for the case of B-DNA (without counter-ions and at low humidity).

As a consequence of these effects, the water-water interaction energy presents two plateaus, where a small humidity increase brings about a strong water-water interaction variation. The two plateaus border at two steep sides, where a large humidity increase brings about only a small water-water interaction variation.

The relative humidity interval, corresponding to a variation from about 6 to about 10 water molecules per nucleotide unit, appears as a critical humidity interval. The "classical grooves" start to be filled in this interval; by a further increase in the humidity, the final saturation at the DNA solvated sites is reached and the "extended" grooves are filled. When the adsorption process is completed (very high humidity), it is hard to visualize a conformational transition in DNA (for example, from the conformation A to B), without the introduction of a third body. Equally difficult is to visualize a conformational transition at very low humidity, since the few water molecules present are fully occupied at solvating the PO′ counter-ion groups. On the other hand, in the above discussed plateau, where the incipient filling up of the classical grooves perturbs the sugar units (this is tantamount to a directional pressure) one can easily hypothesize conformational transitions. Thus combining the evidences on the ion-induced compression effect and the finding on the variations of the interaction energies with relative humidity, we can obtain a preliminary model to explain conformational transition, under specific conditions. The use of the model however, requires simulations of the type here presented for different conformations and with different counter-ions. The mathematical conditions for a conformational transition are now indicated by the following three models.

First, we consider only one type of counter-ions (for example, Na^+). Let E(A) and E(B) be the energy of DNA in conformations A and B, respectively, for DNA fragments of equal number of units. From a simulation with a equal number of water molecules and at a given temperature, one can obtain the water-DNA interaction ES(A, h(i)) and ES(B, h(j)), where the indices A and B refer to the two conformations, h(i) and h(j) to the two relative humidities (clearly different in the two conformations, since we have assumed the same number of water molecules).

A transition can occur for $\Delta E = E\Delta'$, where $\Delta E' = E(A) - E(B)$ and $\Delta E = ES(A, h(i)) - ES(B, h(j))$, assuming nearly equal entropic variations. With this work we provide one of the four needed quantities.

Second, we consider the same situation as above with, however, two different types of ions I(1) and I(2) (for example, Na^+ and Li^+) yielding either I(1)-DNA or I(2)-DNA. Let E(A, I(1)), E(A, I(2)), E(B, I(1)) and E(B, I(2)) be the energies of the fragment with either one of the two types of counter-ions and in the two conformations. Correspondingly let ES(A, I(1)), ..., ES(B, I(2)) be the corresponding simulated solvation energies as in the previous model. The introduction of more than one type of counter-ions increases the probability of finding an equality between ΔE and $\Delta E'$ since one more variable, the type of ion, has been added.

Third, let us assume the same situation as above, with, however, the additional possibility that a fractional rather than total replacement of I(i) with I(j) is considered. This last assumption considerably increases the probability of finding an equality between ΔE and $\Delta E'$. The model can be made more complete by including *a temperature index, namely by considering more than one temperature.*

As an example, let us consider the ES values, stressing however, that simulations on Li^+-B-DNA, Li^+-A-DNA, Na^+-A-DNA are not available. The different solvation energy values can be tentatively guessed by keeping in mind ionic solvation studies[35,36], i.e. the differences in the solvation energies between the A and B single helix conformation and preliminary simulation data on Li^+-B-DNA. We have made no use of the study by Marynick and Shaeffer[37] since their use of a sub-minimal basis set and the neglect of basis set super position corrections deprive their otherwise very interesting computation of quantitative validity. Their very strong attraction for a phosphate group in $(CH_3\text{-}O)_2\text{-}PO'$ with Li^+ and Na^+ (computed as -303 Kcal/mol and -198 Kcal/mol, respectively) and the very short oxygen-cation distances (computed as 1.77 A and 1.98 A, respectively) are too far from the more reasonable data obtained, however, on $(CH_3\text{-}CH_2\text{-}O)_2PO_2^-$ and yielding -152 Kcal/mol, -134 Kcal/mol, 1.93 A and 2.17 A, respectively.[30] Equivalent data, obtained from relatively old electrophoretic mobility studies, indicates that the binding to DNA for ions is $Li^+ > Na^+ > K^+$.[38]

At low humidity (1 water per nucleotide unit) Na^+-A-DNA is estimated to be more attractive to a water molecule by about 8 Kj/mol (relative to A-DNA). Simulations on A and B single helix[2,3] indicate a value between 5 Kj/mol (average value) and 17 Kj/mol (maximum difference), thus our selected value might be somewhat smaller than the correct one. At high humidity we know that the PO_4^- groups in A-DNA are sufficiently crowded, having one less water molecule strongly bound to a PO_4^- group relative to B-DNA. Thus on a sample of 447 water molecules, about 22 water molecules in A-DNA are less bound to the PO_4^- than in B-DNA by an assumed interaction of about 10 Kj/mol per water molecule. As a result the total interaction energy curve for Na^+-A-DNA crosses the total interaction energy curve of Na^+-B-DNA at a relative humidity corresponding to about 160-190 water molecules (see Fig. 14). For Li^+-B-DNA, we have obtained preliminary data for 22 and 447 water molecules (at a temperature of 300K). At a relative humidity corresponding to 22 water molecules Li^+-B-DNA is more attractive to water than Na^+-B-DNA, but the situation is reversed at high humidity (447 water molecules) bringing about a crossing of the Li^+-B-DNA and Na^+-B-DNA total interaction energy curves at about 190-210 water molecules. Finally for Li^+-A-DNA we assume that the total interaction energy to water differs from the Na^+-A-DNA total interaction energy in the same way as found by comparing Li^+-B-DNA with Na^+-B-DNA. Until definitive simulations on Li^+-A-DNA, Li^+-B-DNA and Na^+-A-DNA will be available, the above estimates are likely all the data one can use. The stabilizations due to solvent effect for the A→B conformational transition are reported on Fig. 14 (bottom insert), (the ordinate gives the number of water molecules for either an A or a B double helix sample with twenty-two phosphate units). We consider four cases (all at 300K). In case 1, we consider the stabilization of a DNA conformation with Na^+ counter-ions, whereas in case 2, we consider the stabilization of a DNA conformation with Li^+ counter-ions. In case 1, at low humidity the form A is stabilized by water and the stabilization reaches a maximum for about 3 water molecules per nucleotide unit, then goes to zero at about 7 water per nucleotide unit, and then the B form becomes stabilized. In case 2 (Li^+ counter-ions) the same behavior is predicted, but the crossing from A to B occurs at slightly lower humidity. In cases 3 and 4, we increase not only the humidity, but we also assume that the Na^+ counter-ions of the A form are substituted with Li^+ counter-ions in the B form (case 3) or vice versa (case 4). From our preliminary data we expect that only the form B is stabilized by the solvent, whereas in case 4, the Na^+-A-DNA has a net solvation stabilization up to about 7 water molecules per nucleotide unit; at higher humidity the solvent effect helps the formation of B-DNA.

We stress that we have only referred to ΔES not to $\Delta E'$; in addition no entropic effect has been considered, or equivalently, we have assumed that the entropic contribution to the free energy is ion-independent and conformation independent at a given relative humidity and temperature. The theoretical behavior of Fig. 23, even keeping in mind its tentative nature, explains a large number of experimental findings relative to A-B transition. Clearly, the same type of reasoning can be used when considering the solvent effect of any other conformational transition. By adding to an ionic solution (containing A- or B- DNA) solvents like alcohol-water the number of water molecules available to DNA decreases because the hydrophobic part of the alcohol removes water from DNA[39,40]. Thus, if one can estimate the latter effect, then Fig. 23 provides an explanation, also for transitions, where not only the humidity and the counter-ions are varied, but also additional solvents, like alcohols, are added. Concerning the energy $\Delta E'$ for conformational transitions from A to B a value of about 84 Kj/mol has been proposed by Ivanov et al. [41] later confirmed by and Sukhorukov et al.[42].

Comment on the Second Approximation

The most crucial limitation in this approximation is in the assumption of a rigid

Na^+-B-DNA fragment, and in the position assumed for the Na^+ ions. From theoretical considerations the Na^+ lowest energy minimum (minimum a) in model compounds containing the phosphate group is in agreement with our choice. However, there is a second minimum away from the PO_2^- plane (minimum b) which will become deeper, if more than a single phosphate group is present; finally there is a third minimum along the PO bond direction (minimum c). The three minima are relatively near in energy. Let us consider the available, but indirect experimental data. A double helix chain can be constructed from the X-ray structure of sodium adenylyl (3′-5′) uridine (ApU), where one Na^+ is coordinated to the PO_4^- group and one to the two uracil carbonyl groups.[43] Thus minimum "b" is known to be a possible candidate in polynucleic acids. A double helix chain can be constructed from the X-ray structure of sodium guanylyl (3′-5′) cytidine (GpC), where the Na^+ is coordinated to the 2 free oxygen atoms of a phosphate group.[44] Thus the minimum "a" is a possible candidate. Finally, X-rays studies on crystals[45] of deoxyribose-dinucleotide sodium thymidylyl (3′-5′) thymidylate (pTpT) can be used to model a double helix with the Na^+ coordinated to one free oxygen of one PO_4^- group and two oxygen atoms on two different thymine bases. This position for Na^+ can be considered as one ralated to the minumum "c". It is noted that none of the above structures refers directly to a crystal of a true polynucleic acid, nor to a 50% G-C, 50% A-T double helix structure, as in our fragment[2]. The nearest case is the one of GpC, where the Na^+ is located close to our chosen position. Most recently (when this work was completed) a large DNA fragment has been analyzed as a single crystal[47], but the counter-ions, Mg^{++}, were not detected. Several water molecules however, have been identified and assigned to the DNA fragment atoms.[60] The overall data on the water location nicely follows some of our early and above reported predictions, keeping in mind, however, the deep perturbation of one DNA fragment on the nearby fragments, the presence of impurities, and the undetermined position of the Mg^{++} ions and of the corresponding solvation water molecules[47]). Another very interesting work involving single crystals and NH_4^+ counter-ions and water molecules with a DNA fragments has recently been published.[48]

In view of the above mentioned problems concerning the determination of the counter-ion position we extended our computer experiment as reported below.

Third Approximation: "Free" Ions Simulation

The selected B-DNA double helix fragment is composed of 30 base-pairs (three full B-DNA turns). Periodically, at each turn the following sequence of ten base-pairs is selected: TA*, GC*, AT*, CG*, TA*, GC*, CG*, AT*, TA*, CG*, where the asterisk denotes the h* strand. The coordinates of this B-DNA fragment have been previously discussed.[2,31] The three B-DNA turns, see Figure 15, are hereafter referred to as *top, middle* and *bottom* turn, respectively.

Four-hundred water molecules and twenty Na^+ counter-ions are placed within the middle section of the cylindrical volume, shown in Figure 15. For each water molecule we compute the interaction energy with the full three turns of the B-DNA fragment, with the water molecules and with the Na^+ ions. In addition, since our B-DNA fragment has a periodically repeated sequence for each turn, each water molecule (or each ion) is associated to two "image" water molecules (or two "image" ions) obtained by a coordinates translation (along the Z axis) of the water molecule (or ion) in the central section. In Figure 15, we have shown a water molecule in the central section of the cylindrical volume and its two "images", symmetrically located in the top and bottom sections. Equivalently, for each counter-ion in the central section, we compute the interaction with the atoms in the three DNA turns, with all the water molecules and the corresponding water images and with the other ions and ion-images. Thus we simulate a system composed by the atoms in the three B-DNA turns, by 1200 water molecules and 60 counter-ions. The solvent particles (water molecules and sodium counter-ions) in the central section are *randomly* displaced; the displacement applied to a solvent particle is also imposed on the "image" particles. Each random displacement (or "move") generates a new "configuration" for the solvent (or a new "step" in the "random walk"). An initial set of about $2x10^6$ conformations was disregarded, however, in order to erase any memory of the initial configuration for the solvent. Thereafter, the cartesian coordinates of the solvent particles (for each configuration) and the corresponding interaction energies (water with water, ions and the DNA and ion with ions and DNA) are stored on magnetic tapes, to be used later in a statistical analysis of the Monte Carlo data.

Determination of the Counter-ion Structure

In a recent and preliminary communication[49] we have reported that the sodium counter-ions in a solution with B-DNA form a pattern corresponding to two helices interwinding with the two B-DNA strands. The overall methodological approach is described at length in Ref. 29 and summarized in Ref. 50.

The most direct way to analyze the counter-ion positions is to provide the projections in the X-Y and X-Z planes (see Figure 15 for the axis choice) of the statistical distributions of the counter-ions obtained from the Monte Carlo data. The following techniques has been adopted: the entire cylindrical volume (see Figure 15) has been subdivided into small cubical cells (of 0.2 A side). A counter at each cell is activated, measuring how many times a counter-ion falls within a cell during the Monte Carlo walk. The statistical distribution is graphically visualized by reporting for each cell a number of points proportional to the number of times an ion is present in the cell. The projections of the probability distributions of the counter-ions at one B-DNA turn are reported in Figure 16. At the *bottom-left insert* we provide the X-Y projection of the ion distribution map (spots-like patterns), with an index (1 to 10) for the ten "spots" external to the phosphates and an index (11 to 20) for the ten "spots" internal to the phosphates. A counter-ion corresponds to each "spot"; the size of the "spot" provides a measure of the counter-ion mobility. The mobility of the counter-ions, determined from the probabilty distributions, is large in the x-y directions, and relatively small in the z direction; alternatively

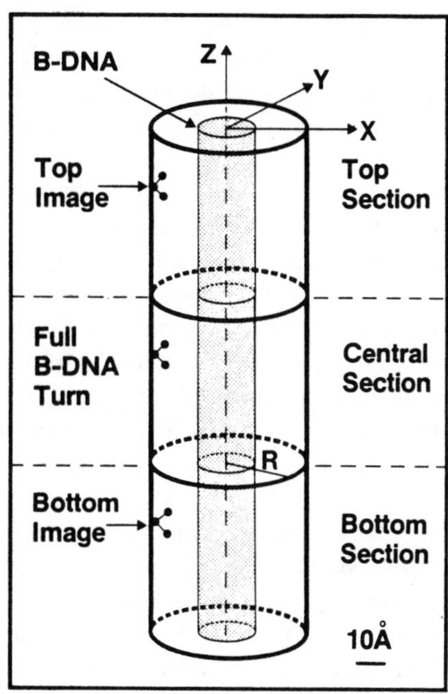

Figure 15. Volume enclosing the B-DNA fragment; "image water" molecules and "image" ions.

stated, a displacement of the counter-ions in the xy plane costs less than an equivalent displacement in the z direction (in energy terms). The phosphates of one strand are indicated by drawing the bonds between the oxygen atoms and the corresponding phosphorous atom and by explicitly indicating the corresponding bases, A, G, C and T. The atoms of the phosphates in the second strand are indicated simply with dots at the nuclear positions and for the corresponding bases we report only a dot for the nitrogen atom position (the one connecting the base with the sugar). *The projections of the twenty counter-ion probabilities form two nearly regular circular patterns.* In the bottom-right insert of Figure 16 we present the same probability density distributions, this time projected into the X-Z plane. In this insert the phosphate groups of the two strands are enclosed into helical envelopes and the base-pairs are identified by reporting the base-pair molecular plane. In the top-right insert we repeat the X-Z projection, this time connecting into an envelope those counter-ion probability distributions which are nearest neighbors; to simplify the diagram we have not enveloped the phosphate groups into two helices as done in the bottom right insert. *The pattern emerging from these three inserts is repeated in the top-left insert, where we draw the two helical envelopes for the phosphates of the two strands, and the two helical envelopes for the sodium counter-ions, one penetrating into the major groove and the other outside to the minor groove.* The counter-ions helices are designated by the letters H and H*. From the density projections it is evident that the H helix is external to the cylindrical volume determined by the phosphate groups, whereas the H* helix is internal to it. The cross section of the imaginary "cable" enveloping the ions of H* is larger than the corresponding crossection of the "cable" enveloping the ions of H. The physical reason is rather obvious:the counter-ions in H* are strongly affected by the base-pairs. Therefore the *exact position* for an ion in H* *is base-pair sequence dependent; the* "irregularities" of the distributions 11 to 20 in the bottom-left insert of Figure 25, are the effect of the base-pair sequence dependence.

The new structure, H and H, obtained in our computer-experiment is physically very pleasant since it optimizes at once several basic energetic requirements:* 1) it keeps the counter-ions as far as possible away one from the other but at the same time; 2) it satisfies the *very strong attraction* between the Na^+ ions and the free oxygen atoms of the phosphates as well as 3) the *strong attraction* between the Na^+ ions and the bound oxygen atoms in the phosphate groups and the *attraction* to the base-pairs, 4) it allows *both* the phosphate groups and the counter-ions to be solvated by the water molecules, thus making use of the solvation energy to stabilize the entire system.

Biologically this structure is very interesting since:1) *it provides for a base-pair recognition mechanism at long range distances due to the base-pair sequence dependent H* structure, 2) it can easily allow for the exchange of one or more sodium counter-ions with different counter-ions (from Na^+ to K^+ or Mg^{++} or Ca^{++} etc. etc.) 3) it can account for rapid structural and conformational reorganization processes, typical of ionic solution with macromolecules as solute and 4) because of the ionic mobility and very strong interactions, it can act either as an important sink (or source) of energy.*

Determination of the Water-Structure Solvating DNA

The statistically most probable distribution of the water molecules (at 300 K temperature) either *bound* or in *the first solvation shell* for an atom (or groups of atoms) in B-DNA is analyzed in Figures 17 to 21. In these figures the probability distribution for the hydrogen atoms is given as a *dotted line* and the one for the oxygen atoms is given as a *full line.* On one axis (ordinates) we report the number of hydrogen or oxygen atoms as a function of R (given in the abscissa, in A units); we recall that R is the distance of the hydrogen (or oxygen) atoms from the Z axis (see Figure 15). The notations B and FS differentiate between *bound* and *first solvation shell* water molecules; the notations h and h* differentiate the two strands.

In Figure 17, we report the analysis for water molecules *bound* to the free oxygen atoms (O1P and O2P), and to the bound oxygen atoms (O5′ and O3′) in the two strands (h and h*). These analyses of water molecules bound at atomic sites are complemented with the *bound* water distributions and the *first solvation shell distributions* at the PO_4^--CH_2 group.

We learn that *at the O1P* sites the water oxygen atoms in the h strand have a maximum at about 11 A, whereas the maximum is at 12 A in the h* strand; the water hydrogen atom distributions (dotted lines) in h differ from those in h*, therefore indicating different orientational arrangements at the two strands. The

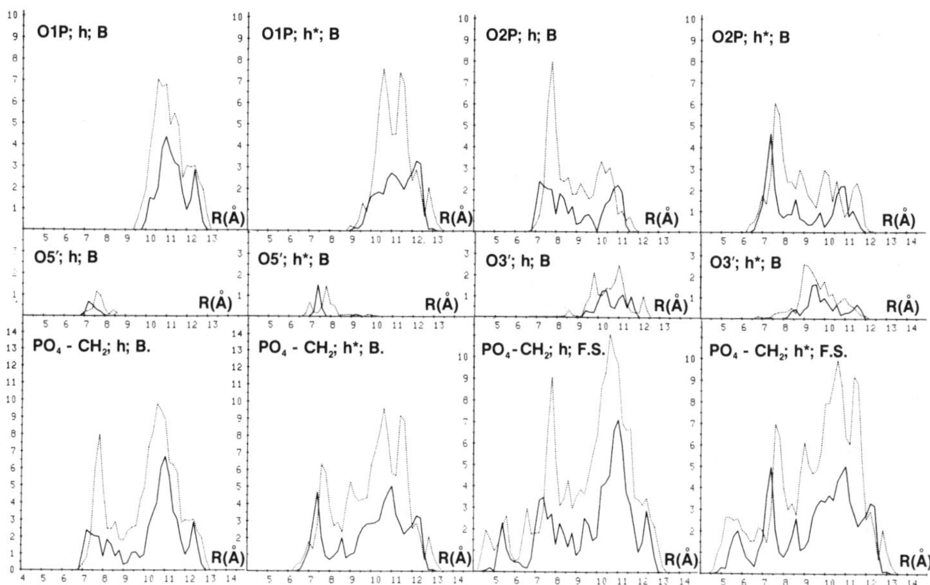

Figure 17. *Top* and *Central* inserts. Distribution of the hydrogen and oxygen atoms of water molecules bound to the oxygen atoms in the PO_4^- groups; *bottom inserts:* equivalent distribution for the PO_4^--CH_2 group either considering bound (first two insert from the right) or first solvation shell water molecules (last two inserts from the right).

same holds for the water molecules at O2P. There is little, if any, water at the O5′ as previously noted and not much water at O3′. This information is iterated by providing the distribution for the water molecules bound at the PO_4^--CH_2 site; as discussed above the first solvation shell can extend much further than the *bound* water distribution, as clearly seen in the figure. The integral of the distribution values of the hydrogen and oxygen atoms as function of R provides the number of water molecules *bound* to the atoms or groups of atoms, reported in this figure. These are given in Table V, both for *bound* and *first hydration shell* water molecules. In the following, we shall (in general) not comment on the features which can be obtained by inspection of the self-explanatory figures. We feel that infrared, Raman, nmr, neutron, X-ray experiments on DNA in solution will now be more easily interpreted, such data being available.

In Figure 18, we report the analysis for the water molecules *bound* to the lone pair nitrogen (N3 and N7) and oxygen (O2, O4 and O6) atoms of the bases. The analysis of the water molecules bound to atoms belonging to the bases is extended in Figure 19 when we report the distributions for the water molecules bound to specific NH_2 groups and the *average* distributions for water molecules bound to NH_2 or oxygen or nitrogen atoms at the bases obtained by considering all the bases at the two strands.

In Figure 20, we compare the distributions of water molecules bound at the O1′ (of

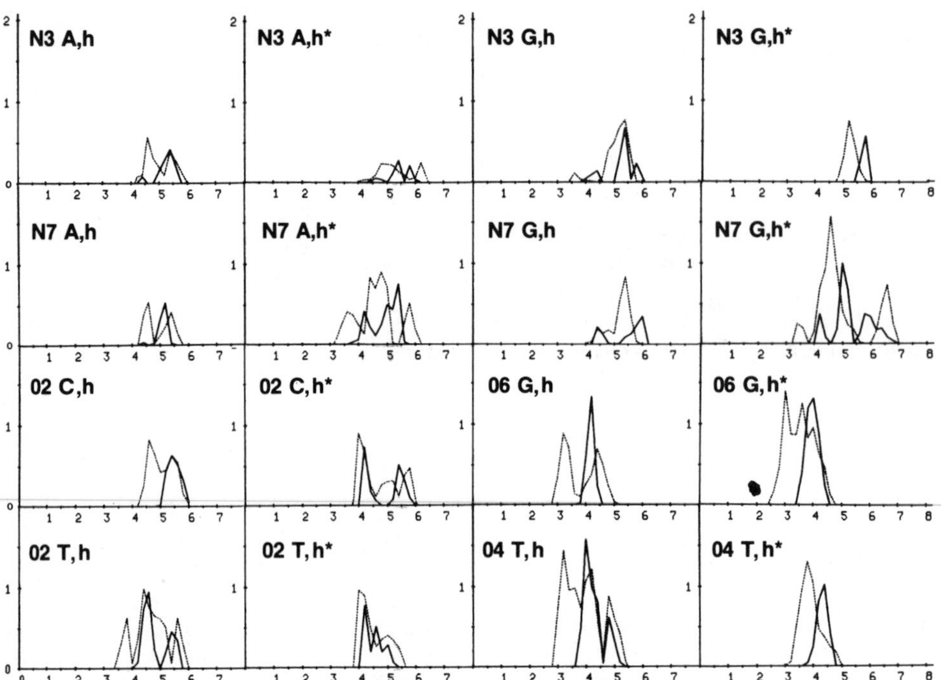

Figure 18. Distribution of hydrogen and oxygen atoms for water molecules *bound* at the nitrogen (N3 and N7) or oxygen (O2, O4 and O6) of bases forming either the h or the h* strand in B-DNA.

sugar) or the first solvation shell of the entire sugar unit. In this figure we also report the distribution of the water bound to the bases (in each strand).

In Figure 21, we complete the analysis of the bases, by reporting the average distribution of the water molecules bound to the A-T and G-C base-pairs, and the average distribution for the base-pairs either in the h or h* strands, and the first hydration shell distribution at the bases (average values for both strands).

These very detailed but coincided graphical presentations of the solvation in B-DNA, are complemented by the data of Table V.

The analysis above reported for 400 water molecules (per B-DNA turn) has been extended to intermediate and low relative humidity; we have now concluded our computer experiment by considering 380, 240, 220, 180, 140, 40, and 20 water molecules per B-DNA turn. *The H and H* structures for the counter-ions have been found at each relative humidity; therefore it is very reasonable to assume that such structures are present in solution.*

Determination of the Water Structure Solvating the Counter-ions

In Figure 22, we report the distribution of the water molecules bound to the ten

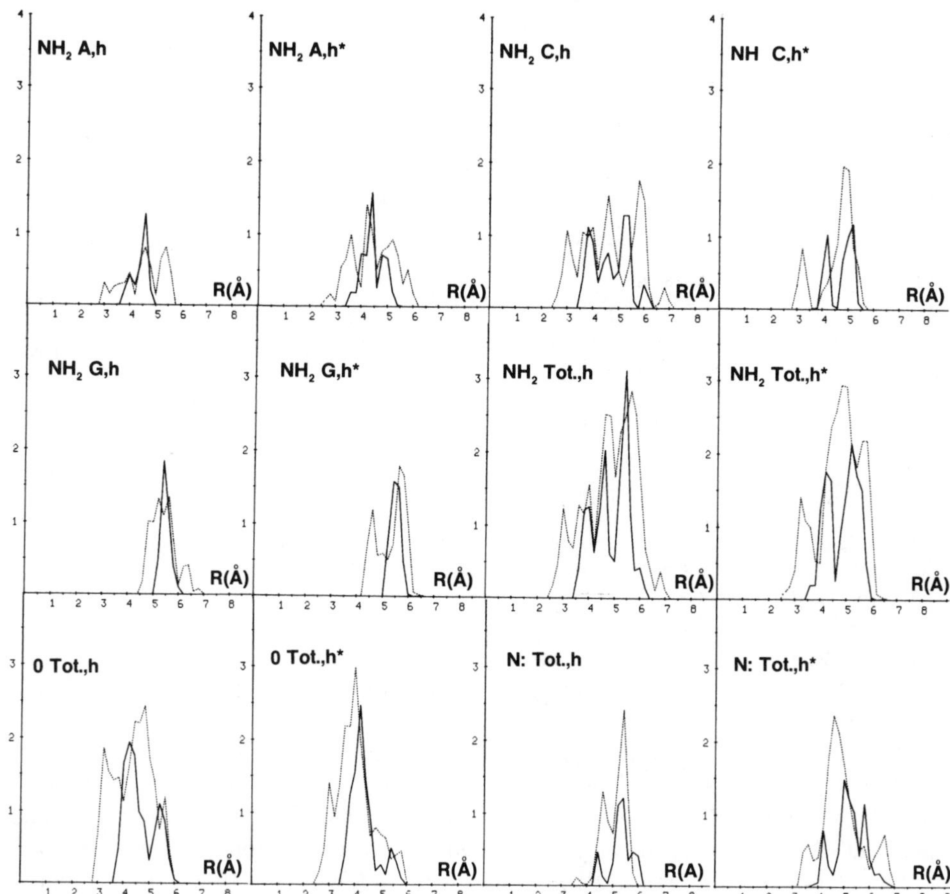

Figure 19. *Top and central* insert: distribution of hydrogen of oxygen atoms for water molecules *bound* at the NH_2 groups of the bases. The four bottom inserts report equivalent quantities but for an average oxygen (either O2, O4 or O6) and an average nitrogen (either N3 or N7) in the bases at the two anti-parallel strands, h and h*.

counter-ions in the H helix (ions 1 to 10) and to the ten counter-ions in the H* helix (ions 11 to 20). In this figure the counter-ion is placed at the *origin* of the axis. The orientation of the water molecule (oxygen nearer, hydrogen farther away) is very typical of a sodium ion in solution[13,35,36]. Each counter-ion is solvated; this finding is expected to be valid also for K^+ counter-ions, and most likely for the Li^+ counter-ions.

In Figure 23, we report the distribution for the water molecules in the two strands either *bound* or in the first hydration shell. From these data we see clearly that the two strands have a different hydration pattern, as also reported in Table V. This very important finding remains hidden when one reports the distributions for the total system of 400 water molecules solvating the ten base-pair turn in B-DNA or when one considers the water molecules in either the first solvation shell or those

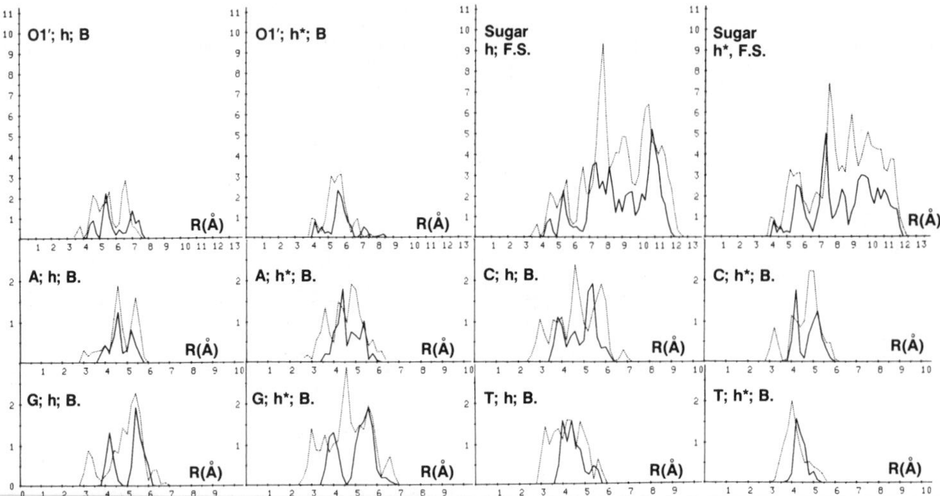

Figure 20. *Top inserts:* water characterization at the sugar units in the two strands:bound water at O1′ atom, and *first solvation shell* water in the sugar units. *Middle and Bottom insert:* distribution for water molecules *bound* to the bases.

bound to B-DNA. In Figure 24, we present the distribution of water molecules bound to B-DNA and to the ions, those in the first hydration shell and those in the grooves.

In the relative humidity range from 400 to 240 water molecules (per B-DNA turn) we find all the time about *four water molecules* per Na^+ counter-ion. By decreasing the relative humidity, namely for the cases of 220, 180, 40 and 20 water molecules (per B-DNA turn) the *average hydration number* for the 20 counter-ions decreases to 3.8, 3.5, 1.5 and 0.8, respectively. Therefore, at low humidity a water molecule solvates either the counter-ions, or B-DNA. This result should warn against extrapolations (relating to B-DNA in solution) of quantum mechanical computations obtained by considering few water molecules and one counter-ion.

In Figure 25, we report the average energy for a water molecule in the volume R and R+dR; the energy is decomposed as water-water, water-DNA, water-ions. Notice how this energy is nearly constant from small to large values of R.

Base-Pair Recognition

The important conclusion from our computational experiment is that the counter-ions and the solvation water molecules form two different patterns at the two DNA strands. We designate the *global system* composed by the h strand, the counter-ions in the H helix and the water molecules bound to h and H as the S *"super-strand"*. The equivalent global system for h* and H* and the solvating water molecules is designated as *S**. As known, the two strands h and h* differ only because they are anti-parallel. This difference, however, is enhanced by the counter-ions distribution and by the water molecules. Any biological process dealing with DNA in water

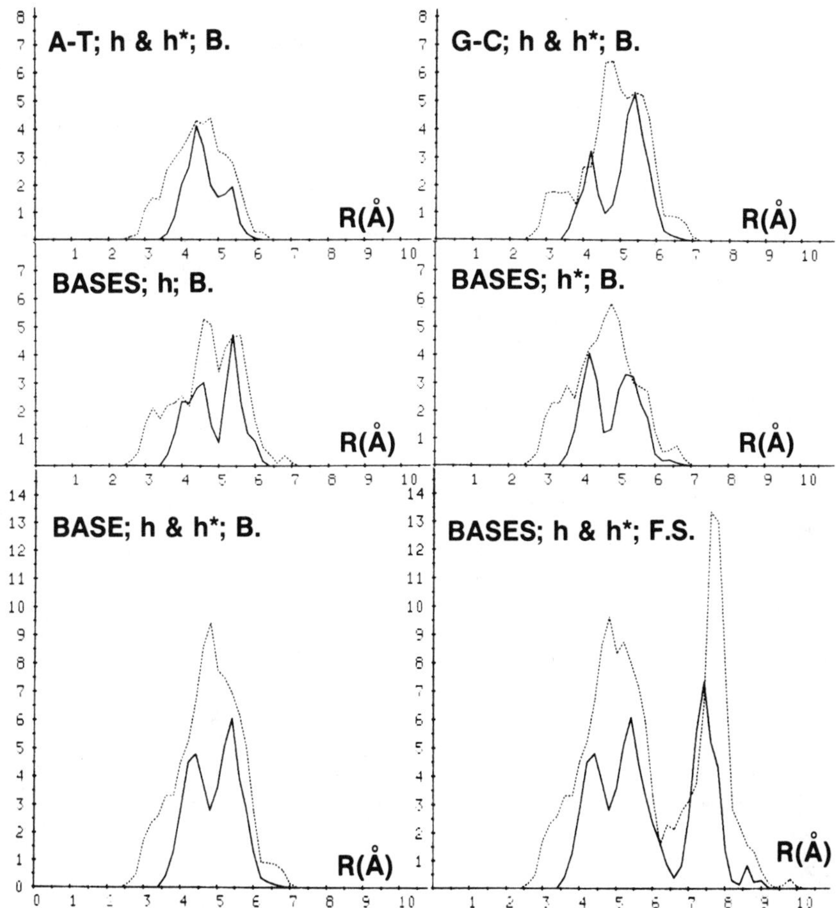

Figure 21. *Top inserts:* Distribution for water molecules at the base-pairs. *Middle inserts:* average distribution for water molecules *bound* in the two strands; *bottom:* average distribution for both strands of water molecules *bound* (left) and in the first solvation shell (right) of the bases.

solution deals with the S and S* super-strands and not only with the h and h* strands. In addition, to a given DNA conformation, there is a corresponding wide spectrum of S and S* conformations. Indeed, the structure of H and H* is dependent upon the *counter-ion charge,* the *ionic radius* and the *counter-ion concentration* for a given *temperature.*

Let us consider a few immediate implications of the above findings. In the following we present a *"base-pair sequence"* recognition mechanism. *Let us consider a molecule,* for example a glycine zwitterion, *approaching DNA in water solution, but still relatively far away from DNA, such that the direct interaction glycine-base-pairs can be assumed as small.* For example, we assume that the C^{α} of glycine is at 17 A on the x axis (with y=0; see Figure 15) and optimally oriented relative to our B-DNA fragment, the 1200 water molecules and the 60 sodium counter-ions. Fur-

Table V
Number of water molecules and its interaction energy (in Kcal/mol) with atoms or groups of atoms of B-DNA and the Na^+ ions.

Atoms or Groups	Number of Water Bound		Average Energy	
	h	h*	h	h*
O1P	3.24	3.36	−37.97	−34.57
O2P	2.92	3.05	−38.33	−41.11
O5′	0.18	0.23	−37.76	−43.52
O3′	0.89	1.12	−37.51	−34.24
O1′	1.19	1.13	−39.00	−37.81
N3 in A	0.49	0.26	−39.07	−34.97
N7 in A	0.46	0.94	−37.97	−37.73
N3 in G	0.72	0.29	−36.70	−39.74
N7 in G	0.60	1.20	−33.03	−41.94
O2 in C	0.67	0.97	−39.91	−37.33
O6 in G	1.08	1.34	−35.91	−40.53
O2 in T	1.00	1.00	−41.08	−38.94
O4 in T	1.62	1.26	−36.97	−29.29
Na^+	4.00	4.40	−39.72	
PO_4^--CH_2	6.43	7.18	−38.04	−37.09
NH_2 in A	1.49	1.77	−39.79	−37.65
NH_2 in C	2.76	2.31	−38.24	−39.18
NH_2 in G	1.94	1.46	−36.23	−39.93
NH_2 in A, G, C	2.13	1.70	−37.81	−39.04
N in A, G	0.57	0.67	−36.40	−38.85
O in C, G, T	1.00	1.06	−38.61	−38.35
A	2.42	2.55	−40.03	−37.46
C	3.43	3.29	−38.58	−38.43
G	3.84	4.15	−35.77	−40.68
T	2.62	2.26	−38.51	−37.68
Grooves	131.82		−30.41	

ther, we assume that glycine is *translated* along the z axis by steps of 0.25 A, re-optimizing its *orientation* at each value of Z.

The interaction of the counter-ions with glycine is rather large (long range interaction of ionic nature), and the interaction of the H* counter-ions with the base-pair has been previously shown to be large and base-pair sequence dependent. Therefore, glycine will recognize the base-pair sequence, via the counter-ions. *The proposed recognition mechanism is a relay-type mechanism: base-pairs to counter-ions, counter-ions to glycine.* A disordered pattern in the ions, rather than the ordered one we have determined, will lead to no base-pair recognition. Only a *special* ordered pattern lead to recognition. Further, the recognition in our model is dependent upon *ion concentration,* ionic *radius* and temperature. This latter comment is of importance in study of the evolution of genetic proto-materials. Among feasible application of this proposed mechanism we mention: 1) recognition of a sequence perturbed by cancer or an anti-cancer intercalating molecule and 2) recognition of amino acids by RNA in protein syntheses.

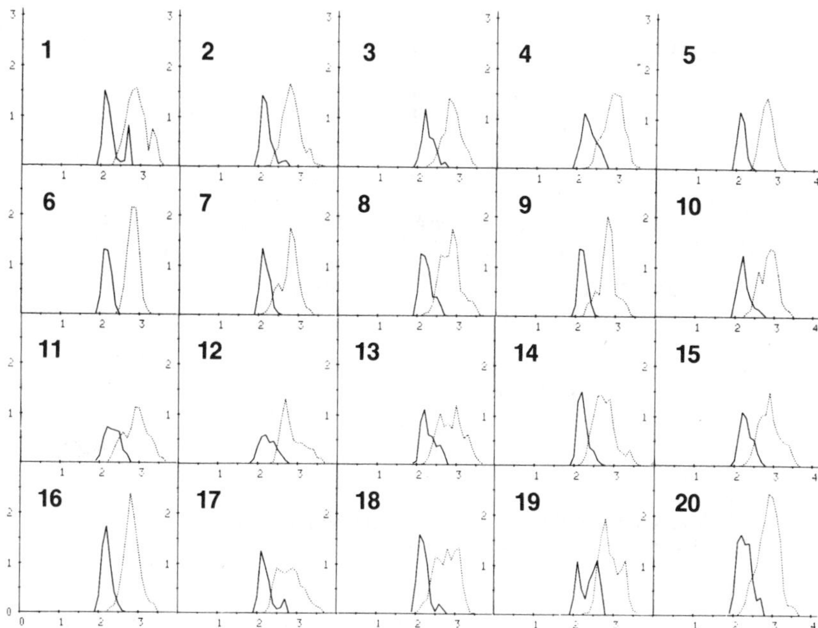

Figure 22. Hydrogen and oxygen atoms distributions for water molecules *bound* to the counter-ions in the H helix[1-10] and in the H* helix[11-20].

In Figure 26 (left insert), we report the interaction energies of GLY (in Kcal/mol) with the atoms of the B-DNA fragment, with the water molecules solvating B-DNA with the counter-ions and the total interaction energy. The interaction of GLY with water is nearly constant and repulsive (screening effect). The interaction of GLY with B-DNA shows a *low* frequency periodicity associated with the B-DNA turn and a high frequency periodicity associated with the nucleotide units. The interaction with the counter-ions shows the low frequency periodicity but with opposite phase as the one for the GLY-B-DNA interaction. The GLY counter-ion interaction is attractive and over compensate (being larger) the repulsive interactions with water and B-DNA. The high frequency spikes in the GLY-B-DNA and in GLY counter-ions are separated by about the same distance as the base-pair to base-pair distance. Clearly the pattern will differ in A-DNA, and it will *locally* differ if a molecule intercalates DNA. Notice that the "recognition spikes" in the total interaction energy are about 10 Kcal/mol, namely a value sufficiently large for being very important in biological mechanisms, but also sufficiently low as to be affected by thermal effects. Notice how one pair has a pattern different from another pair. To our knowledge this figure represents the first quantitative energetic representation of the reading of DNA by a molecule (GLY in our experiment).

DNA Unwinding

Another implication of our findings concerns the unwinding mechanism in the double helix. As known, a double helix structure has a critical temperature and a

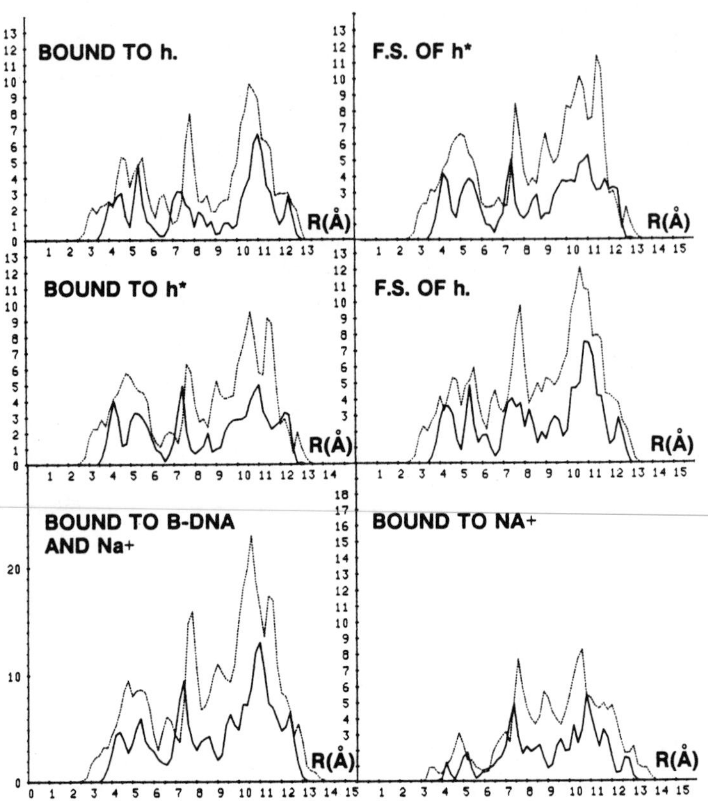

Figure 23. *Top and central* inserts: average hydrogen and oxygen atoms distributions for water molecules in the h and h* B-DNA strands either *bound* or in the *first hydration shells. Bottom inserts:* water molecules bound to B-DNA and to the counter-ions (left) or only to the counter-ions (right).

critical ionic concentration, just beyond of which the two helices extremely rapidly snap apart cooperatively. By a 20 K increase in the temperature (in our simulated system) we obtain a different counter-ions pattern with the counter-ions in the H* helix closer to the base-pairs than at 300 K. We recall that the interaction of an Na^+ ion with the base-pairs, is not only strong when Na^+ is at the perimeter of the base-pair (in the plane containing the base-pair skeleton) but also when Na^+ is *above* the base-pair. In this position the attraction "base-pair to Na^+" is opposed by the hydrogen atoms forming the base-pairs hydrogen bonds. An increase in the system thermal motion (due to temperature) can bring about a separation between two successive base-pair and/or hydrogen bonds breakage within a base-pair. In either case, a sodium ion can approach the bases even further, and oppose the restoration of the original DNA configuration.

The disruption of the double helix structure following progressive removal of counter-ions at constant temperature is easily understood in terms of the large stabilization brought about by the ions to the system "water and DNA"[32]. We note

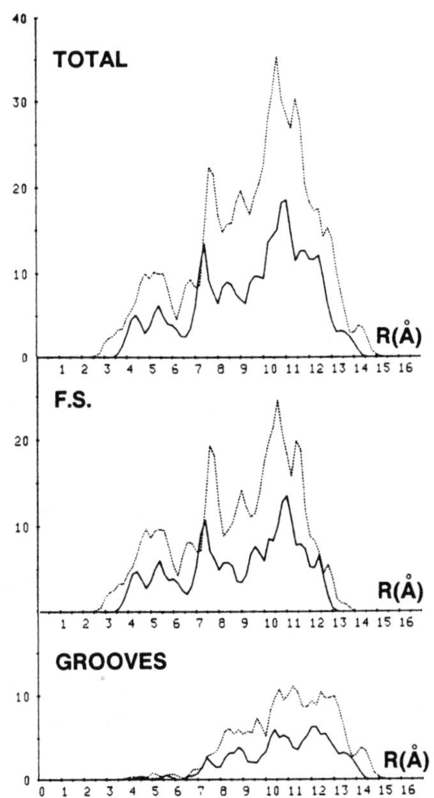

Figure 24. Statistical hydrogen and oxygen atom *total* distribution for 400 water molecules solvating one B-DNA turn and 20 Na+ counter-ions (top) in B-DNA *first solvation shell* (middle) and in the *grooves* (bottom).

that the total energy of the system reported in this work is more stable by about 20 Kj/mol, than the system analyzed in Reference 32.

From our energy data we estimate that the DNA and the counter-ions field extends up to about R=25 A. Therefore, we expect that X-ray crystal studies from single crystals should show evidences of the DNA to DNA perturbation. As a consequence the counter-ion structure in a single crystal is expected to differ from the counter-ion structure of DNA in solution.

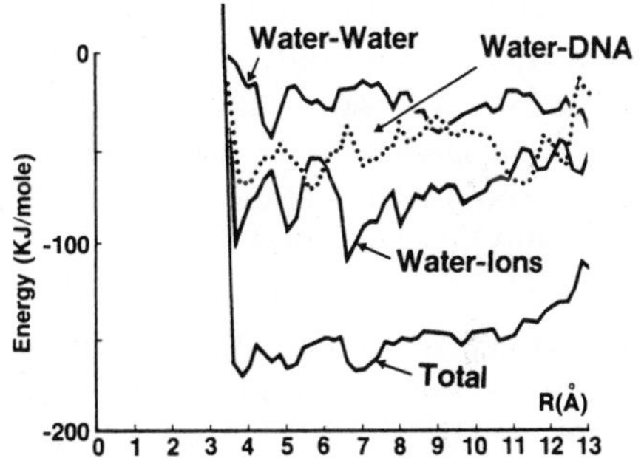

Figure 25. Average interaction energy (Kj/mol) for water with the remaining water molecules (water-water), with the B-DNA fragment (water-DNA), with the ions (water-ions) and the total interaction as function of R.

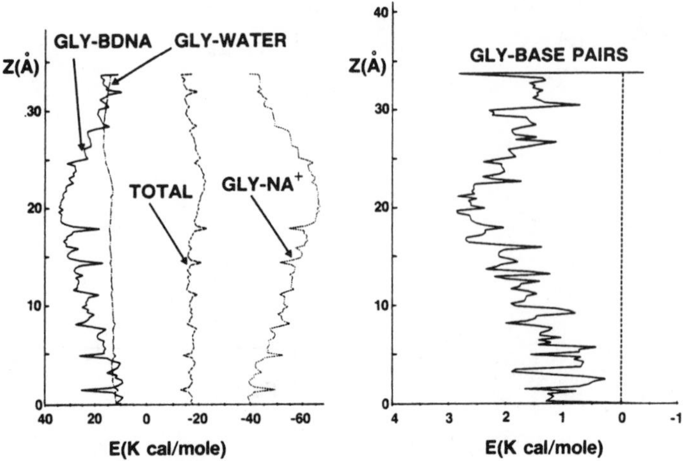

Figure 26. Recognition by GLY of DNA:Interaction energies (in Kcal/mol) of GLY with B-DNA, with the water molecules solvating DNA, with the counter-ions and total interaction; interaction energy of GLY with base-pairs is shown on the right-hand side panel.

Conclusions

Study is in progress for the determination of the counter-ion structure for Li^+, K^+, Mg^{++} and Ca^{++}. In addition, the approximation of the rigid solute is under scrutiny.[51,52] In a recent work[53] on the agar-agar double helix, the rigid structure in the solute has been partially relaxed; the experience gained in this recent work will be transferred to future DNA simulation. These computer experiments are now being extended in order to assess the optimal number of counter-ions for our B-DNA fragment. *From preliminary simulations, at low relative humidity and 300 K,* we obtain that 19 sodium counter-ions (per B-DNA turn) bring about a *net stabilization* in the total interaction energy of about 0.58% relative to the case with 20 sodium-ion (per B-DNA turn). This net stabilization results mainly from about a 5% decrease in the ion-ion repulsion and an increase of about 0.10% in the ion-DNA attraction. If we simulate (at the same low relative humidity and at 300 K, as above) 21 counter-ions per B-DNA turn, then we obtain a net *destabilization* of about 0.64% in the total interaction energy of the system, relative to the case with 20 sodium counter-ions per B-DNA turn; the destabilization is mainly the effect of an increase of the ion-ion repulsion (by about 5%) and a decrease (0.07%) in the ion-DNA attraction. When we consider either 18 or 17 or 10 counter-ions, we obtain no additional stabilization for the 18 ions case (relative to the 19 ions case) and we notice a destabilization for the 17 and 10 ions cases.[54] This type of study should allow us to combine our micro-analyses with the thermodynamical models presented by Record et al.[55] and by Manning.[56]

The model proposed for conformational transitions (see the section Conformational Transitions) can be adapted most easily, from the simulations where the counter-ions are fixed at a predetermined position, to simulations where the counter-ions are mobile in the water solvent. The specific solvation energy of different ions in water, corrected by the specific DNA field effect (H and H* structures) constitutes an appreciable and ion-specific contribution to ES(A) and ES(B). Computer experiments in progress[54] appear to nicely reproduce laboratory ion-induced conformational transitions[39-42,57] and observed melting point at different ionic strength.[58]

In this review, we have omitted a detailed analysis of the approximations adopted, and we have not stressed the generality of our approach for the *"computer experiments",* reported. The interested reader can find such information in references 29 and 50.

One trend is becoming more and more evident:the very approximated nature of the "quantum-mechanical rationalizations" of laboratory experiments, (rather conspicuous in the sixties and still retained, for example, in studies based on approximated electrostatic potentials) is becoming more and more apparent[59] and therefore, there is an increasing reliance on those theoretical formulations, where essential parameters like *temperature, statistical distributions, time, solvents* and *reaction fields* are no longer ignored. Indeed these formulations, proposed about twenty to thirty years ago and generally, long neglected in quantum-biophysics, are complementary to laboratory experiments. This trend is emerging not accidentally, but because more and more attention is given to dynamical aspects in nucleic acid and protein chemistry.

References and Footnotes

1. Scordamaglia, R., Cavallone, F. and Clementi, E., *J. Am. Chem. Soc. 99,* 5545 (1977).
2. Clementi, E. and Corongiu, G., *J. Chem. Phys. 72,* 2979 (1980).
3. Clementi, E. and Corongiu, G., *Biopolymers 18,* 2431 (1979).
4. Clementi, E. and Corongiu, G., *Int. J. Quant. Chem. 116,* 897, (1979).
5. Clementi, E. and Corongiu, G., *Chem. Phys. Letters 60,* 175 (1979).
6. Clementi, E. and Corongiu, G., *Gazz. Chim. It. 109,* 201 (1979).
7. Romano, S. and Clementi, E., *Int. J. Quant. Chem. 17,* 1007 (1980).
8. Clementi, E., Corongiu, G., Jonsson, B. and Romano, S., *FEBS 100,* 313 (1979).
9. Clementi, E., Corongiu, G., Johnsson, and Romano, S., *J. Chem. Phys.* 260 (1980).
10. Lewin, S., *J. Theor. Biol. 17,* 181 (1967).
11. Clementi, E., and Corongiu, G., *Biopolymers 20,* 551 (1981).
12. Lie, G. C., Yoshimine, M. and Clementi, E., *J. Chem. Phys. 64,* 2314 (1976); Matsuoka, O., Yoshimine, M. and Clementi, E., *J. Chem. Phys. 64,* 1351 (1976).
13. Clementi, E., *Determination of Liquid Water Structure,* Lecture Notes in Chemistry, Vol. 2, Springer-Verlag, Berlin, (1976).
14. Falk, M., Hartman, K. A. and Lord, R. C., *J. Am. Chem. Soc. 84,* 3843 (1962).
15. Falk, M., Hartman, K. A. and Lord, R. C., *J. Am. Chem. Soc. 85,* 387 (1963).
16. Falk, M., Hartman, K. A. and Lord, R. C., *J. Am. Chem. Soc. 85,* 391 (1963).
17. Rupprecht, A. and Forslind, B., *Biochim. Biophys. Acta 204,* 304 (1970).
18. Hearst, J. E. and Vinograd, J., *Proc. Natn. Acad. Sci. U. S. A. 47,* 825 (1961).
19. Hearst, J. E. and Vinograd, J., *Proc. Natn. Acad. Sci. U. S. A. 47,* 999 (1961).
20. Hearst, J. E. and Vinograd, J., *Proc. Natn. Acad. Sci. U. S. A. 47,* 1005 (1961).
21. Wolf, B. and Hanlon, S., *Biochemistry 14,* 1661 (1975).
22. Tunis, M. J. B. and Hearst, J. E., *Biopolymers 6,* 1325 (1968).
23. Tunis, M. J. B. and Hearst, J. E., *Biopolymers 6,* 1345 (1968).

24. Kuntz, I. E., Branfield, T. S., Law, G. A. and Purcell, G. V., *Science 163,* 1329 (1969).
25. Privalov, P. L., Ptitsyn, O. B. and Birshtein, T. M., *Biopolymers 8,* 559 (1969).
26. Texter, J., *Prog. Biophys. Molec. Biol. 33,* 83 (1978).
27. Dahlborg, U. and Rupprecht, A., *Biopolymers 10,* 849 (1971).
28. Corongiu, G. and Clementi, E., *Gazz. Chim. It. 108,* 687 (1978); *108,* 273 (1978).
29. Clementi, E., Lecture Notes in Chemistry Vol. 19, *Computational Aspects for Large Chemical Systems* Springer-Verlag, Heidelberg, Berlin, New York, (1980).
30. Clementi and E., Corongiu, G., & Lelj, F., *J. Chem. Phys. 70,* 3726 (1979), (and references, thereby given). For the Na^+ and Li^+, K^+, Mg^{++}, Ca^{++}, atom-atom pair potentials with DNA see G. Corongiu and E. Clementi (to be published).
31. Fieldman, R., *Atlas of Macromolecules* document 13 (1976), Natl. Inst. Health, Bethesda, Maryland, U. S. A. For earlier references see: Arnott, S. and Hukins, D. W. L., J. Molec. Biol., *81,* 93 (1973). Arnott, S. and Hukins, D. W. L., Biochim. Biophys. Res. Commun., *47,* 1502 (1972).
32. Corongiu, G. and Clementi, E., *Biopolymers* (in press).
33. Perahia, M. S. J. and Pullman, B., *Biochim. Biophys. Acta 474,* 349 (1977), and references therein given. The very extended number of papers referenced in 33 are very similar in the technique adopted.
34. Ranghino, G., and Clementi, E., *Gazz. Chim. It. 109,* 170 (1978).
35. Barsotti, R. and Clementi, E., *Theor. Chim. Acta 42,* 101 (1977). See also Marynick, D. A. and Schaffer, III, H. F., *Proc. Nat. Acad. Sci. U. S. A.* 3794 (1975).
36. Clementi, E. and Barsotti, R., *Chem. Phys. Letters 59,* 21 (1978).
37. Marynick, D. S. and Schaeffer, H. F., *Proc. Nat. Acad. Sci. U. S. A. 72,* 3794 (1975).
38. Ross, P. D. and Scruggs, R. L., *Bioplymers 2,* 79 (1964).
39. Frisman, E. V., Slonitsky, S. V. and Vaselkov, A. N., *Int. J. Quant. Chem. 16,* 847 (1979).
40. Frisman, E. V., Vaselkov, A. N., Solnitsky, S. V., Karavaev, L. S. and Vorob'ev, V. E., *Biopolymers 13,* 2169 (1974).
41. Ivanov, V. I., Zhurkin, V. B., Zavriev, S. K., Lysov, Yu. P., Minchenkova, L, E., Minyat, E. E., Frank-Kametskii, N. D. and Schyolkina, A. K., *Int. J. Quant. Chem 16, 1*89 (1979), see in addition Zhurkin, V. B., Lysov, Yu. P. and Ivanov, V. I., *Biopolymers 17,* 377 (1978).
42. Sukhorukov, B. I., Gukowsky, I. Ya., Pekrov, A. I., Gukowskaya, A. S., Mayevsky, A. A. and Guenkova, N. M., *Int. J. Quant. Chem. 17,* 339 (1980); Bunville, L. G., Geiduschek, E. P., Rawitscher, M. A. and Sturdevant, J. M., *Biopolymers, 3,* 213 (1965).
43. Seeman, N. C., Rosenberg, J. M., Suddath, F. L., Kim, J. J. P. and Rich, A., *J. Mol. Biol. 104,* 109 (1976).
44. Rosenberg, J. M., Seeman, N. C., Day, R. O. and Rich, A., *J. Mol. Biol. 104,* 145 (1976).
45. Camerman, N., Fawcett, J. K. and Camerman, A., *J. Mol. Biol. 107,* 601 (1976).
46. Drew, H. R., Takano, T., Tanaka, S., Ikatura, K. and Dickerson, R. E., *Nature 286,* 567 (1980).
47. Drew, H. R. and Dickerson, R. E., *J. Mol. Biol* (in press). We thank the above authors for having sent us a preprint of this paper.
48. Klug, A., Jack, A., Viswamitra, M. A., Kennard, O., Shakked, Z. and Steitz, T. A., (1979), *J. Mol. Boli. 131,* 669 (1979), and references therein given. We thank Dr. Kennard for preprints of their work.
49. Clementi, E. and Corongiu, G., *Int. J. Quant. Chem.* (submitted) (1981).
50. Clementi, E., IBM *J. Res. and Dev. 25,* 315 (1981).
51. Schellman, J. A., *Biopolymers, 13* 217 (1974).
52. Olson, W. K., *Nucleic Acid Geometry and Dynamics* Sarma, R. H., Ed., Pergamon Press, New York, pg. 383 (1980).
53. Fornili, S., Corongiu, G., Palma, U. and Clementi, E., (to be published).
54. E. Clementi (to be published).
55. Record, M. T., Jr., Anderson, F. C. and Lohman, T. M.,*Quart. Rev. Biophys. II, 2,* 103 (1978); Anderson, C. F. and Record, M. T., *Biophys. Chem., 77,* 353 (1980).
56. Manning, G.S., *Quart. Rev. Biophys. II,* 2179 (1978); see also Gueron, M. and Weinsbuck, G. *Biopolymers 19,* 353 (1980).
57. Chan, A., Kilkuskie, R. and Hanlon, S., *Biochem., 18,* 84 (1979); Anderson, P. and Bauer, W.

Biochem., 17, 594 (1978); Pohl, F. and Jovin, T., *J. Mol. Biol. 67,* 375 (1972); Hanlon, S., Chan, A. and Berman, G. *Biochem. Biophys. Acta, 519,* 526 (1978).
58. Schildkraut, C. and Lifson, S. *Biopolymers, 3,* 195 (1965).
59. Clementi, E., *J. Phys. Chem., 84,* 2122 (1980).
60. Dickerson, R. E., Drew, H. R., and Conner, B., in *Biomolecular Stereodynamics, Volume I,* Ed., Sarma, R. H., Adenine Press, NY p. 1-xx (1981).
61. Work partially supported by the National Foundation for Cancer Research.

Proceedings of the Second SUNYA Conversation in the Discipline Biomolecular Stereodynamics
Volume I, ISBN 0-940030-00-4, Ed., Ramaswamy H. Sarma,
Adenine Press, New York, ©Adenine Press

Intercalation of Water Molecules Between Nucleic Acid Bases: Sandwiches and Half-Sandwiches

R. Parthasarathy, T. Srikrishnan and Stephan L. Ginell
Center for Crystallographic Research
Roswell Park Memorial Institute
Buffalo, New York 14263, U.S.A.

Introduction

The intercalation of dyes and drugs into nucleic acids is well known and has been well studied.[1-4] Most of the simple intercalating agents contain positively charged heterocyclic aromatic rings. In this paper, we show that the intercalation between nucleic acid bases is not the prerogative of charged aromatic systems only. Water molecules intercalate between modified pyrimidine bases and form 'sandwich' structures. Modified purines seem to prefer 'half-sandwich structures' in which water molecules stack on top of purines. We have studied the stereochemistry of such water intercalations in nucleic acids and we suggest that water intercalations will be important in the dynamics of nucleic acids and in facilitating the entry of drug and intercalating agents into nucleic acids.

Sandwich Structures[5-6]

Three structures with novel intercalation of water molecules between pyrimidine bases 6.2 to 6.6Å apart have been discovered our laboratory. The water molecule is situated between two bases at base-stacking distances of 3.11 to 3.4Å. These three structures are 5-nitro-1-(β-**D**-ribosyluronic acid) -uracil monohydrate,[7-8] 1-methyl-5-nitrouracil monohydrate[9] and 6-azathymine hemihydrate.[10] The first structure contains a stack of sandwiched bases, the second one has a layered arrangement in which pyrimidines and water molecules in alternate layers stack, and the third structure has local domains of two different sets of sandwiched water molecules between bases, related by centers of inversion and glide planes. Figures 1, 2 and 3 illustrate the edge view of these three types of sandwich formation and Figures 4, 5 and 6 show the top view and the overlap of the bases and the water molecules. The characteristic features of these sandwich structures are: (i) the water molecule participates only in three rather than four hydrogen bonds, (ii) the hydrogen bonding environment around water is planar (sp^2) rather than tetrahedral (sp^3), (iii) there is no hydrogen bonding between the intercalated water molecule and the bases on top or bottom and (iv) there is no stacking of the bases; instead, the water molecule acts as a 'spacer'. The water molecules do not seem to have any specific orientation with respect to the atoms in the bases on top and bottom. The closest pairs of atoms

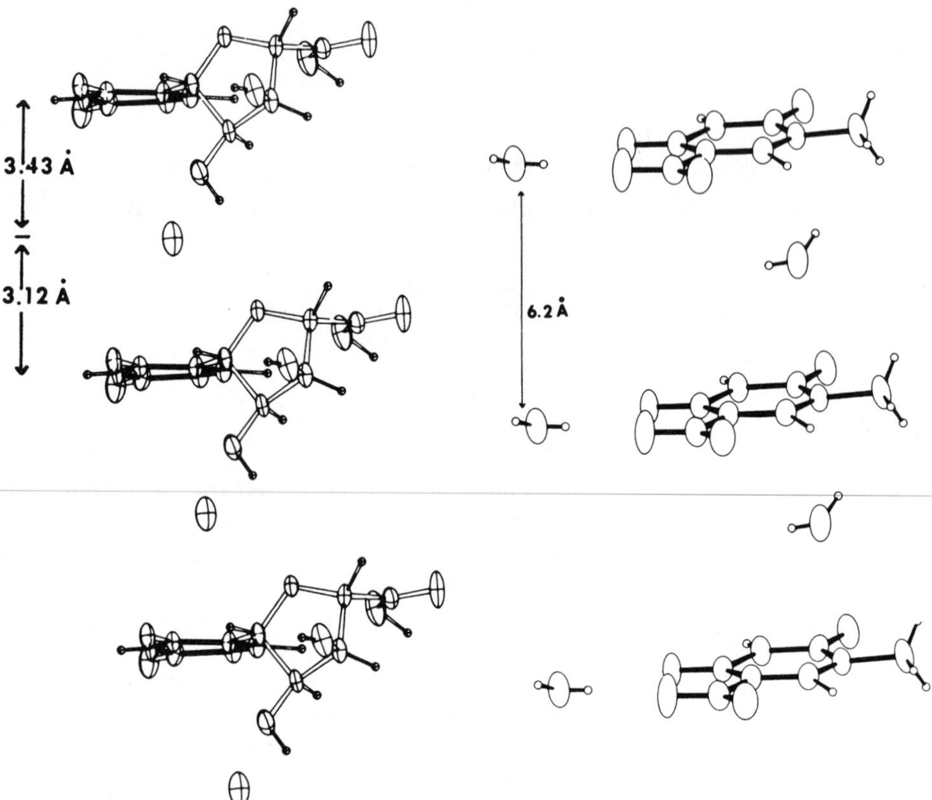

Figure 1. The intercalation of water molecules in the crystal structure of 5-nitro-1-(*B*-**D**-ribosyluronic acid)-uracil monohydrate. The molecules form a vertical column. The 5-nitro group has been omitted for clarity (from Ref. 5).

Figure 2. The intercalation of water molecules in the crystal structure of 1-methyl-5-nitrouracil monohydrate (from Ref. 9).

to the water molecules in the four sandwich structures are, respectively, N(1) and C(6), N(1) and N(1), N(1) and C(5), and C(2) and C(5). The sandwich formation has been observed so far only in pyrimidines.

Half-Sandwich Structures[11]

The purines seem to prefer the half-sandwich structures in which a water molecule stacks on top of a base; there is no base on top of the water molecule. Instead, in some structures, two bases forming a 'window' are located on top of the water molecule. There is no hydrogen bonding between the water molecule and the base on which it stacks (see Figures 7, 8). By re-examining structures reported in the literature, we found that these half-sandwich structures are observed in the crystal structures of guanosine hydrobromide monohydrate,[12] 5′-methylammonium-

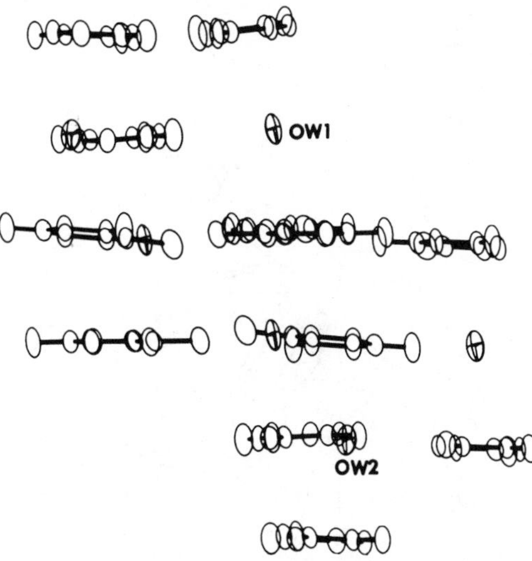

Figure 3. The 'sandwiching' of water molecules in the crystal structure of 6-azathymine hemihydrate. There are four independent molecules (A, B, C, D) of 6-azathymine and two independent water molecules (OW1, OW2) in the asymmetric unit.

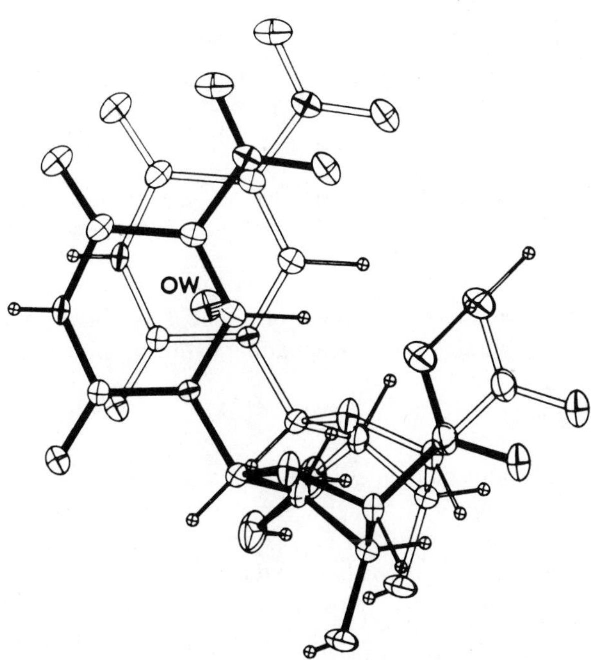

Figure 4. Partial overlap of the bases and the water in between (from Ref. 5).

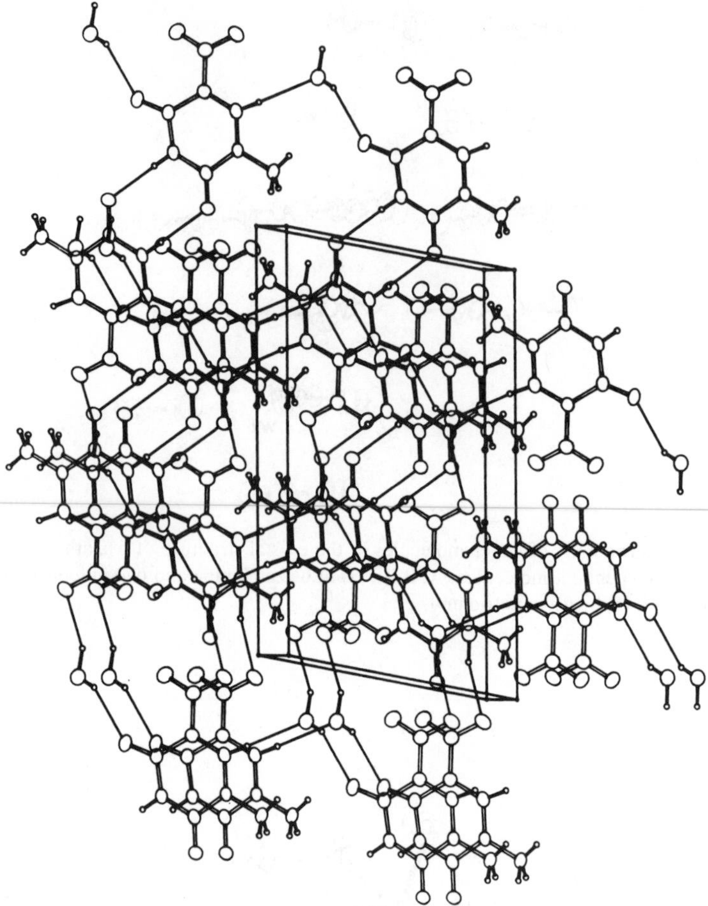

Figure 5. The layer structure of 1-methyl-5-nitrouracil monohydrate. Note the overlap of the bases and the water in between.

5′-deoxyadenosine iodine monohydrate,[13] orotic acid monohydrate,[14] 8-azaguanine hydrochloride monohydrate[15] and 8-azaguanine hydrobromide monohydrate.[16]

Nature of Base Stacking

The intercalation of water molecules between stacked bases is an unique situation, and may seem surprising at first sight. An extensive study of base stacking in crystals[17] indicates that base stacking in a crystal is specific and involves the interaction between polar regions of one base and polarizable regions of the other base. Solution studies[18] of bases and nucleosides show that base stacking takes place in water but not in organic solvents, leading to an interpretation that attributes stacking to "hydrophobic interactions".[18-19] Thermodynamic analysis of the solubility of bases in water and organic solvents shows that transfer of a nucleic acid

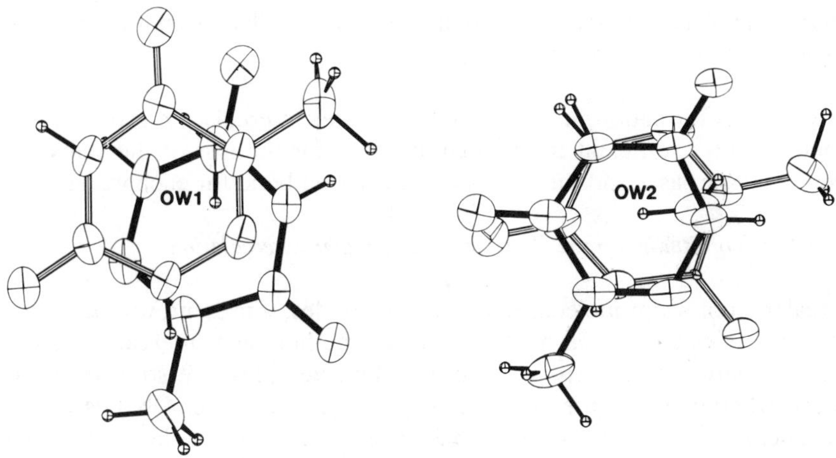

Figure 6. The intercalation of water molecules between 6-azathymine molecules.

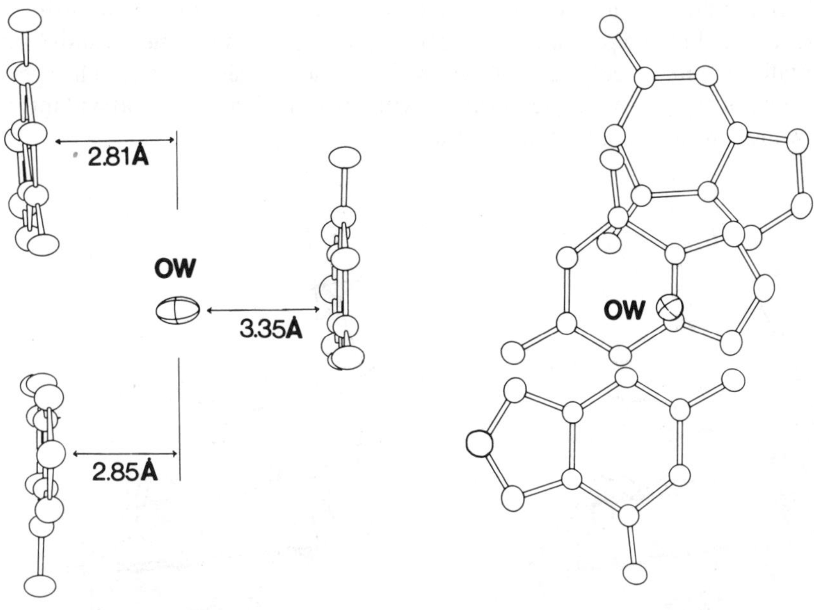

Figure 7. Half-sandwich formation in the crystal structure of 8-azaguanine hydrochloride monohydrate. Note the formation of the 'window' over the water molecule.

Figure 8. Top view of the half-sandwich in the crystal structure of 8-azaguanine hydrochloride monohydrate.

base from an organic environment into water yields positive values for ΔH and ΔS.[20] Hence the overall interaction between nucleic acid bases and water cannot be hydrophobic.[20] Recent work tends to support the idea that hydrophobicity is not primarily involved: rather, the planar overlap of polar aromatic rings leads to increased polarization energy.[21]

The sandwich formation indicates that the water intercalation provides alternate attractive forces sufficient to compensate both for the loss of the stacking interaction between bases and the loss of one hydrogen bond for each water molecule.

Model Building Studies on the Possible Role of Water Intercalation

Intercalation of water molecules in polynucleotides, if present, will be very important. No studies have been made to check whether such intercalations of water molecules between bases is possible in polynucleotides. We studied, using CPK models, whether such intercalation of water molecules in DNA is allowed on stereochemical grounds. We examined whether a water molecule can act as a 'spacer' and stabilize the helical structure of DNA, should a base be turned outside due to dynamical fluctuations or to noncomplementary base opposition. Our model building studies (Fig. 9) indicated that it is possible for a water molecule to replace a pyrimidine and stabilize the DNA helix by hydrogen bonding to a purine base in one strand and to the phosphate oxygen in the other strand. Such a stabilization, if it actually occurs in polynucleotides, will produce torsional angle changes of the back bone at the site of the deletion or swinging out of the base and will introduce a kink, which in turn might facilitate the entry of intercalators.

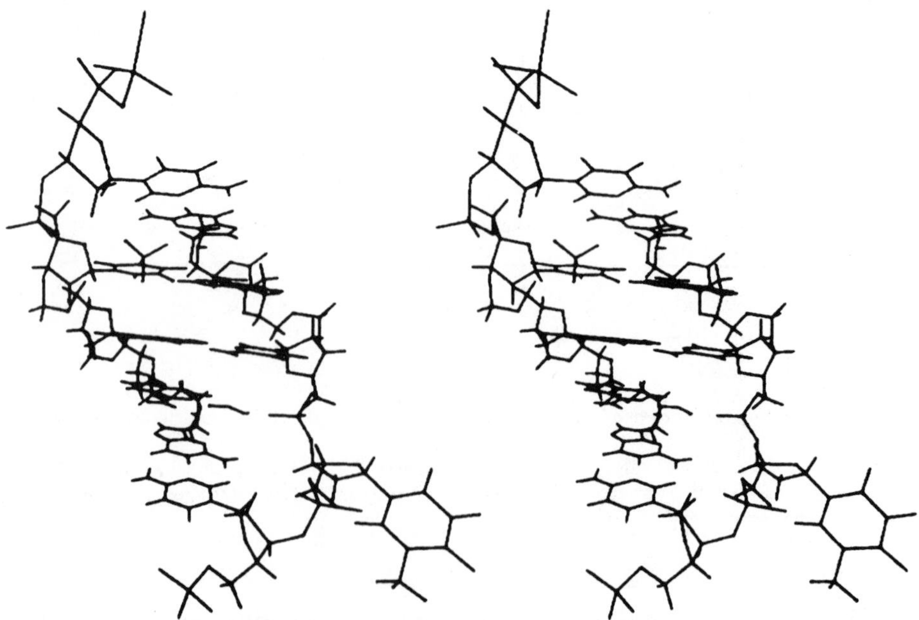

Figure 9. Water molecule acting as a 'spacer' and hydrogen bonded to the base on one strand and phosphate oxygen on the other strand.

Acknowledgement

We are grateful to NIH-GM-24864 for financial support. We thank Dr. R. Garduno and Dr. R. Rein for help in the graphics of Figure 9.

During the conference (April 26-29, 1981), we came to know of the first example of water intercalation in purines, in the crystal structure of hypoxanthine nitrate monohydrate carried out by Dr. Nadrian Seeman of SUNYA (private communication).

References and Footnotes

1. Lerman, L.S. *J. Mol. Biol 3,* 18-30 (1961).
2. Lerman, L.S. *Proc. Natl. Acad. Sci. USA, 49,* 94-102 (1963).
3. Sobell, H.M. in *Nucleic Acid Geometry and Dynamics,* ed. Sarma, R.H., Pergamon Press, New York. pp. 289-323 (1980).
4. Waring, M.J. *Nature 219,* 1320-1325 (1968).
5. Srikrishnan, T. and Parthasarathy, R. *Nature 264,* 379-380 (1976).
6. Parthasarathy, R., Srikrishnan, T. and Ginell, S.L. *Fed. Proc. 39,* 1880 (1980).
7. Srikrishnan, T. and Parthasarathy, R. *Acta Cryst. B34,* 1363-1366, (1978).
8. Takusagawa, R., Koetzle, T.F., Srikrishnan, T. and Parthasarathy, R. *Acta Cryst. B35,* 1388-1394 (1979).
9. Ginell, S .L. and Parthasarathy, R. *Biochem. Biophys. Acta,* submitted (1981).
10. Srikrishnan, T. and Parthasarathy, R. *Astracts, XIIth International Union of Crystallography,* Ottawa, Canada, (1981).
11. Srikrishnan, T. and Parthasarathy, R. *Abstracts, Amer. Cryst. Assn. Meeting, Series 2,* 6, 68 (1979).
12. Tougard, P., and Chantot, J.F. *Acta Cryst. B30,* 214-220 (1974).
13. Saenger, W., *J. Amer. Chem. Soc., 93,* 3035-3041 (1971).
14. Takusagawa, F. and Shimada, A. *Bull. Chem. Soc. (Japan) 46,* 2011-2019 (1973).
15. Kozlowski, D., Singh, P., and Hodgson, D.J. *Acta Cryst. B30,* 2806-2811 (1974).
16. Kozlowski, D., Singh, P., and Hodgson, D.J. *Acta Cryst. B31,* 1751-1753 (1975).
17. Bugg, C.E., Thomas, J.M., Sundaralingam, M. and Rao, S.T. *Bioplymers 10,* 175-219 (1971).
18. Ts'o, P.O.P., Melvin, I.S. and Olson, A.C. *J. Amer. Chem. Soc. 85,* 1289-1296 (1963).
19. Crothers, D.M. and Zimm, B.H. *J. Mol. Biol. 9,* 1-9 (1964).
20. Scruggs, R.L., Achter, E.K. and Ross, P.D. Biopolymers 11, 1961-1972 (1972).
21. Wagner, K.G., Arfman, H., Lawaczeck, R., Opatz, K., Schomberg, I. and Wray, V. in *Nuclear Magnetic Resonance Spectroscopy in Molecular Biology,* Ed. Pullman, D., B. Reidel Publishing Co., Holland, pp 103-110 (1978).

Proceedings of the Second SUNYA Conversation in the Discipline Biomolecular Stereodynamics
Volume I, ISBN 0-940030-00-4, Ed., Ramaswamy H. Sarma,
Adenine Press, New York, ©Adenine Press

Nucleic Acid Junctions: Building Blocks for Genetic Engineering in Three Dimensions

Nadrian C. Seeman
Department of Biological Sciences
State University of New York at Albany
Albany, New York 12222

But man must light for man
The fires no other can,
And find in his own eye
Where the strange crossroads lie.

. . . David McCord

Introduction

The complementary double helical paradigm for nucleic acids[1] is the foundation of our thinking about these molecules which constitute the genetic material of all living organisms. The pre-eminent structural characteristic of double helical nucleic acids is that the positions of all atoms in the molecule bear a well-defined relationship to a linear (although not necessarily straight) axis which exhibits no junctions (branch points). Nevertheless, conformational variability[2] and backbone flexibility[3] permit the formation of junctions which are crucial to the biological role played by nucleic acids. The replicational junction, shown in Figure 1, was implicit in the original Watson-Crick proposal for the mechanism of DNA replication. The Holliday structure,[4] indicated in Figure 2, is a critical intermediate in genetic recombination,[5] while the Gierer cruciform structure,[6] closely related to the Holliday structure, may play an important role in the regulation of gene expression.[7] Other types of junction structures are involved as intermediates in single-strand-displacement recombination and as transcriptional intermediates.

To date, it has not been possible to study the structural and dynamic properties of these junctions in oligonucleotide model systems, where the junction will contribute a significant signal. This is due to the sequence symmetry evident in Figures 1 and 2: The strands there are unlikely to form junction structures in preference to double helices; if they did occasionally form such structures, the process of branch point migration, shown in Figure 2, will result in the rapid resolution of the junction structures into double helices.

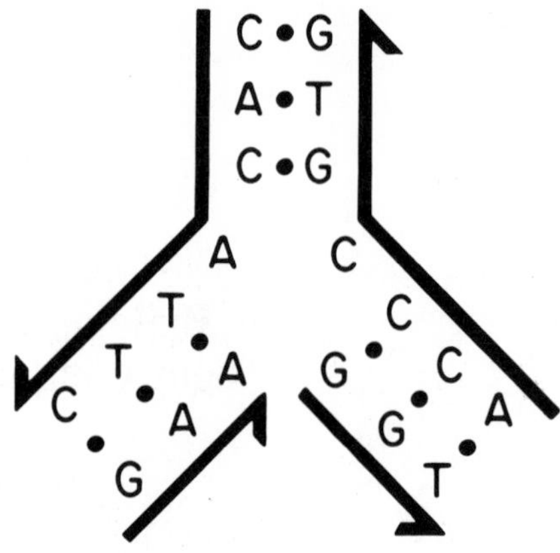

Figure 1. *A Replicational Junction.* The half-arrows indicate the 5′→3′ direction of the individual strands. The outer strands represent the parent double helix, while the inner strands are newly synthesized. This disjoint junction structure shows sequence symmetry between the two newly formed double helical segments. Since this structure is not likely to be as stable as individual double helices, mixing these oligonucleotides would probably generate a hexameric double helix and another trimeric double helix.

Conversely, abandonment of this sequence symmetry results in the possibility of generating junctions which will form preferentially, and which are immobile with respect to the migratory process. This statement rests on two assumptions: First, that Watson-Crick base pairing is the optimal form of association for individual strands of nucleic acids; and second, that individual strands of nucleic acids will

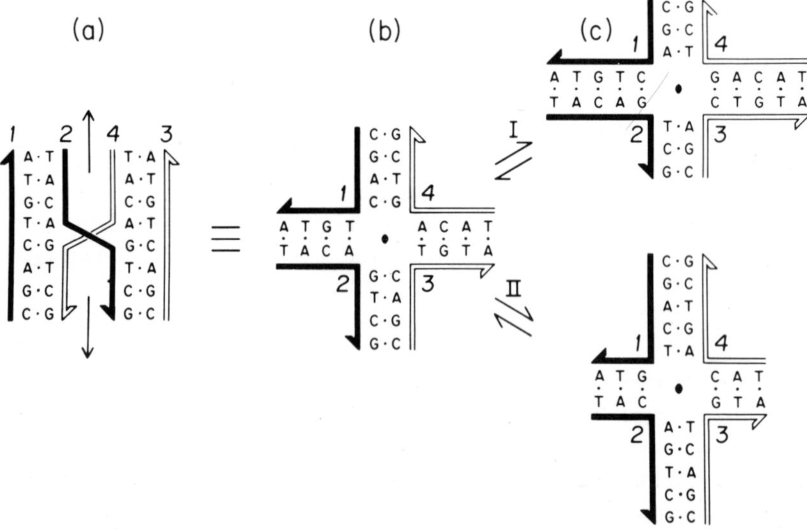

Figure 2. *The Recombinational Junction.* (a) The Holliday Structure[4] as originally proposed. The shaded backbones were initially paired exclusively to each other, as were the unshaded backbones. The half-arrows indicate the 5′→3′ directions of the strands. The full headed arrows indicate the axis of two-fold symmetry. The place where the strands cross is the junction. Migration of the junction corresponds to the movement of this point up or down. This representation is identical to that shown in (b), in which the possible 4-fold backbone symmetry originally suggested by Sobell[14] is more apparent. The two-fold sequence symmetry is indicated by the lens-shaped object in the middle of the structure. Migration of the branch point in either direction is indicated by reactions I and II in the transition to (c). The eventual end product of the repetition of reaction I is a return to the original pairing. The eventual end product of the repetition of reaction II is a newly hybridized pair of double helices.

maximize their Watson-Crick base pairing when they associate. The first assumption is weak near the junction, where the Watson-Crick pairing specificity associated with double helical nucleic acids may be perturbed. Since many forms of pairing between the bases are possible;[8,9] it is clear that the double helical backbone is an important element in generating the Watson-Crick pairing specificity between Adenine and Thymine or Uracil and between Guanine and Cytosine. Nevertheless, as the lengths of the double helical arms which flank the junction increase, so does the strength of this assumption. Below, I eludicate the rules for selecting sequences which generate immobile junctions flanked by double helices. Junctions which display limited migrational mobility are also devisable, as are junctions which involve single strands. Although this discussion is couched primarily in terms of DNA oligomeric fragments, all arguments should be equally applicable to oligomeric RNA fragments and to mixtures of RNA and DNA oligomeric fragments.

Rules for Junction Formation

Several concepts must be defined to facilitate this discussion. First, we must establish a criterion for the number of base pairs that every double helical arm that flanks the junction must contain. This number is termed *nocrit,* and is expected to fall in the range of 3-6 base pairs. A sequence of nocrit bases is termed a *criton.* Unique base pairing patterns which will generate junctions are, of course, a function of the free energy of association of individual strands. In line with our assumptions, however, we will approximate this consideration by treating only the number of Watson-Crick pairing interactions. Therefore, violation of the following rules for sizes smaller than nocrit will be ignored. The sequence complementary to a criton is termed its *anti-criton.* A *bend* is a phosphodiester linkage which is flanked by bases paired to different strands. The concept of a bend will be generalized below when semi-mobile junctions are discussed. The *rank* of a junction is the number of base pairs which directly abut it. Thus, the junction shown in Figure 1 is rank 3, while those shown in Figure 2 are rank 4.

In order to generate uniquely paired structures with non-migratory junctions, the following rules must be obeyed:

1. Every criton in the individual strands forming the junction must be unique throughout all strands, regardless of frame.

2. The anti-criton to any criton which spans a bend in a strand must not be present in any strand, regardless of frame.

3. Self-complementary critons are not permitted. If nocrit is an odd number, this injunction holds for all critons of size (nocrit + 1).

4. In order to eliminate migratory mobility, the same base pair can only abut the junction twice. If it is present twice, those two occurrences must be on adjacent arms.

The first three rules are designed to assure pairing specificity. The fourth criterion limits the maximum rank of junctions. Since there are only four base pairs, A-T, T-A, G-C, C-G, and since each pair can only appear twice, junctions of rank greater than eight are not possible with the conventional bases. Due to the weakness of assumption 1, criterion 4 should be applied to base pairs further from the junction as well. I have written a FORTRAN computer program to generate sequences which fulfill these criteria. By ranking the critons in order of the most rapidly changing bases which they contain, the alogrithm used by this program eliminates many otherwise wasteful steps in the generation of appropriate sequences. Figure 3 demonstrates a junction of rank 4, generated with nocrit equal to 3. A more sophisticated program which takes account of free energy considerations is in preparation.

Immobile junctions of the sort described in this section will be enormously valuable in studying the structural charcteristics of junctions. There are several questions associated with this system which crystallographic and solution studies of immobile junctions will answer: These pertain to the structure of the junction and the dynamics of its motion. It has been suggested by several authors that the dynamics of junction structure are critically related to the process of branch point migration[10-12]. However, in order to get a handle on the direct mechanism of the migratory event, it is necessary to go to a second class of molecules, the semi-mobile junctions; these complexes are capable of limited migration within the confines of a fundamentally immobile structure. An example of a semi-mobile junction is shown in Figure 4. This junction is able to accomplish a flip-flop, as indicated in the figure, but it is incapable of complete resolution. In order to generate such a junction, the flipping bases and the phosphates which flank them must be considered part of the bend. Once this sophistication of the concept of a bend is in effect, the same four rules apply. Clearly, it is not possible to have more flipping bases than nocrit-2, since bends will not be properly spanned: It would be advisable for the number of

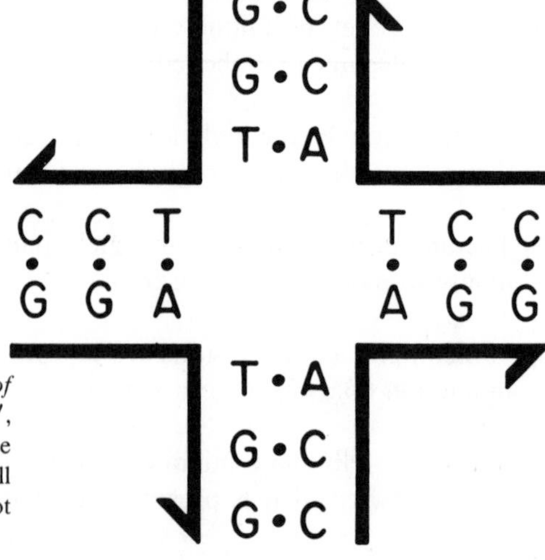

Figure 3. *An Immobile Nucleic Acid Junction of Rank 4.* The directions of the backbones, 5′→3′, are indicated by the half-arrowheads. The sequence fulfills all the rules listed in the text. Note that all the sets of base pairs are in accord with rule 4, not just the base pairs which flank the junction.

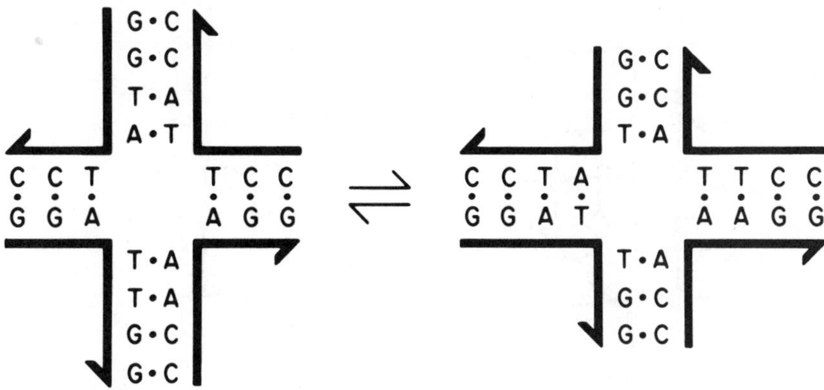

Figure 4. *A Semi-Mobile Junction.* This junction may undergo the reactions indicated, but may not go beyond them and resolve into two linear duplexes. Thus, it constitutes a simple flip-flop. The rules of migration are satisfied for the two states shown, but the rules for non-migration come into play for any further migratory events in either direction.

mobile bases in a semi-mobile junction to be much less than nocrit, to ensure pairing fidelity.

Junctions and Networks

It is evident that the ability to construct rank N junctions (N = 3,4,5,6,7,8) implies that it will be possible to construct highly specific N-connected networks[13] and polyhedra of double helical nucleic acids. Examples of 3,4,5 and 6-connected networks are shown in Figure 5. This may be done by using the sticky-ended ligation technology which is so fruitful today in genetic engineering applications. This involves the use of sequences in which the junction crossroads structure does not end in a 'blunt ended' fashion, as shown in Figure 3. Rather, one strand extends beyond the end of each double helix, so that a single stranded region is dangling off the end. The enormous specificity of Watson-Crick base pairing is then utilized to link up two different pieces of DNA which have complementary 'sticky ends'. Among the 2- and 3-dimensional networks that are possible, some are of course periodic in their connectivity. It is to be hoped that such networks are also periodic spatially. If this is the case, they will grow to large size and will be suitable for diffraction analysis by x-rays, and perhaps neutrons. An example of such a 2-dimensional network is shown in Figure 6. It should be remembered that the relative orientation of successive junctions is a function of both junction structure and the separation of junctions, since the connecting segments between junctions are helical. The hybridized pieces indicated in Figure 6 can be ligated together enzymatically, if the size of the lattice is large enough to allow free diffusion of the ligase. Thus, a covalently connected 2- or 3-dimensional network of nucleic acids can be constructed using these junctions. Insofar as genetic engineering consists of constructing specific structures of genetic material (nucleic acids) by these techniques, this procedure can be termed genetic engineering in two or three dimen-

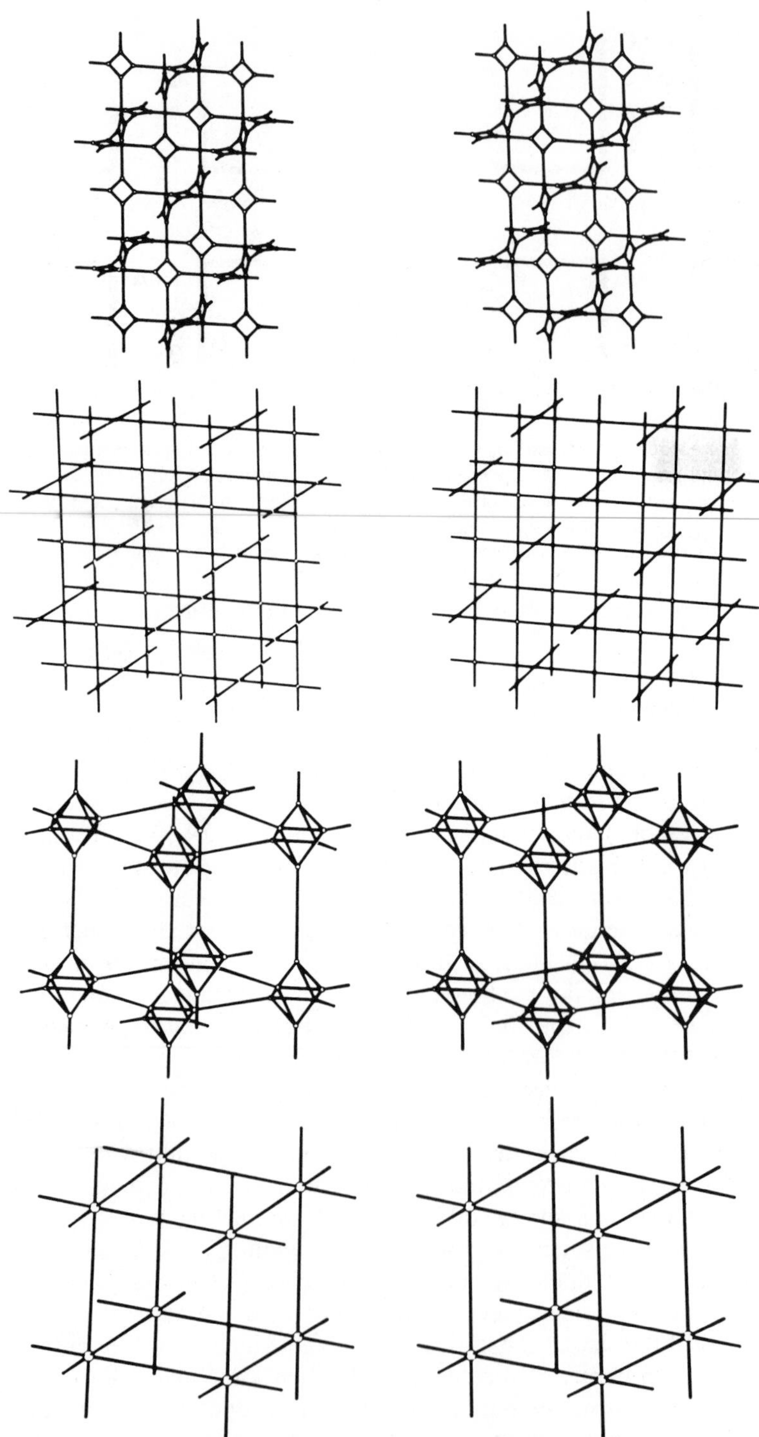

Legend is on the next page.

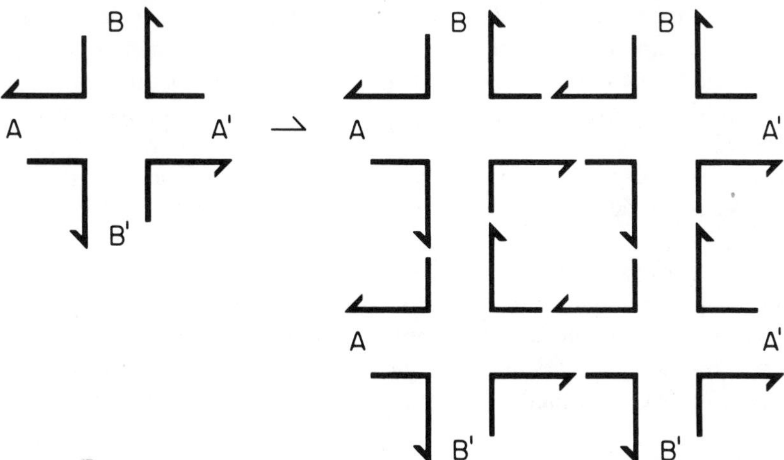

Figure 6. *Formation of a 2-Dimensional Lattice from an Immobile Junction with Sticky Ends.* A is a sticky end, and A′ is complementary to it. A similar relationship exists between B and B′. Four of the monomeric junctions on the left are complexed in parallel orientation to yield the structure on the right. If the inter-junctional spacing is large enough, a ligase would be able to close the overlapped gaps to make the complex on the right a covalently bonded structure. Note that the complex has maintained open valences, so that it can be extended by the addition of more monomers. This procedure is not limited in theory to rank-4 junctions, nor is it limited to two dimensions.

sions. It should be noted, however, that it would not be possible to clone these structures in living organisms, since a round of replication would destroy the three-dimensional properties of the molecules: All known nucleic acid polymerases make complementary copies of template molecules.

The individual junctions can be considered to be macromolecular valence clusters of nucleic acid. The ways in which they can be assembled in 3-dimensional space are limited only by imagination and a few physical properties. It is undesirable to have a length of DNA which is a significant fraction of the persistence length, since that will lead to floppiness in the structure, and possibly to undesirable bonds being formed. Probably 150 A is the limit for which an individual double helical linkage should be allowed. This is 1/4 to 1/3 of the reported persistence lengths for linear duplex DNA. Specificity comes from the identities of the individual sticky ends. If we wish to link two separate junctions together, and they each contain complementary sticky ends, we can be assured that this reaction will proceed with high fidelity. The major constraint on the generation of spatially periodic arrays is the repeat

Figure 5. *N-Connected Networks.* Four different N-connected networks are shown in stereoscopic projection. These are all indicated as forming cubic lattices, although this is certainly not necessary. The dark lines represent double helical stretches of nucleic acid. The large circles represent junction regions of the appropriate rank. The small circles at the ends of each line on the periphery of each figure represent unsatisfied valences. From the top, these are respectively units of 3,4,5 and 6 connected networks.

length of the nucleic acid double helix. Clearly, structures must be appropriately designed so that spatial periodicity is possible. The relationship between spatial periodicity, the lengths of double helix involved, junction flexibility, crystal packing forces and lattice construction algorithms will be treated in detail elsewhere. It is, however, evident that it is possible to mix the ranks of the materials that form these lattices, and it is furthermore evident that every double helical 'line' in these networks can be replaced by a prismatically shaped nucleic acid network.[13]

Those junctions of rank greater than 3 suffer from a potential problem of pseudo-enantiomerism. Although all molecules of nucleic acids are chiral (the sugars are all D-ribose or D-deoxyribose), the valence clusters which these junctions constitute are not guaranteed to assume a unique 'configuration'. For example, if the rank 4 junction assumed a tetrahedral structure, a left and a right handed version of its

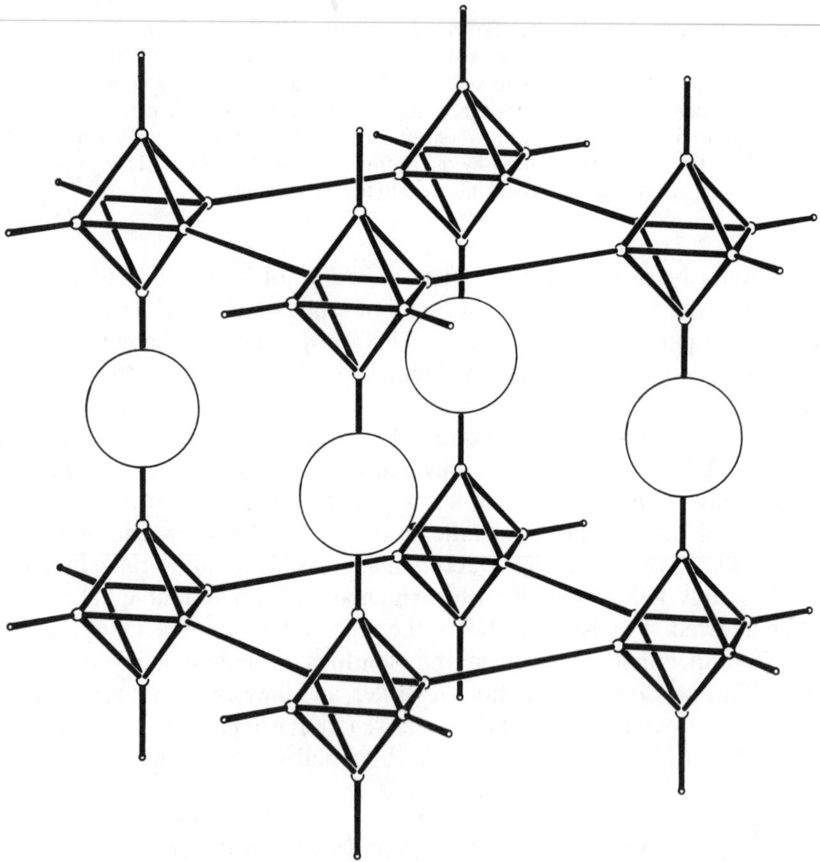

Figure 7. *Schematic Representation of a Cognate Protein in a 5-Connected Lattice.* The dark lines represent double helical stretches of nucleic acid, the small circles at the ends of each line represent unsatisfied valences, and the intermediate size circles represent junctions. The huge spherical objects represent a cognate macromolecule bound to the double helical nucleic acid in a periodic fashion.

sticky-ended specificity could exist. This problem may be obviated by decreasing the backbone symmetry of the system. If one of the strands is selected to be RNA and the other 3 strands are DNA, the equivalence between the two forms would be broken, and one could be selected for.

The mechanical and electrical properties of these periodic networks are of course not known at all. Since nucleic acids are polyanions, which exist in aqueous solutions, clearly this characteristic will be retained in the network structures. It should be noted, however, that methylation of the phosphates will neutralize the polymers, and aminoalkylation of the phosphates will render them positively charged. Large gaps will exist between the individual columns of DNA or RNA or DNA-RNA hybrids. Other macromolecular materials, such as cognate proteins, may be inserted into these gaps. A schematic example of such a complex is indicated in Figure 7.

The essential feature of the system derived above is the construction of nucleic acid molecules which do not conform to the linear model, despite heavy reliance on Watson-Crick base pairing. This unusual system should prove to be a valuable tool in understanding the dynamics of branch point migration, and in elucidating the means by which sequence specific recognition of nucleic acids is effected by proteins.

Acknowledgements

This work has been supported by a Basil O'Connor Starter Grant from The March of Dimes Birth Defects Foundation and Grant GM-26467 from the NIH. I would like to thank Neville R. Kallenbach, Bruce H. Robinson and Ramaswamy H. Sarma for valuable discussions and invaluable encouragement.

References and Footnotes

1. Watson, J.D. and Crick, F.H.C. *Nature 171,* 737-738 (1953).
2. Kim, S.H., Berman, H.M.,Seeman, N.C. and Newton, M.D. *Acta. Cryst. B29,* 703-710 (1973).
3. Sarma, R.H., *Nature 263,* 567-572 (1976).
4. Holliday, R., *Genet. Res. 5,* 282-304 (1964).
5. Broker, T. and Lehman, I.R., *J. Mol. Biol. 60,* 131-149 (1971).
6. Gierer, A. *Nature 212,* 1460-1461 (1966).
7. Lilley, D.M.J. *Proc. Nat. Acad. Sci USA 77,* 6468-6472 (1980).
8. Voet, D. and Rich, A. *Prog. Nucl. Acid Res. and Mol. Biol. 10,* Academic Press, 183-265 (1970).
9. Seeman, N.C., in *Nucleic Acid Geometry and Dynamics* (ed. by R.H. Sarma) Pergamon Press, 109-142 (1980).
10. Thompson, B.J., Camien, M.N. and Warner, R.C. , *Proc. Nat. Acad. Sci. USA 73,* 2299-2303 (1976).
11. Warner, R.C., Fishel, R. and Wheeler, F., *Cold Spring Harbor Symp. Quant. Biol. 43,* 957-968 (1978).
12. Seeman, N.C. and Robinson, B.H. These Proceedings, Vol. 1 (1981).
13. Wells, A.F., *Three Dimensional Networks and Polyhedra,* John Wiley & Sons, New York (1977).
14. Sobell, H.M., *Proc. Nat. Acad. Sci. (USA) 69,* 2483-2487 (1972).

Proceedings of the Second SUNYA Conversation in the Discipline Biomolecular Stereodynamics
Volume I, ISBN 0-940030-00-4, Ed., Ramaswamy H. Sarma,
Adenine Press, New York,

Simulation of Double Stranded Branch Point Migration

Nadrian C. Seeman
Department of Biological Sciences
State University of New York at Albany
Albany, New York 12222

and

Bruce H. Robinson
Department of Chemistry
University of Washington
Seattle, Washington 98195

I travel for travel's sake. The great affair is to move.

Robert Louis Stevenson

Introduction

Recombination between two homologous double helices of DNA to form two newly hybridized double strands of DNA is one of the fundamental processes in the generation of genetic variability in biological systems. The recombinant DNA is believed to arise via the Holliday intermediate,[1] indicated in Fig. 1a. This cross-stranded structure, in which the strands from two DNA double helices have been exchanged, is shown with a hypothetical sequence. This structure is topologically identical to the "crossroads-junction" structure illustrated in Fig. 1b. Both intermediate structures have twofold rotational symmetry, as indicated by the appropriate symbols in the figure. The strands with the same shading in the figures are considered to have been paired before strand exchange was initiated. The place where the strands interchange is termed the branch point, or, alternatively, the junction. This corresponds to the crossover point of Fig. 1a, and to the middle of the intersection in Fig. 1b. The interchange process is thought to occur by migration of the branch point, via rotation of the double helices which constitute it. Either of the two migration reactions (I or II) shown in the transition from Fig. 1b to Fig. 1c may occur. Repetition of reaction I results in a return to the original pairing: strand 1 paired with strand 2 and strand 3 paired with strand 4. Conversely, the repetition of reaction II yields a new hybridization: strand 1 paired with strand 4 and strand 2 paired with strand 3. If the two initial double helices (1-2 and 3-4) are exactly identical, no genetically assayable recombination event will occur, regardless of which reaction pathway is involved. However, if the two double helices are only homologous, with slight sequence differences between them, reaction II will lead to a recombinant product. Thus, resolution of the junction structure via reaction II

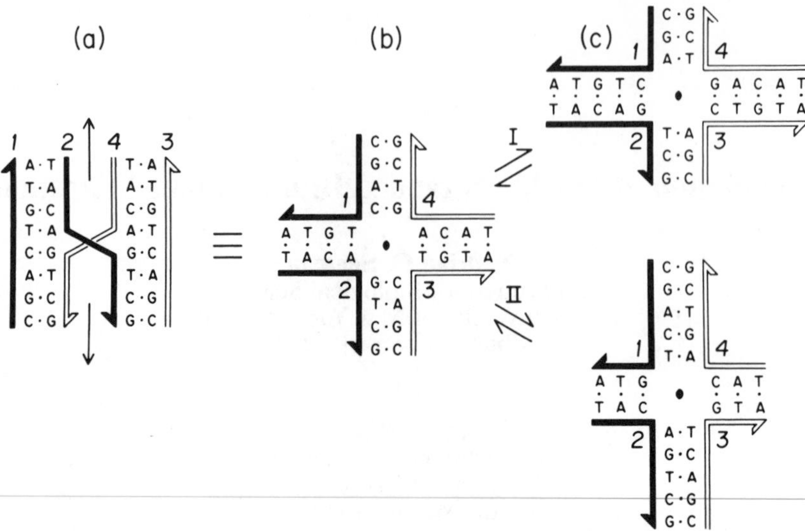

Figure 1. *The Recombinational Junction.* (a) The Holliday Structure[1] as originally proposed. The shaded backbones were initially paired exclusively to each other, as were the unshaded backbones. The half-arrows indicate the 5′→3′ directions of the strands. The full headed arrows indicate the axis of two-fold symmetry. The place where the strands cross is the junction. Migration of the junction corresponds to the movement of this point up or down. This representation is identical to that shown in (b), in which the possible 4-fold backbone symmetry originally suggested by Sobell[3] is more apparent. The two-fold sequence symmetry is indicated by the lens-shaped object in the middle of the structure. Migration of the branch point in either direction is indicated by reactions I and II in the transition to (c). The eventual end product of the repetition of reaction I is a return to the original pairing. The eventual end product of the repetition of reaction II is a newly hybridized pair of double helices.

results in two double helices which are not optimally paired, and are therefore appropriate substrates for repair enzymes.

Numerous attempts have been made to model the structural and dynamic features involved in the mechanism of double stranded branch point migration. Structural models have been proposed by Sigal and Alberts,[2] Sobell,[3] Wilson[4] and ourselves.[5] Dynamic models will be discussed below. The structural model proposed by Sigal and Alberts is physically analogous to the schematic diagram in Fig. 1a. In particular, they have postulated that the two helix axes of the double helices which participate in forming the recombinational intermediate remain essentially unperturbed. Therefore, the structures of two strands which are homologous and parallel to each other (one from each of the double helices), remain undeformed. The other two strands, which are anti-parallel to the first pair of strands, are exchanged between the double helices; clearly these strands must have altered structures near the branch point. The sequence similarities imply twofold spatial symmetry, which is probable because of the virtual identity between the two double helices. This symmetry is implicit in the Sigal-Alberts model,[2] but is explicit in Sobell's.[3] Indeed, Sobell has pointed out that the backbone structures can assume fourfold symmetry. Sobell's fourfold model in projection corresponds to the schematic diagram in Fig.

1b: The helix axes of the four double helices emanating from the branch point in that figure are related by a fourfold rotational symmetry operation, whose axis is coincident with the 2-fold axis indicated.

An intriguing proposal for the structure of the intermediate has also been advanced by Wilson.[4] He proposed his model in order to solve certain topological problems associated with migration in the case of double helices whose ends are not free to rotate. His solution involved forming a tetrahelical structure from the two double helices which had already recombined. Thus, the double helices (1-4) and (2-3) in Fig. 1b would be wrapped around each other, while the double helices (1-2) and (3-4) would remain separate. Stabilization of this tetrahelical structure was supposed to be accomplished by hydrogen bonding between base pairs across the major grooves of the two recombined double helices. Wilson's speculations are of interest, but are beyond the scope of our structural considerations. The details of our structural model are described below. All of these structural proposals are based on molecular model building, supported by chemical intuition. No experimental data yet exist at an adequate resolution to indicate the structural properties and parameters of the branch point; however, electron microscopic observations[6-9] have confirmed the existence of the branch point and of the migratory phenomenon. It is likely that physical studies of immobile and semi-mobile nucleic acid junctions will soon generate some of the missing structural information. However, as will be seen below, it is possible to establish a consistent mechanism for branch point migration in this system, even in the absence of explicit structural data.

The dynamical properties of the branch point are somewhat better characterized than the structural parameters. In particular, the work of Warner and his colleagues[11,12] has produced the rate constant for the migrational event *in vitro.* These kinetic parameters were derived by using the bacteriophage G4 figure-8 replicative intermediate as a model mogratory system. The disappearance of cross-stranded structures was monitored following restriction by Eco R1, which initiated migration. Furthermore, this work has generated estimates for the enthalpy of activation associated with the unit migratory event. The rate is highly temperature dependent, being about 10,000 base pairs/second at 30°C., and 60 base pairs/second at 10°C. Thus, at 30°C., the migrational event takes about 100 μ seconds. As might be expected from such a system, the enthalpy of activation is large, 22 kcal/mole between 20°C. and 30°C., while it is 75 kcal/mole between 0°C. and 20°C.[11,12] These two temperature ranges are referred to throughout this text as the high and low temperature ranges, respectively.

The major kinetic model at this time, independent of an external energy source, is due to Meselson,[13] although it antedates the work of Warner's group. Meselson postulated that rotary diffusion would generate migration, and he suggested that either this random process or base pair opening constitutes the rate limiting step. Both of these processes are much faster at high temperatures than the observed migration time of 100 μ seconds. Therefore, Meselson's assumptions have been criticized,[11,12] since they ignore junction structure.

Here, we present a dynamic structural model for the migrational process. This model subsumes the major features of the preceding Sigal-Alberts and Sobell models, which correspond to special cases of our more general formulation. The kinetic pathway we propose incorporates Meselson's use of rotary diffusion, although we have elaborated his mechanism considerably. Our parameterization of the branch point junction structure results in a large number of conformers which the molecular complex can assume via the process of dynamic equilibrium. The calculations which we have executed utilizing this model involve the electrostatic energies which characterize these different conformers. We have calculated the conformational electrostatic energies and their associated probabilities for the entire range of possible structures concordant with our parameterization. From these probabilities, known quantities are employed by a dynamic model to estimate the rate and the activation energy for the migratory event. The results of these calculations are in good agreement with the data of Warner and his colleagues. We have also shown that no electrostatic driving forces are likely to promote migration.

Besides the lack of models consistent with the data for this system, a second, more subtle problem exists concerning branch point migration. This point deals with the relative rate constants of reaction I and reaction II (Fig. 1) in the case of homologous DNA. The slight differences in free energy between the end products of reaction I and reaction II may or may not be critical from a physico-chemical perspective, because of the large number of residues which are typically involved. Nonetheless, some questions are raised if the completely paired intermediates shown in Fig. 1 are presumed to be obligatory, which is reasonable on stereochemical grounds. In the case of identical double helices, an equal probability is to be expected for reaction I or reaction II at each transition. In the case of homologous double helices, this is not necessarily true. When a mismatch occurs, the system is moving (by reaction II) from a state containing two Watson-Crick base pairs (bridging arms (1-2) and (3-4)), to a state having no Watson-Crick base pairs stabilizing chains (1-4) and (2-3) at the junction point. This would generate a significant kinetic barrier to reaction II, thereby introducing a large bias in favor of reaction I. The calculations which we have performed bear on this problem as well. It is clear from our calculations that no electrostatic torques are available to ameliorate this situation.

The Structural Model for the Junction

One of the fundamental problems in dealing with recombinational intermediates such as the Holliday structure is that no detailed structural data yet exist for the junction, although several models have been proposed, as noted above. We have parameterized the junction under an inclusive formulation which subsumes the major previous suggestions, while retaining the intrinsic twofold rotational symmetry implied by the covalent structures involved. Figure 2 shows our formulation of the junction structure. The difficulty in understanding a complete molecular model of the junction has led us to represent it both schematically (Fig. 2a) and by comparison with a familiar staircase (Fig. 2b). In the staircase, the banisters correspond to the backbone structures, and the steps correspond to the base pairs. The junction region is suggested by the plateau in the middle. The structure has a twofold axis of rotational symmetry perpendicular to the plateau plane. Note that the banisters are alternately anti-parallel, as indicated by the directions of the arrows on them. From this figure, it can readily be understood how each backbone (banister) is part of two separate double helices. Of course, in a true junction, the stairs would be related to each other by helical symmetry, rather than by a simple translation.

The four helix axes, which go up or down the middle of each flight of stairs, are indicated schematically in Fig. 2a, which shows the rest of our parameterization. We have constructed a rhombus, which corresponds to the square plateau of the staircase. The variable parameter of this rhombus is the angle α, as indicated in Fig. 2a. The four phosphates which flank the junction are positioned at the corners of this rhombus. Their separation is taken as that found in B-DNA.[14] Because the phosphates on opposite strands of B-DNA are not at the same height along the helix axis, the phosphorus atoms have accordingly been placed slightly above and below the vertices of the rhombus. Besides the variability of the rhombus angle, we have generated two virtual bonds coincident with two adjacent sides of the rhombus. The two variable angles associated with these virtual bonds, θ_T and θ_B, are indicated in Fig. 2a. The maintenance of the twofold rotational symmetry axis perpendicular to the plane of the rhombus requires that the helices attached to the two sides of the rhombus opposite the virtual bonds also be varied simultaneously about two symmetrically positioned virtual bonds. These virtual bonds allow us to change the angles θ_T and θ_B, which the double helices make with the plane of the rhombus. Although these angles refer to the double helices which are initially above ($\theta_T = 0$) and below ($\theta_B = 0$) the plane of the rhombus, the unique, physically feasible, range of each of these angles is 180°, as is the range of α. In the staircase illustration, for example, $\alpha = 90°$, $\theta_T = 60°$ and $\theta_B = 60°$. Thus, if $\alpha = 36°$, and $\theta_T = \theta_B = 0°$, the Sigal-Alberts proposed structure is approximated. Similarly, if $\alpha = \theta_T = \theta_B = 90°$, the fourfold symmetric backbone structure suggested by Sobell is obtained.

It should be emphasized at this point that the symmetry of the recombinational intermediate is intrinsically twofold, as can be seen from Fig. 1. The twofold rotational symmetry of the sequence is mandatory, except for the occasional mismatches noted above. Higher symmetry is possible for the backbone of the Holliday intermediate for a series of structures whose helix axes are related to each other like the ribs of an umbrella or parasol. These structures arise when $\alpha = 90°$, and $\theta_T = 180° - \theta_B$. As can be seen from the staircase picture, there is no symmetry relationship between the parts of the Holliday structure above and below the branch point. Thus, the only symmetry indicated for this structure, based on the sequence, is the twofold symmetry perpendicular to the junction plane of the structure. We have maintained this symmetry throughout our calculations.

The staircase in Fig. 2b illustrates another point of importance, which was first noted independently by Wilson[4] and ourselves.[5] Two individuals symmetrically walking down the two staircases above the plateau would face each other; however, if they continued below the plateau, down opposite staircases, they would face away from each other. Thus, although two-fold rotational symmetry exists between helices attached to the opposite sides of the rhombus none exists between helices attached to adjacent sides of the rhombus. The molecular equivalent of this asym-

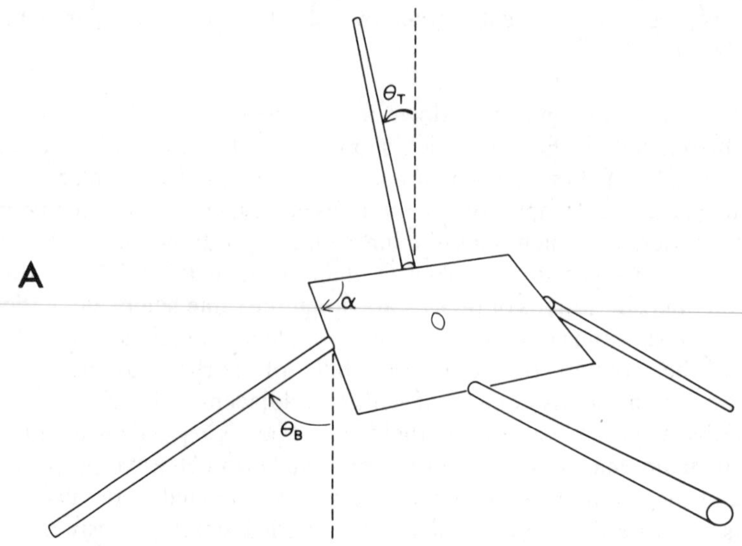

For Legend see opposite page.

metry is that just above the plateau, the two double helices would face each other in, say, the major groove, while just below the plateau, they would face each other in the opposite, or minor, groove.[4,5] Because double helices undergo a half-revolution approximately every five residues, the situation will reverse itself every five base pairs; thus, a half-turn away from the junction, the top double helices will face each other in the minor groove, and the bottom double helices will face each other in the major groove.

The differences between the two juxtapositional geometries of the double helices surrounding the branch point will be reflected in the phosphate-phosphate electrostatic repulsions, since the phosphates are not identically situated with respect to the two grooves of the double helix. For this reason, we have calculated those electrostatic quantities associated with the branch point junction structure. The first of these quantities is the electrostatic energy associated with each of the conformations available to the structure, $E(\alpha,\theta_T,\theta_B)$. The map of these energies is readily transformed into one showing the probability that the structure will assume any of these given conformations, if one neglects all but electrostatic considerations.

The second quantity of interest is the total electrostatic torque, generated by phosphate-phosphate repulsions, about the helix axes, i.e., an electrostatic torque which can facilitate branch point migration in either direction. In Fig. 3a, we see a sample conformation, with $\alpha = 110°$, $\theta_T = \theta_B = 0°$. A productive migrational event will occur when all the helices are rotating either counter-clockwise (as indicated) or clockwise (yielding migration in the opposite direction). We term the total torque about the helix axes in either of these directions the "productive torque". The magnitude of the productive torque indicates its contribution to migration, while its sign indicates the direction of migration. It should be noted that this concept of a productive torque remains valid, even when the double helices which constitute the junction are no longer parallel to each other; it refers to the sum of the torques about the individual helix axes, regardless of the orientations of the axes themselves.

Figure 2. *Representation and Parameterization of the Recombinational Junction Structure.* (a) Schematic representation of the junction. The four double helices which comprise the junction are indicated by their helix axes, which are the four rod-like elements emanating from the central rhombus. This rhombus represents the junction itself, the four phosphates flanking the junction being placed near the corners of the rhombus. The twofold symmetry axis perpendicular to the plane of the rhombus is indicated by the lens-shaped symbol at the center of the rhombus. The three degrees of freedom are indicated, as well: α is the variable angle of the rhombus, θ_T is the degree of freedom for the two double helices indicated above the plane of the rhombus and θ_B is the degree of freedom for the two double helices indicated below the plane of the rhombus. (b) The staircase analogy to the junction structure. In this representation, the rhombus of (a) corresponds to the plateau at the center of the staircase. The banisters correspond to the backbones of the four strands which comprise the junction, while the steps are analogous to the base pairs. The polarity of the backbones is indicated by the arrows on the banisters, indicated the alternately anti-parallel nature of the backbones. The twofold symmetry of the junction is again indicated by the lens-shaped symbol at the center of the plateau. Note that two people walking down the opposite top stairways will face each other, while two people walking down the opposite bottom stairways will have their eyes oriented in opposite directions.

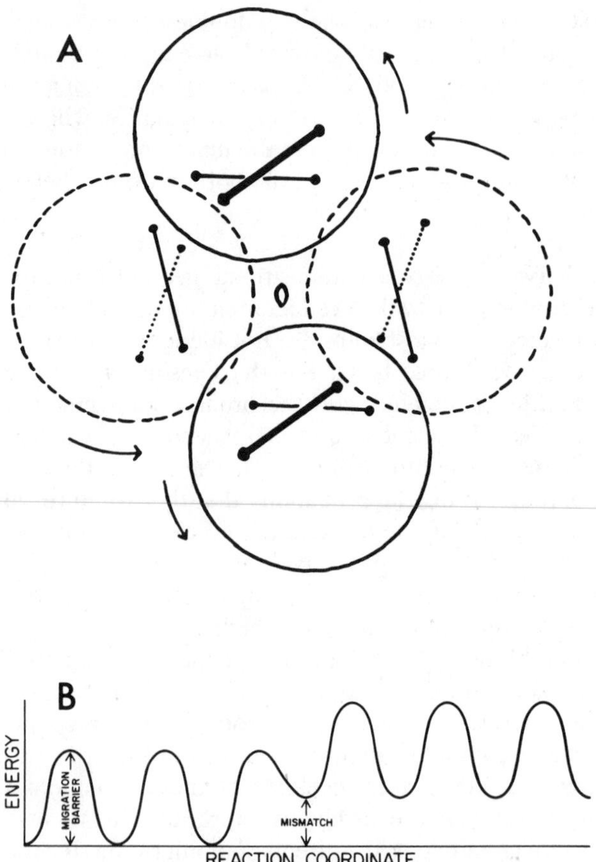

Figure 3. *Nature of Movement Through the Junction.* (a) The directions of rotation to produce a productive migration. Shown is a junction with parameters $\alpha = 110°$, $\theta_T = \theta_B = 0°$. The view is perpendicular to the plane of the rhombus. The twofold symmetry is indicated by the lens-shaped symbol in the center of the figure. Two double helices emanate toward the reader from the plane of the paper these are indicated by solid circles. Two other double helices are below the plane of the paper, emanating in the other direction; these are indicated by dashed circles. Within each circle are two pairs of dots, representing C1′ atoms. Those connected in pairs belong to the same base pair. The two pairs for each double helix nearest the junction are indicated in this fashion. Within the solid circles, the thicker line represents the base pair nearer the reader. Similarly, within the dashed circles, the solid line represents the base pair nearer the reader. In the course of migration toward the reader, the thin lines in the solid circles will move to replace the thick lines; the thick lines in the dashed circles will break and reform to replace the thin lines in the solid circles; while the dotted lines in the dashed circles will move to replace the solid lines in the dashed circles. For these coordinate motions to occur, all four double helices must rotate in the counter-clockwise directions indicated by the curved arrows. Migration in the other direction similarly entails coordinate rotation in the clockwise direction. (b) Schematic reaction coordinate for the migratory process. A single-step migration is separated by a single peak. The migration barrier is indicated in the first step. The result of a mismatch is also indicated: The barrier remains the same, but the relative minimum to which the system returns is now higher. No detailed statement about the structure of the energy function near the region of the mismatch is being made; only the qualitative nature of the phenomenon is being indicated.

The reaction coordinate shown in Fig. 3b complements the structural diagram shown in Fig. 3a. Here, we have plotted the energy of the state of the system along the ordinate, while the effective reaction coordinate is indicated along the abcissa. Migration in one direction (say, counterclockwise) in Fig 3a. corresponds to rightward movement along Fig. 3a, while migration in the other direction (say clockwise) corresponds to movement in the leftward direction in Fig 3b. We will discuss below the reasons that we believe that this diagram represents a projection of the true situation, which can only be adequately represented in two (or perhaps more) dimensions. At the right side of Fig. 3b, we have indicated the nature of the energetics when a mismatch is encountered. There is a raised region, in which the relative minimum is higher than in the rest of the diagram, but the barriers between the minima are not higher. The kinetic barriers on either side of the mismatch will be unaffected, but the equilibrium will be shifted in favor of the direction without the mismatch; this is because the leftward rate constant will be higher than that in the rightward direction at the site of the mismatch. The details of this diagram in the vicinity of the mismatch are necessarily conjectural: we have indicated the effect of the mismatch as being manifest entirely at its site; it is possible that the effect is spread over a short region after the mismatch has been encountered (corresponding to the right side of Fig. 3b).

The productive torques might provide the necessary driving force for branch point migration in the presence of mispairing. It is *a priori* possible that mispairing could be kinetically overcome by an averaging mechanism which jumped the system from one side of the mismatch trauma to the other. A large driving force which involved a great many multiple steps would fall into this category, hurtling the state of the system from the left side of Fig. 3b to the far right. Similarly, an unraveling-reraveling mechanism involving many bases could also effectively make this jump. As will be seen below, our calculations indicate that there is no significant electrostatic component to either of these mechanisms, if they exist.

The quantitative details of our model for purposes of the calculations are as follows: We have placed a single turn of B-DNA[14] on each of the sides of the rhombus, as indicated schematically in Fig. 2. In the initial conformation ($\theta_B = \theta_T = 0$), the vertices of the rhombus correspond to the projections of the phosphorus atoms onto the plateau plane; this specification serves to orient the double helices about their helix axes. The plane bisects the projection on the helix axis of the two phosphorus atoms on opposite strands, thereby fixing the position of the helix axis vertically. The four double helices are covalently joined into four polynucleotide backbones by the phosphate linkages associated with the corners of the rhombus. Thus, these phosphorus atoms occupy positions analogous to the corners of the four banisters (Fig. 2b) which flank the four "double helical" sets of steps.

We have placed a negative charge at the positions of the phosphorus atoms. This charge has been isotropically screened 76%, according to simple polyelectrolyte theory,[15] to yield an effective charge of -0.24. No further screening has been included due to the current lack of a sufficiently detailed structural model. The dielectric constant of water has been taken to be 78.4.[16] Both the shielding and dielectric factors are important scaling factors for the energies and torques about the helix axes. In order to generate torques about the helix axes, it is necessary to use the actual positions of the phosphorus atoms, rather than projecting them onto the helix axes. We have performed our calculations using a single turn of double helical B-DNA[14] for each of the double helices for all allowed conformations of the complex. However, we have used longer lengths for selected regions of conformation space, as indicated below. The stereochemical feasibility of all conformations has been checked. Those conformations which generated impossible steric contacts were omitted from the calculations. The only exception to this exclusion involves bad intra-strand contacts within a single residue of the virtual bonds. Our model makes no statement about the detailed geometry of the linkage there, and a small amount of conformational flexibility at those sites is to be expected. Viable space-filling and Kendrew models can be constructed for these regions which eliminate the bad contacts, with small modifications to the local geometry. We have made no attempt to include flexibility parameters for the DNA, due to computing limitations.

Results of the Electrostatic Calculations

The calculations were performed over the entire ranges noted above. The results for a single turn and for 5 turns of helix at selected conformations have been extrapolated to appropriate lengths for the energies and probabilities which we have computed. The results for α are summarized in Fig. 4. Here we see the value of the projected probability for this parameter. The distribution clearly peaks at 90°. Therefore, conformations in which the rhombus approximates a square are the most favored ones.

We present the most favored section in α, the section at $\alpha = 90°$, in Fig. 5a. We have plotted the angles θ_T and θ_B in two dimensions, while we have contoured the electrostatic energy of the structure in increments of ¼ kT. Note that the innermost contours are of the lowest energy, while the highest contours are on the outside of the figure. The region outside the highest contours is filled primarily with conformations which are sterically inaccessible. The diagonal corresponding to $\theta_T + \theta_B = 180°$ is an axis of twofold rotational symmetry for the function. Because this section happens to intersect that axis, the twofold axis appears as a mirror; however, inspection of the three-dimensional map, $E(\alpha,\theta_T,\theta_B)$ indicates the true nature of the symmetry. It should be remembered that this diagonal corresponds structurally to the fourfold symmetric umbrella-rib-like backbone structures mentioned above. The actual minimum of the energy function is very close to the position $\alpha = \theta_T = \theta_B = 90°$, which corresponds to a square planar configuration of the helix axes. In fact, the minimum deviates slightly from this position ($\theta_T = 95°$, $\theta_B = 85°$), although it is still on the diagonal. This results from the differences between the major and minor groove phosphate structures noted above; i.e., the top and the bottom of the "umbrella" have a slightly different distribution of charge

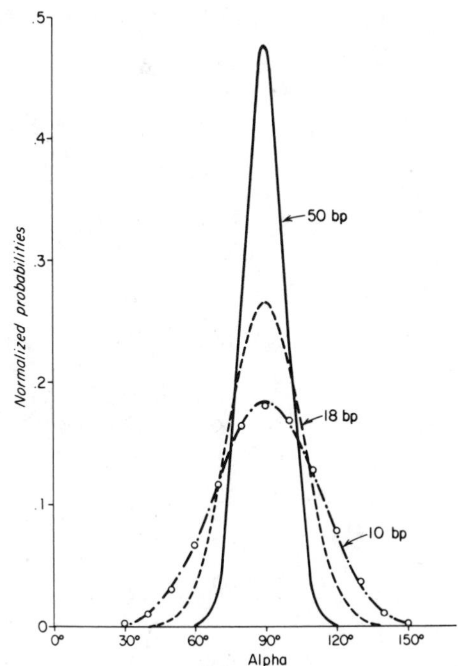

Figure 4. *The Probability Distribution of α for Various Chain Lengths.* Curves are shown for lengths of double helix of 10, 18 and 50 base pairs. The points along the 10 base pair curve are the calculated values. The curve through these points is the best fit to these points via the function indicated in the text. The curve for 50 base pairs is based on calculated values for selected points, from which an appropriate scale factor was derived. The curve for 18 base pairs is based on the function derived in the text. Note that the probability for the rhombus being a square increases with the length of the double helices emanating from the junction.

on their surfaces. This slight deviation results in the small deviation of the minimum energy structure from planarity.

The productive torques about the helix axes are also related to the same diagonal to which the energy is related. Figure 5b indicates the productive torques which have been calculated for the section $\alpha = 90°$. The function is contoured at intervals of 0.006 Kcal/mole-radian. Here, we can see that the diagonal becomes an axis of twofold rotational antisymmetry. Again, since this plane contains the axis, it appears as a plane of antisymmetry, rather than an axis. From the magnitudes of the contoured figure, it is evident that the torques in this section are very small, typically about 0.030-0.070 Kcal/mole-radian. The values of the torques throughout the other sections of the map do not differ from these values markedly: only rarely do values exceed 0.100 Kcal/mole-radian, and then, only for positions of extremely low probability. Thus, if the values of the dielectric constant and the shielding factor are accepted, a single turn of helix clearly does not generate sufficient torque to provide the driving force which would ameliorate the mispairing problem. The torques have been calculated for selected longer lengths, at selected conformations, and they do not increase to significant values.

The structures of the energy and torque maps indicate that the symmetry diagonal is the most useful direction in which to project these functions. The probability plot in Fig. 4 indicates that little information will be lost if the functions are projected in α, as well. We have projected the results in both these directions, as shown in Fig. 6.

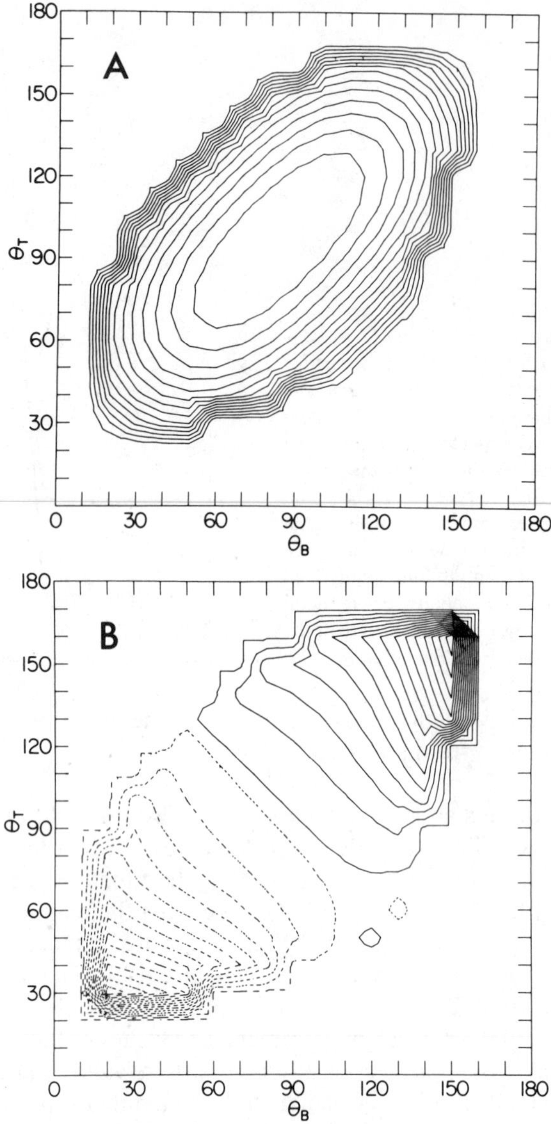

Figure 5. *The Energy and Torque Functions at $\alpha = 90°$.* (a) The energy section is contoured at intervals of ¼ kT. The minimum is at the center of the figure, and higher energies are indicated as one crosses contour lines from that point. Note the twofold axis lying in the plane of the section along the diagonal $\theta_T + \theta_B = 180°$. (b) The productive torque function is contoured at intervals of 6 cal/(mole-radian). The positive contours are indicated by solid lines, while negative contours are indicted by dotted lines. Note that the twofold symmetry axis along the diagonal of (a) is here an axis of anti-symmetry.

This is a plot of the conformational probabilities, versus the distance from the symmetry diagonal. In Fig 6, we show the plot directly calculated, for a single turn of double helix emanating from each corner of the junction, and a curve which simulates this function analytically,

$$P(\delta) = A \cos \left(\frac{\delta}{2}\right)^{\eta} ,$$

where

$$\delta = \Theta_T + \Theta_B - 180° .$$

Extrapolations to longer lengths of double helix, based on calculations at selected values are also shown. It should be noted that the abcissa (δ) corresponds roughly to deviation from fourfold symmetry. In fact, it describes the difference in the values of θ_B and θ_T: those values closest to the center relate to structures whose sets of helix axes are oriented most similarly relative to the axis of 2-fold symmetry. One of the extrapolated curves corresponds to five turns of double helix, while the other is an extrapolation to a element length. The scale factor for the last curve is based on a calculation for the section $\alpha = 90°$ for five turns, which scaled to 1.8% disagreement factor relative to a linear response.

The vanishing of the torques about the helix axes on the symmetry diagonal ($\delta = 0$) results from the cancellation of two equal and opposite forces at this highly probable set of conformations. One must ask, therefore, whether the magnitudes of these cancelling forces are indeed large enough to contribute to an unravelling mechanism for the accomodation of mispairing, as discussed above. They are not: these forces are extremely small, of the order of 0.010 or 0.020 Kcal/mole-radian, at the vanishing points. Thus, we feel that mechanisms involving unravelling are not likely to have a significant electrostatic component to them.

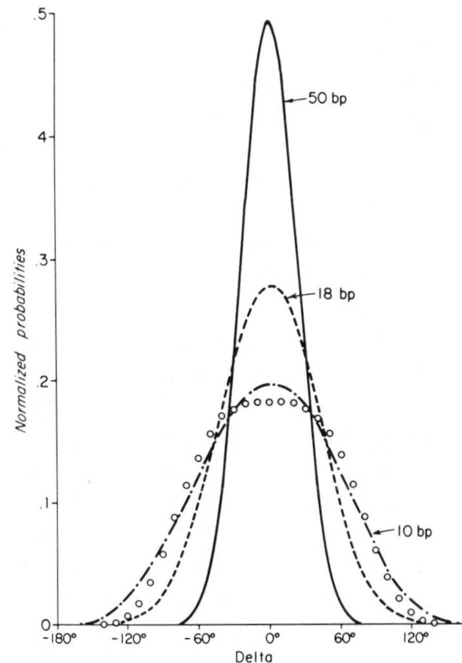

Figure 6. *The Probability Distribution of δ for Various Chain Lengths.* Curves are shown for lengths of double helix of 10, 18 and 50 base pairs. The points along the 10 base pair curve are the calculated values. The curve through these points is the best fit to these points via the function indicated in the text. The curve for 50 base pairs is based on calculated values for selected points, from which an appropriate scale factor was derived. The curve for 18 base pairs is based on the function derived in the text. Since δ is an indication of the deviation of the structure from 4-fold symmetry, the probability of a 4-fold symmetric structure increases as the chain length increases.

Qualitative Nature of the Dynamic Model

With the foregoing results in mind, let us now proceed with the description of a dynamic model which is in agreement with the kinetic and thermodynamic data gathered by Warner and his colleagues.[13,14] There are three obvious elements to consider when developing this model. The first is rotary diffusion about the double helix axes, as advanced by Meselson.[13] He treated rotary diffusion in this direction, which we label ϕ, as the driving force for migration, and with the modifications noted below, we believe that he was correct in this ascription. His calculations, based on this assumption, indicated that the rate of migration should be quite large. He calculated the rotary diffusion constant to imply a migrational rate of about 20 μ seconds per transition for bacteriophage lambda. In fact, Meselson included the entire length of the DNA involved in forming the recombinant complex when calculating his diffusion constant; for his sample case, the paired bacteriophage lambda genomes total about 100 kb in length. For the case of bacteriophage G4, used in Warner, *et al's* experiments, the rate would increase by an order of magnitude, if Meselson's algorithm for calculating the rotational diffusion constant were correct. We have chosen to calculate the diffusion constant corresponding to the torsional correlation length of DNA, which is 580 A.[17] This length will be invariant to the size of the genome being studied. This increases Meselson's migrational rate discrepancy by yet another order of magnitude. On the other hand, Warner and his colleagues have stated that the structure of the junction must be involved in the migratory mechanism, since Meselson's rates, independent of this assumption, are much too fast. Our model takes the thinking of Warner's group into account.

The second feature of the dynamic model to consider is the structure of the junction, particularly relative to the efficacy of rotary diffusion for a given conformer. From molecular models, it appears that if the junction structure does not closely resemble the Sigal-Alberts structure, rotary diffusion forces will not be effectively transduced into migratory forces. Rather, they will be dissipated by the viscous drag of the other arms of the junction. Thus, we must take into account the ability of the junction structure to assume a Sigal-Alberts-like structure, as well as the migrational properties of this structure. We assume that flexural diffusion can account for the achievement of this structure. However, because of the results of the electrostatic calculations described above, the Sigal-Alberts-like structures are not readily available regions of conformation space, since they are characterized by high values of δ. Thus, since these structures are far removed from fourfold symmetric regions of the energy map, their probability of occurrence is very low. The third quantity to consider is the need for base-pair opening at the junction, and the thermodynamics of this process. As will be seen below, the high energy of activation for this process makes it relevant to the low-temperature kinetics of migration (<20°C.), but it is not rate limiting in the high temperature pathway. Thus, at low temperatures, a third, intermediate step, unpairing of the bases is involved. This temperature dependence of opening may be due to the structure of the junction above 20°C.; it may already be an "activated" intermediate, in which the base pairs which adjoin the junction are at least partially open.

Our qualitative model of the high temperature kinetic process is illustrated in Fig. 7, where we show the nature of the migratory transition state potential surface. In this two dimensional diagram, the horizontal direction, ϕ, corresponds to rotation about the helix axes, while the vertical direction corresponds to differential rotation about the virtual bonds perpendicular to the helix axes, i.e., along δ. A rotation of 36° in ϕ will advance the branch point by a single transition in a given direction. As will be noted from the figure, conformations near fourfold symmetry are associated with large barriers to this advancement; those conformations far from fourfold symmetry have a much lower barrier to rotation. Meselson's single step mechanism is indicated by the double arrow at the bottom of the the diagram; this arrow extends horizontally from side to side, over the indicated barrier. The model which we propose indicates that the pathway that the reaction must take is (at least) a two step reaction, first in the direction vertically, and then a second step which also involves crossing a rotation barrier in the horizontal direction. At low temperatures, this barrier includes resistance not only to rotation, but also to the opening of

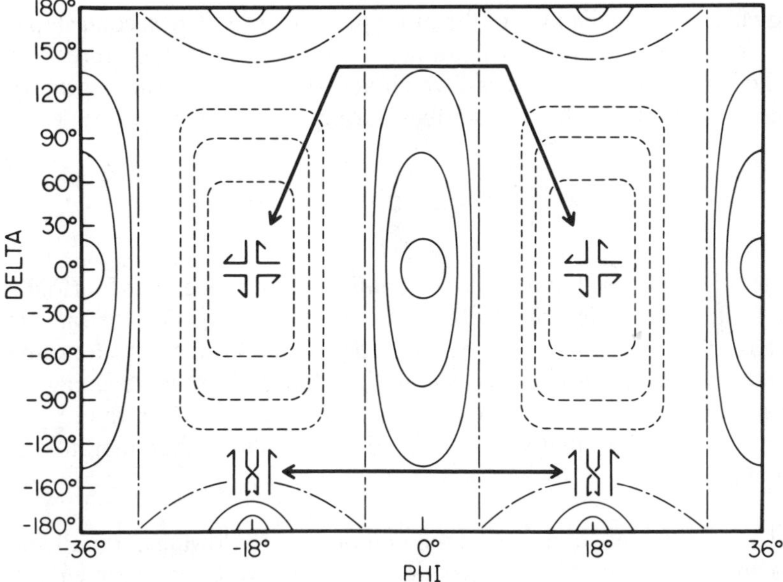

Figure 7. *A Qualitative Picture of the Reaction Coordinate in Two Dimensions.* ϕ is the direction of migration, and δ is the deviation of the helix axes from 4-fold symmetry. 36° steps in ϕ indicate a unit migrational event. An arbitrary zero of energy is indicated by the lines of alternating dashes and dots. Positions negative with respect to this energy are indicated by dashed contours, while positions positive with respect to this energy are indicated by solid contours. Note that the function is mirror symmetric about $\delta = 0$. This line corresponds to the most probable four-fold symmetric structures. Sigal-Alberts-like structures correspond to structures with δ near 180. Meselson's mechanism[13] is indicated at the bottom of the figure: migration over a barrier between two Sigal-Alberts-like structures. The mechanism proposed in this paper is indicated by the thick line at the top of the figure: movement towards a Sigal-Alberts-like structure ($\delta > 140°$) and then movement over the barrier. The barrier between the two four-fold symmetric structures arises because the rotary diffusion forces are not capable of generating migrational forces unless the structure is Sigal-Alberts-like.

the base pairs. Our model states that the migration will not occur unless the angle δ is greater than a specific minimum angle, which is called δ_c. Since the potential energy surface is symmetric about $\delta = 0$, δ may also be less than $-\delta_c$, for migration to occur. This pathway is indicated by the other double arrow, at the top of the diagram in Fig. 7. Implicit in this pathway is the assumption that the barrier between the wells is higher than that for the pathway we have selected. Note that although we have indicated this path as containing two distinct steps, for purposes of clarity, mixing the two steps is certainly permitted by the surface.

Thus, we propose that three processes are involved in the migratory event. The first of these involves transforming a structure in which the difference in orientation of the two sets of helix axes is relatively small, to one in which it is large. The larger this difference, the closer a Sigal-Alberts-like structure is approximated. In order for rotary diffusion to be effective in promoting migration, we feel that it is necessary for the helix axes, and, *a fortiori,* the bases, near the junction to be approximately parallel. Otherwise, the torsional diffusion forces about any pairs of symmetric arms emanating from the junction will be dissipated by the viscous drag perpendicular to the helix axes of the other pairs of arms. The second step involves base pair opening, which is not limiting the rate at high temperatures. The third step is the rotary diffusion process, which actually generates the migratory event. The following reaction scheme describes our postulated mechanism:

$$\text{S4L} \underset{k_{-1}}{\overset{k_1}{\rightleftharpoons}} \text{SAL} \underset{k_{-2}}{\overset{k_2}{\rightleftharpoons}} \text{SALO} \overset{k_3}{\longrightarrow} \text{P.}$$

S4L is the initial material in a region of conformation space near fourfold symmetry, SAL is the material in a Sigal-Alberts-like structure, SALO is the same material with its base pairs open, and P is the migrated product. k_1, k_2 and k_3 are respectively the forward rate constants for becoming Sigal-Alberts-like, for opening the bases and for rotation to a migrated position. k_{-1} and k_{-2} are the back rates for the first two reaction steps. Note that k_3 is the rate of diffusion for either migratory reaction (I or II) of Fig. 1.

The first step of the reaction may be thought of in the following alternative fashion: There is only a fraction of junctions which are arranged so that δ is greater than δ_c, and only this fraction is able to react. Therefore, the rate is proportional to this fraction, called F, which represents those helices able to undergo branch point migration. The second and third steps are then considered independently of the first step, as a two step mechanism. Furthermore, we assume that the overall rate is proportional to the initial rate of base pair activation and branch point migration. We neglect the base pair closing rate. The combined effective rate for these two processes is just

$$\frac{k_2\, k_3}{k_2 + k_3} \, .$$

This effective rate does not consider the distance of migration or the possibility of multiple branch point steps during migration. The total number of steps depends upon the time during which the helices are correctly positioned to allow migration and the rate at which the migration can proceed by torsional diffusion. The root mean square average number of steps (N) taken by migrating molecules is

$$N = (k_3 t^*)^{\frac{1}{2}},$$

where t* is the mean time during which those conformations with $\delta > \delta_c$ will maintain themselves in this migrating state. The total migration rate constant, K is, thus,

$$K = \frac{FN(k_2 k_3)}{(k_2 + k_3)} .$$

A more detailed treatment of this model and the types of assumptions necessary to arrive at this simplified treatment will be published elsewhere.[18]

Quantitative Details of the Dynamic Model

The rate of branch point migration, as well as the associated thermodynamic constants may be calculated from a quantitative model which quantitates the description outlined in the previous section. The first quantities which must be determined are the torsional diffusion constant about the helix axes, D^-, and the flexural diffusion constant perpendicular to the helix axes, D^+. As noted above, the values for D^- and D^+ are estimated from the diffusion coefficients for the rotation of two non-interacting helices, one persistence length long, and freely hinged at the junction. The flexural persistance length is approximately 600Å while the torsional persistence length is 580Å.[17] The radius of the individual helices has been taken as 12Å. The diffusion coefficient D^+ is related to the effective friction factor, f by

$$D^+ = kT/f .$$

here k is Boltzmann's constant and T is the absolute temperature. The effective friction factor is given in terms of f_r^+ and f_t^+, which are the friction factors for pure rotation and pure translation for a cylinder moving respectively about and along an axis perpendicular to the helix axis. The relation is:

$$f = 2\left(f_r^+ + \left(\frac{L_p}{2}\right)^2 f_t^+\right) .$$

The factor of 2 accounts for two helices in a single unit, which are hydrodynamically noninteracting to a first approximation. L_p is the persistence length of double helical DNA, 600Å.[19] The translational term f_r^+ and f_t^+ contributes to the frictional factor, since the rotation of a helix about its end at the junction, is coupled to a translation. The terms f_r^+ and f_t^+ are calculated by Broersma's equations.[20] At 25°C.,

$D^+ = 7500$ rad^2-sec^{-1}. Since this diffusion coefficient is for a full persistence length, and since direct hydrodynamic interaction has been neglected, this is the smallest value expected. The diffusion coefficient[19] D^- is 33000 rad^2-sec^{-1}.

The process of flexural diffusion where the individual strants are repelled by electrostatic forces may be treated by the Smoluchowski equation, appropriately incorporating a restoring potential.[21] The distribution in terms of δ, P(δ) is defined by

$$\frac{dP}{dt} = D^+ \frac{d}{d\delta} \left(\frac{d}{d\delta} - \frac{Q}{kT}\right) P = AP ,$$

where Q is the longitudinal torque on the helices, generated by the electrostatic potential. This should not be confused with the productive torque, which is about the helix axes. This longitudinal torque is generated from the conformational energy profile calculated above. The equilibrium distribution of helices with respect to δ, $P_o(\delta)$, calculated earlier, is described in terms of the single adjustable parameter, η, since $P_o(\delta) = A \cos^{\eta}(\delta/2)$.

The value of A is adjusted so that Po(δ) is normalized to unity. As noted above, the scaling was nearly linear with the number of base pairs (NBP), suggesting a simple relation between η and NBP,

$$\eta = \frac{(NBP-4)}{2} .$$

The application of the Boltzmann condition to the Smoluchowski Equation gives the equation that

$$\eta \log \left(\cos \left(\frac{\delta}{2}\right)\right) = \frac{U}{kT} .$$

where U is the potential due to the electrostatic repulsions, and the torque, Q, is

$$\frac{Q}{kT} = \eta \tan \left(\frac{\delta}{2}\right) .$$

which is a very tractable term and easily used in the computations.

We will define t* to be the "mean first passage" time. This the mean time for all junctions with $\delta > \delta_c$ to relax to $\delta = \delta_c$. Junctions with angles less than δ_c cannot contribute to this time. Once a helix has attained δ_c, it is removed, so that it can no longer contribute to the passage time. This quantity may be computed from the diffusion equation. In order to carry out the computation, the operator, $d/d\delta$ is represented in terms of a finite difference form and P(δ) is replaced by a vector in δ. This transformation from differential operators to quantized difference forms is known as the transition, or rate matrix approximation to the differential equation.[22]

The Smoluchowski Equation is now in a general form of $dP/dt = AP$, and therefore the solution is $P_t(\delta) = P_o(\delta) \exp(At)$, where P(0) is the distribution at time $t = 0$. The mean first passage time, t* is computed as

$$t^* = \int_{\delta_c}^{\pi} A^{-1} P_o(\delta) d\delta \; .$$

The matrix A is augmented so that junctions which move below angle δ_c cannot return. This assures that only junctions above the critical angle contribute to t*. P_o is renormalized over the range from δ_c to 180°. $P_o(\delta)$ is taken to be the equilibrium distribution.

Results of the Dynamic Calculations

Fig. 8 shows the values of the rate constant, K based on the above equations over the temperature range from 0 to 40°C. More complex equations, discussed elsewhere, were also used to evaluate the rate constant for migration, and the results are quite similar. The cutoff angle δ_c was taken to be 140 degrees, when the electrostatic repulsion term contributes 5.5 Kcal/mole to the total enthalpy for the migratory process. The value of η was approximately 7 at 0°C. This value of η corresponds to slightly less than two turns of double helical B-DNA. Hence, two turns represents the effective length necessary to describe the electrostatic repulsion at the junction. The electrostatic calculation, which assumed rigid helices, did not account for systematic bending of the helices, which would occur when δ becomes large, and would, of course, vanish at $\delta = 0$. The persistence length of 600Å implies that there is a 5° root mean square flexure between base pairs. Electrostatic repulsion would likely force the helices to be somewhat distorted from linearity. A systematic 5° flexure would bend the helices by 90° in just under two turns. The rms number of base pairs migrated per activation is about 2.5, over the entire temperature range. While this number is on the order of a single transition, our model does suggest that multiple transitions do occur. As is shown in Fig. 8, the diffusion processes are rate limiting at temperatures greater than 20°C. k_2 does not contribute over this temperature range. This result produces a simplified

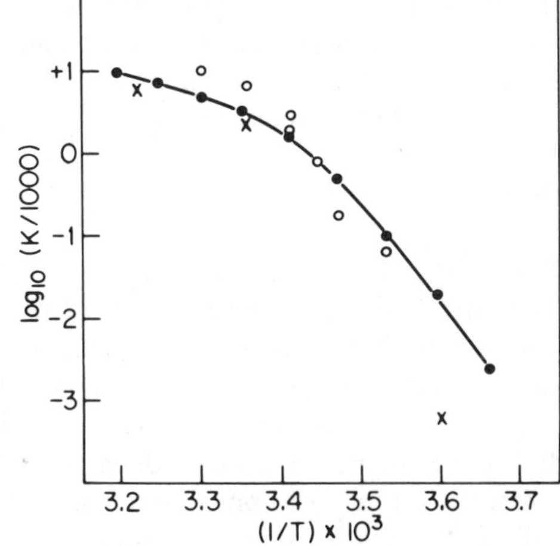

Figure 8. *Calculated and Observed Arrhenius Plots.* The X's represent experimental points from Thompson, Camien and Warner[11] and the open circles represent experimental points from Warner,Fishel and Wheeler.[12] The filled circles which are connected by lines represent points calculated from the model presented in the text.

model which contains some features which are qualitatively similar to the model suggested by Meselson.[13] In this temperature range, the calculated enthalpy of activation is 16.5 Kcal/mole. This quantity is the sum of three terms: Both the flexural and torsional rotation contribute 5.5 Kcal/mole, as noted by Warner et al.[12] Another 5.5 Kcal mole comes from the electrostatic repulsion term. This is in reasonably good agreement with the experimental result of Warner, et al of 20 ± 10 Kcal/mole. At temperatures lower than 20° C., the processes specific to the pairing status of the junction become important. The rate of 9 x 10^8 sec^{-1} at 20°C and enthalpy of 50 Kcal/mole were chosen to show the dramatic change in slope which occurs around 20°C. This value of the enthalpy is on the order of the enthalpy of opening the two base pairs. The enthalpy for the base pair opening process in linear, duplex DNA, as measured by tritium exchange, is around 18 Kcal/mole.[23] This should be considered a lower bound on this quantity, since the amount of opening to allow tritium exchange may be considerably less than that necessary to permit migration. The slope in the region from 0 to 20°C. gives an enthalpy of 60 Kcal/mole. This comes from two 25 Kcal/mole base pair openings plus the flexural diffusion and the electrostatic repulsion.

Thus, it seems likely, as noted by Warner, Fishel and Wheeler,[12] that base pair opening comes into play at temperatures of 20°C. and below, while it is not rate limiting above that temperature. Although this seemed unlikely with the earlier dynamic model, it is consistent with our model. This implies that Warner and his colleagues are correct in stating that the structure of the junction is crucial to the kinetics of this system. It is neither necessary nor forbidden that a structural change of some sort takes place in this structure at the transition temperature of 20°C. The very high enthalpy of activation of base pair opening is sufficient to account for the change in the rate limiting step at the transition temperature. However, the high temperature dependence of the rate of the process may have structural implications for the effective detailed conformation of the junction. Nevertheless, it is quite clear that the dynamic structure of the junction is dramatically different above and below 20°, with respect to the migratory process. Our model is able to justify Warner, Fishel and Wheeler's suggestion[12] that opening the base pairs which adjoin the junction is rate limiting below 20°C., while this appears not to be the case above that temperature.

Discussion

The results obtained above indicate that the model which we have used is able to qualitatively emulate the behavior of the system observed by Warner and his colleagues.[11,12] The success of our electrostatic structural model and our three-step kinetic model implies that the dynamics of double stranded branch point migration can be modeled by the means we have selected. The structure of the branch point during the migratory process does not appear to be static: rather, it is best represented by a range of conformations. Some of these conformations are of substantially different electrostatic energies, and the transitions between them appear to be critical factors in the dynamics of the migratory process. It is interesting that the two major structures previously proposed, *viz,* the Sigal-Alberts structure analogous to Fig. 1a, and the Sobell structure analogous to Fib. 1b, are both representative of conformational classes relevant to the migratory process. Those structures analogous to Fig. 1b appear to be the most probable structures from the standpoint of electrostatics; on the other hand, those structures are the least likely to be involved in the migratory process. Those structures furthest from fourfold symmetry are closest to the Sigal-Alberts model; these structures have gained electrostatic energy through flexural diffusion in the course of their excursions away from the most stable region of conformation space. The Sigal-Alberts-like structures appear to be critical intermediates in the migratory process, corresponding to those structures with $\delta > \delta_C$. Along with Meselson, we conclude that the high temperature migratory process is diffusion limited however, rather than allowing rotation about the helix axes to be rate limiting, we suggest that diffusion perpendicular to the helix axes is rate limiting in the system. For the low-temperature system, clearly a process such as base-pair opening is rate-limiting instead. Nonetheless, diffusional processes approximating those about the virtual bonds are an apparantly necessary step in reaching a state of sufficient energy to overcome the transition barrier. The process is then limited by the slower diffusion perpendicular to the helix axes, rather than the faster diffusion about them. Therefore, flexural diffusion must be rate-limiting in the migratory process at high temperatures. However, since the electrostatics near the junction determine the structure, the activation parameters and the reaction potential surface, we conclude, as did Warner and his colleagues, that the structure about the branch point is the determining factor in the rate of branch point migration. For this reason, the importance of immobile and semimobile junctions[10] in understanding this phenomenon at the detail affordable by high resolution physical techniques is paramount.

Acknowledgements

This work has been supported by a Basil O'Connor Starter Grant from The March of Dimes Birth Defects Foundation and Grant GM-26467 from the NIH. We would like to thank Dr. J.M. Schurr for valuable discussions and information about torsional correlation lengths. We would also like to thank Ms. Kathleen A. McDonough for assistance in model building. Ms. Diane B. Robinson and Mr. Ryland Loos gave graphic assistance, and Mr. Robert Speck made the photographs. Computational facilities were kindly provided by the SUNY/Albany Computer Center and by the University of Washington Computer Center.

References and Footnotes

1. Holliday, R., *Genet. Res. 5,* 282-304 (1964).
2. Sigal, N. and Alberts, B., *J. Mol. Biol 71,* 789-793 (1972).
3. Sobell, H.M., *Proc. Nat. Acad. Sci. (USA) 69,* 2483-2487 (1972).
4. Wilson, J.H., *Proc. Nat. Acad. Sci. (USA) 76,* 3641-3645 (1979).
5. Seeman, N.C., Robinson, B.H., and McDonough, K.A., abstr., Second Basil O'Connor Symposium, Key Biscayne, October, 1979.
6. Broker, T. and Lehman, I.R., *J.Mol.Biol. 60,* 131-149 (1971).

7. Wolgemuth, D.S. and Hsu, M.T., *Nature 287,* 168-171 (1980).
8. Kim, J.S., Sharp, P. and Davidson, N., *Proc. Nat. Acad. Sci. (USA) 69,* 1948-1952 (1972).
9. Lee, C.S., Davis, R.W. and Davidson, N., *J. Mol. Biol 48,* 1-22 (1970).
10. Seeman, N.C. These Proceedings, Vol 1. (1981).
11. Thompson, B.J., Camien, M.N. and Warner, R.C., *Proc. Nat. Acad. Sci. USA 73,* 2299-2303 (1976).
12. Warner, R.C., Fishel, R. and Wheeler, F., *Cold Spring Harbor Symp. Quant. Biol. 43,* 957-968 (1978).
13. Meselson, M. *J. Mol. Biol. 71,* 795-798 (1972).
14. Arnott, S. and Hukins, D.W.L., *Biochem. Biophys. Res. Comm. 47,* 1504-1509 (1972).
15. Manning, G.S., *Quart. Rev. Biophys. 11,* 179-246 (1978).
16. *Handbook of Chem. and Phys.,* CRC Press, Cleveland, p. E-61 (1975).
17. Allison, S.A. and Schurr, J.M. *Chem. Phys. 41,* 35-46 (1979).
18. Robinson, B.H. and Seeman, N.C. manuscript in preparation (1981).
19. Barkley, M.D. and Zimm, B.H., *J. Chem. Phys. 70,* 2991-3007 (1979).
20. Broersma, S., *J. Chem. Phys. 32,* 1626-1631 (1960).
21. Chandrasekhar, S., *Rev. Mod. Phys. 15,* 1-89 (1943).
22. Montroll, E.W. and Shuler,K.E., in *Advances in Chemical Physics,* vol 1, Academic Press, New York, pp. 361-393 (1958).
23. Bird, R.E., Lark, K.G., Curnutte, B. and Mansfield, J.E., *Nature 225,* 1043-1045 (1970).

Proceedings of the Second SUNYA Conversation in the Discipline Biomolecular Stereodynamics Volume I, ISBN 0-940030-00-4, Ed., Ramaswamy H. Sarma, Adenine Press, New York, ©Adenine Press

The Nature of the Mobility of the Sugar and its Effects on the Dynamics and Functions of RNA and DNA

M. Sundaralingam and E. Westhof
Department of Biochemistry
College of Agricultural and Life Sciences
University of Wisconsin
Madison, Wisconsin 53706

"On all great subjects, much remains to be said."
Macaulay

Introduction

Nucleic acids are sugar-phosphate polymers with heterocyclic side-chain bases. Their most important biological properties reside in the sequence of the heterocyclic bases which contains the genetic information. The sugar-phosphate backbone appears, therefore, as a "passive" carrier helping in the information storage and decoding processes. Cells use two kinds of nucleic acids for storage and processing. The genetic information is stored in DNA, where the sugar is a 2′-deoxyribose; while the information is processed by RNA, where the sugar is a ribose.

The DNA molecules should be flexible to be compactly packed in the chromosomes of the cell nucleus and, at the same time, responsive to the environment (interactions with proteins, metal ions, etc.) so that the information can be retrieved. Thus, the double helical Watson-Crick base paired secondary structure of DNA folds into the coiled and supercoiled tertiary structures of chromosomes. Also, DNA should be adaptable to specific conformations so that the information reading and processing can be done orderly and faultlessly.

On the other hand, the complex ribosomal machinery which translates the genetic information into cell components requires molecules which can help assemble various protein components into a stable functional structure. This role is played by ribonucleic acids which characteristically adopt complex tertiary structures. This is well illustrated by the tRNAs[1] where a complex secondary structure assembles to form a compact three-dimensional structure which fulfills the needs for the translation of the genetic code. Similarly, it is expected that the highly complex secondary structures of the ribosomal RNAs[2] will fold into stable tertiary structures to act as scaffolds for the specific attachment and arrangement of the numerous ribosomal proteins.

We will show that *nucleic acids owe their adaptive flexibility to their sugar-phosphate backbone which astutely exploits the peculiar properties of the five-membered furanose ring through the 3′-5′ phoshodiester linkage.* The presence or absence of the 2′-hydroxyl group in the furanose ring adds further subtleties to the conformational properties of the sugar-phosphate backbone of RNA and DNA. Most of the interesting properties of the furanose rings originate in their capacity of undergoing pseudorotation.[3] This leads to the inherent mobility of five-membered rings in contrast to the relatively "rigid" chair forms of six-membered rings.

In nucleic acids, depending on the nature of the sugar (ribose *versus* deoxyribose) and also of the base (purine *versus* pyrimidine), the furanose ring adapts preferential puckered states, which are conserved and amplified upon their incorporation into a polynucleotide chain.[4,5,6]. The preferred conformations of the various polynucleotide torsion angles will, in this way, be modulated by the furanose substituents. Also, and most importantly, because of the 3′-5′ linkage, the inherent flexibility of the furanose ring is transmitted to the backbone differently depending on the sugar pucker.[7] This sugar pucker dependent flexibility is an important key to the understanding of the conformational dynamics of ribo- and deoxyribonucleic acids.

Conformational Variety of the Nucleotide Building Blocks

Both from the standpoint of the chemistry and biology of nucleic acids, the 5′-nucleotide can be regarded as the structural unit of nucleic acids. The main conformational variables for this unit are the sugar pucker, the sugar-base glycosyl linkage, and the exocyclic C4′-C5′ linkage (Fig. 1). In addition, the C5′-O5′ and C3′-O3′ linkages are amenable to rotation over a singular conformational range centered on the *trans* or skewed *trans* orientation. Although the sugar can potentially move over a large region of pseudorotation space, it prefers to populate the sectors in the ${}^{3}E$ range (comprising the closely related states ${}_{2}E$, ${}^{3}_{2}T$, ${}^{3}E$, ${}^{3}_{4}T$) and in the ${}^{2}E$ range (comprising the states ${}_{1}E$, ${}^{2}_{1}T$, ${}^{2}E$, ${}^{2}_{3}T$, ${}_{3}E$). These two sectors, referred to as ${}^{3}E$ and ${}^{2}E$ for short, are separated by energy barriers. The sugar-base glycosyl linkage can adopt orientations in two broad domains, the *anti* and *syn* conformations, which are also separated by energy barriers. And, finally, the exocyclic C4′-C5′ torsion angle can adopt one of the three staggered conformers g^{+}, t, or g^{-}. *These conformations do not all occur independently of each other. Certain combinations are known to be preferred over others, while certain others are mutually incompatible. Thus, the conformations of the nucleotide building blocks are dominated by correlated changes about the various bonds. This concept of correlated motions has proved invaluable in the interpretation of solid-state and solution data on nucleic acids and their constituents.*[5]

Pseudorotation of Five-Membered Rings

In crystals, the furanose ring of nucleo*sides*, nucleo*tides*, and oligomers is not planar and, as a rule, has an unique puckered state. The various puckered states of five-membered rings are best understood in the framework of the pseudorotation

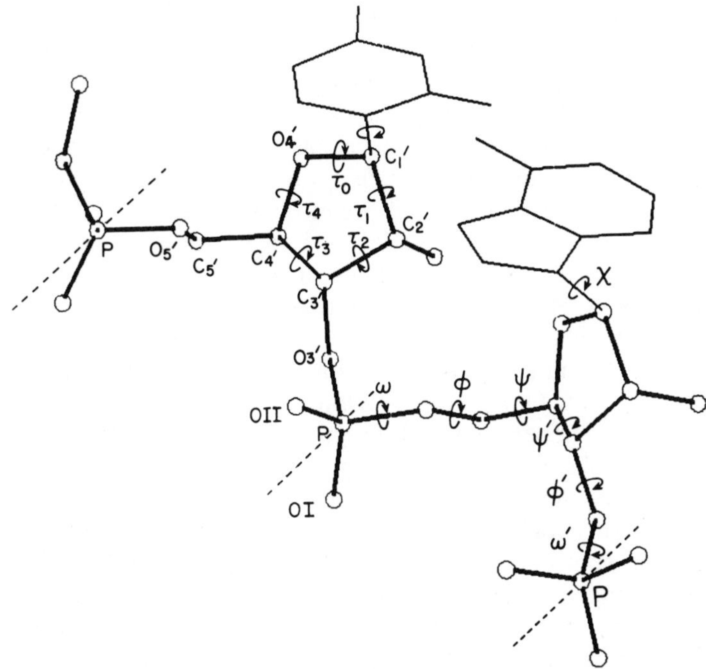

Figure 1. Atom numbering and torsion angle notation for the sugar-phosphate backbone. The pair (ω, ω') corresponds to the P-O bonds, the pair (ϕ, ϕ') to the C-O bonds, and the pair (ψ, ψ') to the C-C bonds of the backbone; it may be regarded that the unprimed notations are for the 5′-end and the primed ones for the 3′-end. Viewing down a bond, the torsion angle is positive when the far bond moves in a clockwise direction. The *gauche*$^+$ domain is centered on $+60°$, the *trans* on 180°, and the *gauche*$^-$ on $-60°$. The torsion angle ψ' is determined by the sugar pucker and depends on both the angle of pseudorotation P and the amplitude of pucker τ_m. The torsion angle about the glycosyl bond is denoted by χ and is defined by the atom sequence O4′-C1′-N9(N1)-C8(C6) in purine (pyrimidine). The *anti* range is centered on $\chi = 0° \pm 90°$ and the *syn* range on $\chi = 180° \pm 90°$. From reference (18).

concept where each pucker is described by two parameters, the phase angle of pseudorotation, which determines the puckered atom(s), and the amplitude of pucker, which gives the maximum puckering. It is worth remembering how pseudorotation of a five-membered ring was introduced. In order to explain the measured values of the entropy and of the heat capacity of cyclopentane, Aston et al.[8] had to invoke a puckering of the cyclopentane ring. Later, Kilpatrick et al.[9] concluded that "the puckering of the ring is not of a definite type, but that the angle of maximum puckering rotates around the ring", and they showed that, in cyclopentane pseudorotation occurs because all puckered states are equally more favorable than the planar state. Besides, this motion of the atoms (perpendicular to the direction of rotation and thus without angular momentum), there is an "ordinary vibration in which the amount of puckering oscillates about a most stable value."[9] Thus, pseudorotation is an inherently dynamic concept leading to a continuum of puckered states.

In a five-membered ring, there are two puckering modes which retain a symmetry element: the envelope conformation (C_S or E with one atom displaced from the other four containing the plane of symmetry) and the twist conformation (C_2 or T with two atoms oppositely displaced from the other three and with a two-fold axis of symmetry) (Fig. 2). In cyclopentane, because of symmetry, the five envelope and the five twist puckers are energetically equivalent, i.e. the potential energy is independent of the phase angle, and pseudorotation is free. This is because the opposing forces of torsional strain and bond angle strain balance out. The introduction of an endocyclic substituent, like in *tetrahydrofuran,* leads to restricted pseudorotation between the envelope and twist puckers which have the endocyclic substituent on the symmetry element.[10] The energy barrier hindering the rotation of the phase of puckering depends mainly on the difference between the torsional barriers (C-C *versus* C-O in tetrahydrofuran). The addition of an exocyclic substituent leads to a further energy difference between potential minima because puckers with the substituent in either equatorial or axial orientation do not have the same energy content. In the substituted furanose rings of nucleotides, nonbonded interaction between endocyclic and exocyclic substituents and between exocyclic substituents (eclipsing and steric repulsions) have to be considered. Thus, in ribofuranose rings, puckerings involving the C2′ and/or C3′ atom relieve most the torsional strain and eclipsing interactions between exocyclic substituents. These puckers are, therefore, usually observed in crystal structures. In deoxyribofuranose rings, however, the absence of an exocyclic substituent relieves some interactions between endo- and exocyclic substituents and between exocyclic substituents leading to preferential stabilization of puckerings which involve the nonsubstituted atoms.

As a rule, in crystals of the monomers, ribofuranose rings have no preference for either ^{2}E or ^{3}E puckers while deoxyribofuranose rings prefer puckers in a broad C2′-*endo* range.[6]

According to Kilpatrick et al.,[9] the various puckers are characterized by the atomic out-of-plane displacements z_j of the five-ring atoms in such a way that the amplitude of pucker (q, Å) of each atom is modulated by the phase angle of pseudorotation (ϕ):

$$z_j = q \cos\left(\frac{4\pi}{5}(j-1) + \phi\right), \; j = 1, \ldots 5, \text{ with } \Sigma z_j = 0 \qquad (1)$$

Because of the closed structure, the torsion angles around the ring are linear combinations of the out-of-plane displacements. Consequently, the torsion angles transform like the out-of-plane displacements and an equation similar to (1) was proposed[11]:

$$\tau_j = \tau_m \cos\left(\frac{4\pi}{5}(j-1) + \underline{P}\right), \; j = 1, \ldots 5, \text{ with } \Sigma\tau_j = 0. \qquad (2)$$

Equation (2) renders best the fact that interconversion between the puckered states occur through smooth and correlated changes of the ring torsion angles. Through

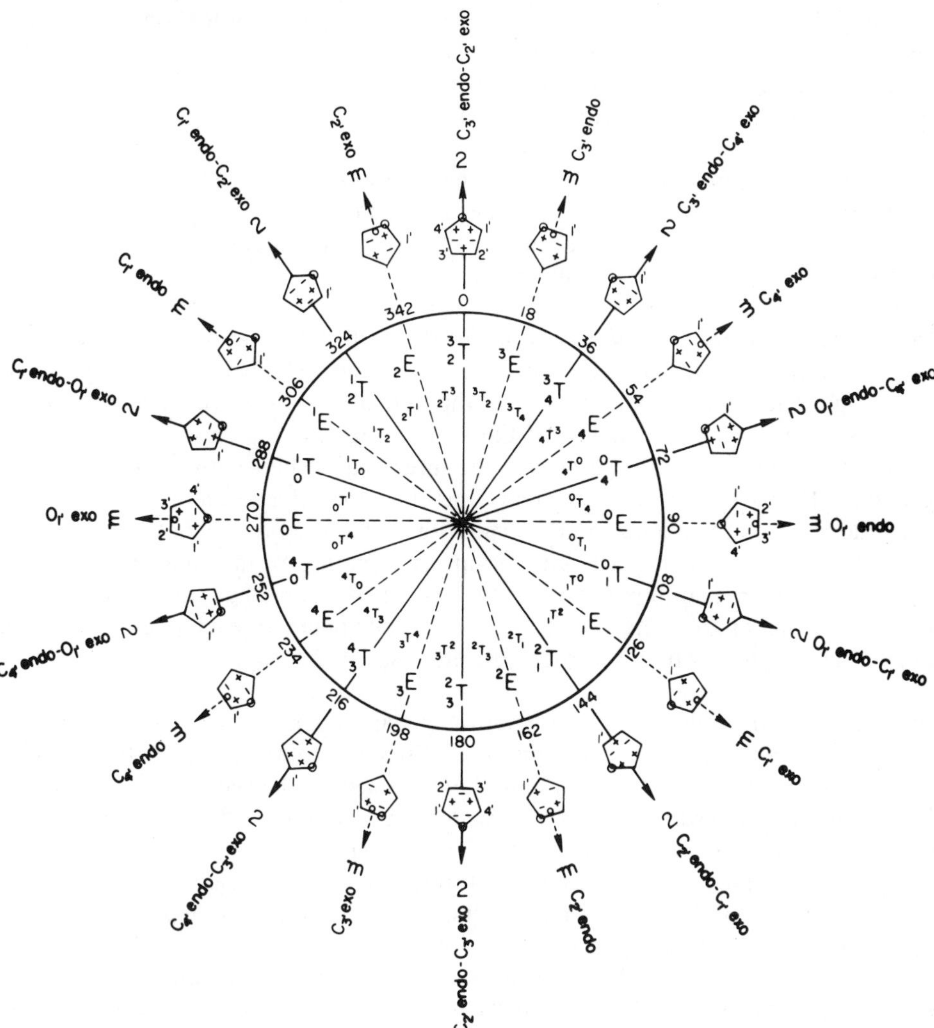

Figure 2. The pseudorotation wheel showing the relationship between the phase angle of pseudorotation P and the ten envelop (C_S-symmetry, E-type pucker) and ten symmetric twist (C_2-symmetry, T-type pucker) conformations for the furanose ring. The pseudo mirror symmetry in the envelope conformations is indicated by the arrows in dashed lines and the pseudo twofold axis in the symmetric twist conformations by the solid arrows. The non-symmetric twist conformation between the envelope and symmetric twist conformations are also indicated. The signs of the torsional angles of the sugar ring are indicated within the furànose ring.

The origin of P is taken as the C3′-*endo*-C2′-*exo* twist conformation (${}^{3}_{2}T$), where the atoms C3′ and C2′ are equally displaced from the mean least-squares plane defined by the other three sugar atoms C1′, O4′, C4′, but in opposite directions (*endo:* toward the C5′ atom; *exo:* away from the C5′ atom). The envelope conformation C3′-*endo* (${}^{3}E$) where only the C3′ atom is displaced from the mean least-squares plane defined by C2′, C1′, O4′, C4′ has a P value of 18°. The corresponding twist conformation C2′-*endo-C3′-exo* (${}^{2}_{3}T$) and envelope conformation C2′-*endo* (${}^{2}E$) have P values of 180° and 162°, respectively. The preferred sectors for the P values are 0°-36° and 144°-198° and they are generally referred to as C3′-*endo* (${}^{3}E$) and C2′-*endo* (${}^{2}E$) puckerings, respectively. From reference (3).

modulation of the phase angle of pseudorotation *(P)*, the ring torsion angles can take in a concerted fashion all values between $-\tau_m$ and $+\tau_m$. Since the torsion angles never become zero simultaneously, the ring does not need go through the planar state when interconverting between puckers.

Recently, Cremer and Pople[12] devised another method for extracting pseudorotation parameters from atomic coordinates where a mean plane is chosen so as to guarantee that the sum of the displacements is zero with the direction of the displacements parallel to the plane normal.

In principle, the pseudorotation concept allows the generation of the various puckers around the pseudorotation circuit necessary to analyze the puckering effects in model building, structural, and theoretical studies. At first glance, owing to the simple cosine dependence of the z coordinate, the Cremer-Pople method seems most appropriate for describing the interrelationships between pucker and geometry. However, the ring being not symmetric, the orientation of the reference plane depends on the pucker, and the variation of the z coordinate is consequently more complex. In other words, the disymmetric nature of the ring prevents full cancellation of angular momentum. In the Cremer-Pople method, this cancellation is assured by an appropriate choice of the reference plane for each pucker, but the puckering parameters thereby absorb the effects.

In an improved method of the torsion angle approach,[13] exact pseudorotation parameters are extracted disregarding small distortions of the ring, i.e., the parameters correspond exactly to the amplitude and phase of the second order Fourier wave of period $4\pi/5$ which alone gives rise to pseudorotation. The geometry of five-membered rings observed in the crystal depends on the puckered state adopted.[14] This geometrical dependence has been parametrized from an analysis of numerous crystal structures of nucleic acid monomers.[15] A method has also been developed to calculate the coordinates of the five atoms of a furanose ring with any puckering on the pseudorotation circuit.[16] The variations with *P*, at a common amplitude of puckering, of the endocyclic bond angles for cyclopentane and for a furanose ring are shown in Fig. 3.

The Preferred Conformations of the Nucleotides Follow the Preferred Conformations of the Furanose Ring

With one common bond between the furanose ring and the sugar-phosphate backbone the $3' \rightarrow 5'$ sugar-phosphate linkage is most appropriate for taking advantage of the furanose pseudorotational preferences and flexibility. It is thus expected that the sugar pucker will influence not only the nature of the preferred conformations about the glycosyl bond (χ), the exocyclic C4′-C5′ bond (ψ), and the internucleotide phosphodiester bonds (ω', ω) but also the conformational mobilities and correlations about these bonds. For instance, C3′-*endo* puckers favor low χ angles in the *anti* range, while C2′-*endo* puckers favor higher χ angles[4] (Fig. 4). In addition, the χ range available in the C2′-*endo* domain is broader, it extends from the usual *anti*

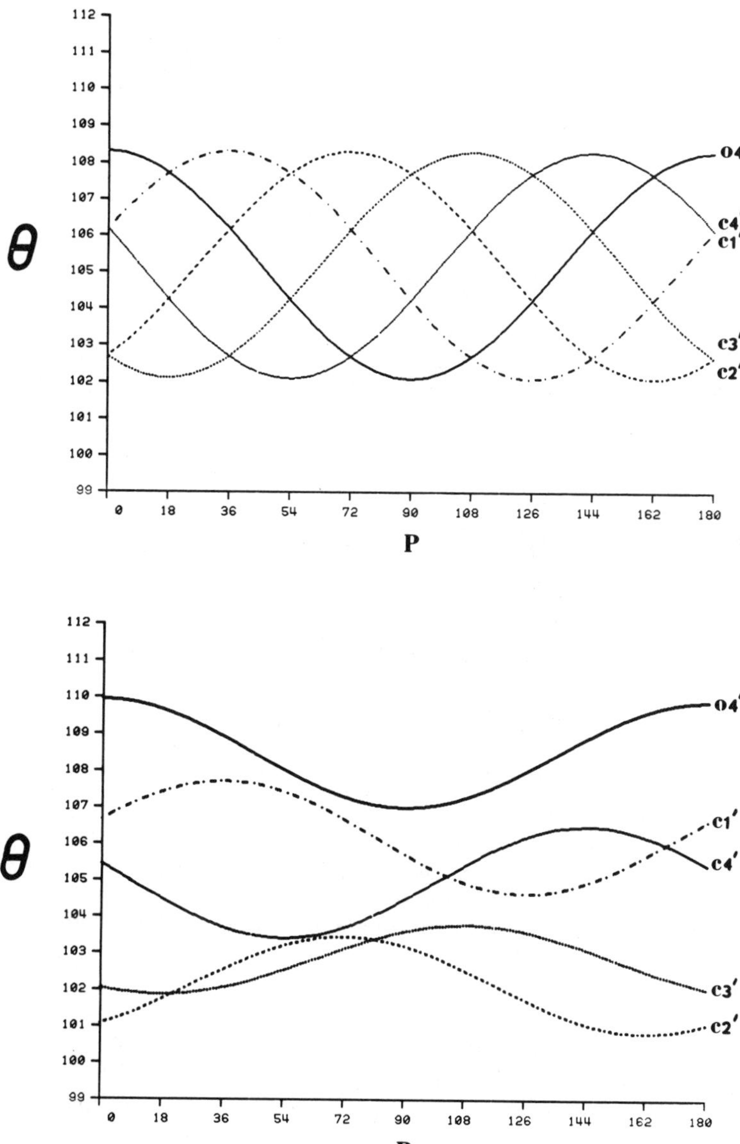

Figure 3. Variation with the phase angle of pseudorotation, *P*, of each endocyclic angle (θ) of a cyclopentane ring *(top)* and of a ribose ring *(bottom)* at the mean value of amplitude of puckering, $\tau_m = 38.7°$. From reference (15).

domain to the high *anti* domain. Of the three staggered conformers about the exocyclic C4′-C5′ bond, the g^+ rotamer is highly preferred ($\psi \sim 60°$), independently of the base or sugar pucker.[5] With the sugar pucker in the C2′-*endo* domain, the other two rotamers are observed, in the preference order $t > g^-$, but the g^- rotamer is rare with a C3′-*endo* pucker.[4,17] Also, the O4′-*endo* pucker tends to favor the *t*

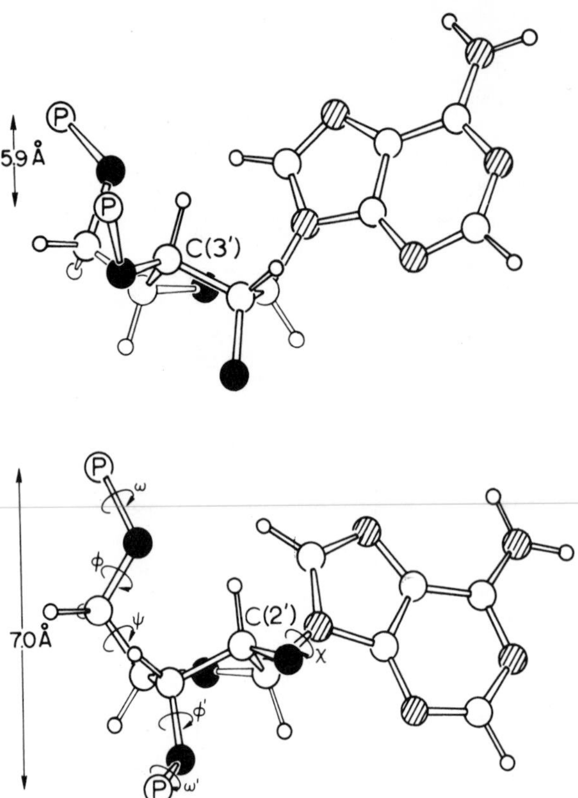

Figure 4. The two preferred conformations for ribonucleotides: C3′-*endo*-low *anti*-g^+ *(top)* for the pucker, the glycosyl bond orientation, and the rotamer about C4′-C5′, respectively; and C2′-*endo*-high *anti*-g^+ *(bottom).* Notice the low *anti* orientation of the base goes with the shorter interphosphate separation, while the high *anti* orientation of the base with the larger P-P distance. From reference (18).

conformer.[18] The phosphodiester bonds P-O3′ (ω') and P-O5′ (ω) are, respectively, one and two bonds away from the furanose ring, and it is, therefore, expected that they respond differently to the puckering parameters with ω' the most sensitive. Both ribose and deoxyribose polymers have a preferred (ω', ω) helical domain in the (g^-, g^-) region.[4]

The Allowed Conformations of the Nucleotides Follow the Allowed Conformations of the Furanose Ring

In the C2′-*endo* and C3′-*endo* puckers, the *anti* and *gauche*$^+$ conformations of the base and of the exocyclic bond, respectively, bring these groups together and, by electrostatic and van der Waals short-range interactions, additionally stabilize these favored conformations of the nucleotide monomers.[19] Besides the two preferred puckered states of the furanose ring, there are other allowed puckered states for the sugar in the pseudorotational cycle. Because these allowed states lose the above short-range favorable interactions by either increasing the separation between the base and the C5′-substituent (O4′-*endo* sector) or by increasing steric interactions (O4′-*exo* sector), they usually occur simultaneously with other allowed or less preferred conformations of the other bonds of the nucleotide (Figs. 5,6).

The O4′-Endo Sector. The O4′-*endo* pucker, which separates maximally the base from the C5′-exocyclic group, destabilizes the g^+ rotamer about the C4′-C5′ bond and can promote the *t* conformer. With the *t* conformation around the C4′-C5′ bond, steric interactions between the base and the 5′-phosphate group are relieved

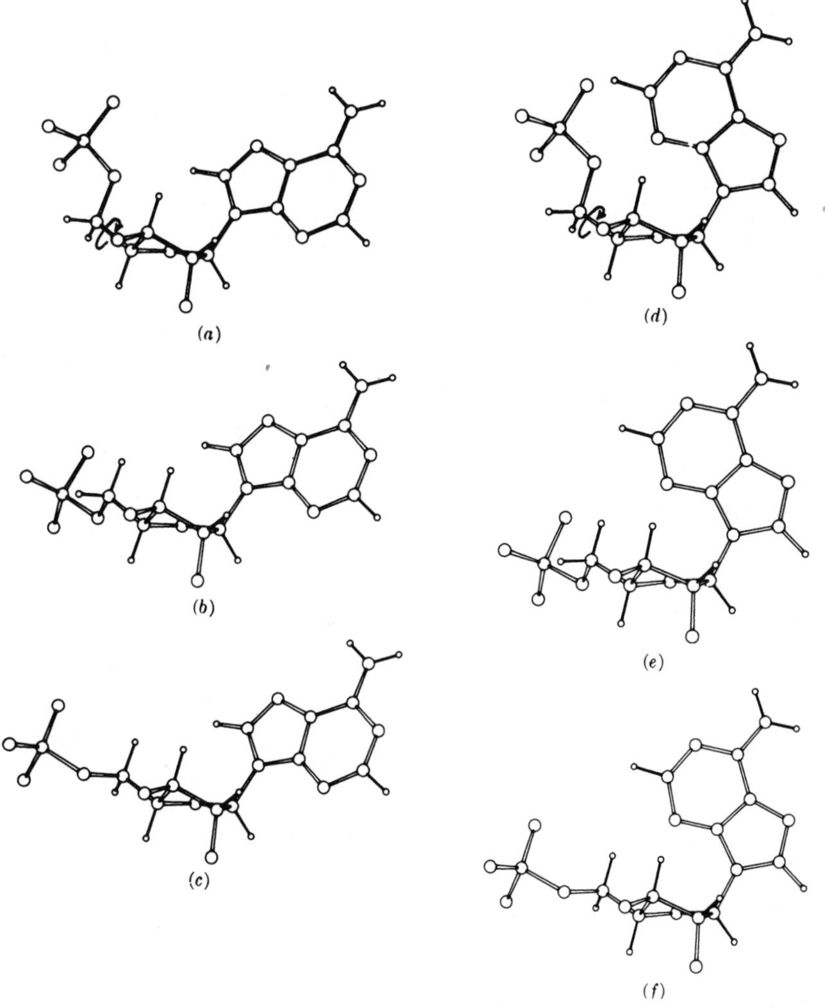

Figure 5. Perspective views showing the *anti* (χ = 10°), left side, and the *syn* (χ = 210°), right side, conformations in 5′-AMP with C(3′)-*endo* sugar puckering for the three staggered conformations about the C(4′)-C(5′) bond: (a) *anti-g*$^+$; (b) *anti-t;* (c) *anti-g*$^-$; (d) *syn-g*$^+$; (e) *syn-t;* (f) *syn-g*$^-$. From reference (52).

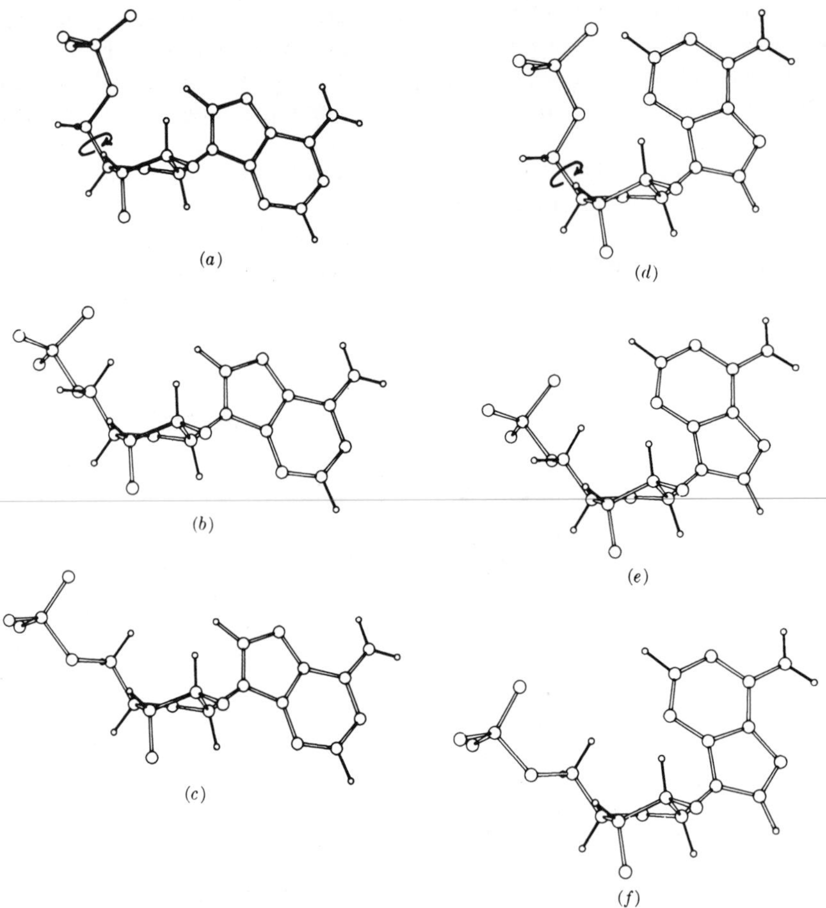

Figure 6. Perspective views showing the *anti* (χ = 50°), left side, and the *syn* (χ = 230), right side, conformations in 5′-AMP with C(2′)-*endo* sugar puckering for the three staggered conformations about the C(4′)-C(5′) bond: (a) *anti-g*$^+$; (b) *anti-t;* (c) anti g$^-$; (d) *syn-g*$^+$ (e) *syn-t;* (f) *syn-g*$^-$. From reference (52).

and the base-sugar glycosyl bond is relatively free and the base can adopt the *syn* conformation.[20]

The O4′-Exo Sector. The O4′-*exo* pucker brings close together the base and the C5′-exocyclic group and the resulting steric repulsion destabilizes the g$^+$ rotamer about ψ, tends to flatten the furanose ring, and prevents the base to take the *syn* orientation.[20]

The Preferred Secondary Structures of Nucleic Acids Incorporate the Preferred Nucleotide Conformations

The structures of the right-handed helices of the A- and B-DNA polymorphs, as derived from X-ray diffraction of fibers, incorporate the two preferred conformational combinations of nucleotides, C3′-*endo*-low *anti*-g^+ and C2′-*endo*-high *anti*-g^+, respectively, in the helical repeat[18] (Fig. 7). The internucleotide phosphodiesters adopt the helical conformations (g^-, g^-) for the (ω', ω) torsion angles (Fig. 8). The recently determined structure of single crystals of the dodecamer d(C-G-C-G-A-A-T-T-C-G-C-G) revealed that the sugar conformation and the sugar-phosphate backbone torsion angles are around their preferred values.[21]

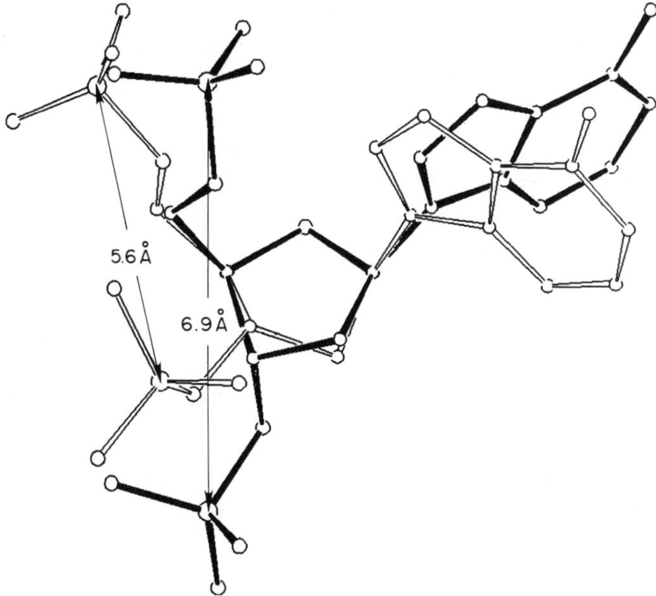

Figure 7. Overlay of the two principal conformations for deoxyribonucleotides: C3′-*endo*-low *anti*-g^+ in light bonds and C2′-*endo*-high *anti*-g^+ in dark bonds. Notice again the marked difference in the interphosphate distance for the two nucleotides. From reference (45).

Right-handed Double Helices in Transfer RNA. An analysis of the polynucleotide chain conformation in the crystal structure of yeast $tRNA^{Phe}$ reveals that the nucleotide residues of the double-helical stems exhibit conformations favored by the ribomononucleotides themselves, C3′-*endo*-low *anti*-g^+.[22] The internucleotide phosphodiesters in the helical stems are all in the double *gauche*$^-$ conformation. These preferred conformations are also adopted by many nucleotides in the loops, thereby conferring helical character to segments of singe-stranded regions. The nucleotides which lead to the formation of loops and bends usually assume less favored alternative conformations for the phosphodiester P-O bonds and, often, for the other torsion angles. Rotations around the internucleotide P-O bonds are a principal source of flexibility in polynucleotide chains.[18,23]

The Left-handed Z-helices: Alternating Preferred and Allowed Nucleotide Conformations

The crystal structure determinations of the deoxyoligonucleotides $d(CG)_3$[24] and d(CG)$_2$[25] have demonstrated the possibility of left-handed structures for DNA. Recently, the occurence of left-handed helical regions containing a long stretch of alternating CG in a plasmid DNA has been claimed.[26] In the Z-helices,[24,25] the repeating motif is a dinucleoside diphosphate rather than a nucleoside monophosphate (Figs. 8 and 9). The two kinds of Z-helices which have been presented incorporate an alternation in the stereochemical features of the sugar pucker, ^2E-^3E; of the sugar-base glycosyl conformation, *anti-syn;* of the exocyclic C4′-C5′ bond, g^+-t; and of the phosphodiester groups, (g^+, g^+) - (g^-, t) for the Z_1 helix and (g^+, g^+) - (g^-, t) for the Z_2 helix. The main difference between the Z_1 and Z_2

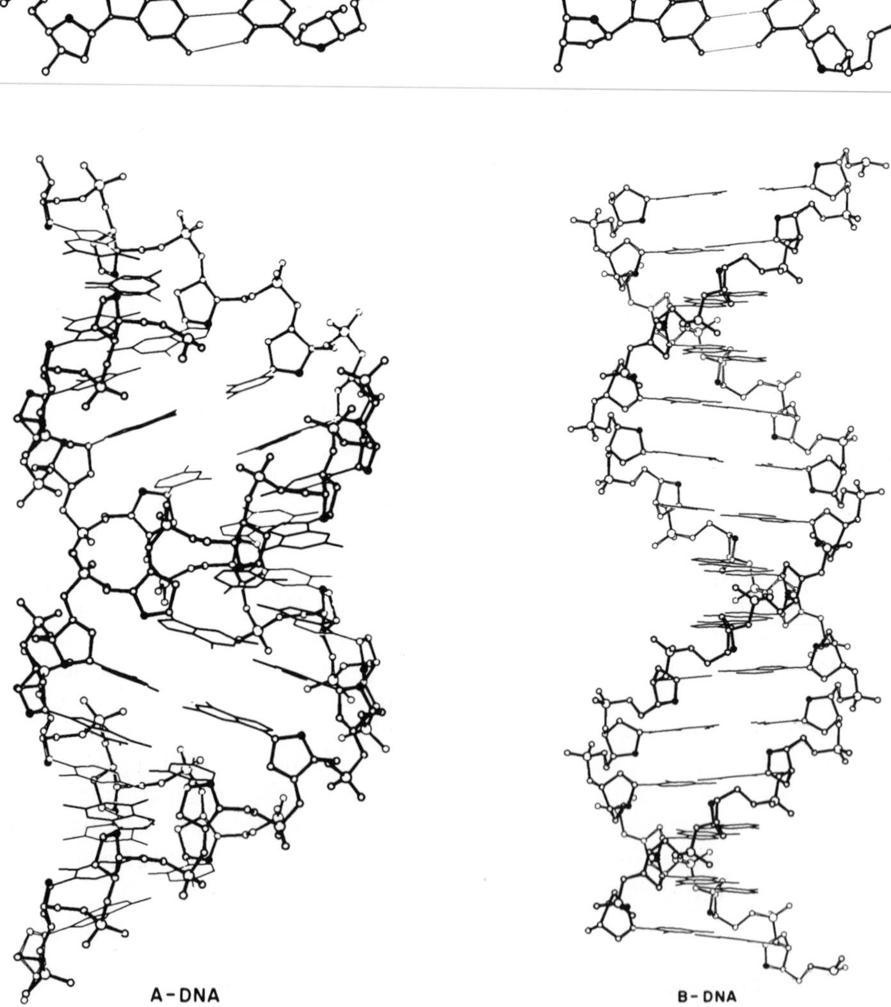

See balance of Figure and Legend on opposite page.

left-handed helices is in the conformation of the second phosphodiester group. The stereochemical features follow the alternating pyrimidine purine base sequence with the C residues assuming the preferred "right-handed" conformational combination ^{2}E-*anti-gauche*$^{+}$ and the G residues the less favored "left-handed" combination ^{3}E-*syn-trans.*

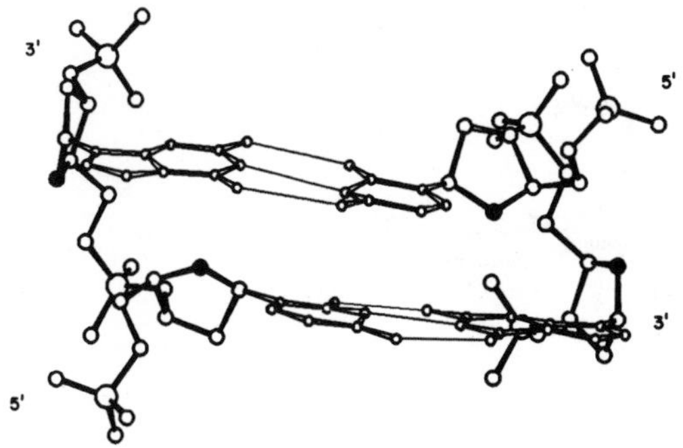

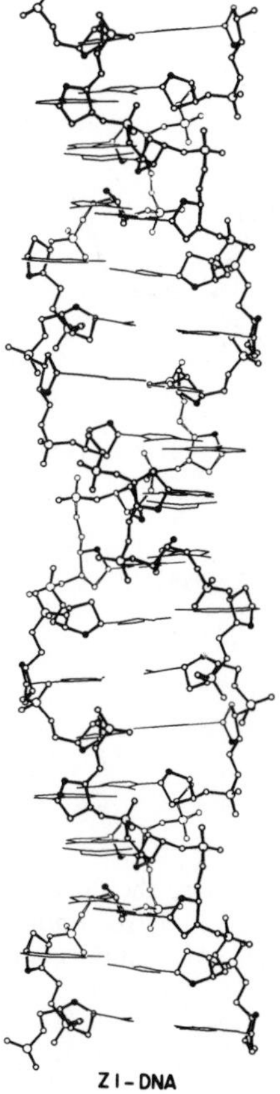

Figure 8. Drawings of an A-DNA helix, a B-DNA helix, and a Z-DNA helix after published coordinates.[53,54] Because of the antiparallelism of the chains, the furanose ring oxygens of O4′ point in opposite directions in the right-handed A- and B-forms. In the left-handed Z-helix, the furanose ring oxygens O4′ point alternatively up and down so that, despite the antiparallel orientation of the chains, each base pair has a local parallel orientation. The repeating units are shown on the top.

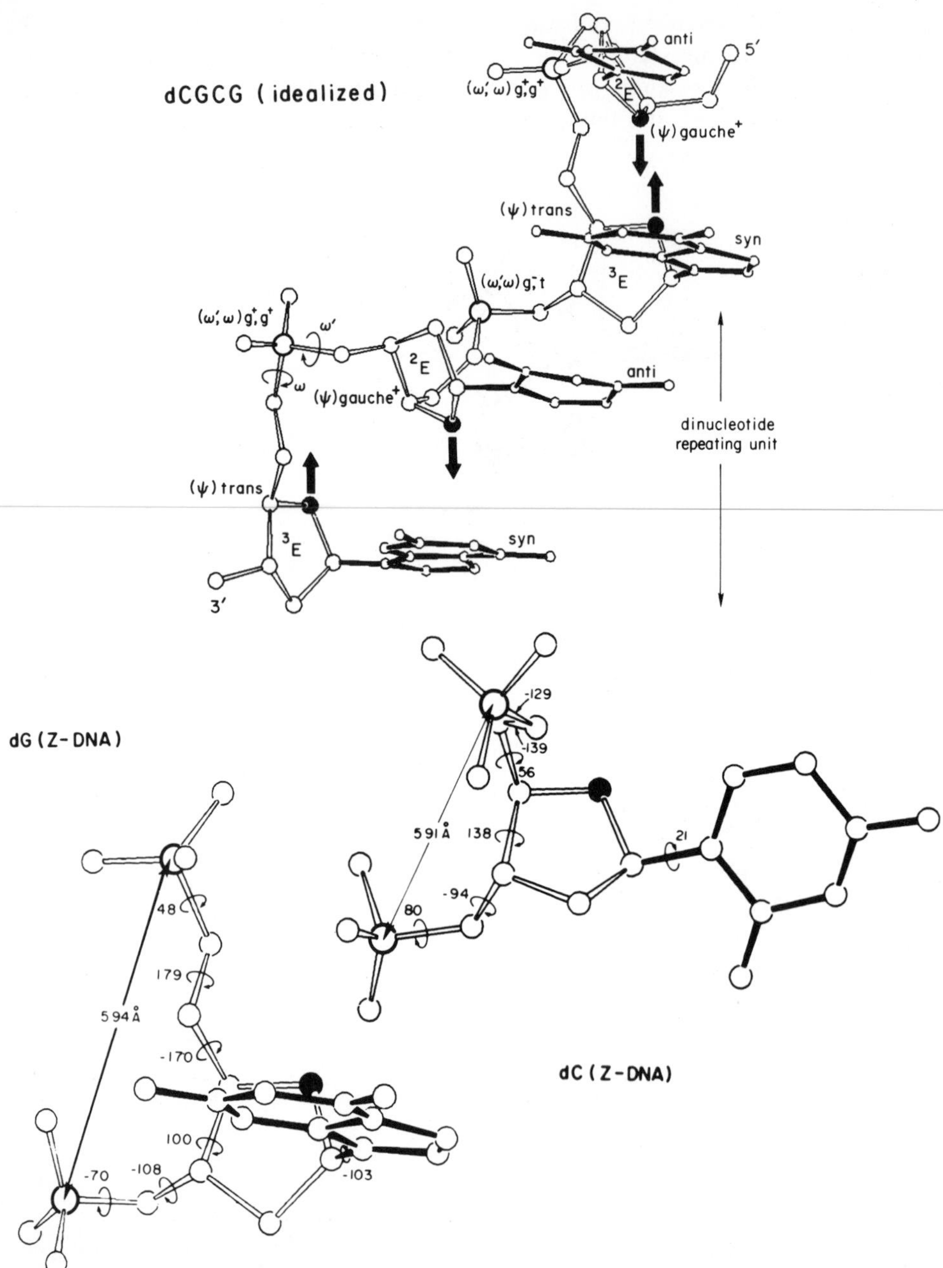

Figure 9. *(Top)* One chain of the Z-helix tetramer dCGCG[54] showing the dinucleotide diphosphate repeating unit and the conformational domains of the main torsion angles. *(Bottom)* Drawings illustrating the conformational differences between the guanine nucleotide and the cytosine nucleotide of the Z-helix.

Influence of Pseudorotation on the Polynucleotide Backbone Flexibility

Since the furanose bond C3′-C4′ is part of the polynucleotide backbone in nucleic acids, the flexibility of the furanose ring can be transmitted to the other backbone bonds. But, more importantly, we will show now that the propagation of the pseudorotational mobility of the furanose ring to the sugar-phosphate backbone depends on the sugar puckering parameters, i.e., there is a sugar-pucker dependent conformational flexibility of nucleic acids. We have calculated the variation of ψ' with P and τ_m from the coordinates of the five-membered furanose rings deduced by the method shown above. The exocyclic substituents were placed at their appropriate average values. The calculated values of ψ' were then fitted to the function:

$$\psi' = (\tau_m + \alpha) \cos (\underline{P} + \beta) + \gamma \tag{3}$$

for four selected values of τ_m and the results are:

$$\psi' = (\tau_m + 3.7) \cos (\underline{P} + 144) + 119.1 \tag{4}$$

or, with a slightly better fit:

$$\psi' = (\tau_m + 3.8) \cos (\underline{P} + 148) + 119.1 \tag{5}$$

These equations are plotted in Fig. 10. As a general rule, in the C3′-*endo* domain ($P = 18°$), changes in ψ' are brought about by changes in the amplitude of pucker and, in the C2′-*endo* domain ($P = 162°$), by changes in the phase angle of pseudorotation. The variation in ψ' at constant τ_m given by (5) are given in the following Table.

Amplitude of pucker τ_m	*Pseudorotation Phase Domain*	ψ' *Variation*	$\Delta\psi'$
39°	$0. \leq P \leq 36°$	$80° \leq \psi' \leq 86°$	6°
	$144° \leq P \leq 180°$	$135° \leq \psi' \leq 155°$	20°

It is known that the furanose ring in nucleic acids has not only preferred domains for the phase angle of pseudorotation but also preferred amplitudes of puckering, with an average value for ribose around 39° and for deoxyriboses around 35°.[6] A study of monomers has shown that the variation in τ_m is minimal, while that in P is much larger especially in the C2′-*endo* domain.[6,7]

The observed spread in phase angle of pseudorotation indicates that, in the preferred domains, the potential energy profiles do not depend strongly on P. Thus, thermal agitation will induce fluctuations in P, and probably τ_m as well, in the two preferred domains. Specifically, furanose rings, when in the ^2E domain could oscillate between ^2_1T ($P = 144°$) and $_3E$ ($P = 198°$) pucker. From the table, it can be seen that, in the ^2E domain, since the linkage torsion angle ψ' is strongly coupled with P,

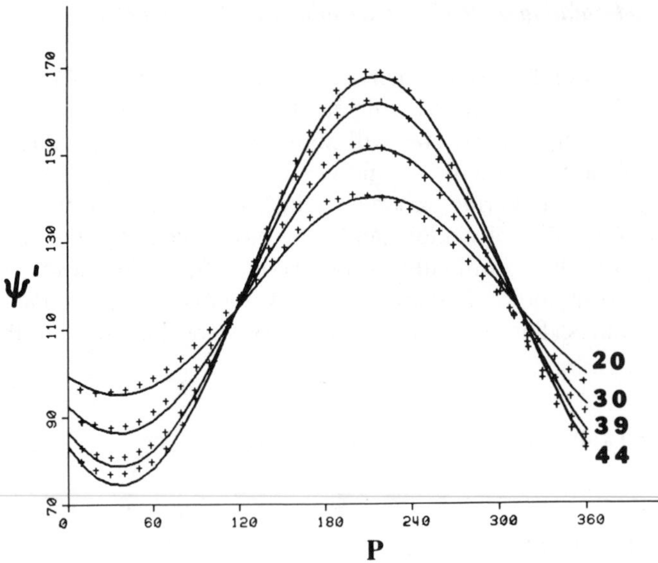

Figure 10. Dependence of the sugar-phosphate backbone torsion angle ψ' with the phase angle of pseudorotation, P, for various values of the amplitude of pucker τ_m (20°, 30°, 39°, 44°). The crosses represent the values of ψ' calculated from the sets of furanose ring coordinates for various P and τ_m obtained by Merritt and Sundaralingam.[16] The continuous lines are the best fit given by eq. (5). It can be seen that a flattening of the furanose ring decreases the range available to ψ', regardless of P. A flat furanose ring (i.e. ψ' independent of P) and the C1′-envelope conformations, $_1$E ($P = 126°$) or ^{1}E ($P = 306°$), (i.e. ψ' independent of τ_m) both give $\psi' \simeq 120°$. Thus, in nucleic acids incorporating furanose rings with C1′-envelope conformations, the orientation of the bases relative to the sugar-phosphate backbone can be altered not only by changes in the glycosyl bond torsion angle but also by mere variations in τ_m. In other words, the C1′-envelope puckered nucleic acids will have strongly dampened conformational transmissions or correlations between side-chain base and sugar-phosphate backbone.

fluctuations in P will be transferred to the other backbone torsions. On the other hand, a furanose ring in the ^{3}E domain, when it oscillates between $^3_2$T ($P = 0°$) and $^3_4$T ($P = 36°$), will transmit minimally these fluctuations to the rest of the backbone because the linkage torsion angle does not depend much on P in that pucker region. *This leads to a sugar-pucker-dependent flexibility in nucleic acids: when incorporated in a polynucleotide, a C3′*-endo *puckered furanose ring gives a small degree of freedom to ψ' in sharp contrast to a C2′*-endo *puckered furanose ring which allows a larger degree of freedom to ψ'. In other words, C3′*-endo *sugars confer "rigidity" to the polynucleotide chain while C2′*-endo *sugars confer "flexibility".*

In RNAs where the ribofuranose ring favors the ^{3}E domain, the thermally-induced fluctuations in the sugar will be damped. Helical RNAs do not accomodate readily C2′-*endo* puckered riboses. However, as exemplified by the crystal structure of tRNA^{Phe}, C2′-*endo* puckered riboses are commonly found in the loops and bends of the complex tertiary structure, usually in association with less preferred conformers about the neighboring bonds.[22] The concomitant increase in flexibility of these regions due to the more extended C2′-*endo* nucleotide might have some important biological function. In this respect, it is interesting that the purely geometrical origins for the occurrence of C2′-*endo* pucker in single-stranded tertiary structural regions are associated with an increased flexibility and adaptability of the sugar-phosphate backbone over that of the helical regions.

On the other hand, deoxyribonucleic acids can adopt both C2′-*endo* and C3′-*endo* puckered furanose rings and can, therefore, adapt their flexibility and the extent to which thermally-induced fluctuations in the five-membered rings are transferred to the backbone. In the B-polymorph, where the sugar can pseudorotate between $^{2}_{1}T$ and $_{3}E$, flexibility of the sugar-phosphhate backbone is expected. On the other hand, in the A-polymorph, even without the presence of the 2′-OH group, the conformational variation will be more restricted and the flexibility of the sugar-phosphate backbone dampened. X-ray diffraction analysis of oriented fibers[27] and circular dichroism studies[28] of nucleic acid polymers in solution have shown that the range in the winding angle between adjacent base pairs varies between 30° and 33° in the A-type polymers (C3′-*endo* puckers) and between 33° and 45° in th B-type polymers (C2′-*endo* puckers). The source of the greater variation of the B-form might come not only from the sugar pucker but also from coupled changes in the other torsion angles.

Correlated Motions in DNA: Mechanics of Twisting and Bending

The phosphodiester torsions (ω', ω) are correlated with the sugar pucker (ψ') and the torsion about C4′-C5′ (ψ) in the helix forming domains.[29] More specifically, there is a strong anti-correlation between the torsion angle ω about P-O5′ and the torsion angle ψ about C4′-C5′ on the 5′-side of the nucleotide unit, and between the torsion angle ω' about P-O3′ with the torsion angle ψ' about C3′-C4′ on the 3′-side. In DNA helices incorporating C2′-*endo* puckered deoxyribose rings, the P-O3′ bond is roughly parallel to the helix axis and rotations about ψ' will lead to lateral movements of the bases in a plane perpendicular to the helical axis, i.e. resulting in unstacking of the bases. However, correlated motions about the (ω', ψ') pair of bonds will change the relative orientation of the bases and ultimately determine the winding angle and the number of residues per helical turn.[7] Thus, thermally-induced fluctuations in the furanose ring pseudorotational parameters will lead to thermal fluctuations in the number of residues per helical turn.

The free energy required to change the DNA winding angle determines the torsional stiffness (shear modulus) of the DNA considered as an elastic rod. Recent studies conclude that this free energy is small and lead to an amplitude of thermal fluctuations between 3° and 5°.[30] In the framework presented here these large fluctuations in the winding angle are due to thermal motions in the furanose ring coupled mainly with changes in ω'. The persistence length of DNA is mainly determined by the bending stiffness (Young modulus) of DNA considered as an elastic rod. Since the P-O5′ bond is roughly parallel to the base pairs in B-DNA type helices, it is expected that rotations about ω will lead to a tilt of the bases and bending of DNA.

Therefore, since ω is correlated with ψ, correlated motion about (ω,ψ) pair of bonds lead to smooth bending of a helical segment. If these correlated motions are modulated over the helical repeat, a variety of super helices can be formed.[31]

Pseudorotational Mobility and Nucleic Acid Flexibility

In tetrahydrofuran and cyclopentane, the lowest energy path for interconversion between the various envelope and twist conformations is the pseudorotation path at constant amplitude of puckering with a small or no activation energy, respectively.[9,10] In furanose rings, the pseudorotational mode of interconversion between the preferred C2′-*endo* and C3′-*endo* puckers requires the ring to pass through intermediate puckered states like C1′-*exo,* O4′-*endo,* and C4′-*exo.*[6] The alternative pseudorotational path, through O4′-*exo*, leads to close contacts between the C5′ exocyclic group and the base, and would require not only the *t* or g^- conformation about ψ, but also a flattening of the ring by 10-20°.[15] Interconversion by inversion through the planar state is a path of high energy in nucleic acid monomers because of the eclipsing interactions involving the endocyclic and exocyclic substituents. In polynucleotides, the adoption of puckered states involving O4′, C1′, or C4′ might disrupt the hydrogen bonding and/or the stacking interactions depending on the amplitude of the puckers. Other modes of interconversion for the $^2E \Leftrightarrow {}^3E$ transition might therefore occur in polynucleotides, like pseudorotation with variable amplitudes of puckering or inversion through the planar state,[15] besides pseudorotation at constant amplitude of puckering (Figs. 11-12).

As mentioned above, the most probable intermediate puckered states spread apart the base and the phosphate groups and, consequently, tend to destabilize the short-range interactions and promote rotational motions around the C4′-C5′ bond and the base-sugar glycosyl bond. Thus, rotations about the C4′-C5′ bond and the glycosyl bond are coupled with the C2′-*endo*⇔C3′-*endo* interconversion.[19] In other words, the pseudorotational mobility of the sugar facilitates rotation around the exocyclic bonds by lowering the barriers. The energy barrier for the $^2E \Leftrightarrow {}^3E$ interconversion is therefore a main determinant of nucleic acid flexibility. *Since this energy barrier is much less in deoxyribose rings than in ribose rings, it is expected that* $^2E \Leftrightarrow {}^3E$ *interconversions, anti⇔syn transitions, rotations about C4′-C5′, and consequently, about the phosphodiester linkages would occur relatively more easily in DNA than in RNA systems.*

Purine nucleotides have energy barriers for the *anti⇔syn* transitions lower than pyrimidine nucleotides, mainly because they are linked to the sugar at the five-membered imidazole ring (internal angle ~106°) and not at the six-membered pyrimidine ring (internal angle ~120°). *Thus, the most flexible systems would be the purine deoxyribonucleotides with the* ^2E *pucker (*^2E*-dR-Pu), which would tend to exist as (*$^3E \Leftrightarrow {}^2E$*) - (*anti⇔*syn* - ($g^+/t/g^-$)*) combinations, and the least flexible systems the pyrimidine ribonucleotdes with the* ^3E *pucker (*^3E*-R-Py), which would assume mainly the* ^3E-anti-g^+ *combinations.*

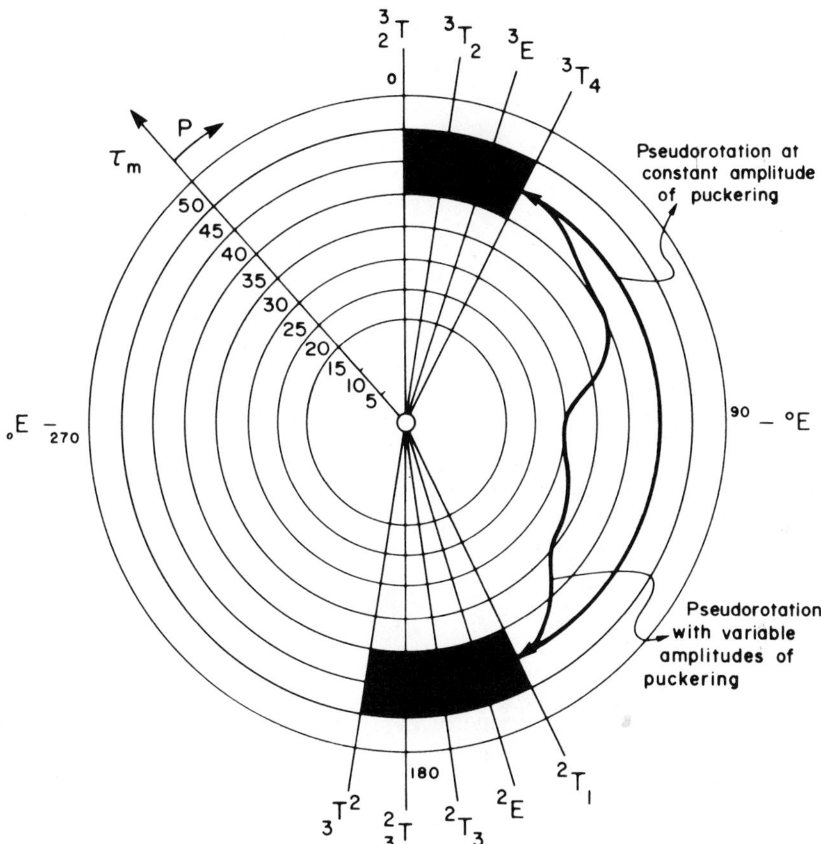

Figure 11. Pseudorotational wheel (or P, τ_m) plot of furanose rings showing two pseudorotational pathways between the preferred pseudorotational domains: one with constant amplitude of puckering and one with variable amplitudes of puckerings. While the first path is the most probable, it is conceivable that the amplitude of puckering varies with the phase of pseudorotation in polynucleotides undergoing A⇔B transitions. The pseudorotation pathway through $P = 270°$ (O4′-*exo*) is not shown, since it is energetically less favored. The inversion through the planar state, which goes through the center of the pseudorotation wheel, is expected to be higher in energy but cannot be ruled out. From Reference (15).

Nucleic Acid Polymorphism: Effect of Humidity, Salts and Solvents

X-ray diffraction studies of oriented fibers have shown that nucleic acids can exit in several kinds of double-helical structures (for a review, see Reference (32)). At low humidities (~75%) and low ionic strength, the A-form of Na, K, and Rb DNA is observed. If excess salt is present (>3% w/w), the A-form of Na, K, and Rb DNA transforms into the B-form upon an increase in the relative humidity to 92%. Li DNA remains in the B-form down to 66% relative humidity.[33] RNA-DNA hybrids

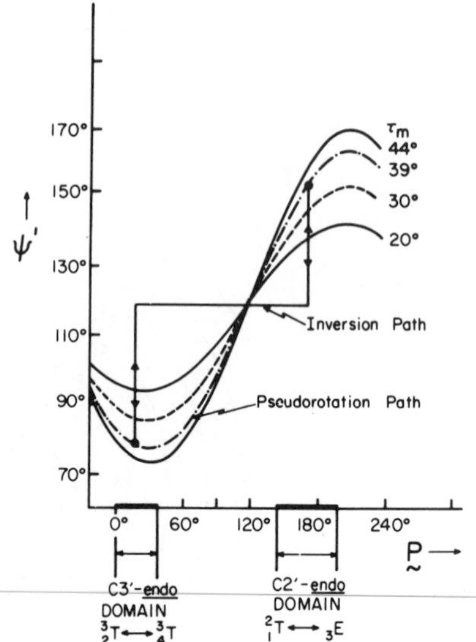

Figure 12. Variation of ψ' during $^2E \Leftrightarrow {}^3E$ interconversion through the pseudorotation path or through the inversion path. During inversion, the amplitude of pucker decreases steadily toward the planar ring ($\psi' \simeq 120°$) wherefrom it will again increase. In the pseudorotation path, the ring stays puckered during interconversion but the pucker type varies.

always adopt the A-form.[34] Diffraction patterns, attributed to the C-form of DNA, have been obtained on partial drying of Li DNA fibers, or at low hydration and intermediate salt concentration with DNA.[35] The C-form is a structure closely related to the B-form with an increase in the winding angle of 2.6°.

Circular dichroism studies of nucleic acids in alcoholic solutions have revealed a sharp transition at 70% ethanol (v/v) from the B-form to the A-form.[36] The CD spectral changes induced by increasing the salt concentrations or by decreasing the temperature were interpreted as indicating a B- to C-transition.[37,39] Precise measurements with closed circular DNA molecules[40] indicate that DNA unwinds with increasing temperature at the rate of $13 \pm 2x10^3$ deg. bp^{-1} K^{-1} Also, at constant concentration of electrolytes, the winding angle increases in the order $Na^+ < K^+ < Li^+ < Rb^+ < Cs^+ < NH_4^+$. It is interesting that, except for Li^+, this follows the radii of the hydrated ions: $Li^+ < Na^+ < K^+ < Rb^+ < Cs^+$.[38] This has been correlated with the width of the narrow groove, where the hydrated ions are supposed to bind, which increases with unwinding of the helix.[28]

We have shown above how the pseudorotational mobility of furanose rings, driven by thermal fluctuations, can lead to correlated conformational fluctuations in backbone torsion angles and ultimately to variations in winding angle. These phenomena might be at the origin of the "breathing" modes[41,42] and of the dynamical motion on a nanosecond timescale observed by fluorescence[43] and NMR spectroscopy.[44] Depending on the amount of salt or type of solvent, the conformational equilibria or the preferential angular domains of the torsion angles will be displaced, leading to the diversity of observed secondary structures.[45]

Transitions in Nucleic Acids: A⇔B and B⇔Z

The geometrical arrangements of the exocyclic groups of the furanose ring in nucleic acids depend on the puckering parameters of the five-membered ring. The network formed by the water molecules, the counterions, the oxygen atoms of the sugar-phosphate backbone, and the polar atoms of the base will tend to stabilize some puckered states and conformational domains of the other torsion angles. For example, the strong preference of ribonucleic acids for A-type structures with C3′*endo* sugar pucker might reflect a preferential interaction between water molecules in the minor groove and the axially oriented 2′-hydroxyl groups of the ribose sugars.[45] By allowing interconversion from one puckering to the other, the pseudorotational mobility of the sugar is at the heart of the A⇔B transition. The intermediate puckers in the O4′-*endo* sector, by promoting conformational adjustments in the sugar-phosphate backbone and the sugar-base linkages, lead to transient hydration schemes and, finally, to the A⇔B transition.[45]

Correlated conformational changes in the sugar-phosphate backbone can, as well, bring about not only *anti⇔syn* interconversion but also a switch in the helical sense of helical polynucleotides. This might be important for understanding the right-left transition observed by circular dichroism with poly d(G-C).[46] We assume that a Z-helix similar to that observed in the crystal structure of $d(CG)_3$ gives rise to the inverted Cotton effects in the circular dichroism of poly d(G-C) at high salt. *The concerted motions described below might be triggered by pseudorotational interconversion of the guanosine sugars from* ^2E *to* ^3E. *The intermediate puckers encountered in the pseudorotational cycle promote the* syn *disposition of the base around the glcosyl bond and the* trans *orientation about the C4′-C5′ bond (Fig. 13). Crankshaft motions imposed simultaneously on both sugar-phosphate chains will tend to switch the phosphodiester bonds of the CpG fragment from* (g^-, g^-) *to* (g^+, g^+)*, thereby bringing about the left-handed backbone* where the phosphates of adjacent chains are on the same side of the helix with the *syn* guanine base and the *anti* cytosine base on the opposite side.[45] The phosphodiester bonds of the GpC fragment do not require such large changes because the cytosine base stays in the *anti* orientation.

Phosphodiester Conformations and Helix Chirality

All right-handed helical structures determined so far incorporate the characteristic double *gauche*$^-$ orientation for the two phosphodiester P-O bonds together with the preferred conformations of the nucleotide. The enantiomerically related area in the (ω', ω)-map, (g^+, g^+), is expected to give the corresponding left-handed helical structures if the polynucleotide is constituted of hypothetical L-sugars in the preferred conformations (Fig. 14). However, the (g^+, g^+) phosphodiester conformations with the preferred conformations for the D-nucleotide is stereochemically unfavorable in a polynucleotide[47] and requires less favored conformations for some torsion angles of the nucleotide, like the *trans* rotamer about C4′-C5′[48] (see also the Z-helices).

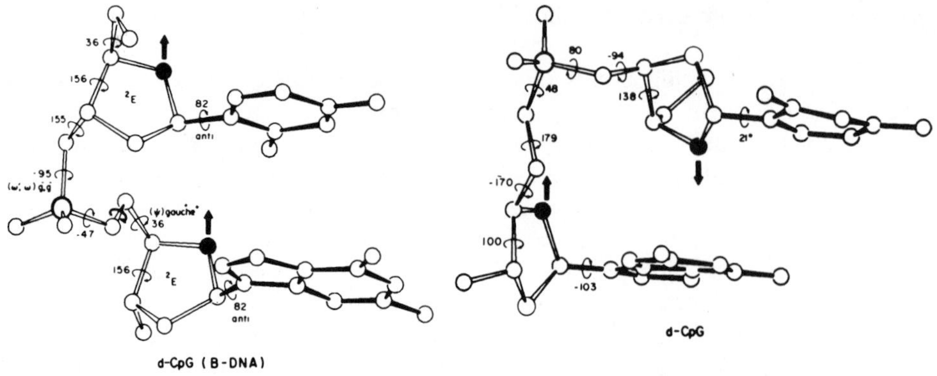

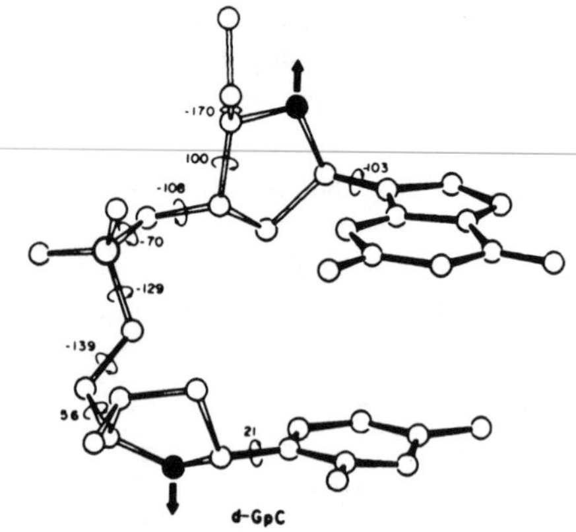

Figure 13. Drawings showing the dinucleotide fragments of B-DNA (left) and Z-DNA (right) illustrating the torsion angle changes in going from the B- to the Z-helix. Notice that in the CpG-fragment of the Z-helix, the O4′-oxygen atoms point towards each other and, in the GpC-fragment, away from each other.

In contrast, a study of the helical parameters, n and h, as a function of the P-O torsion angles (ω', ω) has revealed that P-O bond rotations can switch the helical handedness since small changes in (ω', ω) separate the right-handed from the left-handed domains.[29,49] This led to the suggestion that small changes in the P-O bond rotations, coupled with changes in the pseudorotation parameters of the sugar and an increase in the sugar base glycosyl torsion to the high *anti* range, can switch the helical sense. These principles were applied to the Ikehara polymers where the base-sugar cyclization (8,2′-anhydro-arabinoadenosine and 6,2′-anhydro-arabinouridine) produces the combined effects of the high *anti* glycosyl torsion ($\chi \sim 125°$) and the C4′-*exo* sugar pucker to stabilize the left-handed base stack and left-handed helix.[49-51]

Conclusions

The discovery of base complementarity led not only to the intertwined anti-parallel double helical structure for DNA, but immediately suggested a simple mechanism for the replication of the hereditary molecule. However, the structural principles of the nucleic acids which involve the conformational properties of the sugar and its intricate role in governing the stereochemistry of the nucleotide building blocks

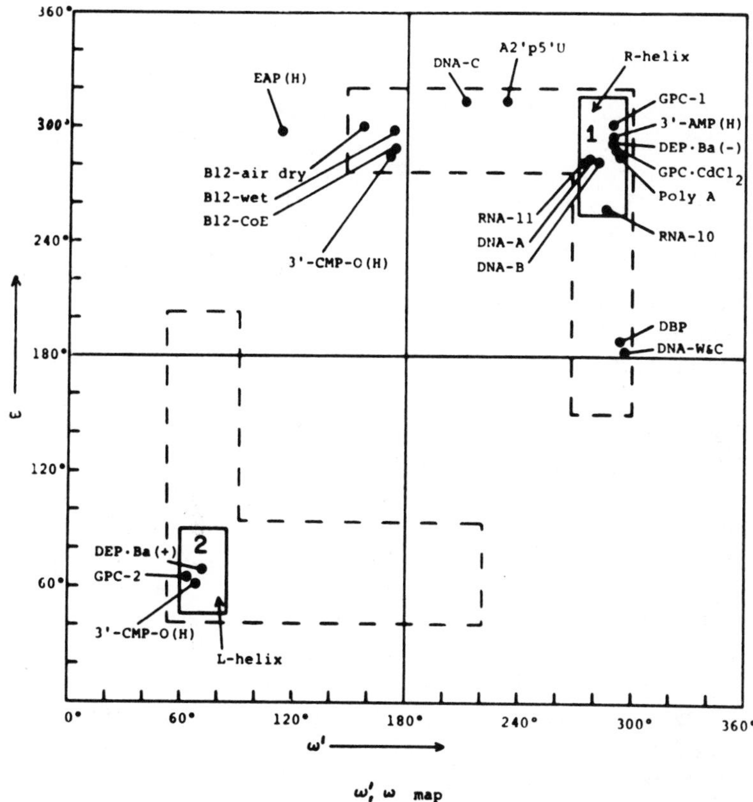

Figure 14. The (ω', ω) map showing the preferred and allowed internucleotide phosphodiester conformations (after reference (4)). In the upper right-hand corner lie the (g^-, g^-) phosphodiester conformations typical of the right-handed double helices, while in the lower left-hand corner are the enantiomerically related (g^+, g^+) phosphodiester conformations which were predicted for left-handed helices and have since been found in the Z-helices (see also text).

and the folding schemes of the sugar-phosphate backbone were elucidated subsequently. These were based on single crystal x-ray diffraction studies and conformational analysis, of the nucleic acid constituents and systems containing molecular fragments of the nucleic acids. A vocabulary of notations and nomenclature was also devised for the description and rationalization of the conformational manifestations of the nucleotide and of the polynucleotide chain about the sugar ring, sugar-base glycosyl bond and the sugar-phosphate backbone bonds. It was found that rotations around the various single bonds of the polynucleotide chain were restricted and that the nucleotides themselves showed preference for certain conformational combinations. The internucleotide phosphodiester conformational map (ω', ω) early depicted the preferred domains for both the right- and left-handed helices as well as for loops and kinks. These structural principles have been amplified and placed on a firmer basis with the availability of crystal structural data on di-, tri-, tetra-, hexa-, and dodecanucleotides, nucleoside di- and triphosphates, nucleotide coenzymes and metal- and drug-intercalated nucleotide complexes and conformational energy calculations.

The known stereochemical rules played a crucial role in the interpretation of the medium resolution (3 to 2.5Å) electron density maps of the transfer RNA structures elucidated in different laboratories; tRNA, which is endowed with a rich tertiary structure containing a variety of novel base-pairing schemes between parallel and antiparallel chains, hairpin loops and bends, has thus provided clear support of the conformational priorities already established for the nucleotides and internucleotide phosphodiesters. In addition, further information has been gleaned on the nature of turns in polynucleotide chains, and the forces involved in nucleic acid-nucleic acid recognition interactions.

A property of the furanose ring of the nucleic acid sugars is their inherent mobility, best visualized by the pseudorotation concept. Despite the hindrance to free pseudorotation due to the presence of ring substituents there is sufficient conformational latitude in and between the pseudorotation domains, ^3E and ^2E, to endow flexibility and polymorphism to nucleic acids. Thermal fluctuations capitalize on the pseudorotation property of the furanose ring. Because the furanose ring is linked to the base and the phosphate, these fluctuations are correlated to the rotational mobility of the base and the phosphate, leading to phenomena such as premelting changes, breathing modes, or opening of base pairs. Such concerted changes in the sugar-phosphate backbone torsion angles are instrumental in the polymorphic transitions between right-handed nucleic acid helices, e.g. B⇔A, and between right- and left-handed helices, e.g. B⇔Z. Thus, *the sugar can be regarded as the pulsating heart of the genetic molecules.* The absence of the 2′-hydroxyl group relieves the steric effect between it and the substituents on the C3′ and C1′ atoms, changes the pseudorotational preferences of the furanose ring, and enhances the mobility of the 2′-deoxyribose system compared to that of the ribose. Consequently, the sugar-phosphate backbone, as well as the sugar-base glycosyl linkage, can exhibit greater conformational latitude and fluctuations in DNA than in the relatively rigid RNA. The 2′-hydroxyl group provides subtle conformational variations between ribose and 2′-deoxyribose by enhancing the relative populations of the ^3E and ^2E puckers. Through pseudorotation, the sugars of both RNA and DNA possess the inherent property of the ^2E and ^3E puckers, conferring *Yin-Yang* character to the genetic molecules.

Acknowledgements

We gratefully thank the National Institutes of Health (Grants GM-17378 and GM-18455), the American Cancer Society (Grant CH-128), and the College of Agricultural and Life Sciences of the University for their support of this research.

References and Footnotes

1. *Transfer RNA: Structure, Properties, and Recognition,* Eds. P. R. Schimmel, D. Söll, and J. N. Abelson (Cold Spring Harbor Laboratory, 1979).

2. *Ribosomes: Structure, Function, and Genetics,* Eds. G. Chambliss, G.R. Craven, J. Davies, K. Davis, L. Kahan, and M. Nomura (University Park Press, Baltimore, Md., 1980).
3. Altona, C. and Sundaralingam, M., *J. Amer. Chem. Soc. 94,* 8205 (1972).
4. Sundaralingam, M., *Biopolymers 7,* 821 (1969).
5. Sundaralingam, M., in *Conformations of Biological Molecules and Polymers,* vol. V, E.D. Bergmann and B. pullman, Eds. (Academic Press, New York, 1973), p. 417.
6. Altona, C. and Sundaralingam, M., *J. Amer. Chem. Soc. 95,* 2333 (1973).
7. Sundaralingam, M. and Westhof, E., *Biophys. J. 32,* 250 (1980).
8. Aston, J.G., Schumann, S.C., Fink, H.L., and Doty, P.M., *J. Amer. Chem. Soc. 63,* 2029 (1941).
9. Kilpatrick, J.E., Pitzer, K.S., and Spitzer, R., *J. Amer. Chem. Soc 69,* 2483 (1947).
10. Pitzer, K.S. and Donath, W.E., *J. Amer. Chem. Soc. 81,* 3213 (1959).
11. Geise, H.J., Altona, C., and Romers, C., *Tetrahedron Lett.,* 1383 (1967).
12. Cremer, D. and Pople, J.A., *J. Amer. Chem. Soc. 97,* 1354 (1975).
13. Rao, S.T., Westhof, E., and Sundaralingam, M., *Acta Cryst. A, 37,* 421 (1981).
14. Sundaralingam, M., *J. Amer. Chem. Soc. 87,* 599 (1965).
15. Westhof, E. and Sundaralingam, M., *J. Amer. Chem. Soc. 102,* 1493 (1980).
16. Merritt, E.A. and Sundaralingam, M., in preparation.
17. Jack, A., Ladner, J.E., Rhodes, D., Brown, R.S., and Klug A., *J. Mol. Biol. 111,* 315 (1977).
18. Sundaralingam, M., *Int. J. Quantum Chem. Quant. Bio. Symp. 1,* 81 (1974).
19. Sundaralingam, M., in *Structure and Conformations of Nucleic Acids and Protein-Nucleic Acid Interactions,* M. Sundaralingam and S.T. Rao, eds. (University Park Press, Baltimore, Md., 1975), p. 487.
20. Yathindra, M. and Sundaralingam, M., *Biopolymers 13,* 2061 (1974).
21. Drew, H. R., Wing, R. E., Takano, T., Broka, C., Tanaka, S., Itakura, K., and Dickerson, R.E., *Proc. Natl. Acad. Sci. U.S.A. 78,* 2179 (1981).
22. Sundaralingam, M., in *Biomolecular Structure, Conformation, Function, and Evolution,* R. Srinivasan, ed. (Pergamon Press, Oxford and New York, 1980), p. 259.
23. Sundaralingam, M., Mizuno, M., Stout, C.D., Rao, S.T., Liebman, M., and Yathindra, M., *Nucleic Acids Res. 3,* 2471 (1976).
24. Wang, A.H.J., Quigley, G.J., Kolpak, F.J., Crawford, J.L., van Boom J.M., van der Marel, G., and Rich, A., *Nature 282,* 680 (1979).
25. Drew, H.R., Takano, T., Tanaka, S., Itakura, K., and Dickerson, R. E. *Nature 286,* 567 (1980).
26. Klysik, J., Stirdivant, S.M., Larson, J.E., Hart, P.A. and Wells R. D., *Nature 290,* 672 (1981).
27. Arnott, S., *Biophys. J. 32,* 249 (1980).
28. Ivanov, V.I., Minchenkova, L.E., Schyolkina, A.K., and Popetayev, A. I., *Biopolymers 12,* 89 (1973).
29. Yathindra, N. and Sundaralingam, M., Nucl. Acids Res. 3, 729 (1976).
30. Vologodskii, A.V., Anshelevich, V.V., Lukashiu, A.V., and Frank-Kamenetskii, M.D., *Nature 280,* 294 (1976).
31. Sundaralingam, M. and Westhof, E., *Int. J. Quantum Chemistry: Quantum Biology Symposium 6,* 115 (1979).
32. Arnott, S., In *Organization and Expression of Chromosomes,* Dahlem Conferenzen, Berlin. Allfrey, B.G., Bantz, E.K.F., McCarthy, B.J., Schinke, R.T., and Tissieres, A., Eds., p. 209 (1976).
33. Wilson H.R., *Diffraction of X-rays by Proteins, Nucleic Acids, and Viruses.* (Edward Arnold Publishers Ltd, London, 1966).
34. Spencer M., Fuller, W., Wilkins, M.H.F., and Brown, G.L., *Nature 194* 1014 (1962).
35. Arnott, S. and Selsing, E., *J. Mol. Biol. 98,* 265 (1975).
36. Brahms, J. and Mommaerts, W.G.H.M., *J. Mol. Biol. 10,* 73 (1963).
37. Studdert, D.S., Patroni, M., and Davis, R.C., *Biopolymers 11,* 761 (1972).
38. Tunis-Schneider M.J.B. and Maestre, M.F., *J. Mol. Biol. 52,* 521 (1970).
39. Chan, A. Kilkuskie, R. and Hanlon, S., *Biochemistry 18,* 84 (1979).
40. Depew, R.E. and Wang, J.C., *Proc. Natl. Acad. Sci. USA 72,* 4275 (1975).
41. Printz, M.P. and von Hippel, P., *Biochemistry 18,* 394 (1968).
42. Teitelbaum, H. and Englander, S.W., *J. Mol. Biol. 92,* 55 (1975).
43. Wahl, Ph., Paoletti, J., and Le Pecq, J.B., *Proc. Natl. Acad. Sci. USA 65,* 417 (1970).
44. Hogan, M. and Jardetzky, O., *Proc. Natl. Acad. Sci. USA 76,* 6341 (1979).
45. Sundaralingam, M. and Westhof, E., *Int. J. Quantum Chemistry: Quantum Biology Symposium* (1981), in press.

46. Pohl, F.M. and Jovin, T.M., *J. Mol. Biol. 67,* 375 (1972).
47. Yathindra, N. and Sundaralingam, M., *Proc. Natl. Acad. Sci. USA 71,* 3325 (1974).
48. Jayaraman, S., Yathindra, M., and Sundaralingam, M., *Biopolymers* submitted for publication.
49. Sundaralingam, M. and Yathindra, M., *Int. J. Quantum. Chemistry: Quantum Biology Symposium 4,* 285 (1977).
50. Ikehara, M. and Tezuka, T., *J. Amer. Chem. Soc. 95,* 4054 (1973).
51. Dhingra, M.M., Sarma, R.H., Uesugi, S., Shida, T., and Ikehara, M., *Biochemistry* (in press).
52. Yathindra, N., and Sundaralingam M., *Biopolymers 12,* 297 (1973).
53. Arnott, S. and Hukins, D.W.L., *Biochem. Biophys. Res. Commun. 47* 1504 (1972).
54. Wang, A.H.J., Quigley, G.J., Kolpak, F.J., van der Marel, G., van Boom, J.H., and Rich, A., *Science 211,* 1171 (1981).

Proceedings of the Second SUNYA Conversation in the Discipline Biomolecular Stereodynamics
Volume I, ISBN 0-940030-00-4, Ed., Ramaswamy H. Sarma,
Adenine Press, New York, ©Adenine Press

Understanding the Motions of DNA

Wilma K. Olson
Department of Chemistry
Rutgers University
New Brunswick, New Jersey 08903

Introduction

The molecular motions that characterize double helical DNA are puzzling. At the macroscopic level, the polynucleotide duplex is an extremely stiff structure that attains ideal Gaussian behavior only when extremely long.[1-10] At the local level, however, the nucleotide repeating units are unexpectedly mobile.[11-21] While the fluctuations of the laterally attached bases from their preferred stacked, hydrogen-bonded arrangements are limited, the sugar and phosphate moieties alternating along the polynucleotide backbone are free to vary over rather broad rotational ranges. Furthermore, under certain experimental conditions, the familiar right-handed B-DNA double helix is able to adopt an unusual left-handed Z-form with local torsion angles radically altered from their normally preferred values.[22-26] Moreover, both right- and left-handed duplexes are now found to coexist and to interconvert in naturally occurring systems.[27,28]

Until these recent developments, the motions responsible for the pronounced stiffness of long DNA chains were presumed to be limited in scope.[3,5,6,9] The flexibility of a chain molecule is usually correlated with the conformational mobility of its constituent repeating units.[29] Because a stiff chain is typically composed of less freely rotating units than a flexible one, the local motions of the stiff DNA molecule were initially treated as highly restricted. Indeed, this "rigid" nucleotide concept is consistent with the majority of available X-ray data[30,31] and is also in agreement with the mean-square dimensions[4,5] and cyclization rates[6,10] of DNA over a broad range of chain lengths. The limited conformational variations of such a model, however, are not easily reconciled with the relaxation properties of double helical DNA. The double helix is apparently free to adopt a wide variety of rotational arrangements in solution. In order to account for the limited flexibility of the duplex as a whole, the internal rotations of the chain backbone must vary in a highly correlated fashion.

As part of an overall effort to comprehend the unusual flexibility of DNA, we have carried out a combined geometric and potential energy analysis to identify conformations of the polynucleotide chain that satisfy both the long-range stiffness and the local mobility of the double helix. We find the stacking interactions between consecutive bases in each strand and the hydrogen bonding interactions of complementary bases in opposite strands to favor certain torsional combinations over others and also to introduce long-range correlations between rotations about various bonds of the sugar-phosphate backbone. In this work, we focus particular attention upon some of the strongly correlated second-neighbor rotations of the polynucleotide backbone. The second-neighbor torsions flanking any *trans* bond in the polynucleotide chain may vary over wide ranges of conformation space provided that one angle rotates in a positive sense while the other varies simultaneously in a negative sense. Such fluctuations not only satisfy the local stacking and hydrogen bonding requirements of the nucleotide repeating units but also maintain the preferred direction of the double helix. Similar correlated (crankshaft) motions[32] apparently occur in the conformational transitions of alkane[33] and peptide[34] chains and also account for the nmr properties of model phospholipids[35] at low temperatures. The second-neighbor motions described below additionally provide feasible pathways for the interconversion of DNA between its various helical forms, including possible continually "stacked" trajectories between the B and Z structures.

Two-Dimensional Analysis

The conformational flexibility of DNA is dominated by the interactions of its constituent bases. The stacking of adjacent bases is apparently responsible for the limited flexibility of the chain as a whole. The local helical bending consistent with the macroscopic dimensions,[3,5,9] fluorescence depolarization,[7] and cyclization equilibria[6] of linear DNA is no more than 5-10°. The combined effects of base stacking and hydrogen bonding are also related to the limited (5-10°) twisting of adjacent residues about the helical axis. According to hydrogen exchange studies of DNA

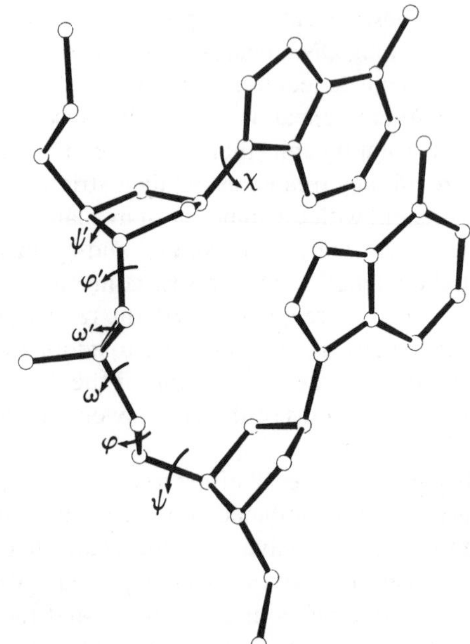

Figure 1. Computer generated representation of a pdApdAp fragment of the reference B-DNA double helix showing chemical bonds and internal torsions. Rotations are defined relative to *cis* = 0° and are fixed at the following values in all units of both strands: $\chi = 78°$, $\psi' = 137°$, $\phi' = -133°$, $\omega' = -157°$, $\omega = -42°$, $\phi = 136°$, $\psi = 42°$.

"breathing,"[36,37] less than 5% of the hydrogen bonds between complementary bases are broken at any one time. As outlined below, conformational models that account for these limited motions are not confined to "rigid" geometries.

A systematic survey of DNA flexibility, however, is complicated by the multi-dimensionality of its conformation space. The orientations of consecutive bases in each dinucleoside monophosphate fragment of the chain molecule are a function of the seven intervening acyclic torsions (defined for one nucleotide repeating residue in Figure 1) and the puckering (related to the angle ψ') of the two pentose rings. More than 10^6 spatial arrangements can be generated for every dimer if these few parameters are varied at only rough (30°) angular increments. Information obtained from direct procedures of this sort is often hidden by the sheer enormity of the resulting data.

The conformational flexibility of DNA is more easily comprehended in terms of the various two-dimensional surfaces within its conformation space. The structural changes that accompany simultaneous variations of two torsion angles and the interdependence of these rotations are readily detected from simple contour maps. A two-dimensional analysis is also a convenient starting point for more complicated higher-order treatment of chain flexibility.

Two dimensional analyses of polymer flexibility are usually confined to adjacent angle pairs.[38] A chain molecule like the polynucleotide containing large, interacting side groups, however, is subject to longer-range interactions between non-adjacent bond rotations. Several such correlations have already been noted in theoretical[39-47] and experimental[48-51] studies of oligonucleotides in recent years. In view of these observations, all pairwise combinations of torsional variables must be included in the analysis of DNA flexibility. Three of the 36 possible pairs of dimer variables are discussed in detail below. A more comprehensive review of DNA flexibility will be presented elsewhere.

The rotational flexibility of double helical DNA is described here in terms of various base stacking and potential energy surfaces. Acceptable motions are taken to be ones that maintain the preferred parallel alignment of adjacent bases without engendering severe steric contacts. Base stacking is defined, as before,[5,6,52] in terms of the angle Λ between the normal to adjacent base planes and the average distance Z between these planes. Steric contacts are monitored by a simple 6-12 potential[53,54] with all states more than 10 kcal/mole above the (reference B-DNA helix) energy minimum excluded. All surfaces are drawn with respect to the recently refined B-DNA fiber model as reference.[24,55,56] The residue torsion angles describing the reference B-DNA double helix are illustrated for the single-stranded pdApdAp fragment in Figure 1. The sugars are fixed in an idealized C2′-endo puckered conformation with pseudorotation parameters[57] $\tau_m = 32.5°$ and $P = 0.89\pi$ rad. This puckering is somewhat altered from that ($\tau_m = 37.8°$; $P = 0.87\pi$ rad.) used in the 10-fold fiber model.[58] Because of this slight difference, the theoretical double helix constructed from the conformers in Figure 1 is characterized by 10.5 residues per helical turn and residue step height h = 4.1Å. The features of local structure in this reference conformer, nevertheless, are typical of conventional helical models. The planes of bases attached to each strand of the resulting complex are oriented nearly perpendicular to the helix axis at angles $\eta = 83.4°$ and are tilted by angles $\Lambda = 3.9°$ from perfectly parallel alignment with their nearest neighbors. The mean distance Z between neighboring base planes is 3.9Å. The two bases forming each Watson-Crick complex (not illustrated here) are twisted in a slight propellor fashion at an angle $\tau = 8.2°$. The three atoms comprising each hydrogen bond of the base pair are nearly collinear ($\gamma = 0.7°$ and 9.6° for ∠ H6(A)-N6(A)-O4(T) and ∠ N1(A)-H3(T)-N3(T), respectively, of the A•T base pair), and the respective hydrogen-bond heavy-atom distances are 2.9 and 2.8Å.

Consecutive Internal Rotations – The Phosphodiester Pair

The conformational flexibility of B-DNA in the $\omega'\omega$ (phosphodiester) plane of conformation space is described by the base stacking and potential energy contour maps in Figure 2. The combinations of these two angles that maintain adjacent residues in a parallel, closely-spaced arrangement typical of normal base stacking are located in the upper right-hand quadrant of Figure 2(a) between the solid Z = 3 and 4Å contour lines and within the dashed $\Lambda = 45°$ boundary. A small extension of this domain is noted in the upper left-hand portion of the figure where $\omega' > 0°$. According to this diagram (where base overlaps are ignored), each angle is free to vary over more than 120° of conformation space provided that it follows the prescribed (shaded) path. When the O3′-P bond ω' is rotated in a positive direction from its *t (trans)* to g^- *(gauche⁻)* range, the P-O5′ ω angle must be simultaneously altered in a positive sense from its g^- to g^+ *(gauche⁺)* range. As evident from Figure 2(b), however, many of the phosphodiester conformations that maintain this base "stacking" are subject to severe steric contacts. A large number of conformations involving steric repulsions of 10 kcal/mole or more energy above the reference 6-12 potential minimum are found in the shaded central band of the figure. Only a small fraction of the allowed pathway of Figure 2(a) is also found in a low energy (unshaded) area of Figure 2(b). The sterically allowed helical motions are thus confined to a region centered about the reference tg^- phosphodiester conformation (noted by × in Figure 2(a)) and to a smaller $g^- g^+$ area located near $\omega'\omega = -100°, 10°$. The former "rigid" nucleotide domain alone was used in previous theoretical treatments of DNA flexibility.[5,6] The energy barrier separating the "rigid" phosphodiester domain from the g^-g^+ stacking region, however, is not insurmountable and can be lowered by minor (<5-10°) variations of torsion angles other than ω' and ω. The $g^- g^+$ phosphodiester motions allowed without variation of other angles in the B-DNA backbone are illustrated in Figure 3. This torsional combination is found to overwind consecutive bases and to invert the sugars relative to their alignment in the reference helix. Such conformational changes are at once apparent from the sequential illustrations in the figure. The phosphodiester rotations in Figure 3(a-c) are varied in gradual steps over the stacking pathway of Figure 2(a) from the (−157°, −42°) reference conformer to the (−110°, −10°), (−90°, 30°), and (−35°, 70°) slates, respectively. The three g^-g^+ combinations are illustrated

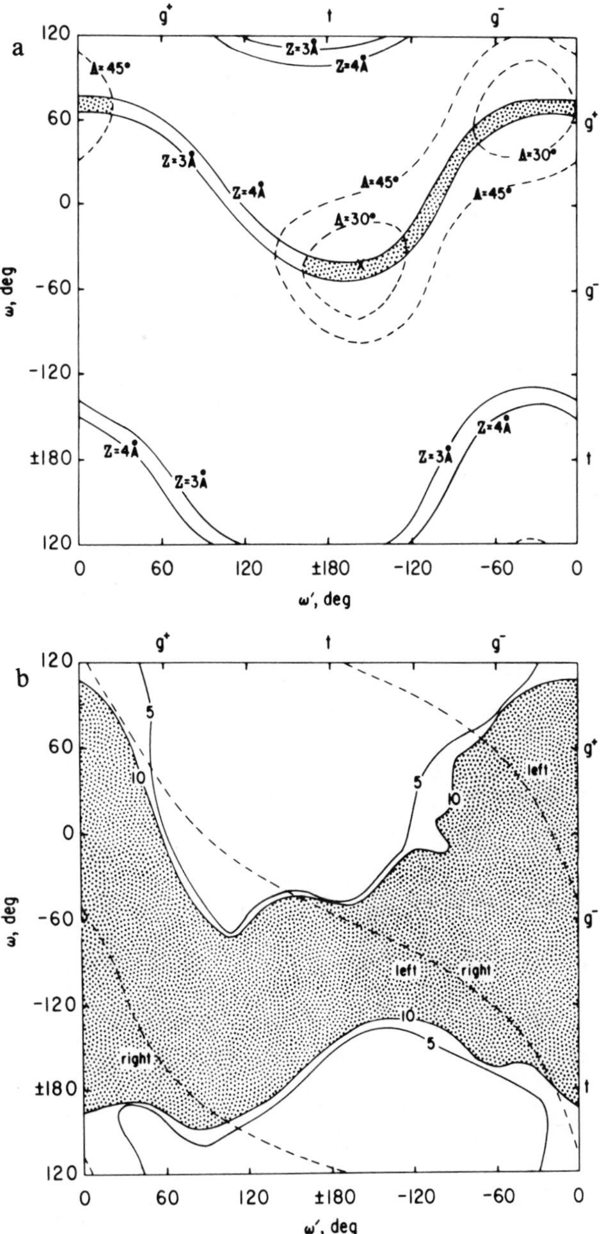

Figure 2. ***(a)*** Composite contour diagram of the base stacking angles Λ (dashed curves) and distances Z (solid curves) as functions of the phosphodiester angles relative to the B-DNA reference helix (noted at ×). The allowed base stacking pathway, where $\Lambda \leq 45°$ and $3\text{Å} \leq Z \leq 4\text{Å}$, is shaded. ***(b)*** Contour diagram in the $\omega'\omega$ plane of nonbonded (steric) potential energy[53,54] relative to the B-DNA reference. The unshaded areas correspond to allowed conformations below the 10 kcal/mole contour and the shaded portion to disallowed states above this boundary. The dashed lines divide the grid into fields of helical handedness.

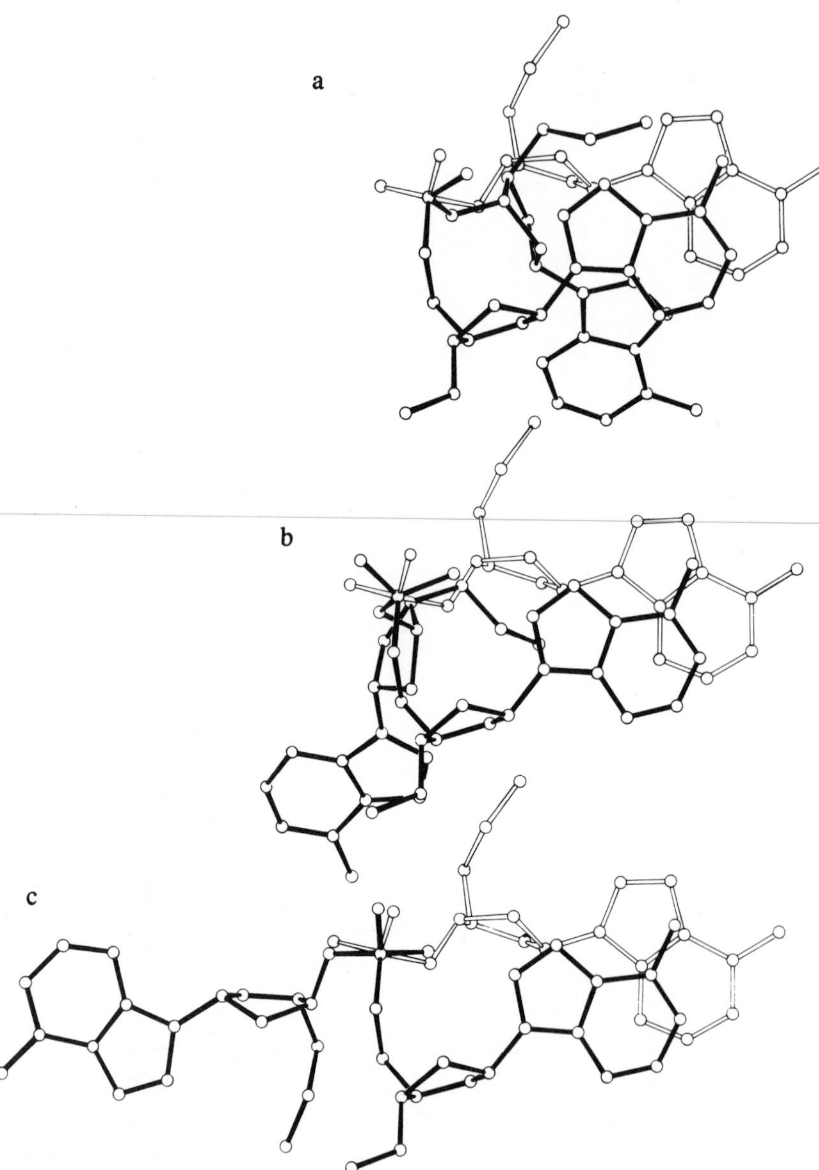

Figure 3. Comparative molecular representations of flexibility along the $\omega'\omega$ stacking pathway. Successive rotational (winding) states at (a) $\omega'\omega = (-110°, -10°)$ (b) $\omega'\omega = (-90°, 30°)$,and (c) $\omega'\omega = (-35°, 70°)$ are superimposed upon the 5′-terminus of the reference helix. Motions are defined relative to the 3′-terminus of the reference helix (open bonds).

by the shaded structures and the reference state by the open ones in the figure. The reference conformer is superimposed upon the 5′-terminus of each g^-g^+ arrangement to emphasize the gradual changes in stacking and sugar orientation. Only the 3′-terminus of the reference state is distinguishable in the illustrations. Base stacking is right-handed and sugar orientation (as measured by the vectors drawn to O1′ from the midpoint of the C2′-C3′ sugar bonds[59]) nearly parallel in the reference B-DNA helix. As the phosphodiester torsions are each increased the attached bases are wound first into a left-handed overlapping stack (Figure 3(a)) and then overwound to nonoverlapping but parallel open arrangements. These angle changes are also responsible for a gradual tilting of adjacent sugar rings to an inverted form reminiscent of that observed in Z-DNA. The extreme sugar inversion illustrated in Figure 3(c), however, is disfavored by major steric contacts between atoms on either side of the central phosphodiester. If compensated by other factors, an extreme conformation of this type is sometimes seen.[60] The unfavorable interactions associated with the inverted g^-g^+ combination are more likely relieved by variations of ω' or ω alone or in concert with other torsions in the chain.

Long Range Angle Correlations – The P-O5′ and C5′-C4′ Bonds

Variations of nonadjacent bonds markedly enhance the rotational mobility of the B-DNA helix. Compared to the limited flexibility of consecutive torsion angles, variations of alternate bonds offer more rotational freedom to the double helix. According to the contour diagrams in Figure 4, the phosphodiester ω angle can adopt nearly all possible rotational values without destroying the parallel stacking of the polynucleotide helix or entertaining any serious steric conflicts. Moreover, this parameter can vary up to 130° without disrupting the linear hydrogen bonding between Watson-Crick base pairs of complementary strands.[61] These changes in ω, however, require corresponding large variations in the ψ torsion that follows it in the chain sequence. According to Figure 4(a), the latter parameter can span nearly 240° without disrupting the parallel alignment of adjacent bases. In contrast to the consecutive phosphodiester torsions that vary in the same rotational sense in Figure 2, the second-neighbor ω and ψ angles in Figure 4 move in the opposite rotational sense. As ω decreases in value from the g^- helical reference (noted at × in Figure 4(a)), ψ simultaneously increases to values greater than its g^+ reference. Because the intervening ϕ rotation is held fixed in a t arrangement, the g^-g^+ and all other rotational combinations of the angle pair remain essentially free of steric overlaps. Indeed, the two small steric barriers in Figure 4(b) disappear upon minor (5-10°) adjustments of the remaining torsions of the nucleotide repeating unit. The observed populations of $\omega\psi$ conformers therefore must reflect the relative energies of the various stacked arrangements together with electrostatic forces, solvent interactions, and intrinsic torsional potentials of the chain backbone. As evident from the dimeric representations in Figure 5, the reference $\omega\psi = g^-g^+$ state involves more extensive base-base overlap than other allowed combinations of these angles. Intrinsic torsional factors of chemical bonding additionally favor *gauche* arrangements of O-P-O-C (ω), O-C-C-C (ψ), and O-C-C-O (ψ-120°) sequences over *trans* conformations.[62-66] Only the $\omega\psi = g^-g^+$ conformation satisfies the *gauche* preferences of all three of these internal rotations. Variations of ω and ψ away from the g^-g^+ domain unwind the bases of adjacent residues and gradually open the chain to a higher energy nonoverlapping parallel arrangement. In contrast to the clockwise $\omega'\omega$ motions illustrated in Figure 3, the bases described by the $\omega\psi$ pathway in Figure 5

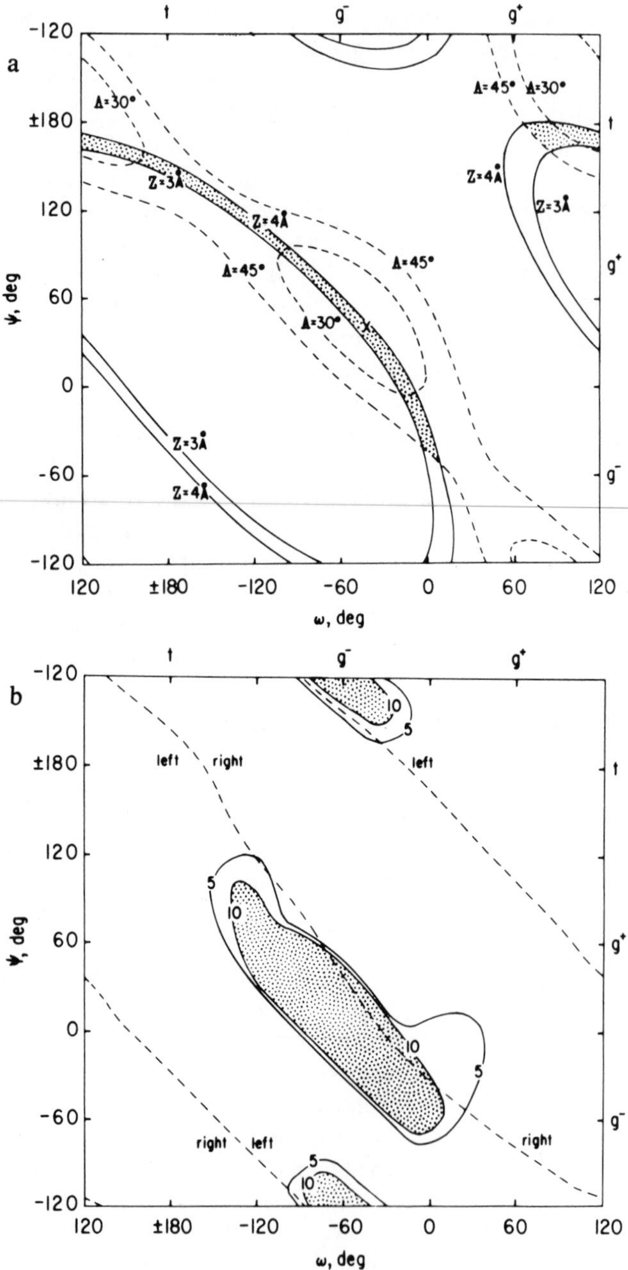

Figure 4. Composite contour surface describing the base stacking of the $\omega\psi$ angle plane. (b) Allowed and disallowed combinations of ω and ψ. See legend to Figure 2.

open in a counterclockwise fashion. The angle pair varies from the $(-42°, 42°)$ B-DNA reference to the $(-32°, 82°)$ and $(120°, 170°)$ states in Figures 5(a) and (b), respectively. As the bases open to the *tt* domain, the sugars also tilt in a perpen-

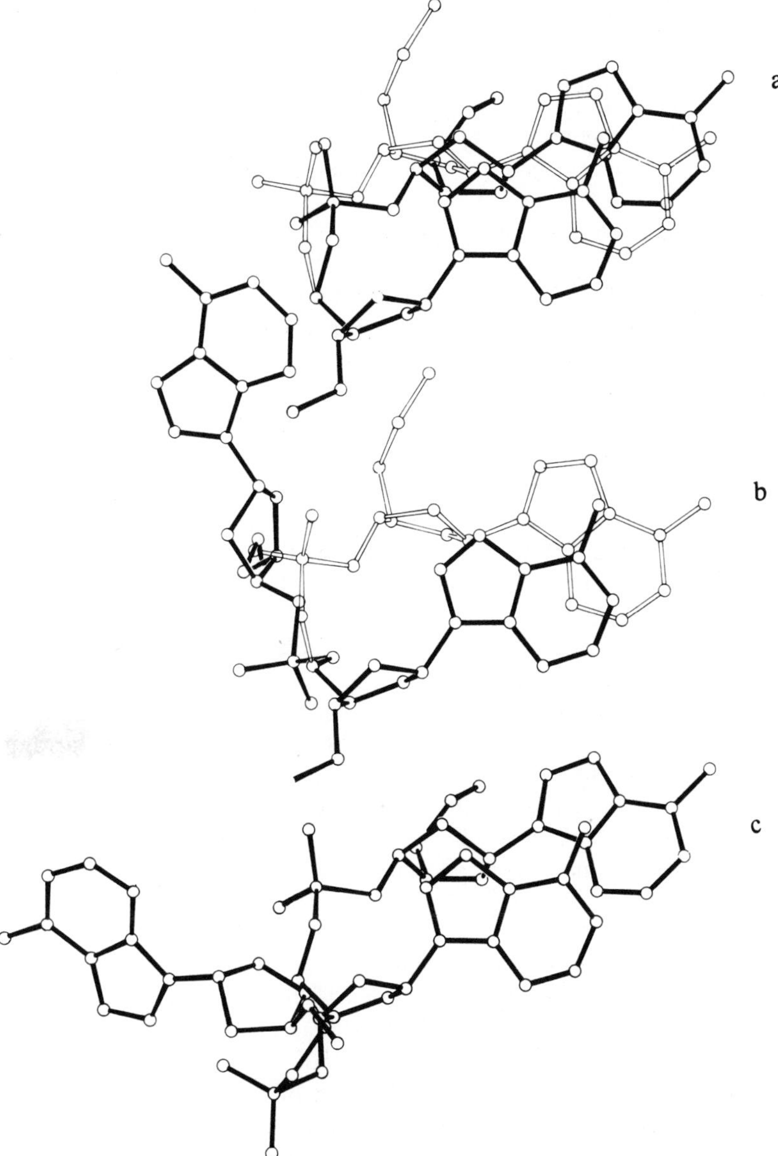

Figure 5. Comparative (unwinding) motions along the $\omega\psi$ stacking pathway. (a) $\omega\psi = (-32°, 82°)$; (b) $\omega\psi = (120°, 170°)$; (c) The open state resulting from sequential winding and unwinding of adjacent dimers. The wound dimer adopts the $\omega'\omega$ state of Figure 3(c) and the unwound dimer the open $\omega\psi$ combination illustrated in (b) of this figure. See legend to Figure 3.

dicular fashion. However, none of the allowed $\omega\psi$ rotational combinations in Figure 4 invert the sugars to a Z-type arrangement. The *tt* combination of ω and ψ, instead, describes a novel all-*trans* helix similar to the E-DNA model recently proposed for fibrous poly d(IIT)•poly d(CCA).[67-69] Without the simultaneous incidence of overwinding in neighboring repeating units, the highly unwound conformation illustrated in Figure 5(b) will disrupt the B-DNA double helix. When coupled with an overwound conformation in the preceding repeating unit, however the unwound base moves, as seen in Figure 5(c), to the outside of the DNA duplex. Such arrangements facilitate hydrogen exchange, promote the *syn-anti* transition of the base about its glycosyl bond (χ), and provide a potential model of frameshift mutation.

Other Torsional Interactions – The C-O3' and P-O5' Bonds

Like the alternate $\omega\psi$ angle pair, the second-neighbor ϕ' and ω torsions of B-DNA can also adopt a broad range of stacked arrangements. According to the distance and angle contours in Figure 6, ϕ' can vary almost 300° and ω nearly 270° without disruption of the preferred parallel base alignment. As noted with ω and ψ in Figure 4(a), the ϕ' and ω angles in Figure 6(a) rotate in the opposite rotational sense along the shaded stacking pathway. The ϕ' rotation moves between t and g^- conformations while the ω rotation varies simultaneously from g^- to t states. These compensating rotations presumably reflect the fixed *trans* torsion that intervenes between the angle pair. So-called 1-3 crankshaft motions[32] of this sort similarly occur on either side of *trans* bonds in the conformational transitions of both alkanes[33,35] and peptides.[34] In these systems, as well as in the polynucleotides, the intervening *trans* torsion aligns the second-neighbor bonds in approximately parallel fashion. The positive rotation of one angle coupled with the negative rotation of the other maintains the direction of the chain backbone. Such motions in the polynucleotide helix simply shift the bases to alternate stacking arrangements without bending the chain backbone. The isolated variation of a single rotation angle, in contrast, usually kinks the chain backbone and, in the case of double helical DNA, disrupts both base stacking and hydrogen bonding. As outlined below, single rotations about any bonds perpendicular to base planes do not destroy an existing parallel stack.

The somewhat broader range of $\phi'\omega$ flexibility in Figure 6(a) compared to that of ω and ψ in Figure 4a apparently stems from the more nearly parallel alignment of the former angle pair. The $\omega' = -157°$ torsion that separates the ϕ' and ω bonds on B-DNA deviates less from a perfect *trans* (180°) arrangement than the $\phi = 136°$ torsion between the ω and ψ angles. With an intervening 180° torsion second-neighbor bonds can vary in concert over their entire rotation range. Close contacts between sugar and phosphate moieties, however, prevent such free rotation of the $\phi'\omega$ angle pair. As evident from Figure 6(b), more than half of the total $\phi'\omega$ angle space involves serious steric contacts. The allowed motions of ϕ' and ω thus fall in a relatively small domain centered at the reference conformation (again noted by $\times$ in Figure 6a). The two angles fluctuate over a 40-50° range and at no time disrupt the hydrogen bonds between complementary strands.[61] As evident from Figure 7, these rotational changes unwind the DNA backbone in a clockwise fashion similar to the $\omega\psi$ angle pair. The above cited steric limitations upon ϕ' and ω prevent extreme unwinding of the helix in this manner. The maximum unwinding corresponds roughly to the extreme open stack in Figure 7(b). These limited variations,

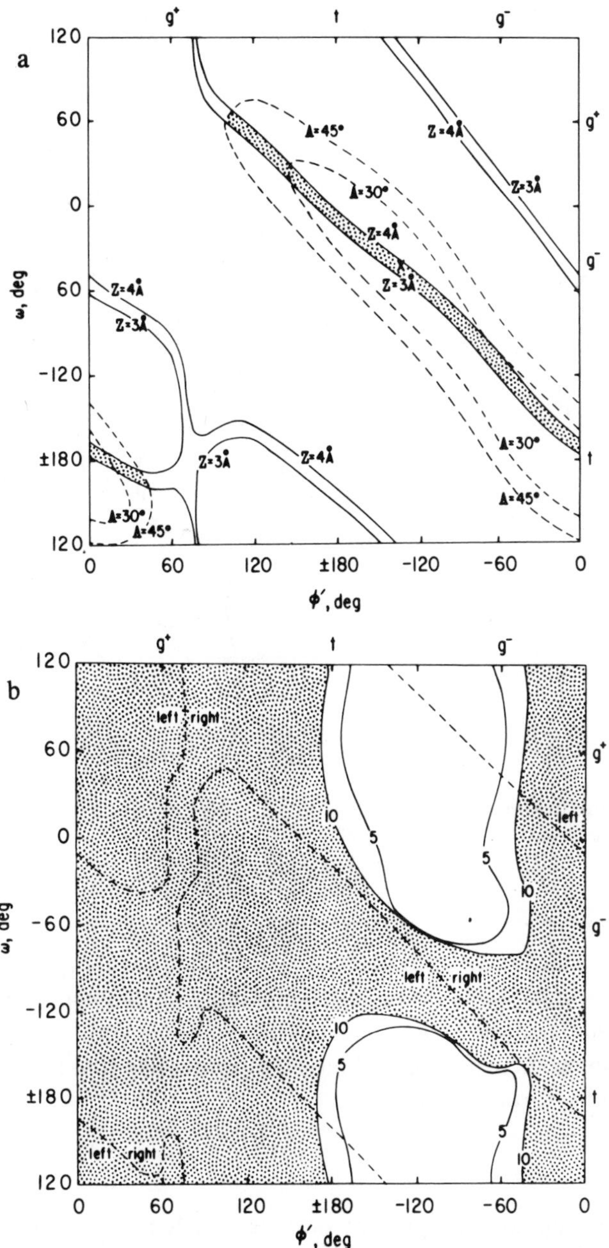

Figure 6. (a) Composite contour surface describing the base stacking of the $\phi'\omega$ angle pair. (b) Allowed and disallowed combinations of ϕ' and ω. See legend to Figure 2.

nevertheless, correspond to the observed conformational differences of ϕ' in B- and Z-DNA.[24,26] The transition from B- to Z-DNA requires an approximately 60° increase in ϕ' analogous to that illustrated in Figures 7(a) and (b). The angle

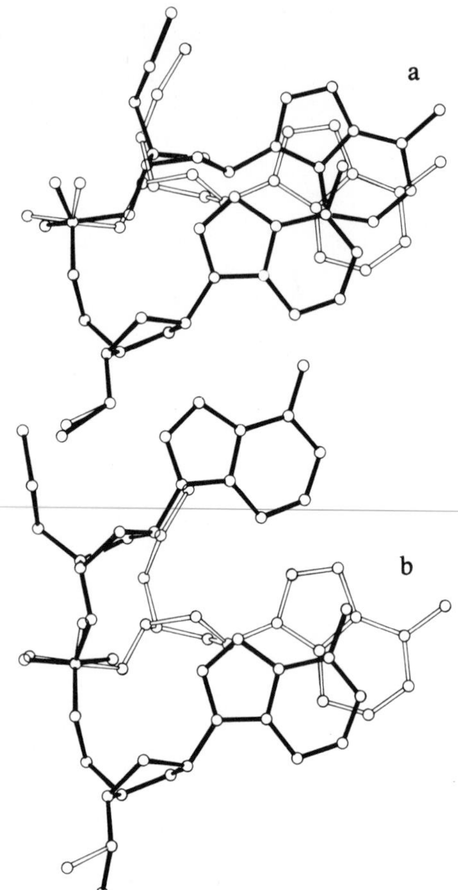

Figure 7. Comparative (unwinding) motions along the $\phi'\omega$ stacking pathway. (a) $\phi'\omega = (-100°, -70°)$; (b) $\phi'\omega = (-30°, -150°)$. See legend to Figure 3.

increases in the figure from its $-133°$ reference successively to $-100°$ and $-30°$. At the same time, the related ω rotation decreases from $-42°$ to $-70°$ and $-150°$, respectively. Combination of these unwinding motions with a neighboring overwinding will also produce an exposed base conformer like that illustrated in Figure 5(c) using ω and ψ variations.

Discussion

The two-dimensional analyses presented above clarify the unusual conformational mobility of double helical DNA. Internal bond rotations of the nucleic acid backbone can fluctuate alone or in pairwise fashion over wide angular ranges without serious distortion of base stacking or disruption of the chain trajectory. The extreme flexibility of the chain repeating units accordingly promotes the well-known stiffness of the polynucleotide as a whole. The various correlations of torsional parameters also corroborate angular effects noted in previous X-ray and potential energy studies. The rotation angle combinations observed in most known nucleic acid structures fall within the allowed base stacking pathways. The continuous sequence of angular changes on these routes provide, in addition, likely models of the transitions between the major conformational domains.

The observed rotational correlations of the double helical structure clearly reflect the strong tendency of complementary nucleic acid bases to adopt stacked, hydrogen-bonded parallel arrangements. Rotational changes of individual bond angles usually kink the chain backbone and disrupt these favorable associations. Torsions about bonds that are perpendicular to the base stacking planes, however, can maintain the parallel alignment of adjacent bases. The variations of such angles simply twist the DNA helix so that successive bases oscillate in their stacking planes. Because none of the acyclic backbone bonds of the B-DNA fiber model are perpendicular to the stacking planes, large fluctuations of these rotations cannot maintain the preferred direction of the double helix. The greatest rotational flexibility of the B-DNA system defined in Figure 1 resides in the ω' angle about the O3′-P bond. As evident from Figure 2(a), this bond, which is aligned at an angle $\theta_{O3'\text{-}P} = 55°$ with respect to the base stacking axis, can vary over a range of 70-80° without serious disruption of normal planar stacking. Rotations about bonds oriented more nearly parallel to the base stacking plane readily distort the helical structure. The C3′-O3′ torsion (ϕ') where $\theta_{C3'\text{-}O3'} = 69°$, for example, varies less than 20° in Figure 6(a) without disruption of planar stacking. Unlike "freely rotating" acyclic torsions such as ω', variations of the ring torsion ψ' alter the initial perpendicular alignment of the C4′-C3′ bond with respect to the attached B-DNA base and thus disrupt normal base stacking. This distortion stems from the changes in ring puckering that accompany variations of ψ'. If the puckering changes induce additional variations of the glycosyl torsion χ, motions of ψ' will not destroy the stacking of a B-DNA dimer unit. Such changes in χ in a polymer, however, induce further rotational changes in the backbone angles of adjacent dimers.

The concerted variations of two rotatable bonds counteract the intrinsic bending associated with isolated torsional changes. Particularly strong compensating effects occur between second-neighbor torsions separated by an intervening *trans* bond. Provided the parameters vary in the opposite rotational sense, second-neighbor rotations can move "freely" over their complete angular range. Adjacent torsions, in contrast, vary over more limited areas without disruption of normal base stacking. This weaker interdependence apparently reflects the fixed (non-linear) valence angles that orient consecutive bonds along the polynucleotide backbone. The strongly correlated second-neighbor $\phi'\omega$ and $\omega\psi$ motions stem from the nearly parallel alignment of the respective bond pairs (C3′-O3′ and P-O5′; P-O5′ and C5′-C4′). Compared to the O-P-O valence angle of 101.4° between consecutive ω' and ω torsions, the pseudovalence angles defined by the second-neighbor ϕ, ω and $\omega\psi$ bond pairs are respectively 152.3° and 139.2°. As the pseudovalence angles approach 180° and the orientations of given bond pairs become parallel, the related torsion angles become more strongly correlated. Indeed any pair of polynucleotide torsions will vary in a strongly interdependent fashion if aligned in parallel. According to the matrix of pseudovalence angles in Table I, other strongly corre-

Table I
Valence and Pseudovalence Angles, in Degrees, of Rotatable Backbone Bonds of the B-DNA Double Helix

	(ψ') C4'-C3'	(ϕ') C3'-O3'	(ω') O3'-P	(ω) P-O5'	(ϕ) O5'-C5'	(ψ) C5'-C4'	(ψ') C4'-C3'
(ψ') C4'-C3'	---	111.9	137.2	115.0	126.0	110.8	174.5
(ϕ') C3'-O3'	---	---	119.0	152.3	115.1	111.5	117.3
(ω') O3'-P	---	---	---	101.4	122.8	104.9	139.6
(ω) P-O5'	---	---	---	---	119.0	139.2	119.3
(ϕ) O5'-C5'	---	---	---	---	---	110.0	121.7
(ψ) C5'-C4'	---	---	---	---	---	---	111.4

lated backbone rotation angle combinations within each dimeric segment of the B-DNA helix include the $\psi'\omega'$ second neighbors ($\angle = 137.2°$) the $\omega'\psi'$ fourth neighbors ($\angle = 139.6°$), and the $\psi'\psi'$ sixth neighbors ($\angle = 174.5°$). Each of these angle pairs can fluctuate over fairly broad rotational ranges without disruption of base stacking or kinking the chain backbone. As noted above, variations in ψ' require additional changes in χ to maintain parallel base alignment.

The long-range rotational correlations in the polynucleotide backbone also avoid the viscous drag and inertial effects[70] that accompany isolated torsion angle variations. Such crankshaft-like motions, instead, move only a short section of the chain through the surrounding medium. These conformational changes, however, involve the simultaneous passage of two, possibly unfavorable, energy barriers. The relatively low C-O and P-O torsional barriers should not preclude such concerted motions in the nucleic acids. Low activation barriers apparently account for the second-neighbor rotations seen in dynamical simultations of polypeptide transitions.[34] The favorable stacking and hydrogen bonding interactions of adjacent base moieties additionally enhance long-range rotational correlations in the polynucleotide chain.

The long-range stiffness of DNA ultimately derives from the sequence of heterocyclic bases. The bases force the sugar-phosphate backbone to move in a highly correlated fashion that preserves the preferred stacking interactions. The local mobility of the chain backbone, however, reflects the structural complexity of the sugar-phosphate residues. The six-bond backbone repeating unit offers an unusually large number of acceptable stacking arrangements. The ease of motion stems from the relatively low (10-20 kcal/mole) barriers that separate the various conformational energy minima. Occasional passages across these barriers account for the

Table II
Conformational Pathways of DNA Helical Transition

Helix Form	Backbone, Torsion Angle Sequence (ψ'	ϕ'	ω'	ω	ϕ	ψ)	Torsional Route
WC	g^+	t	g^-	t	t	t	
↑↓							$\omega\psi$
A	g^+	t	g^-	g^-	t	g^+	
↑↓							$\psi'\omega'$
B	t	t	t	g^-	t	g^+	
↑↓							$\omega\psi$
E	t	t	t	t	t	t	
↑↓							$\phi'\omega$
Unidentified	t	g^-	t	g^+	t	t	
↑↓							ω'
Z	t	g^-	g^+	g^+	t	t	

unique breathing and premelting of DNA. Regular variations over particular pathways eventually convert the B-DNA structure into alternate helical forms.

Some potential DNA transitions are outlined in Table II. The B-DNA helix is easily converted to the all-*trans* E-form following a direct $\omega\psi$ pathway like that illustrated in Figures 2 and 3. Transformation of B- to A-DNA is also readily accomplished along an analogous second-neighbor pathway involving ψ' and ω'. The A-helix, in turn, can be converted to the Watson-Crick (WC) model recently found to match the diffraction pattern of hybrid poly(dI)•poly(rC) fibers.[67] The transformation from A to WC, like that from B to E, is accomplished via an $\omega\psi$ pathway. The WC-helix can be further converted to the E-form by variations of the $\psi'\omega'$ angle pair.

The smooth transition of the right-handed B-DNA helix to the unusual left-handed Z-structure requires several pairs of correlated angle changes. Four of the six backbone torsions plus one of the glycosyl bonds must vary as the dCpdG dimer interconverts between the two forms. One potential route via the E-helix appears in Table II. As seen from the pathway, the principal differences between the B- and Z-backbone angles involve ordinary second-order correlations of the $\phi'\omega$ and $\omega\psi$ angle pairs together with a single variation of ω'. The ω' change may occur alone or possibly in concert with the ψ' rotation of the succeeding unit. Both ω' and ψ' rotations involve bonds that are perpendicular to the base stacking planes. Because the $\phi'\omega$ motions require an intervening *trans* bonds, the ω' change must necessarily follow $\phi'\omega$ changes. The $\omega\psi$ step, however, may occur at any point in the scheme.

The first two steps of the proposed B→Z pathway in Table II unwind the original dCpdG base stack while the last step rewinds the dimer. The *anti* to *syn* glycosyl transition of the G base most likely occurs when the base resides on the DNA surface.[71] As explained by Figure 5(c), such arrangements require winding of the dCpdG units and concomitant unwinding of the neighboring dGpdC residues. Such compensating motions prevent the dissociation of complementary strands and also account for the observed B→Z torsional differences in the dGpdC units.[61]

Acknowledgements

This research was sponsored by the U.S. Public Health Service under grants CA25981 and GM20861 and the donors of the Petroleum Research Foundation to grant AC11586. Computer time was supplied by the Rutgers University Center for Computer and Information Services. W.K.O. is also the recipient of a U.S.P.H.S. Research Career Development Award (GM155).

References and Footnotes

1. Bloomfield, V.A., Crothers, D.M. and Tinoco, I., Jr., *Physical Chemistry of Nucleic Acids,* Harper and Row, New York, 1974, Chapter 5.
2. Eisenberg, H., in *Basic Principles in Nucleic Acid Chemistry,* Vol. I, Ts'o, P.O.P., Ed., Academic Press, New York, 1974, pp. 171-264.
3. Schellman, J.A., *Biopolymers, 13,* 217-226 (1974).
4. Godfrey, J.E. and Eisenberg, H., *Biophys. Chem., 5,* 301-318 (1976).
5. Olson, W.K., *Biopolymers, 18,* 1213-1233 (1979).
6. Olson W.K., in *Stereodynamics of Molecular Systems,* Sarma, R.H., Ed., Pergamon Press, New York, 1979, pp. 297-314.
7. Barkley, M.D. and Zimm, B.H., *J. Chem. Phys., 70,* 2991-3007 (1979).
8. Robinson, B.H., Lerman, L.S., Beth, A.H., Frisch, H.L. Dalton, L.R. and Auer, C., *J. Mol. Biol. 139,* 19-44 (1980).
9. Schellman, J.A., *Biophys. Chem., 11,* 321-328 (1980).
10. Shore, D., Langowski, J. and Baldwin, R.L., *Proc. Natl. Acad. Sci., U.S.A.,* in press.
11. Wahl, P., Paoletti, J. and LePecq, J.-B., *Proc. Natl. Acad. Sci., U.S.A., 65,* 417-421 (1970).
12. Klevan, L., Armitage, I.M. and Crothers, D.M., *Nucleic Acids Res., 6,* 1607-1616 (1979).
13. Davanloo, P., Armitage, I.M. and Crothers, D.M., *Biopolymers, 18,* 663-680 (1179).
14. Early, T.A. and Kearns, D.R., *Proc. Natl. Acad. Sci., U.S.A. 76,* 4165-4169 (1979).
15. Hogan, M.E. and Jardetzky, O., *Proc. Natl. Acad. Sci., U.S.A., 76,* 6341-6345 (1979).
16. Shindo, H., *Biopolymers, 19,* 509-522 (1980).
17. Shindo, H., Wooten, J.B., Pheiffer, B.H. and Zimmerman, S.B., *Biochem., 19,* 518-526 (1980).
18. Bolton, P.H. and James, T.L., *J. Am. Chem. Soc., 102,* 25-31 (1980).
19. Bolton, P.H. and James, T.L., *Biochem., 19,* 1388-1392 (1980).
20. Hogan, M.E. and Jardetzky, O., *Biochem., 19,* 3460-3468 (1980).
21. Millar, D.P., Robbins, R.J. and Zewail, A.H., *Proc. Natl. Acad. Sci., U.S.A., 77,* 5593-5597 (1980).
22. Wang, A.H.-J., Quigley, G.J., Kolpak, F.J., Crawford, J.L., van Boom, J.H., van der Marel, G., and Rich, A., *Nature, 282,* 680-686 (1979).
23. Drew, H., Takano, T., Tanaka, S., Itakura, K. and Dickerson, R.E., *Nature, 286,* 567-573 (1980).
24. Arnott, S., Chandrasekaran, R., Birdsall, D.L., Leslie, A.G.W. and Ratliff, R.L., *Nature, 280,* 743-745 (1980).
25. Gupta, G., Bansal, M. and Sasisekharan, V., *Biochem. Biophys. Res. Commun., 95,* 728-733 (1980).
26. Wang, A. H.-J., Quigley, G.J., Kolpak, F.J., van der Marel, G., van Boom, J.H. and Rich, A., *Science, 211,* 171-176 (1981).
27. Klysik, J., Stirdivant, S.M., Larson, J.E., Hart, P.A., and Wells, R.D., *Nature, 290,* 672-677 (1981).
28. Wells, R.D., Klysik, J., Stirdivant, S.M., Larson, J., and Hart, P.A., this volume.
29. Flory, P.J., *Pure and Applied Chem., 26,* 309-326 (1971).
30. Sundaralingam, M., in *Structure and Conformation of Nucleic Acids and Protein-Nucleic Acid Interactions,* Sundaralingam, M. and Rao, S.T., Eds., University Park Press, Baltimore, 1975, pp. 487-524.
31. Sundaralingam, M. and Westhof, E., *Intl. J. Quantum Chem. Biol. Symp., 6,* 115-130 (1979).
32. Helfand, E., *J. Chem. Phys., 54,* 4651-4661 (1971).
33. Helfand, E., Wasserman, Z.R., and Weber, T.A., *Macromolecules, 13,* 526-533 (1980).
34. Pear, M.R., Northrup, S.H., McCammon, J.A., Karplus, M. and Levy, R.M., *Biopolymers, 20,* 629-632 (1981).

35. Huang, T.H., Skarjune, R.P., Wittebort, R.J. Griffin, R.G., and Oldfield, E., *J. Am. Chem. Soc., 102,* 7377-7379 (1980).
36. Teitelbaum, H. and Englander, S.W., *J. Mol. Biol., 92,* 55-78 and 79-92 (1975).
37. Kallenbach, N.R., Mandal, C. and Englander, S.W., in *Stereodynamics of Molecular Systems,* Sarma, R.H., Ed., Pergamon Press, New York, 1979, pp. 271-282.
38. Olson, W.K. and Flory, P.J., *Biopolymers, 11,* 25-56 (1972).
39. Perahia, D., Pullman, B., Vasilescu, D. Cornillon, R. and Broch, H., *Biochim. Biophys. Acta, 478,* 244-259 (1977).
40. Broch, H. and Vasilescu, D., *Biopolymers, 18,* 909-930 (1979).
41. Kumar, N.V. and Govil, G., *Indian J. Biochem. Biophys., 16,* 414-420 (1979).
42. Broyde, S. and Hingerty, B., *Nucleic Acids Res., 6,* 2165-2178 (1979).
43. Sasisekharan, V. and Gupta, G., *Curr. Sci., 49,* 43-48 (1980).
44. Olson, W.K., *Macromolecules, 13,* 721-728 (1980).
45. Srinivasan, A.R. and Olson, W.K., *Nucleic Acids Res., 8,* 2307-2329 (1980).
46. Sundaralingam, M. and Westhof, E., *Biophys. J., 32,* 280-281 (1980).
47. Yathindra, N., in *Biomolecular Structure, Conformation, Function and Evolution,* Vol. 1, Srinivasan, R., Ed., Pergamon Press, Oxford, 1981, pp. 379-401.
48. Viswamitra, M.A., Kennard, O., Jones, P.G., Sheldrick, G.M., Salisbury, S., Falvello, L. and Shakked, Z., *Nature, 273,* 687-688 (1978).
49. Hingerty, B.E., Brown, R.S. and Jack, A., *J. Mol. Biol., 124,* 523-534 (1978).
50. Holdbrook, S.R., Sussman, J.L., Warrant, R.W. and Kim, S.-H., *J. Mol. Biol., 123,* 631-660 (1978).
51. Stout, C.D., Mizuno, H., Rao, S.T., Suaminathan, P., Rubin, J., Brennan, T., and Sundaralingam, M., *Acta Crystall, B34,* 1529-1544 (1978).
52. Olson, W.K., *Biopolymers, 15,* 859-878 (1976).
53. Suter, U.W., *J. Am. Chem. Soc., 101,* 6481-6496 (1979).
54. Srinivasan, A.R. and Olson, W.K., manuscript in preparation.
55. The B-DNA reference helix was generated following established procedures[56] using as input the torsion angles listed in reference 24.
56. Olson, W.K., in *Fiber Diffraction Methods, ACS Symposium Series,* No. 141, French, A.D. and Gardner, K.H., Eds., American Chemical Society, Washington, D.C., 1980, pp. 251-265.
57. Altona, C., and Sundaralingam, M., *J. Am. Chem. Soc., 94,* 8205-8212 (1972).
58. Arnot, S., personal communication.
59. Kim, S.-H., Berman ,H.M. Seeman, N.C and Newton, M.D., *Acta Crystall., B29,* 703-710 (1973).
60. Such factors are apparently responsible for the unusual g^+g^- phosphodiester combination reported in this volume by Broyde and Hingerty from minimization studies of carcinogen-modified DNA.
61. Olson, W.K., unpublished data.
62. Newton, M.D., *J. Am. Chem. Soc., 95,* 256-258 (1973).
63. Govil, G., *Biopolymers, 15,* 2303-2307 (1976).
64. Hayes, D.M., Kollman, P.A. and Rothenberg, S., *J. Am. Chem. Soc., 99,* 2150-2154 (1977).
65. Abe, A. and Mark, J.E., *J. Am. Chem. Soc., 98,* 6468-6472 (1976).
66. Olson, W.K., *J. Am. Chem. Soc.,* submitted.
67. Chandrasekaran, R., Arnott, S., Banerjee, A., Campbell-Smith, S., Leslie, A.G.W. and Puigjaner, L.C., in *Fiber Diffraction Methods, ACS Symposium Series,* No. 141, French, A.D. and Gardner, K.H., Eds., American Chemical Society, Washington, D.C., 1980, pp. 483-502.
68. Arnott, S., Chandrasekaran, R., Bond, P.J., Birdsall, D.L., Leslie, A.G.W., and Puigjaner, L.C., in *Proceedings of the Seventh Aharon Katzir-Katchalsky Conference on Structural Aspects of Recognition and Assembly in Biological Macromolecules,* Nof Ginossar, Israel, February 1980.
69. Leslie, A.G.W., Arnott, S., Chandrasekaran, R. and Ratliff, R.L., *J. Mol. Biol., 143,* 49-72 (1980).
70. Pear, M.R., Northrup, S.H. and McCammon, J.A. *J. Chem. Phys., 73,* 4703-4704 (1980).
71. Ciancia, L. and Olson, W.K., manuscript in preparation.

Proceedings of the Second SUNYA Conversation in the Discipline Biomolecular Stereodynamics
Volume I, ISBN 0-940030-00-4, Ed., Ramaswamy H. Sarma,
Adenine Press, New York, ©Adenine Press

NMR Relaxation Studies of the Structural and Dynamic Properties of DNA

David R. Kearns, Nuria Assa-Munt, Ronald W. Behling, Thomas A. Early,[50] Juli Feigon, Joseph Granot
Department of Chemistry
University of California—San Diego
La Jolla, California 92093

Wolfgang Hillen
Institut fur Organische Chemie und Biochemie
Technische Hochschule
6100 Darmstadt, Petersenstr. 22
Darmstadt, Germany

and

Robert D. Wells
Department of Biochemistry
University of Wisconsin
Madison, Wisconsin 53706

Introduction

Nuclear magnetic resonance spectroscopy has long been recognized as one of the most powerful techniques for investigating the structural and dynamic properties of molecules in solution. The application of NMR to studies of DNA molecules has in the past been hampered both by difficulties in obtaining suitable samples and by a lack of adequate NMR methods and instrumentation for studying large macromolecules in solution.[1] During the past couple of years, the sample preparation problem has been solved by a variety of new approaches (cloning,[2,3] chemical synthesis of short DNA oligonucleotides,[4-6] enzymatic[7,8] and physical techniques for cleaving high molecular weight DNA to low molecular weight samples), and we now have up to gram quantities of short (10-200 bp) DNA for NMR and other physical studies. There has also been major progress in the field of NMR. Highfield spectrometers operating at 500 MHz and 600 MHz have become available and a variety of new pulse methods (heteronuclear spin exchange,[9] 2-dimensional relaxation experiments,[10-13] multiple-quantum spectroscopy[14]) have been invented. By taking advantage of the recent developments in these two quite different areas, NMR can now be used to study some of the long-standing, but unsolved problems concerning the structural and dynamic properties of DNA molecules in solution.[15] In the present work, we have examined samples ranging in length from 12 to ~100 bp in length with the following questions in mind. What are the precise solution state structures of various DNA fragments and how does sequence affect conformation? Over what distances can a conformational perturbation introduced at one site be propagated along the DNA? What are the dynamic properties of DNA and how are they affected by sequence and various environmental factors? How do various metal ions and other polycations bind to DNA? The answer to these and many other interesting questions may soon be provided by NMR.

The work described in this paper is concerned with the use of ^{1}H and ^{31}P relaxation measurements to explore the properties of selected DNA molecules in solution. Experiments on a "simple" alternating copolymer, poly(dA-dT), serve to illustrate a number of 1-dimensional relaxation techniques and provide specific information about some of the properties of this molecule. Studies on a 12 base pair restriction fragment are used to explore some of the interesting questions regarding structure and sequence effects on conformation. In the latter part of the paper, ^{31}P relaxation measurements are used to investigate the binding of divalent metal ions to DNA, Relaxation measurements are specifically emphasized because they can be most directly related to inter-nuclear separations (and hence structure) and to the frequencies of the internal and overall molecular motions.

Proton Relaxation Studies of DNA

Theory of Proton Relaxation in Multispin Systems. A variety of relaxation techniques (non-selective spin-lattice relaxation rate measurements, steady state nuclear Overhauser effects (NOEs),[16] off-resonance decoupling[17,18]) have been developed to investigate the structures of small molecules in solution; however, in proton NMR studies of DNA molecules which have rotational correlation times longer than 5 nsec, a somewhat different approach is required.

For a multi-spin system of protons, the longitudinal spin lattice relaxation of I_{zi}, the z-component of magnetization of the i-th spin, due to dipolar interactions with other spins j is given by:[19]

$$\frac{dI_{zi}}{dt} = -\sum_j \rho_{ij}(I_{zi}-I_{oi}) - \sum_j \sigma_{ij}(I_{zi}-I_{zj}) \tag{1}$$

where

$$\rho_{ij} = \frac{3}{2}\gamma^4\hbar^2\frac{1}{r_{ij}^6}[J_1(\omega) + 4J_2(2\omega)] \tag{2}$$

and,

$$\sigma_{ij} = \frac{1}{2}\gamma^4\frac{\hbar^2}{r_{ij}^6}[J_o(0)-6J_2(2\omega)] \tag{3}$$

and I_{oi} is the equilibrium value of I_{zi}, r_{ij} is the distance between spins i and j, $J_0(0)$,

$J_1(\omega)$ and $J_2(2\omega)$ are the usual spectral densities, and γ and $\hbar$ have their standard meaning.[20] The σ_{ij} terms govern cross relaxation between spins i and j, and in a large, slowly tumbling macromolecule where $\omega\tau_c > 1$, the rate of transfer of spin polarization between protons is faster than the rate of exchange of polarization with the lattice. This phenomenon is known as spin diffusion and it has very pronounced effects on the measurement of spin-lattice relaxation rates, depending on how the system is initially prepared. Two cases deserve special consideration.

Non-Selective T_1 Measurement: In a non-selective measurement of the spin-lattice relaxation rate, all protons in the system are initially polarized by application of short, high power, non-selective 180° pulse to the system. Immediately after application of the pulse, (t=0), $I_{zi} = I_{zj}$. The *initial* decay of the spin polarization is given by

$$\frac{dI_{zi}}{dt} = -(\Sigma_j \rho_{ij})(I_{zi} - I_{oi}) \quad . \tag{4}$$

However, in slowly tumbling molecules where $\sigma_{ij} > \rho_{ij}$ there is rapid spin diffusion (resulting from the faster cross-relaxation processes) and Eq. (4) no longer holds. Therefore, as Kalk and Berendsen have shown, all N interacting protons tend to have the same T_1 values, given by the following average:[19]

$$\frac{1}{T_1^{ns}} = \frac{1}{N}\sum_i(\sum_j \rho_{ij}) \quad . \tag{5}$$

In the special case of a slowly tumbling isotropic rotor with a rotational correlation time τ_c and $\omega\tau_c > 1$[16]

$$J_1(\omega) = \frac{\tau_c}{5(1+\omega^2\tau_c^2)} \cong \frac{1}{5\omega^2\tau_c} \;, \quad J_2(2\omega) \cong \frac{1}{5}\left(\frac{1}{4\omega^2\tau_c}\right)$$
$$\rho_{ij} \sim \frac{1}{H^2\tau_c} \tag{6}$$

thus

and the average T_1 value will increase with the square of the applied magnetic field, H. Similar results are obtained for an anisotropic rotor, provided the relevant correlation times satisfy the condition $\omega\tau_c > 1$.[21] Because the measured T_1 is an average of a number of interactions in a non-selective relaxation measurement, information about the different local interproton interactions is combined into one single quantity making it impossible to sort out the contribution from any specific proton-proton interaction.

Selective T_1^s Measurements: Quite different results are obtained by selective inversion of the "i" spins. Since only the i spins are inverted, $-I_{zi} = I_{oj} = I_{oi}$ immediately after application of the pulse. In this case the *initial* rate of relaxation of the "i" spins is given by

$$\frac{1}{T_1^s} = \sum_j (\rho_{ij} + \sigma_{ij}) \tag{7}$$

which depends only on those protons j which directly interact with the i spin. Therefore, by measuring the initial rates in a *selective* spin-lattice relaxation measurement it is possible to circumvent the problems which occur in the non-selective relaxation measurement due to rapid spin diffusion. Note further that for a slowly tumbling molecule where $J_0 > J_1, J_2$, $(\rho_{ij} + \sigma_{ij}) > \rho_{ij}$ and consequently for initial rates

$$\frac{1}{T_1^{ns}} < \frac{1}{T_1^s} \sim \tau_c$$

From Eqs. (2) and (3) we see that initial values of the selective spin-lattice relaxation rates vary with the inverse 6th power of the distance between the observed proton and any protons which interact with it. In addition, the rates are dependent on the overall motion of the molecule and any internal motions which affect the interproton interaction. Detailed treatments of the appropriate spectral densities $J_n(\omega)$ for various models of internal motion are to be found in the literature,[22-25] but the simplest results are obtained for an isotropic rigid rotor (Eq. (6)) or for an anisotropic rigid rotor, in which case the result of Woessner may be used.[21]

In the special case where only a single proton j interacts strongly with proton i, the selective relaxation rate for i can be used to obtain a reasonably accurate value of r_{ij}, provided the appropriate values for rotational correlation time(s) can be obtained. When more than one proton strongly interacts with proton i, bi-selective relaxation measurements can be used to determine the individual σ_{ij}s. Because of the 6th power dependence of the relaxation rates on r_{ij}, relatively imprecise values for the correlation times can still yield reasonably accurate distances. For example, a 3-fold error in the correlation times leads to a 20% error in r_{ij}. In short DNA where the overall tumbling of the molecule is sufficiently fast so that it dominates the effects of internal motions, it should be possible to calculate the relevant $J_n(n\omega)$ values to better than a factor of 2, and in some cases, even more accurately. In this case r_{ij} values can be determined even more accurately. Since the overall molecular tumbling rates for DNA molecules can be independently measured using other methods, the NMR measurements can be directly used to evaluate the structural parameters. While internal motions are likely to have relatively little effect on the relaxation behavior of short DNA (~10 bp), in the larger DNA where overall tumbling is slow, high frequency, large amplitude internal motions will become important.

Combining Eq. (7) with Eqs. (2) and (3) we obtain

$$\frac{1}{T_1^s} = \frac{1}{2}\gamma^4\hbar^2\sum_j \frac{1}{r_{ij}^6}(J_0(0)+3J_1(\omega)+6J_2(2\omega)) \quad . \tag{8}$$

For most DNA samples we expect $J_0(0)$ to dominate the $J_1(\omega)$ and $J_2(\omega)$ terms and this considerably simplifies analysis of the selective relaxation rates, since the contributions from low amplitude, high frequency internal motions are suppressed. This point will be elaborated in more detail when we consider the ^1H relaxation measurements on short DNA restriction framents.[26,27] When the number of interacting protons is small (3 or 4), analysis of Eq. (1) shows that the relaxation behavior of any resonance following selective inversion will be non-exponential.[16] Furthermore, as the spin polarization diffuses away from the initial site, one expects to observe transient nuclear Overhauser effects in other resonances.

Some important characteristics which distinguish selective and non-selective spin-lattice relaxation measurements on DNA molecules when $\omega\tau_c \gg 1$ are summarized below.

Experimental Characteristics

Non-Selective T_1	*Selective T_1*
Varies with H^2	Independent of magnetic field, H
Longer than T_1^s	Shorter than T_1^{ns}
Uniform values for all interacting protons	Different initial values for each proton
Exponential decay	Non-exponential decay
Sensitive to high frequency internal motions, but insensitive to slow overall motions	Sensitive to overall motions of the molecule, but insensitive to low amplitude internal motions

Spin-Spin Relaxation Rates: The spin-spin relaxation rate, R_2, for the "i" spins in a multi-spin system is given by the following expression:[19,20]

$$R_2 = \frac{1}{T_2} = \frac{\gamma^2}{4}\hbar^2\sum_j \frac{1}{r_{ij}^6}[5J_0(0) + 9J_1(\omega) + 6J_2(2\omega)] \quad . \tag{9}$$

Since R_2 depends only on the nearest neighbor proton interaction, it, too, can be used to directly monitor the local proton environments. We now compare these general theoretical predictions with the experimental results obtained on double helical poly(dA-dT).

Investigation of the Poly(dA-dT) Structure by Relaxation Methods

Non-Exchangeable C-H Protons: The results of a non-selective T_1 measurement on a sample of poly(dA-dT) ~70 bp long are shown in Fig. 1 along with assignments of the various resonances. Other selective and non-selective relaxation measurements have also been carried out on poly(dA-dT) samples and these results are summarized in Table I. The relaxation rates measured by the different pulse techniques exhibit the behavior predicted. In particular, we note that the non-selective T_1 values are about the same for all non-exchangeable protons in the molecule and they vary approximately with the square of the applied magnetic field (Table I). These observations are clearly indicative of rapid spin diffusion and this is confirmed by an examination of T_1 values measured following selective polarization. In this case the relaxation rates are much shorter and vary from proton to proton. Because of the relatively high molecular weight (~70 bp) of the particular sample used in these experiments and the possible contribution from internal motions, a direct quantitative comparison of these results with theory is complex, and we defer discussion of the extraction of structural data from relaxation rates until the next section.

Relaxation Measurements on the Hydrogen Bonded T-imino Proton: We have recently shown that the soft pulse method can be used to make spin-lattice and spin-spin relaxation measurements on exchangeable protons of DNA and RNA in H_2O[28] and some of the results obtained on poly(dA-dT) are shown in Fig. 2. In these spectra, the single resonance located at 13.2 ppm is due to the slowly exchanging hydrogen bonded T-imino protons in Watson-Crick base pairs (see Fig. 3). In the spin-spin relaxation (T_2) measurements the decay is exponential at all temperatures studied, and these results are shown in Fig. 4. According to theory, the main contribution to T_2 arises from the $J_0(0)$ spectral density term which in turn is expected to be proportional to the correlation time for end-over-end tumbling of the molecule. The ~2-fold decrease in the R_2 with increasing temperature can, therefore, be attributed to a decrease in the molecular tumbling rate.[27]

In the selective spin-lattice relaxation measurements, the decay of the polarization was found to be non-exponential at temperatures below 25°C and an example of this for the 1°C data is shown in Fig. 5. A summary of results obtained at other temperatures is shown in Fig. 4. The non-exponential decay at the lower temperatures can be understood as follows. Since the 180° pulse selectively polarizes only the lowfield T(N-H) proton, the initial decay can be identified with Eq. (7) in which case the relaxation is due primarily to a through-space dipolar interaction with an amino proton of adenine located 2.6Å away, and to a lesser degree through interaction with the A-H_2 proton (Fig. 3). However, as a result of rapid cross-relaxation those protons which interact with the T-imino proton also become polarized and the decay becomes slower as the spin polarization becomes more uniformly distributed among the interacting protons. At longer times, the decay is more appropriately described by Eq. (5) for a non-selective T_1 measurement, because all three (or possibly 4 counting the A-H_2 proton) interacting protons are polarized.

The effect of temperature on the T_2 and on the fast and slow components of the T_1 spin-lattice relaxation rates are shown in Fig. 4. At the lower temperatures, the behavior of the fast component of the spin-lattice relaxation R_1(fast) can be de-

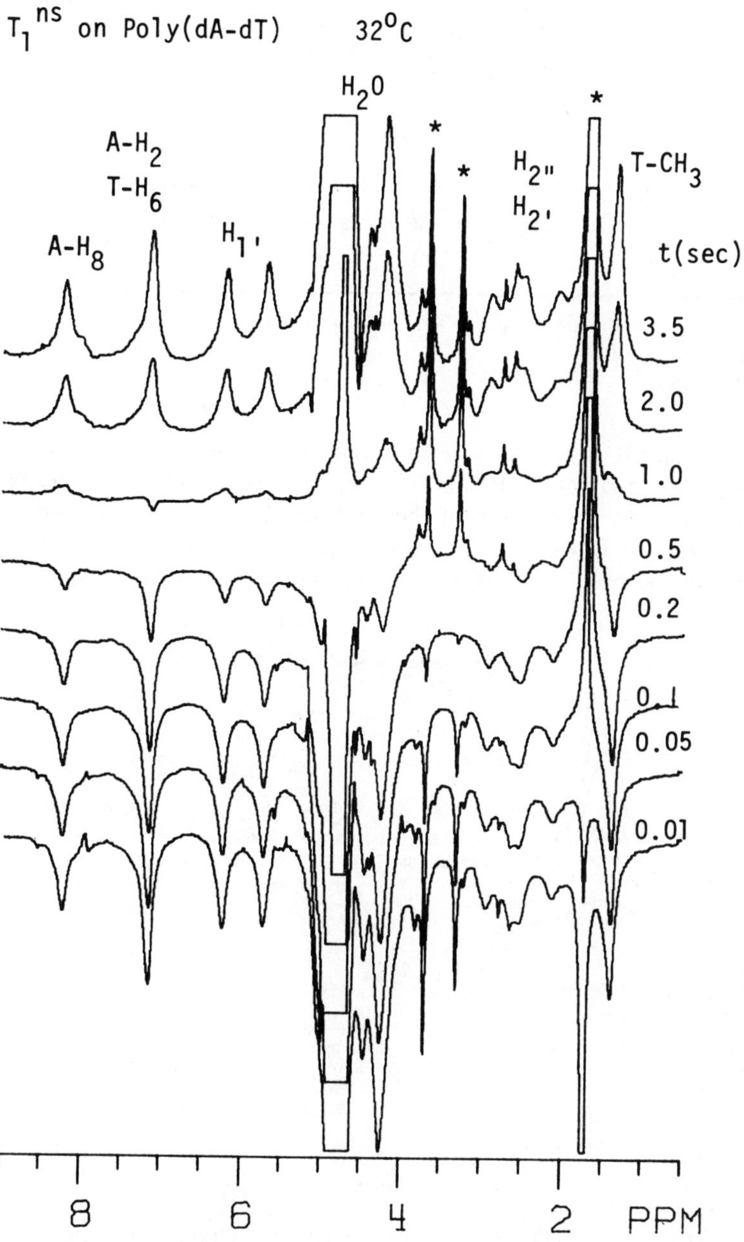

Figure 1. A non-selective spin-lattice relaxation measurement on poly(dA-dT) obtained using the standard inversion-recovery (180°-t-90°) pulse method. The sample (~70 bp in length) was dissolved (5 mg/0.1 ml) in 0.25 M NaCl, 25 mM cacodylate buffer at pH 7.

Table I
A Summary of the ^1H Relaxation Times of poly(dA-dT)[i]

PROTON	Relaxation Times (sec) T_1(ns) 300 MHz	T_1(ns) 200 MHz	T_2(ns) 200 MHz	T_1(s) 200 MHz	$T_1(s)/T_2$*	T_1(ns)(300)/T_1 (ns)(200)**
A(H_8)	1.59	.93	.056	.095	1.7	1.7
A(H_2)	1.75	1.04				1.7
T(H_6)	1.75	1.04				1.7
T($H_{1'}$)	1.69	.81	.050	.12	2.4	2.1
A($H_{1'}$)	1.80	.84	.046	.12	2.6	2.1
T(CH_3)	1.46	.92	.078	.15	1.9	1.6

[i]The measurements were carried out at 38°C on 70 bp material.
*Theoretical value ~2.5.
**Theoretical value ~2.25 if pure DD.

scribed by Eq. (8). Since the J_0 term is expected to be the major contributor, the decrease in R_1(fast) between 1 and 22°C is expected to parallel that observed for R_2. Since the slow component of the spin-lattice relaxation rate can best be described by Eq. (5), R_1(slow) is primarily determined by the magnitudes of $J_1(\omega)$ and $J_2(2\omega)$ which are predicted to vary inversely with the correlation time τ_c for molec-

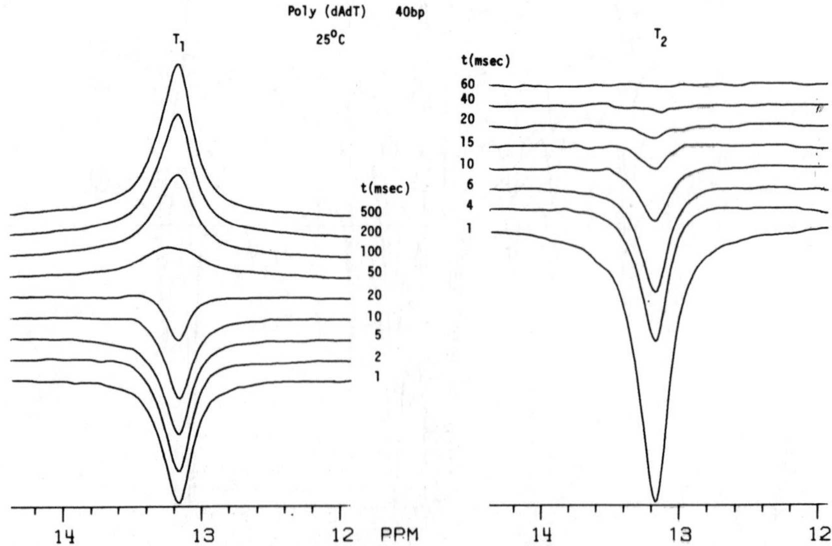

Figure 2. Examples of spin-lattice and spin-spin relaxation measurements on a ~40 bp sample of poly (dA-dT) obtained by the soft pulse method. In the T_1 measurement the 180°-t-90° pulse sequence was selectively applied at 13.2 ppm. The sample was dissolved (5 mg/0.1 ml) in 0.31 M NaCl and 15 mM cacodylate buffer, pH 7.

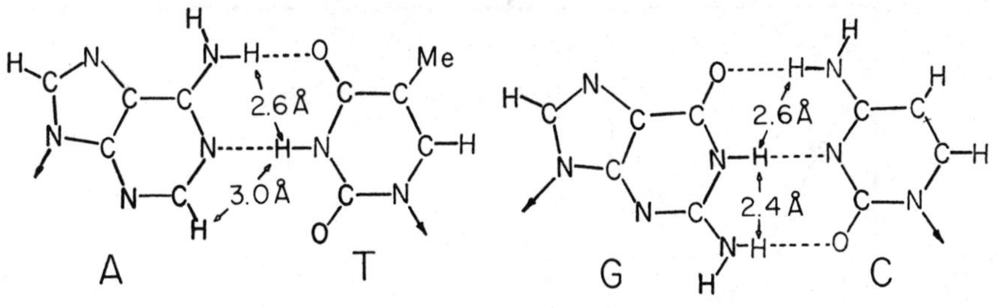

Figure 3. The standard Watson-Crick Base Pairs

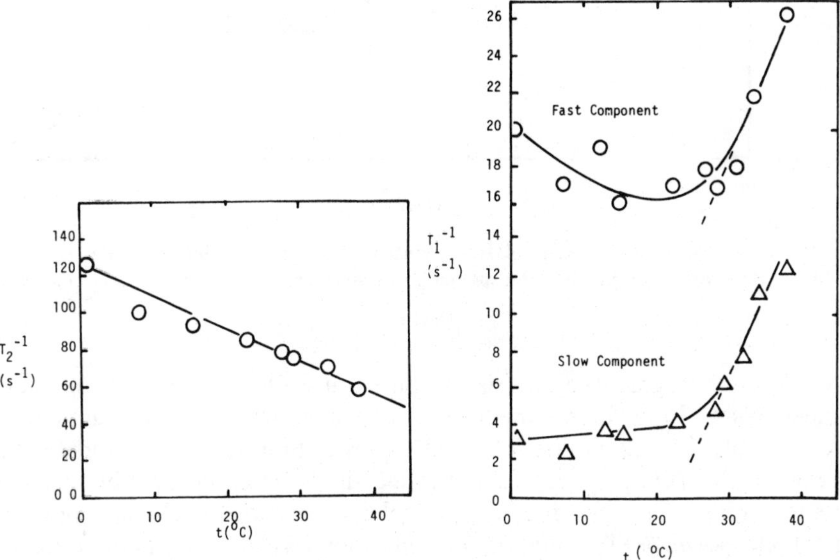

Figure 4. The temperature dependence of the spin-lattice and spin-spin relaxation rates obtained on a ~40 bp sample of poly(dA-dT). The contributions from the slow and fast components of the spin-lattice relaxation are separately plotted . For experimental conditions see Fig. 2.

ular tumbling (in the $\omega\tau_c > 1$ limit). The slight increase in R_1 (slow) with temperature might be attributed to a temperature effect on τ_c.

Above 25°C (Fig. 4) there is a dramatic increase in the spin-lattice relaxation rates due to the exchange of the T-imino protons with solvent protons. By 35°C the exchange mechanism totally dominates the spin-lattice relaxation. Completely analogous results have been previously observed for the relaxation behavior of the T-imino resonances in DNA restriction fragments varying in length from 12 to 69 bp.[27] Interestingly enough the activation energy of ~17 kcal for the high temperature exchange in poly(dA-dT) is close to the value ~16 kcal obtained in our restriction fragment studies.[27] Furthermore, the value of the exchange rate for poly(dA-dT) (k $\simeq$25 sec^{-1} at 38°C) corresponds well with the value of 18 sec^{-1} observed in the

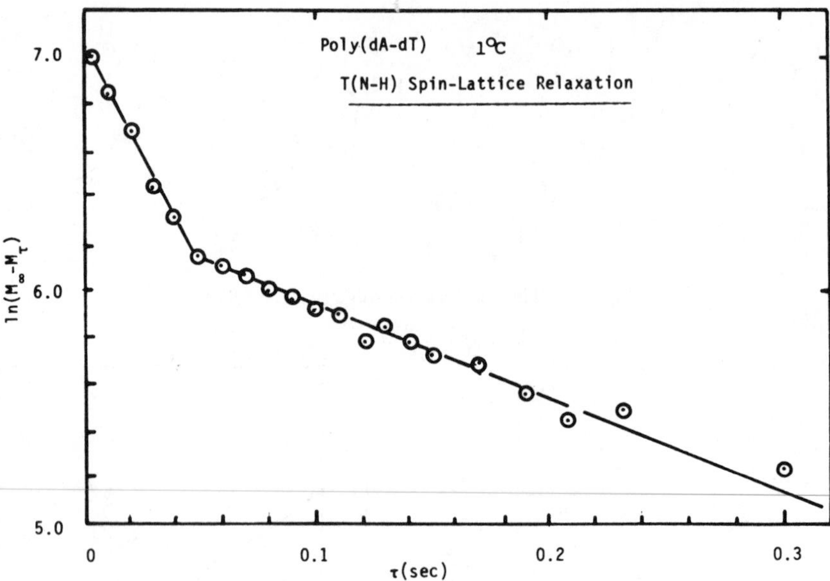

Figure 5. The results of a spin lattice relaxation measurement on the imino resonance (13.2 ppm) in poly(dA-dT) illustrating the highly non-exponential decay at 1°C. For experimental conditions see Fig. 2.

DNA restriction fragments at this same temperature. On this basis we conclude that the mechanism of the "high temperature" exchange process is the same in native DNA as in poly(dA-dT). Studies of the DNA restriction fragments indicate that the exchange of the T-imino proton occurs via a single base pair opening mechanism without opening of neighboring base pairs,[27] and the same mechanism presumably applies to the poly(dA-dT) system. Our results, therefore, do not support a "travelling-loop" mechanism for exchange which had been proposed for homopolymers or alternating systems such as poly(dA-dT) or poly(dG-dC).[29]

Although the temperature dependence of the decay of the spin-polarization in this simple DNA exhibits surprisingly complex behavior, most of the features can be qualitatively understood. The following analysis suggests that many aspects may even be quantitatively understood. According to theory,[26] the proton-nitrogen dipolar interaction makes a significant contribution to the spin-spin relaxation rates of the T-imino protons, but *not* to their spin-lattice relaxation rate. As a result of this large nitrogen contribution, we predict that R_2 will be 5.7 times larger than R_1. In the low temperature regime where magnetic interactions dominate the relaxation rates we find experimentally that $R_2/R_1 = 125/20 = 6.3$. The magnitudes of the magnetic contributions to the initial value of R_1 and to R_2 are also in accord with the values observed for a 43 bp DNA restriction fragment.[27]

While we restricted our initial studies of poly(dA-dT) to three different types of relaxation experiments (measurement of selective T_1 and non-selective T_1 and T_2) it is evident from these limited results that relaxation measurements are a rich source of information about the structure of this DNA in solution. However, the determination of the precise solution state structure of this molecule in solution will have to await more extensive studies on shorter fragments, analogous to those discussed in the next section, and to the application of more sophisticated two dimensional relaxation experiments.[30]

^1H Relaxation Measurements on a 12 Base Pair DNA Restriction Fragment

One reason for carrying out NMR studies on short DNA helices, in addition to the improved spectral resolution, is that these molecules are small enough so that an isotropic rotor model can be used to describe the overall motion of the molecule.[26] More importantly, however, the correlation time for overall rotation of small molecules is so short that internal motions are likely to be of minor importance in affecting the relaxation rates. Thus, the relaxation data can reasonably be used to extract structural information without having to deal with the formidable problem of developing the correct model to describe internal motions in the DNA.[31] When internal motions have to be introduced, as in the treatment of the relaxation behavior of the ^{31}P relaxation data, [32-35] then considerable uncertainties arise from unknown structural factors, amplitudes and directions of the internal motions.[31] While it remains to be proven, we believe that internal molecular motions make relatively little contribution to the R_2 and selective R_1 ^1H relaxation rates of short (12 bp or less) DNA.[26] This proposal may be checked by carrying out measurements over a wider range of temperatures, and by purposely increasing the viscosity of the sample to enhance the contributions from internal motions to the relaxation mechanisms. We now discuss the results of some relaxation experiments we have carried out on short DNA restriction fragments.

We have examined the proton NMR relaxation properties of three different DNA restriction fragments containing 12, 43, and 69 bp respectively,[26,27] The most extensive studies were carried out on the 12 bp fragment and the 600 MHz proton NMR spectrum of this molecule is shown in Fig. 6 along with assignments.

Extraction of Structural Information from Relaxation Rate Measurements on the 12 bp Fragment. Although assignments of some of the non-exchangeable resonances to specific base pairs are uncertain, all resonances in the 12-mer spectrum can be assigned to specific types of protons (e.g. A-H_2) and, therefore, relaxation measurements on these resonances can provide information about the structure of this molecule in solution. The results of a Hahn spin-echo experiment on the non-exchangeable protons at 500 MHz are shown in Fig. 7 and these results along with other spin-lattice relaxation rate measurements are summarized in Table II. In a non-selective T_1 measurement on the non-exchangeable protons, we observed that virtually all of the protons had the same relaxation rate (about 0.8 sec^{-1} at 300 MHz) in contrast with the larger and variable relaxation rates observed in selective relaxation rate T_1 measurements in which just the aromatic resonances were initially polarized. This comparison shows that, even with this short DNA molecule,

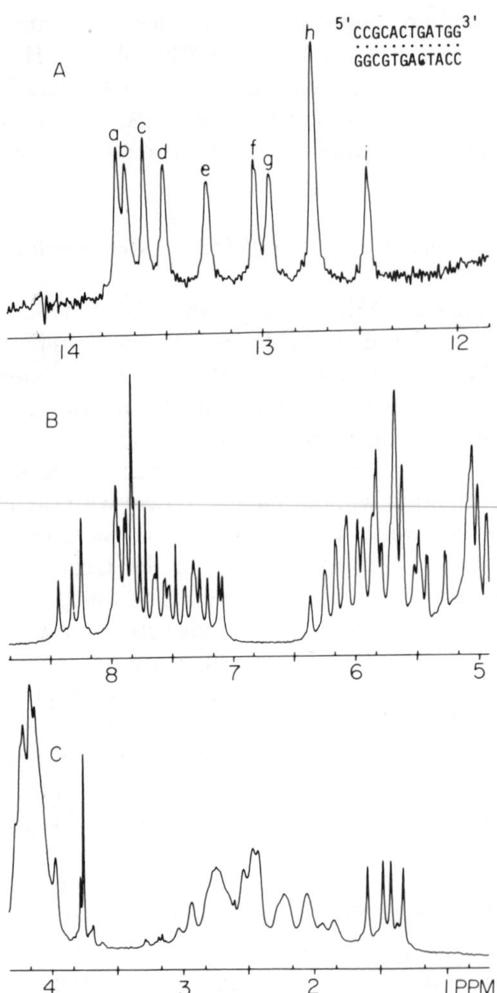

Figure 6. The 600 MHz proton NMR spectrum of a 12 base pair DNA restriction fragment with the sequence shown. (A) shows the lowfield spectrum taken in H_2O, (400 scans), (B) and (C) show the nonexchangeable proton spectrum in D_2O (800 scans). The probe temperature was 20°C. This sample contained material dissolved in 0.12 ml of solution containing 350 mM NaCl, 85 mM phosphate at pH = 7.3. The assignments of the resonances in the spectra are as follows. In the lowfield spectrum the four resonances between 13.8 and 13.3 ppm are due to the T-imino protons of the four A•T base pairs and the remaining resonances between 13.3-12.0 ppm are due to G-imino protons in G•C base pairs. In B, the resonances between 8.5 and 7.0 ppm are due to the base protons (approximately in the order A-H_8, G-H_8, C-H_6, A-H_2 and T-H_6) and resonances between 6.5-5 ppm are a mixture of $H_{1'}$ sugar and C-H_5 resonances and other ribose protons. In the highfield spectrum (C) the large collection of resonances around 4 ppm corresponds to ribose protons ($H_{5'}$, $H_{5''}$, $H_{3'}$, and $H_{4'}$). The $H_{2'}$ and $H_{2''}$ protons are located between 3.5 to 1.7 ppm and the four thymine methyl protons are responsible for the four highest field peaks in the spectrum.

cross-relaxation among most protons is rapid compared with transfer of spin polarization to the lattice. One important goal of these studies is to use the relaxation data to extract information about the DNA structure in solution. The 12 bp

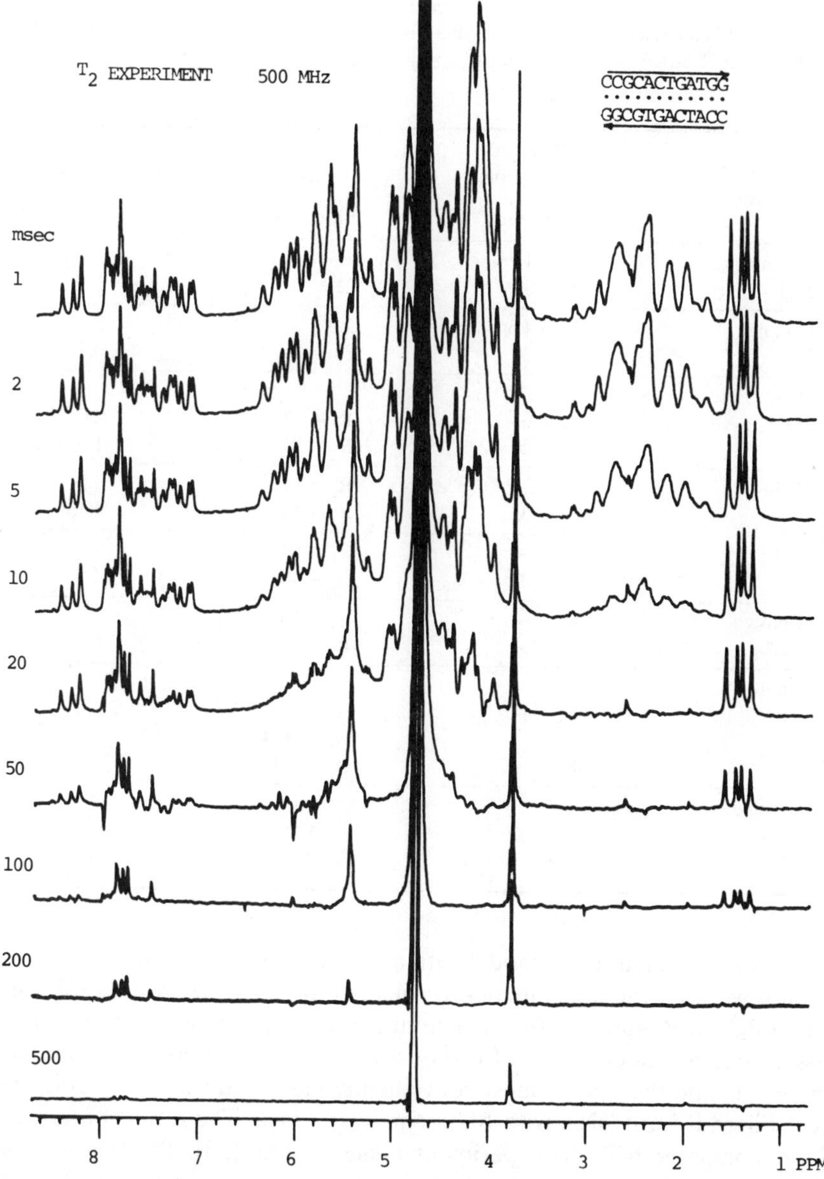

Figure 7. A spin-spin relaxation rate measurement (at 500 MHz) on non-exchangeable protons in the 12 base pair fragment at 78°C in D_2O. The Hahn spin echo method (90°-τ-180°-τ) was used and τ delays are indicated to the right of each spectrum. The peak at 3.75 ppm is due to the buffer.

DNA fragment is ~40Å long and ~25Å thick and theory predicts that the rotational correlation time for end over end tumbling will be about 7 nsec, compared with about 4 nsec for the axial reorientation.[26] When these values are used with the Woessner expressions for a rigid *anisotropic* rotor to calculate the spectral densities, we find that the results obtained are virtually identical with those obtained

Table II

A Comparison of Semi-Selective Proton Spin-Spin (R_2) and Spin-Lattice (R_1) Relaxation Rates Measured for a 12 Base Pair Restriction Fragment at 21°C with Rates Computed Using an Isotropic Rotor Model ($\tau_c = 7$ nsec):

Proton of Interest	Observed Relaxation Rates, sec^{-1}	Calculated Relaxation Rates, sec^{-1}	Ratio Calc./Obs.
Thymine N-H	$R_1 = 1.7$ $R_2 = 20$	1.5 21	0.88 1.05
Guanine N-H	$R_1 = 4.5$ $R_2 = 26$	3.0 25	0.66 0.97
Purine H-8	$R_1 = 6$ $R_2 = 20 \pm 2$	$R_1 = 12$ $R_2 = 33$	2.0 1.6±0.2
C(H-6)	$R_1 = 7.2$ $R_2 = 35$	$R_1 = 14$ $R_2 = 40$	2.0 1.15
C(H-5)	$R_1 = -$ $R_2 = 33$	$R_1 = 14$ $R_2 = 40$	1.2
Ribose (H-1′)	$R_1 = -$ $R_2 = 33$	$R_1 = 10.5$ $R_2 = 29$	0.90
Ribose 2′ (or 2″)	$R_1 = -$ $R_2 = 62$	$R_1 = 18(25)$ $R_1 = 9(14)$ $R_2 = 53(70)$	0.86 (1.1)
Adenine (H-2)	$R_1 = <0.5$ $R_2 = <10$	$R_1 = 0.14$ $R_2 = 0.56$	>0.4 >.13

using an isotropic rigid rotor model with a 7 nsec correlation time. For simplicity, therefore, we used the latter model to calculate the expected values for both the selective R_1 and R_2 values.[26] In order to make such a calculation, however, it was necessary to first choose values for the various relevant inter-proton distances in the molecule and these were most conveniently measured using a Dreiding model for a standard B form DNA with 10 base pairs per turn. The results of the calculations are compared with the experimental values in Table II. The fact that relaxation rates calculated using this set of assumptions agree rather well with the observed values suggests that the distances which we used in the calculations are not seriously in error. In fact, relatively minor changes in some of the distances would permit us to obtain perfect agreement between theory and experiment. Because of the 6th power dependence of the relaxation rates on the interproton distances, a 20% error in an important interproton distance would lead to a 300% change in the predicted relaxation rate. On the basis of the present, albeit limited data, we find no evidence that the time averaged conformation of the 12 bp fragment deviates very significantly from B-form geometry, although at the present stage of refinement some interproton distances could be in error by 10%. A more accurate determination of the structure will require a variety of bi-selective spin-lattice relaxation measurements and two dimensional spin exchange measurements such as those recently used in our study of the structure of poly(C).[30]

A Search for Sequence Effects on the Conformation: In the spectra of the 12-mer, each type of base proton gives rise to four separate resonances (one from each of the four A•T base pairs). In particular, resonances from the four separate methyl protons, A-H_2, T-H_6 and the A-H_8 protons are separately resolved. By examining the partially relaxed spectra obtained in a Hahn spin-echo experiment, the relative relaxation rates of the four protons within each group can be accurately compared and these results are shown in Fig. 7 for spectra obtained at 500 MHz. An expanded version of the 50 msec data, shown in Fig. 8, reveals the following interesting features. First, one or possibly both of the A-H_8 resonances located at about 8.2 ppm decay slightly faster than the two lower field A-H_8 resonances. Virtually no differences are discerned in the relaxation behavior of the four T-H_6 resonances which are located between 7.0-7.3 ppm. Expansion of the 200 msec spectrum (not shown) further reveals that of the four very slowly relaxing A-H_2 resonances (located between 7.4-7.9 ppm) the highest field resonance has a slightly longer relaxation rate. The four T-methyl protons have identical relaxation rates. These data demonstrate tha the time averaged conformation of the base pairs surrounding the four different A•T base pairs in this 12 bp fragment are nearly identical. The NMR

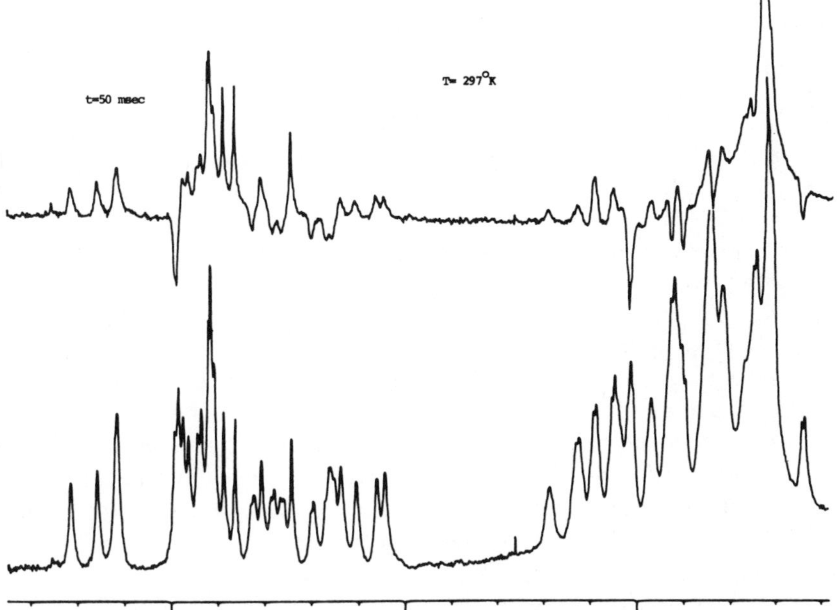

Figure 8. An expanded version of the 1 and 50 msec relaxation data shown in Fig. 7 illustrating small sequence effects on the relaxation properties of some resonances. The inverted peaks in the spectrum arise from C-H_5 and C-H_6 resonances which are phase modulated via a scalar interaction.

studies of this molecule provide, therefore, little evidence for sequence effects on conformation, although studies on other DNA clearly indicate that pronounced sequence effects can exist.[36,37]

There is another feature in the T_2 relaxation measurement that deserves mention. In the 50 msec spectrum there are a number of negative peaks between 5-8 ppm. These inverted resonances arise from a phase modulation of the spin-echo due to scalar interactions between the C-H_6 and C-H_5 protons.[38] This immediately permits us to identify the resonance positions for all of the C-H_6 and C-H_5 protons in the molecule. While this has been done by more tedious decoupling experiments, the spin-echo experiment has obvious advantages. Furthermore, this technique may be of considerable value in detecting very small coupling constants which would otherwise be unresolved. Theoretically[38] the maximum intensities in the inverted peaks occur at a delay time of $\tau = 1/2J$, which, in the present case, should be between 60-70 msec.[39] By a careful examination of the appropriate time range, the coupling constants could be more precisely determined.

Although our relaxation studies of the 12 bp fragment are far from complete, they clearly indicate that the 1-dimensional relaxation technique described here, and 2-dimensional relaxation techniques described elsewhere, may finally provide the necessary tools for precisely determining the structures of short DNA molecules in solution. Now that large quantities of short DNA of any desired sequence can be chemically synthesized, we anticipate very rapid progress in this field.

Use of ^{31}P NMR Relaxation Measurements to Investigate Metal Ion Binding to DNA

In the preceeding section, we described the use of proton relaxation techniques to probe DNA structure in solution. ^{31}P NMR has also been used for this purpose, but because of the physical proximity of the phosphorus to the $H_{5'}$, $H_{5''}$ and $H_{2'}$ sugar protons, the ^{31}P relaxation data provide only limited information about the DNA conformation.[32-35] Furthermore, because the zero quantum transitions involve cross-relaxation between the ^1H and ^{31}P nuclei, the leading term in the expression for the ^{31}P spin-lattice relaxation rate resulting from dipolar interactions with protons involves the spectral density $J_0(\omega_P-\omega_H)$. This term is primarily sensitive to high frequency internal motions, and it is virtually insensitive to the slower overall motions of the molecules.[32] Consequently, extraction of structural data from the ^{31}P spin-lattice relaxation rate measurements requires the development of a model for internal motions in the backbone, and it now appears that there may not be a unique solution to this problem.[31] The data are, however, useful for obtaining information about the time constants for internal motion and a number of papers on this subject have appeared during the past two years.[32-35]

While ^{31}P NMR relaxation measurements may be of somewhat limited use in structural studies, they provide an excellent method for investigating the interaction of DNA with paramagnetic divalent metal ions.[40-42] ^{31}P NMR measurements of metal ion binding are sensitive only to those paramagnetic ions which are directly coordinated with the phosphate groups (i.e. inner-sphere) or bound through an intervening water molecule (outer-sphere complex). We have investigated the interaction of DNA with the paramagnetic Mn^{2+} ion, and from these measurements we are able to quantitatively evaluate the extent of inner-sphere binding to the phosphate groups, set limits on the amount of outer-sphere binding, and determine the kinetic and some of the thermodynamic properties of the inner-sphere metal-DNA complexes.[43,44] According to theory, the measured paramagnetic contribution of Mn^{2+} to the longitudinal (T_{1p}) and transverse (T_{2p}) ^{31}P relaxation times are related to the intrinsic bound state values (T_{1M} and T_{2M}, respectively) through the relation[42]

$$\frac{1}{T_{ip}} = \frac{f}{T_{iM} + \tau_M} \qquad i = 1,2 \tag{10}$$

where f is the ratio of the concentration of the metal bound ligand to that of the total ligand concentration and τ_M is mean lifetime of the metal ligand complex.

$$T_{1M}^{-1} = \frac{C^6}{r^6} \cdot \frac{3\tau_c}{1+\omega_I^2\tau_c^2} \tag{11}$$

$$T_{2M}^{-1} = \frac{C^6}{2r^6}\left(4\tau_c + \frac{3\tau_c}{1+\omega_I^2\tau_c^2}\right) + \frac{2}{3}S(S+1)(A/\hbar)^2\tau_e \tag{12}$$

$$\tau_c^{-1} = \tau_R^{-1} + \tau_e^{-1} \quad , \quad \tau_e^{-1} = \tau_M^{-1} + \tau_{1S}^{-1} \tag{13}$$

where τ_R is an isotropic rotational correlation time, τ_{1S} is the longitudinal electron spin relaxation time, r is the metal-phosphorus internuclear distance in Å, and C is a constant equal to 601 for Mn^{2+}—^{31}P interaction. The frequency dependence of τ_{1S} is given by

$$\tau_{1S}^{-1} = B\left(\frac{\tau_v}{1+\omega_S^2\tau_v^2} + \frac{4\tau_v}{1+4\omega_S^2\tau_v^2}\right) \tag{14}$$

where B is a constant related to the zero field splitting and τ_V is the correlation time for modulation of the zero field splitting. The effect of Mn^{2+} on ^{31}P longitudinal and transverse relaxation rates of DNA (25.2 mg/ml), measured at 40.5 MHz, is shown in Fig. 9. The observed linear increase in the relaxation rates with Mn^{2+} concentration under conditions where $[DNA(P)] \gg K_D$[44] indicates that all the Mn^{2+} is bound to the DNA, and from these data we calculated bound state relaxation rates of $1/fT_{1p} = 3.1x10^4\ sec^{-1}$ and $1/fT_{2p} = 3.8x10^5\ sec^{-1}$.[43] The variation with temperature of the net paramagnetic contributions to the relaxation rates were measured between

1-40°C and a maximum was observed for T_{1p}^{-1} at $10^3/T$ ~3.3 suggesting that $\tau_c \sim \omega_I^{-1}$ = 3.9 nsec at this temperature. Measurements at 81 MHz yielded T_{1p}^{-1} (40.5 MHz)/T_{1p}^{-1} (81 MHz) = 2.5 and assuming that the correlation time is frequency independent, we obtained $\tau_c = 4 \pm 1$ nsec. The frequency independence of the correlation time is consistent with τ_c being much shorter than the electron spin relaxation time of the DNA-bound Mn^{2+}[25,44]

The variation of T_{2p}^{-1} with temperature (data not shown) is typical for an exchange limited relaxation, i.e. $\tau_M \gg T_{2M}$ and $1/fT_{2p} = \tau_M^{-1}$ = 3.8 x 10^5 sec^{-1}. Using this value for τ_M and Eq. (10), T_{1M}^{-1} is calculated to be 3.4 x 10^4 sec^{-1}. This value, together with the correlation time determined above and Eq. (11), can be used to calculate the Mn^{2+}—P distance. Assuming each Mn^{2+} ion interacts only with one or two phosphates at a time, the effective internuclear distance is found to be 4.5±0.5 or 5.0±0.5Å, respectively. This value is intermediate between the value expected for an outer-sphere complex (5.8-6.0Å) or for an inner-sphere complex (3.0-3.3Å)[45,47] and can be used to set limits on the fraction of Mn^{2+} ions which participate in inner-sphere association. A maximum value of 16% is obtained by assuming that there is no outer-sphere binding. However, if we assume that all of the Mn^{2+} ions are bound either as inner-sphere or as outer-sphere complexes, then a minimum value of 4% for the inner-sphere binding can be deduced from the observed relaxation data. A value of ~15% is estimated from the T_2 data.[43]

The competition of Mn^{2+} and Mg^{2+} was also studied by measuring the effect of Mg^{2+} on the paramagnetic contribution to the ^{31}P longitudinal relaxation rate of DNA (25.5 mg/ml) in the presence of Mn^{2+} (40 and 90 μM), and these results are shown in Fig 10. At low Mg^{2+} levels ($\beta \equiv [Mg^{2+}]_T/[DNA(P)] \lesssim 0.15$), T_{1p}^{-1} increases

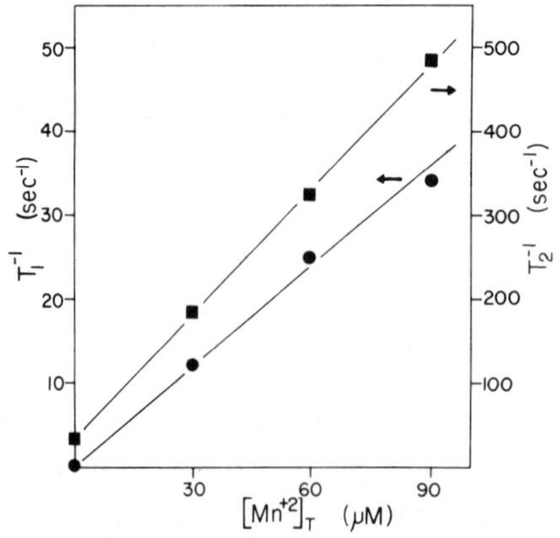

Figure 9. Effects of Mn^{2+} on the longitudinal (circles) and transverse (squares) relaxation rates of the phosphorus nuclei of DNA (25.2 mg/ml) at 27°C, 40.5 MHz.

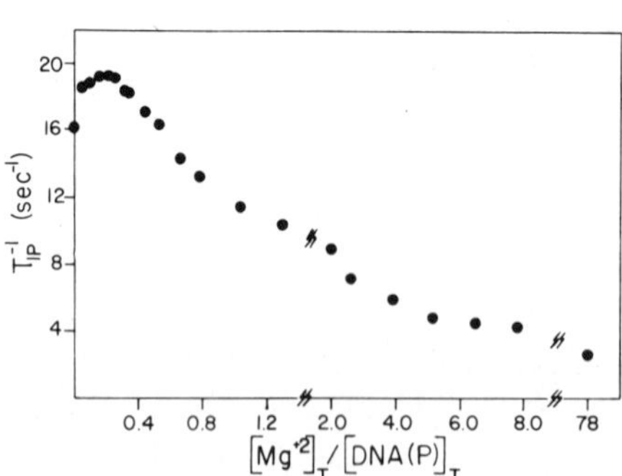

Figure 10. Effect of Mg^{2+} on T_{1p}^{-1} of the phosphorus nuclei of DNA (25.5) mg/ml) in the presence of 40 μM Mn^{2+}.

by up to ~20%, and this is attributed to displacement of Mn^{2+} ions from terminal phosphate groups. In the DNA studied (~150 bp) terminal phosphates have low abundance (~0.7%) but a higher affinity for divalent metal ions than the internal phosphates.[45,48] Displacement of Mn^{2+} ions from the terminal sites by excess of Mg^{2+} ions results in an increased relaxation rate of the *internal* phosphates due to the binding of the released Mn^{2+} ions at the latter sites. In the second region ($0.15 \lesssim \beta \lesssim 0.25$), T_{1p}^{-1} remains approximately constant and the added Mg^{2+} ions appear to bind to DNA without displacing Mn^{2+} ions bound to the bulk of the internal phosphates.

The competitive effects in the third region can be used to determine the relative affinities of DNA for Mn^{2+} and Mg^{2+}, and an analysis of the data shown in Fig. 10 indicates that the binding constant for Mn^{2+} is about 3.5 times greater than for Mg^{2+}, in agreement with results obtained by other techniques.[49]

The above studies provide the first accurate assessment of the amount of inner-sphere binding exhibited by DNA-Mn^{2+} complexes, and our results show that it is a small (~15%) percentage of the total amount of Mn^{2+} bound. Most of the Mn^{2+} ions are, therefore, bound to DNA as outer-sphere complexes in which the Mn^{2+} ions retain their waters of hydration, or less likely, are in association with the bases.[44] Evaluation of this latter possibility will require studies of the effect of Mn^{2+} on DNA proton relaxation rates. Entirely analogous results were obtained with Co^{2+} binding. It is interesting to note that Gueron and Leroy[48] have carried out extensive studies of Mn^{2+} binding to transfer RNA with entirely similar results. The measured average Mn^{2+}—P distance for tRNA was estimated to be about 4.6Å, compared with 4.5Å obtained here for DNA, and the correlation time for the ^{31}P paramagnetic relaxation of Mn^{2+}—tRNA was found to be 5×10^{-9} sec, compared with 4×10^{-9} sec for Mn^{2+}—DNA. At 30°C, $\tau_M = 2.5 \times 10^{-6}$ sec was obtained for Mn^{2+}—tRNA, whereas we find 2.6×10^{-6} sec for Mn^{2+}—DNA. Evidently the mode of binding of Mn^{2+} to DNA and tRNA is very similar. In the present study, the ^{31}P relaxation measurements only provided information about metal ions closely associated with the phosphate groups. By using proton relaxation measurements on water, the interaction of paramagnetic metals with DNA can be monitored via its effect on water proton relaxation. The results of these experiments are described elsewhere,[44] and relaxation measurements on the DNA protons (now in progress) will eventually provide a rather complete picture of the metal binding properties of DNA.

Acknowledgements

The support of the National Science Foundation (PCM 7911571), the U.S. Public Health Service (GM 22969) and the American Cancer Society (CH-32) is most gratefully acknowledged. Some measurements were carried out at the Southern California Regional NMR Facility (supported by National Science Foundation Grant No. CHE 7916324A1).

References and Footnotes

1. Kearns, D.R., *Ann. Rev. Biophys. Bioeng. 6,* 477 (1977).
2. Hillen, W., Klein, R.D. and Wells, R.D., *Biochemistry 20,* 3748 (1981).
3. Klysik, J., Stirdivant, S.M., Larson, J.E., Hart, P.A. and Wells, R.D., *Nature* (1981), in press.
4. Broka, C., Hozumi, T., Arentzen, R. and Itakura, K., *Nuc. Acids Res. 22,* 5461 (1980).
5. Haasnoot, A.G.C., den Hartog, J.H.J., de Rooij, J.F.M., van Boom, J.H. and Altona, C., *Nature 281, 235 (1979).*
6. *Miller, F.S., Cheng, D.M., Dreon, N., Jayaraman, K., Kan, L.-S., Leutzinger, E.E., Pulford, S.M. and Ts'o, O.P., Biochemistry 19,* 4688 (1980).
7. Rill, R.L., Hilliard, Jr., P.R., Bailey, J.T. and Levy, G.C., *J. Am. Chem. Soc. 102,* 418 (1980).
8. Tatchell, K. and Van Holde, K.E., *Biochemistry 16,* 5295 (1977).
9. Muller, L. and Ernst, R.R., *Mol. Phys. 38,* 963 (1979).
10. Meier, B.H. and Ernst, R.R., *J. Am. Chem. Soc. 101,* 6441 (1979).
11. Nagayama, K., Kumar, A., Wüthrich and Ernest, R.R., *J. Chem. 40,* 321 (1979).
12. Jeener, J., Meier, B.H., Bachmann, P. and Ernst, R.R.. *J. Chem. Phys. 71,* 4546 (1979).
13. Nagayama, K., Bachmann, P., Wuthrich, K., and Ernst, R.R., *J. Magn. Reson. 31,* 133 (1978).
14. Emid, S., Bax, A., Konijnendijk, J. and Smidt, *J., Physica 96B,* 333 (1979).
15. Wells, R.D., Goodman, T.C., Hillen, W., Horn, G.T., Klein, R.D., Larson, J.E., Müller, U.R. Neuendorf, S.K., Panayotatos, N. and Stirdivant, S.M. in *Progress in Nucleic Acid Research and Molecular Biology,* Vol. 24, Academic Press, Inc., New York, pp. 168-267 (1980).
16. Noggle, J.H., and Schirmer, R.E., *The Nuclear Overhauser Effect: Chemical Applications,* Academic Press, New York, 1971.
17. Leipert, T.K., Noggle, J.H., Freeman, W.J. and Dalrymple, D.L., *J. Magn. Reson. 19,* 208 (1975).
18. James, T.L. and Matson, G.B., *J. Magn. Reson. 33,* 345 (1979).
19. Kalk, A., and Berendsen, H.J.C., *J. Magn, Reson. 24,* 343 (1976).
20. Abragam, A., *Principles of Nuclear Magnetism,"* The University Press, Oxford, 1978.
21. Woessner, D.E., *J. Chem. Phys. 37,* 647 (1962).
22. London, R.E. and Avitabile, J., *J. Chem. Phys. 65,* 2443 (1976).
23. Wittebort, R.J. and Szabo, A., *J. Chem. Phys. 69,* 1722 (1978).
24. Levine, Y.K., Birdsall, N.J.M., Lee, A.G., Metcalfe, J.C., Partington, P. and Roberts, G.C.K., *J. Chem. Phys. 60,* 2890 (1974).
25. Ribeiro, A.A., King, R., Restivo, C. and Jardetzky, O., *J. Am. Chem. Soc. 102,* 4040 (1980).
26. Early, T.A., Kearns, D.R., Hillen, W. and Wells, R.D., *Nuc. Acids Res. 8,* 5795 (1980).
27. Early, T.A., Kearns, D.R., Hillen, W. and Wells, R.D., *Biochemistry 20,* 3756, 3764 (1981).
28. Early, T.A., Feigon, J. and Kearns, D.R., *J. Magn. Reson. 41,* 343 (1980).
29. Teitelbaum, H. and Englander, S.W., *J. Mol. Biol. 92,* 55 (1975).
30. Broido, M.S. and Kearns, D.R., *J. Magn. Reson. 41,* 496 (1980).
31. Lipari, G. and Szabo, A., presented at the 25th Biophysical Society Meeting, February 23-25, Denver, Colorado, abstract p. 307a, 1981.
32. Bolton, P.H. and James, T.L., *J. Am. Chem. Soc. 102,* 25 (1980).
33. Klevan, L., Armitage, I.M. and Crothers, D.M., *Nuc. Acids Res. 6,* 1607 (1979).
34. Bolton, P.H. and James, T.L., *J. Phys. Chem. 83,* 3359 (1979).
35. Hogan, M.E. and Jardetzky, O., *Biochemistry 19,* 3460 (1980).
36. Pohl, F.M. and Jovin, T.M., *J. Mol. Biol. 67,* 375 (1972).
37. Pohl, F.M., Ranade, A. and Stockburger, M., *Biochim. Biophys. Acta 335,* 85 (1973).
38. Freeman, R. and Hill, H.D.W., in *Dynamic Nuclear Magnetic Resonance Spectroscopy,* Jackman, L.M. and Cotton, F.A. (Eds.), Academic Press, New York, pp. 131-162, 1975.
39. Cheng, D.M. and Sarma, R.H., *Am. Chem. Soc. 99,* 7333 (1977).
40. Luz, A. and Meiboom, S., *J. Chem. Phys. 40,* 2686 (1966).
41. Swift, T.J. and Connick, R.E., *J. Chem. Phys. 37,* 307 (1962).
42. Dwek, R.A., in *NMR in Biochemistry,* Oxford University Press (Clarendon) London and New York, 1973.
43. Granot, J., Feigon, J. and Kearns, D.R., *Biopolymers,* (1981), in press.
44. Granot, J. and Kearns, D.R., *Biopolymers,* (1981), in press.

45. Bean, B.L., Koren, R. & Mildvan, A.S., *Biochemistry 16,* 3322 (1977).
46. Mildvan, A.S., *Accts Chem. Res. 10,* 246 (1977).
47. International Tables for X-Ray Crystallography, Vol. 3, Physical and Chemical Tables, The Kynoch Press, Birmingham, 1962.
48. Gueron, M. & Leroy, J.L., in *ESR and NMR of Paramagnetic Species in Biological and Related Systems,* Bertini, I. & Drago R.S. (Eds.), D. Reidel Publishing Co., pp. 327, 1979.
49. Clement, R.M., Strum, J. and Daune, M.P., *Biopolymers 12,* 405 (1973).
50. Present address: Department of Chemistry, Colorado State Universtiy, Fort Collins, Colorado, 80523.

Proceedings of the Second SUNYA Conversation in the Discipline Biomolecular Stereodynamics
Volume I, ISBN 0-940030-00-4, Ed., Ramaswamy H. Sarma,
Adenine Press, New York, ©Adenine Press

Backbone Conformational Changes in DNA Restriction Fragments of Known Sequence. ^{31}P Relaxation Studies

Phillip A. Hart,* Charles F. Anderson,*
Wolfgang Hillen† and Robert D. Wells†
*School of Pharmacy and †Department of Biochemistry
University of Wisconsin
Madison, Wisconsin 53706

Introduction

The view that double helical DNA manifests considerable internal conformation flexibility is supported by a substantial body of recent data some of which are discussed in the sequel. Whether these motions are associated with premelting changes or "breathing" modes is uncertain. Recently we have observed temperature dependent phosphorus relaxation times in unique 43, 69, 84 and 180 base-pair (bp) DNA restriction fragments that reflect changes in conformation and/or internal mobility in the premelting temperature range. Those unique observations and their rationalization are the substance of this report.

Internal motion on the nanosecond time scale in calf thymus DNA has been invoked to rationalize the depolarization of fluorescence of intercalated ethidium bromide.[1] A theoretical model for this internal motion has been developed[2] and it is concluded that twisting and bending of the intact double helix contribute to the internal motion, but that the twisting mode dominates. Recent ESR studies of intercalated spin labels[3,4] indicate a time scale for internal motion comparable to that reported earlier[1] based on the double helix modeled as a number of rotatable coaxial disks.

The above methods monitor dynamics at the complementary base-pair sites. Although the fluorescent or spin labels are presumably sufficiently dilute so that gross perturbation of the DNA structure is minimized, it is probable that local structure (vicinal to the probe) is perturbed,[5] consequently, it is uncertain whether the motions sensed by these probes are truly (quantitatively) characteristic of unperturbed double-stranded DNA.

Various DNA's unperturbed by added ligands have been studied via ^{1}H, ^{13}C and ^{31}P NMR. In most cases, the relaxation times of these nuclei are consistent with rapid (10^{-7} - 10^{-10} sec) motions of the bases, the sugars and the phosphodiester backbone. Thus, the proton linewidths of salmon sperm or calf thymus DNA that had been subjected to S1 nuclease and deoxyribonuclease digestion suggest different internal mobilities of hydrogen bonded aromatic protons and sugar protons when the average number of bp is greater than 200.[6] The correlation times for internal motion reported are in the 10^{-7}-10^{-8} sec range and were estimated by comparing actual linewidths with linewidths computed for rigid ellipsoids.[7] No model was proposed for the internal motion. Phosphorus, carbon and hydrogen relaxation measurements on DNA fragments 150, 300 and 600 (average) base-pairs long have been interpreted under the assumption of independent and isotropic internal and overall motion; it was concluded that internal motion in the nanosecond range involves the coupling of sugar pseudorotation and phosphodiester motions.[8,9] A similar study of calf thymus DNA[10,11] gave phosphorus and carbon relaxation data that were consistent with long-range bending and internal wobbling about the P-O bonds on the sub-nanosecond time scale. In this case, a two-correlation time model was used for computation,[12] but explicit anisotropic overall motion was not included. An earlier ^{31}P relaxation study of 140 base-pair DNA indicated greater internal mobility than had previously been suggested but no specific model for this motion was proferred.[13]

In all of the above studies the assumption was made that all relaxation mechanisms were predominantly dipolar. That assumption is undoubtedly valid for proton-proton and carbon-proton (direct) interactions. The assumption is weaker in the phosphorus-proton case; however, it is based on phosphorus relaxation studies of deoxyribooligonucleotides that showed large phosphorus proton nuclear Overhauser effects.[14] Ths assumption is generally accepted at fields of 2.4 T and below, but recent studies suggest that the major competing relaxation mechanism for phosphorus (chemical shift anisotropy) may intrude at fields lower than had been considered previously.[15,16] This intrusion may be particularly troublesome when the extreme narrowing condition is violated. The authors of this latter work did not emphasize internal motion in their analyses. They adopted a rigid anisotropic model for overall motion and assumed a distribution of correlation times. It is possible, however, that they assumed too large a contribution from phosphorus chemical shift anisotropy; their computed correlation times for overall motion do not agree with those determined by electric birefringence[17] and computed phosphorus-proton NOE's are an order of magnitude smaller than those measured routinely.[8,11]

It is noteworthy that DNA of intact core particles from calf thymus chromatin has phosphorus relaxation characteristics similar to those of the separated 140 bp DNA.[18,20]

Substantial effort has been devoted during the last several years to the construction of a useful model relating dipolar relaxation rates and NOE's to the dynamics of small oligoribonucleotides (Hart and Anderson, unpublished) and to the development of appropriate experimental methods for the purpose of conformational analysis of the phosphodiester backbone[21] (Hart and Yang, unpublished). It is clear from these studies that computed phosphorus relaxation parameters reflect the choice of rotamer distributions throughout the backbone as well as the choice of redistribution rates and that the parameter values are model dependent. The model dependence extends to the choice of overall motional modes as well as internal modes and to the number of interacting protons. The set of three P-H interactions usually chosen is too small; at least five interactions are necessary in the ribo series and (see below) six in the deoxyribo series. However, it has been exceptionally difficult to know how to model the probable non-independence of internal *vs.* overall motion in the relatively small and conformationally-mobile systems. Systems in which one can safely assume that overall and internal motions are reasonably independent are defined sequence, DNA restriction fragments 25 to 200 bp long. The availability of these fragments in amounts that would allow moderately precise phosphorus relaxation studies prompted the initiation of an effort to characterize the phosphorus relaxation properties of these polymers. If one adopts the viewpoint that the overall motion of double helical DNA fragments shorter than one persistence length is suitably analogous to that of a "rigid rod", one can calculate relaxation times for nuclei positioned on that structure in a known way.[7] To simulate the observed relaxation properties one needs only a good model for internal motion and the mathematical means to superimpose that model upon anisotropic (and independent) overall motion. This report details a preliminary study of the length and temperature dependence of phosphorus spin-lattice relaxation times (T_1) in 43, 69, 84 and 180 base-pair DNA restriction fragments. A qualitative rationalization of the observations is outlined using a specific model for the internal motion.

Preparation of the Restriction Fragments

The DNA fragments used in this study were isolated from a *Hae* III digest of pVH51 by RPC-5 column chromatography.[22] Each of the fractions corresponding to the 43, 69, 84 and 180 basepair fragments contained 2 mg of material and was reduced in volume to one ml. Each reduced-volume sample was then dialyzed five times against a solution 5 mM in sodium cacodylate, 0.1 mM in EDTA, pH 7.0. In one case, the dialysis procedure was followed by Chelex chromatography but that step had no affect on phosphorus relaxation properties so the Chelex step is not routine. Following the last dialysis, the DNA solutions were lyophilized, dissolved in 99.9% D_2O and the D_2O solutions were lyophilized. The residues were dissolved in 100 μl of 99.9% D_2O and these solutions were transferred to 5 mm cylindrical micro cells for NMR analysis. The samples were not degassed.

Phosphorus NMR Measurements

Phosphorus nuclear magnetic resonance was detected at 36.44 MHz using a Bruker HX90E spectrometer modified for quadrature detection, operating in the pulsed mode and interfaced with a Nicolet 1080 computer. Phosphorus chemical shifts were measured relative to external 0.95 M H_3PO_4 in D_2O, the resonance frequency of which was known as a function of temperature. Recorded temperature dependent chemical shifts (Fig. 2) reflect the shift of the entire signal, not of an arbitrarily selected point on it, thus reflect an average change of the phosphorus ensemble.

Phosphorus spin-lattice relaxation times (T_1) were determined by the inversion recovery method. Full proton irradiation was used to ensure single exponential relaxation of phosphorus. Intensities were measured by use of a line integral routine for *each* tau value and the areas were subjected to a non-linear least square fitting procedure that allows correction for delays $<5\ T_1$. The reported variances are those that reflect the goodness of fit of the regression curves. Probe temperature was measured before and after each run using a YSI bridge and thermocouple. The thermocouple was housed in a 5 mm NMR tube and the assembly was positioned so that the thermocouple occupied the sample location precisely. Temperatures are accurate to $\pm$ 0.1 degree and did not vary during data accumulation for a given T_1 value. No corrections were made for imperfect 180° or 90° pulses nor were corrections made for observing pulse inhomogeneity. Therefore, the reported T_1's may not be accurate. However, as instrument conditions have been carefully controlled, the T_1's are precise and reproducible. Phosphorus-hydrogen nuclear Overhauser effects were measured as reported previously.[21]

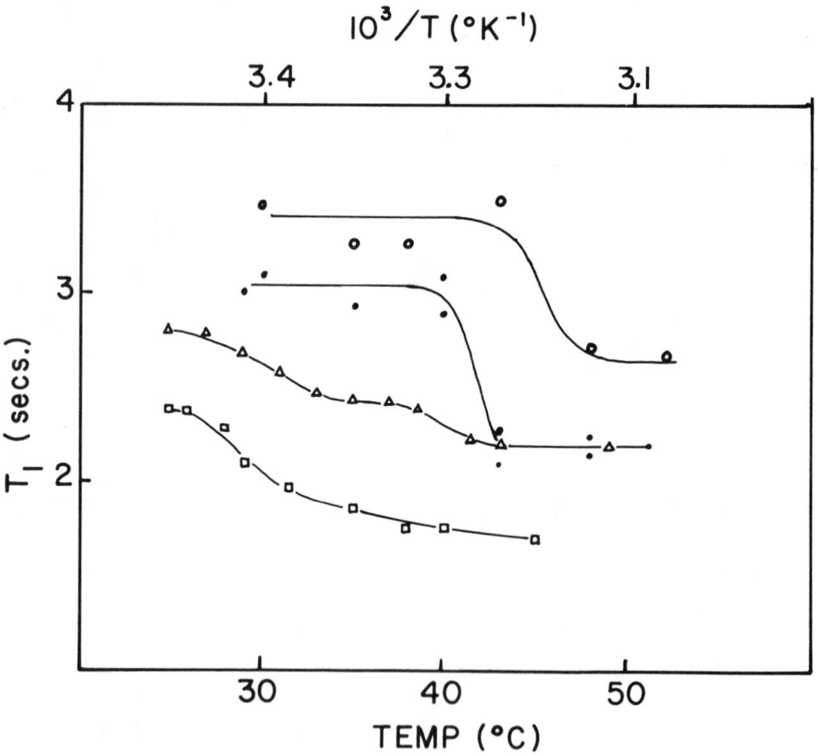

Figure 1. Phosphorus spin-lattice relaxation times (T_1) of the 180 (○○○), 84 (•••), 69 (△△△) and 43 (□□□) bp restriction fragments. Concentrations are all 20 mg/ml, 50 mM sodium cacodylate buffer, pH7. Each T_1 value is the result of a non-linear regression to five or six points. Each point corresponded to the area under the curve resulting from 200-400 transients at a given delay (τ) following the 180° pulse. S/N for short and long τ values was $\pm$8:1, but was much lower near the null. No significant differential relaxation of resolved resonance groups was observed.

There are two trends manifested by the relaxation data represented by Fig. 1 that require rationalization. 1) At lower temperatures (30°C) the observed T_1's depend upon the number of base pairs, decreasing as the double helical fragments shorten. 2) For a given restriction fragment, the observed T_1's are temperature dependent over fairly narrow temperature ranges which are well below the T_m's (70-80°C) of these fragments. Furthermore, while the temperature dependent changes appear to be cooperative in the larger fragments, cooperativity appears to decline in the smaller fragments and the transition temperature decreases. In what follows the fragment length dependence will be shown to be a consequence of plausible changes in overall correlation times and the per fragment temperature dependent changes will be associated with a backbone conformational change.

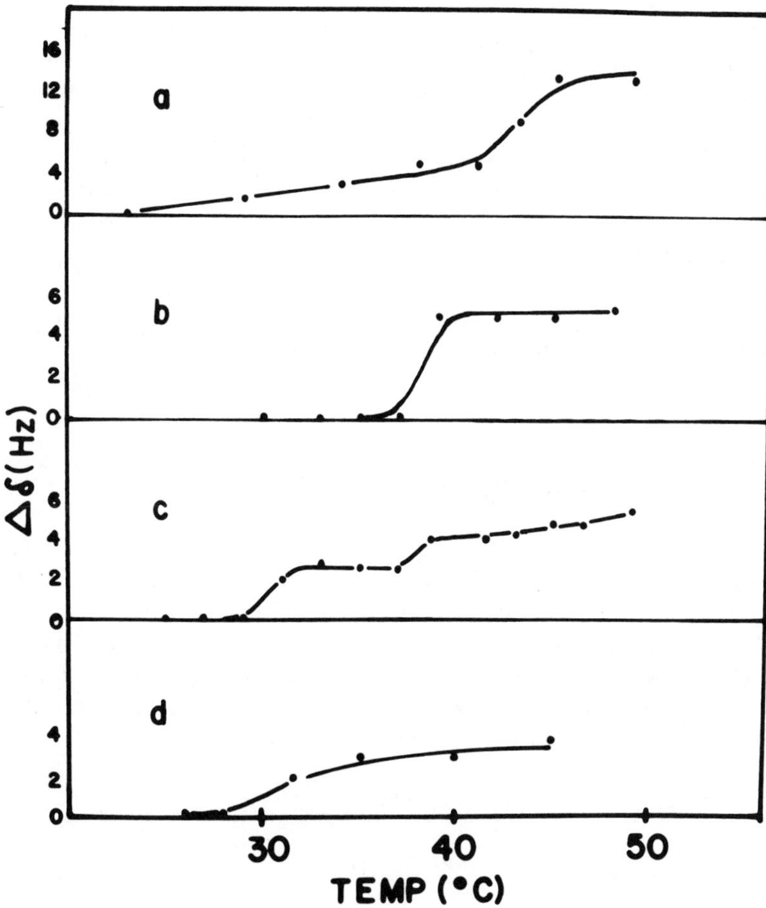

Figure 2. Temperature induced chemical shifts of the restriction fragments. a) 180 bp, b) 84 bp, c) 69 bp and d) 43 bp. Shifts are downfield (in Hz) with increasing temperature relative to the resonance frequency of external 0.95 M H_3PO_4 the temperature dependence of which was known. As nearly as possible the shifts correspond to average frequency changes of the ensemble of resonances; internal changes of resolved groups of resonances were ignored.

T_1 Dependence on Fragment Length

Fig. 1 shows that at 30°C measured phosphorus T_1's corresponding to given fragments decrease as the fragments decrease in length. This observation immediately suggests that the effective correlation times are themselves dependent on helix length. In general, spin lattice relaxation times follow a biphasic dependence upon effective correlation time of overall motion. A T_1 vs. correlation time plot is concave upward with a minimum at $\sim 10^{-9}$secs. The correlation times for the 43-180 bp fragments are certainly slower than 10^{-9} secs, therefore a decreasing effective correlation time (lower molecular weight) will be associated with a smaller value of T_1. Four separate models have been used to simulate the experimental data. The first model (Model I) treats the fragments as rigid rods (symmetric tops). Phosphorus relaxation is considered to arise *via* pure dipolar P-H interactions modulated by anisotropic overall diffusion without significant internal motion. The second model (Model II) treats the fragments similarly, however, modulation of the anisotropic phosphorus chemical shift is allowed to contribute to phosphorus relaxation (chemical shift anisotropy). The third model (Model III) includes all the features of the first two as well as an *arbitrary* mode of internal motion and Model IV represents an attempt to specify the internal motion in geometric terms.

Model I Simulation

Rigid symmetric top diffusional motion can be quantified approximately using equations (1) and (2) quoted by Barkley and Zimm;[2]

$$D_{||} = kT/8\pi\eta b^2 L \tag{1}$$

$$D_{\perp} = (3kT/8\pi\eta L^3)\{\ln(2L/b) - 1.57 + 7[1/\ln(2L/b) - 0.28]^2\} \tag{2}$$

where $D_{||}$ and $D_{\perp}$ are diffusion constants for motion about the helix axis and an orthogonal axis respectively, k is Boltzmann's constant, T is the absolute temperature, η is the solvent viscosity, b is the helix radius and L is one-half the actual helical length. More elaborate calculations have been published by Tirado and de la Torre[23] involving end-effect corrections but the level of the present analysis does not require them. The three characteristic correlation times for overall motion of the symmetric top can be gotten[7] from equations (3a, b, c).

$$1/\tau_0 = 6\,D_{\perp} \tag{3a}$$

$$1/\tau_1 = D_{||} + 5\,D_{\perp} \tag{3b}$$

$$1/\tau_2 = 4\,D_{||} + 2\,D_{\perp} \tag{3c}$$

The computed value of $D_{\perp}$ for a 140 base-pair DNA fragment aqrees sufficiently well with the measured value[17] that one can be confident that in the helix length range of the present study the actual slow diffusion is reasonably approximated. No experimental determination of $D_{\|}$ has been made to which one can compare the computed counterpart. Given the correlation times as a function of the B-DNA helical parameters for the rigid, cylindrical DNA fragments, one is able to compute phosphorus dipolar T_1's using the fomulation of Woessner[7] adapted by Shindo[15] (equation (4)) for DNA simulations.

$$(1/T_1)^{dd} = 1/20 \sum_{k=1}^{n} \frac{\hbar^2 \gamma_H^2 \gamma_P^2}{r_{P,k}^6} \sum_{i=0}^{2} C_i^k Q(J)_i \tag{4}$$

The Q(J) of (4) are sums of spectral densities,

$$Q(J) = J_0(\tau_i, \omega_H - \omega_P) + 3J_1(\tau_i, \omega_P) + 6J_2(\tau_i, \omega_H + \omega_P) \tag{5}$$

in which

$$J_m(\tau_i, \omega_m) = 2\tau_i/(1 + \omega_m^2 \tau_i^2) \tag{6}$$

In these equations $r_{P,k}$ is the distance between phosphorus and the kth hydrogen, and τ_i are the three symmetric top correlation times which are associated with the geometric factors

$$C_0^k = 1/4(3 \cos^2 \beta_k - 1)^2 \tag{7a}$$

$$C_1^k = 3 \sin^2 \beta_k \cos^2 \beta_k \tag{7b}$$

$$C_2^k = 3/4 \sin^4 \beta_k \tag{7c}$$

where β_k are the angle between the kth P-H vector and the helix axis. Conventionally, three P-H interactions are used in calculations of this type, however it is clear from P-H distance calculations of the kind reported earlier[21] using B-DNA geometry,[24] that both Np-2′ hydrogens and the pH-3′ hydrogen must be included. Shindo[15] recognized the need to include pN-H3′, however he did not include the Np-2′ hydrogens. For present computations the two, 2′ hydrogens are included. Their distances from phosphorus were computed using a geometry program that affords P-H distances and direction cosines given bond lengths and appropriate dihedral angles. B-DNA geometry is assumed. Table I gives the phosphorus T_1's computed for a rigid cylinder. There is a significant discrepancy between the computed and the experimental values for the longer fragments, however the discrepancy is not so great for the shorter fragments.

Model II Simulations

It is possible that P-H dipolar interactions are not exclusive contributors to phosphorus relaxation in these systems. Modulation of the anisotropic phosphorus chemical shift by both overall and internal motion is also possible, as mentioned above.

Table I
Experimental and Calculated Spin-Lattice Relaxation Times

	τ_1	dd:CSA*	Fragment Length† 180	84	69	43
Obs. T_1 (sec)‡	---	---	3.47±0.2	3.05±0.13	2.61±0.14	2.12±0.23
Computed T_1	∞	1:0	15.80	7.18	5.86	3.39
	∞	0:1	25.00	11.75	9.74	6.16
	$5x10^{-8}$	1:0	3.62	2.87	2.64	2.03
	$8x10^{-8}$	1:1	3.66	2.55	2.28	1.63

*Proportion of dipole-dipole and chemical shift anisotropy contributions.
†Parameters used: 2L = 3.4(#bp), b = 12.5Å, η = .01 poise.
‡30°C.

The CSA contribution can be estimated using equation (8) adapted by Shindo[15] from the work of Hull and Sykes[25] and Huntress[26]

$$\left(\frac{1}{T_1}\right)^{CSA} = 6/40\ \gamma_p^2\ B_0^2\ \sigma_z'^2 \sum_{i=0}^{2} C_i J_i(\tau_i, \omega_p) \tag{8}$$

where B_0 is the magnetic field strength, $\sigma z'$ is the z component of the chemical shift tensor, the spectral densities are described above and the C_i's are geometric factors that express the coordinate transformation of the principal axes of the chemical shift tensor to the molecular frame of reference.

Adaptation of equation (8) to the rigid cylinder yields the values entered in Table I. It is evident that inclusion of the CSA mechanism by way of equation (9)

$$\left(\frac{1}{T_1}\right)^{obs} = \left(\frac{1}{T_1}\right)^{dd} + \left(\frac{1}{T_1}\right)^{CSA} \tag{9}$$

will not improve the T_1 simulations, therefore, as proposed often by many authors, internal motion must be invoked to rationalize the observations.

Model III Simulations

For purposes of simulation *via* Model III, we choose to represent the complicated effects of internal motion by a *single* correlation time (τ_{int}) *without* reference to specific internal geometric changes. This internal correlation time is then entered in the arguments of the spectral densities by forming harmonic averages with the correlation times that characterize overall motion (3a, b, c). Thus, the spectral densities, (6) are redefined:

$$J_m(\tau_i, \omega_m) = J_m\{\tau_i[\tau_{int}/(\tau_i + \tau_{int})], \omega_m\} \tag{10}$$

The internal correlation time then was treated as an adjustable parameter and various sets of T_1^{dd} and T_1^{CSA} were calculated and combined according to (9). The simulations that gave the best approximations to the experimental T_1's are shown in Table I. Evidently, inclusion of CSA results in poorer simulations. Incorporation of internal motion similarly to the above is done rather commonly[7-11,27] but in all cases the results must be considered qualitative because the geometric and dynamic models are unspecified. It is clear that the computed T_1's for the larger fragments (longer correlation times) are more sensitive to the internal motion parameter than are those of the shorter fragments. This analysis suggests that, while a CSA contribution should be included as a mechanism of phosphorus relaxation at fields of 2.1 T, it may not be necessary to weight that contribution so heavily as it was by Shindo.[15]

The chain length dependence can, therefore, be modeled qualitatively using a single arbitrary internal correlation time superimposed upon the symmetric top model.

Model IV Simulations

To superimpose rigorously internal motion upon the model for overall motion adopted here, it is necessary to choose a plausible conformational representation of internal motion. It is likely in the present case that motion about all bonds of the phosphodiester backbone should be allowed. It is then necessary to compute the time dependence of direction cosines and internuclear P-H separations. It is practically necessary to assume that internal and overall motions are statistically independent. Any model of internal phosphorus motion chosen to represent the fragments of the present study must also be compatible with the observation that only small changes in these fragments' circular dichroism spectra are seen in the 30-50°C temperature range. Thus, internal phosphodiester conformational change must be uncoupled from relative complementary base-pair changes.

A specific conformational model (though probably not the only one) that fits the above criteria is diagrammed in Fig. 3. It represents a concerted jump between two conformational states involving the noted changes of dihedral angles. Careful manipulation of molecular models shows that this conformational change produces minimal perturbation of the remainder of the double helical system and may be analogous to the P-O bond wobbling proposed by Bolton and James.[10,11] What follows is a method for arriving at spectral densities for internal motion based on the two-state model that is generalizable to more complicated models.

The method followed here is after that of Tsutsumi[28] assuming pure dipolar relaxation of phosphorus. As in any rigorous relaxation formulation, construction of the proper autocorrelation function based on the specific dynamic model is critical. A convenient general formulation of the dipolar autocorrelation function used in the present treatment is

$$G_k(t) = \sum_{i,j} f_k(i)\, P[i(t), j(0)]\, f_k(j) \qquad (11)$$

in which i and j are different conformations, and P is a two-time joint probability.

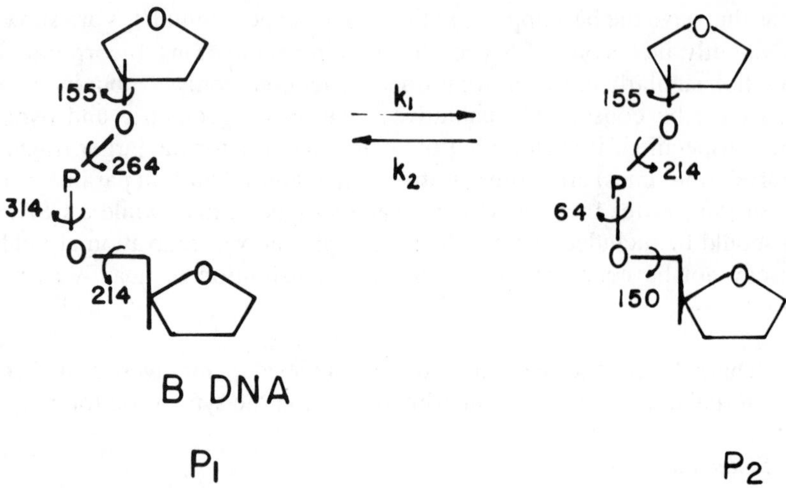

Figure 3. Model chosen to represent backbone conformtional change. The angles labeling 1 correspond to B-DNA backbone geometry; k_1 and k_2 are rate constants.

The f_k are the five orientation factors required by the symmetry of the dipole-dipole interaction and have the form

$$f_1 = (1-3n^2)/r^3 \quad (12a)$$

$$f_2 = (l^2-m^2)/r^3 \quad (12b)$$

$$f_3 = mn/r^3 \quad (12c)$$

$$f_4 = nl/r^3 \quad (12d)$$

$$f_5 = lm/r^3 \quad (12e)$$

in which 1, m and n are direction cosines relative to the helix prinicpal axes and r is a P-H distance. A separate set of the geometric factors must be computed for each conformatiion considered times the total number of P-H interactions.

The time dependent joint conformational probability is derived from the kinetic equations required by the two-state model (Fig. 3). The conventional coupled first order differential equations for this model can be found in standard works and have the particular solutions

$$P_1(t) = P_1^E + C_1 \exp[-k_2 t/P_1^E] \quad (13a)$$

$$P_2(t) = P_2^E + C_2 \exp[-k_1 t/P_2^E] \quad (13b)$$

in which p^E is an equilibrium probability and C_1 and C_2 are determined by the boundary conditions. As will be shown below the conditional conformational probabilities $P[i(t) | j(0)]$ are required to express the joint probabilities needed in the autocorrelation function. Accordingly, the boundary conditions chosen for the evaluation of C_1 and C_2 in (13) are $P_1(0) = 0$ and 1 and P2(0) = 0 and 1. Thus,

$$P(1|1) = P_1^E + (1-P_1^E)\exp[-k_1 P_2^E t] \tag{14a}$$

$$P(1|2) = P_1^E + P_1^E \exp[-k_1 P_2^E t] \tag{14b}$$

$$P(2|1) = P_2^E - P_2^E \exp[-k_1 P_2^E t] \tag{14c}$$

$$P(2|2) = P_2^E + (1-P_2^E)\exp[-k_1 P_2^E t] \tag{14d}$$

and since $P(i,j) = P(i|j)\, Pj^E$, the autocorrelation function for internal motion is

$$\sum_{k=1}^{5} \sum_{i,j} f_{ik} P(i|j) P_j^E f_{jk} \qquad i,j = 1,2 \tag{15}$$

This autocorrelation function can be combined with that for overall motion[28] and the combined function can be Fourier transformed to give a general form of the spectral densities,

$$J(\omega) = [f_{1k}, f_{2k}] \begin{bmatrix} (P_1^E)^2, & P_1^E P_2^E \\ P_1^E P_2^E, & (P_2^E)^2 \end{bmatrix} \begin{bmatrix} f_{1k} \\ f_{2k} \end{bmatrix} [\tau_R/1 + (\omega\tau_R)^2] + \tag{16}$$

$$[f_{1k}, f_{2k}] \begin{bmatrix} P_1^E P_2^E, & P_1^E P_2^E \\ P_1^E P_2^E, & P_1^E P_2^E \end{bmatrix} \begin{bmatrix} f_{1k} \\ f_{2k} \end{bmatrix} \{(\tau_R^{-1} + \tau_{int}^{-1})/1 + \omega^2[\tau_R^{-1} + \tau_{int}^{-1})^{-1}]^2\}$$

in which τ_R is, successively, τ_0, τ_1 and τ_2 of the symmetric top and $\tau_{int} = P_2^E/k_1 \equiv P_2^E\tau_1$ (see Fig. 3 for definition of k_1). It is not necessary to cast (16) in the form shown for the two-state model, however that particular form is convenient for models of greater complexity and can be generalized easily for computation purposes.

Once the appropriate spectral densities have been computed they can be combined in standard ways[7,12,28] to give T_1, T_2, and NOE. The adjustable parameters are C_1 and the equilibrium probabilities. The variations of T_1 with C_1 and population are depicted in Figs. 4 to 6 and the variation of the P-H NOE with C_1 is depicted in Fig. 7 and 8. All computations correspond to T = 30°C. Figure 4 demonstrates that sensitivity to the internal motion parameter is lost by the shorter fragments. In

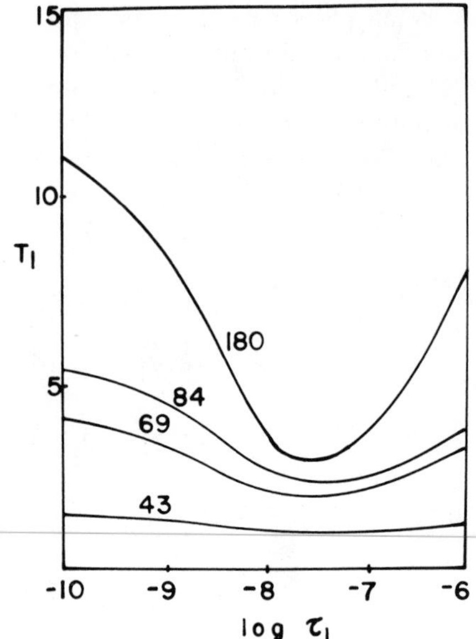

Figure 4. Computed variation of T_1 with τ_1 for the four restriction fragments, $P_1 = 0.9$.

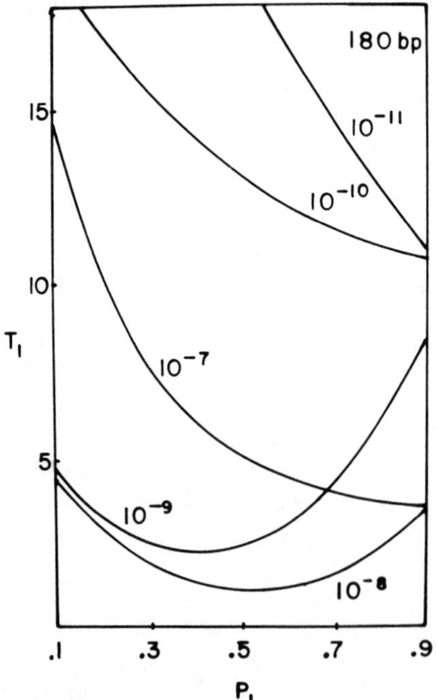

Figure 5. Computed variation of T_1 with P_1 for the 180 bp fragment for various values of τ_1. A more complete dependence of T_1 on τ_1 may be deduced from this figure than can be gotten from Figure 4.

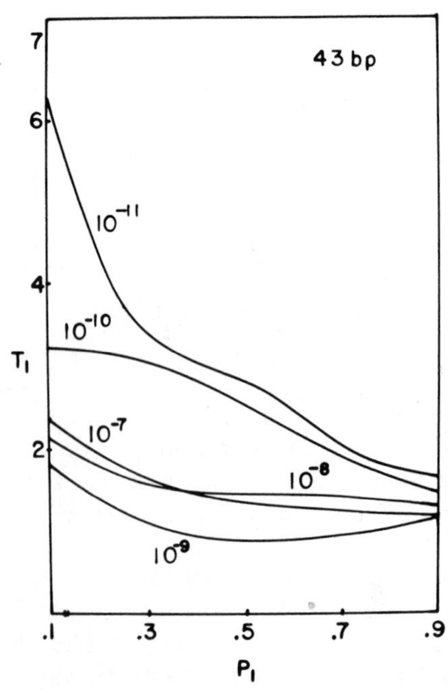

Figure 6. Computed variation of T_1 with P_1 for the 43 bp fragment for various values of τ_1.

contrast, Figs. 5 and 6 show that the dependence of T_1 on conformer population remains for all fragments. The computed NOE's of Figs. 7 and 8 are the sum of pairwise P-H NOE's that would be expected upon broad-band proton irradiation. The conclusion to be drawn from these figures is that the NOE's will depend on both conformation and the internal motion parameter for all four fragments. The apparent conformation dependence reflects the modification of τ_1 by P_2^E to give τ_{int}. This modification appears in the correlation time factor of (16). Geometric information is lost in two-spin NOE calculations.

The Model IV analysis clearly reveals the limits placed on allowable conclusions about these systems based on the model. Considering the near equality of the experimental NOE's (Table II), Fig. 8 suggests just one region for τ_1 ($\sim 5 \times 10^{-9}$) if one assumes conformational homogeneity among the fragments at 30°C. Figures 5 and 6 suggest that better agreement with experimental T_1's would be found if P_1 were 0.1 to 0.2 instead of 0.3 as required by Fig. 8. That merely suggests the need for a more detailed NOE computation. If the assumption of conformational homogeneity among fragments is not allowed, one is faced with a far more difficult and ambiguous fitting process.

Estimates of relaxation contributions *via* modulation of the anistropic chemical shift have not been included in the Model IV computations. The above numbers would be changed somewhat if they were, but not the conclusions.

Temperature Dependence of T_1 for a Given Fragment

If one concludes that thermal activation of overall motion is minimal over the rather narrow range of temperature studied, the Model IV simulations combined

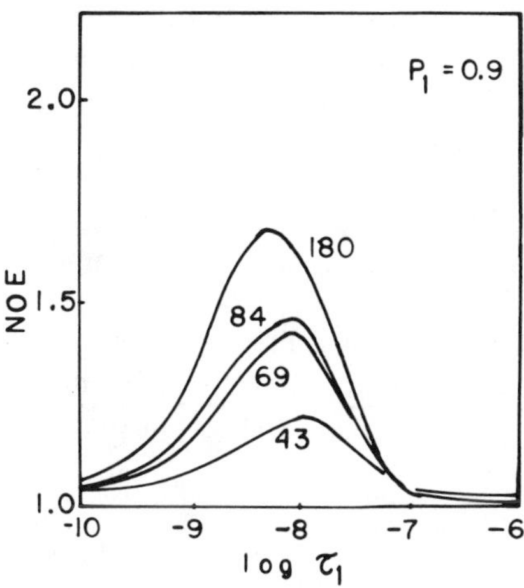

Figure 7. Computed total P-H NOE's. Eight pairwise interactions were weighted according to their relative nuclear separations and summed $P_1 = 0.9$.

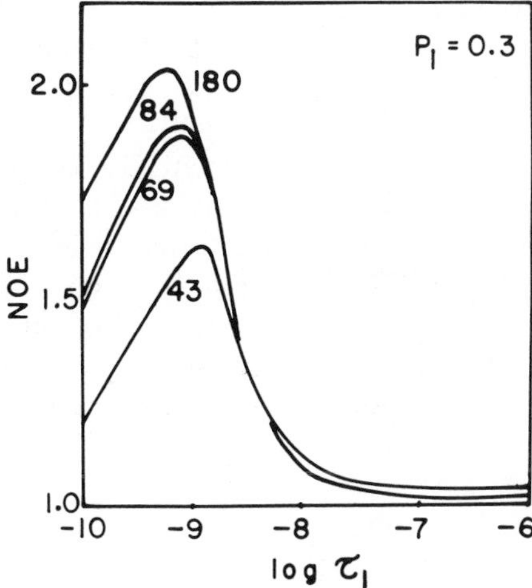

Figure 8. Computed P-H NOE is $P_1 = 0.3$.

Table II
Experimental Phosphorus-Proton Nuclear Overhauser Effects
As a Function of Temperature*

TEMP	NOE(%) 180 bp	84 bp	43 bp
25	20	26	21
29	20	21	23
35	22	24	20
40	21	20	24
45	24	20	20
51	24	23	24

*Broad-band proton irradiation was emlployed.

with the experimental NOE's allow fairly definitive qualitative rationalization of the per fragment T_1 variations with temperature. The constancy (within experimental error) of the P-H NOE's over the relevant temperature range suggests that no significant changes in τ_1 are responsible for the T_1 changes. Furthermore, again assuming only minimal changes in τ_R, increasing the temperature would *decrease* τ_1 and according to Fig. 4 and the above estimated τ_1 values, T_1 should *increase* or remain unchanged. The observed decrease in T_1 is qualitatively rationalized if one assumes a conformational change. One such change could be a change in P_1/P_2 that, according to Fig. 5 and 6 could result in a decrease in T_1. The appearance of some new conformation could also fortuitously lead to the T_1 decrease.

Evidence in support of a conformational change is represented by Fig. 2, a plot of chemical shift against temperature for the four restriction fragments. While these data are crude because of the extreme difficulty of measuring chemical shifts of the broad, noisy spectra of this study, there are clear indications of general temperature dependent chemical shifts that parallel (approximately) the temperature dependent relaxation times. Gorenstein and co-workers[29,30] have proposed that changes in phosphorus chemical shifts of phosphodiesters are correlated with P-O torsion angle changes coupled with O-P-O bond angle changes. If the chemical shift trends of Fig. 2 can be interpreted as conformation changes, those data support both the point-of-view that conformational change is reflected in the T_1 *vs.* temperature plots of Fig. 1 and, also, that the changes are cooperative. Cooperativity is supported because the measured chemical shifts represent shifts of the entire resonance envelope composed of many dispersed resonances. By contrast, the 43 bp fragment that shows no sharp temperature dependent T_1 change, also shows no sharp chemical shift change with temperature.

It is conceivable that all the observations discussed above reflect a cooperative, temperature dependent, disaggregation. The solutions studied are quite concentrated and certainly could contain aggregates. However, Hogan and Jardetzky[9] studied solutions of a 140 bp (average) fragment that was nearly an order of magnitude more dilute than the samples reported here and reported a T_1 that was between the values of the present 180 and 84 bp fragments. We conclude that if aggregation is problematical in the present samples, it must be as well in the more dilute solution studied by Hogan and Jardetzky. In connection with the latter work, it may be wondered why those authors did not observe T_1 transitions of the kind reported here. The explanation is very likely based on the fundamental differences between the fragments used in each study. The present samples are of unique length and sequence. The fragments used by Hogan and Jardetzky are of average nominal length and mixed sequence. Thus, in the averaged length and sequence samples, it appears that the thermal transitions are so dispersed as to be unnoticeable.

Summary

The unique feature of this study is the observation of the temperature dependence of the combined internal and overall motion and its dependence upon fragment-length, which probably indicates a cooperative conformational change of the phosphodiester backbone over a temperature range well below the T_m. It is, therefore, possible that the dramatic salt-induced conformational transitions seen in double stranded dGdC oligomers or polymers,[31,36] dGdC-containing restriction fragments[37] and dAdT polymers,[38] have a more subtle counterpart in the present observation. These solution studies, as well as others not cited here, have been interpreted to indicate salt-induced left-handed double helical structures or alternating phosphodiester conformations. These interpretations are supported by recently-reported x-ray analyses of dGdC oligomers crystallized under high salt conditions that manifest various left-handed helical Z-forms.[39-43] The presumed conformational changes reported here may represent early conformational changes ("premelting" changes[44]) on a continuum from, say B-DNA to Z-DNA or to other forms yet unobserved. Regardless of whether that is true, it is probable that further study of unique sequence and unique length DNA polymers will reveal a diversity of environment-sensitive conformations. The measurement of ^{31}P relaxation rates should provide a sensitive means of detecting these conformations and their transitions, even though unique and rigorous quanitative analyses may be difficult to obtain.

Acknowledgements

This work was supported by grants from the NSF (PCM 77-19927 and PCM 15033) and the NIH (CA 20279). W.H. was supported, in part, by the Max Kade Foundation and the Deutsche Forschungsgemeinschaft.

References and Footnotes

1. Wohl, Ph., Paoletti, J. and LePecq., J.-B., *Proc. Natl. Acad. Sci., U.S.A., 65,* 417-421 (1970).
2. Barkly, M.D. and Zimm, B.H., *J. Chem. Phys., 70,* 2991-3007 (1979).
3. Hurley, I., Robinson, B.H., Scholes., C.P., and Lerman, L.S., in Sarma, R.H., ed., *Nucleic Acid Geometry and Dynamics,* Pergamon (New York), p. 253-271 (1979).
4. Robinson, B.H., Forgacs, G., Dalton, L.R. and Frisch, H.L., *J. Chem. Phys., 73,* 4688-4692 (1980).
5. Hogan, M.E. and Jardetzky, O., *Biochemistry, 19,* 2079-2085 (1980).
6. Early, T.A. and Kearns, D.R., *Proc. Natl. Acad. Sci., U.S.A., 76,* 4165-4169 (1979).

7. Woessner, D.E., *J. Chem. Phys., 37,* 647-655 (1962).
8. Hogan, M.E. and Jardetzky, O., *Proc. Natl. Acad. Sci., U.S.A., 76,* 6341-6345 (1979).
9. Hogan, M.E. and Jardetzky, O., *Biochemistry, 19,* 3460-3468 (1980).
10. Bolton, P.H. and James, T.L., *J. Phys. Chem., 83,* 3359-3366 (1979).
11. Bolton, P.H. and James, T.L., *J. Am. Chem. Soc., 102,* 25-31 (1980).
12. Woessner, D.E., *J. Chem. Phys., 36,* 1-4 (1962).
13. Klevan, L., Armitage, I.M. and Crothers, D.M., *Nucl. Acids Res., 6,* 1607-1615 (1979).
14. Davanloo, P., Armitage, I.M. and Crothers, D.M., *Biopolymers, 18,* 663-680 (1979).
15. Shindo, H., *Biopolymers, 19,* 509-522 (1980).
16. Shindo, H., McGhee, J.D. and Cohen, J.S., *Biopolymers, 19,* 523-537 (1980).
17. Hogan, M.E., Dattagupta, N. and Crothers, D.M. *Proc. Natl. Acad. Sci., U.S.A., 75,* 195-199 (1978).
18. Hanlon, S., Glonek, T. and Chan, A., *Biochemistry, 15,* 3869-3875 (1976).
19. Cotter, R.I. and Lilley, D.M.J. FEBS Letters, 82, 63-67 (1977).
20. Kallenbach, N.R. Appleby D.W. and Bradley, C.H., *Nature, 272,* 134-138 (1978).
21. Hart, P.A., *Biophysical J., 24,* 833-848 (1978).
22. Hillen, W., Klein, R.D. and Wells, R.D., *Biochemistry,* in press (1981).
23. Tirado, M.M. and de la Torre, J.G., *J. Chem. Phys., 73,* 1986-1993 (1980).
24. Arnott, S. and Hukins, D.W.L., *J. Mol. Biol., 81,* 93-106 (1973).
25. Hull, W.E. and Sykes, B.D., *J. Mol. Biol., 98,* 121-153 (1975).
26. Huntress, N.T., Jr., *J. Chem. Phys., 48,* 3524-3533 (1968).
27. Tsutsumi, A., Quaegebeur, J.P. and Chachaty, C., *Mol. Phys., 38,* 1717-1735 (1979).
28. Tsutsumi, A., *Mol. Phys., 37,* 111-127 (1979).
29. Gorenstein, D.G., Findlay, J.B., Momii, R.K., Luxon, B.A. and Kar, D., *Biochemistry, 15,* 3796-3803 (1976).
30. Gorenstein, D.G., *J. Am. Chem. Soc., 97,* 898-900 (1975).
31. Pohl, F.M. and Jovin, T.M., *J. Mol. Biol, 67,* 375-396 (1972).
32. Pohl, F.M., Jovin, T.M., Baehr, W. and Holbrook, J.J., *Proc. Natl. Acad. Sci., U.S.A., 69,* 3805-3809 (1972).
33. Pohl, F.M., Ranade, A. and Stockburger, M., *Biochem. Biophys. Acta., 335,* 85-92 (1973).
34. Pohl, F.M., *Nature, 260,* 365-366 (1976).
35. Patel, D.J., Canuel, L.L. and Pohl, F.M., *Proc. Natl. Acad. Sci., U.S.A., 76,* 2508-2511 (1979).
36. Simpson, R.T. and Shindo, H., *Nucl. Acids Res., 8,* 2093-2103 (1980).
37. Klysik J. Stirdivant S.M. Larson J.E., Hart, P.A. and Wells, R.D., *Nature 290,* 672-679 (1981).
38. Shindo H., Simpson R.T. and Cohen, J.S., *J. Biol. Chem., 254,* 8125-8128 (1979).
39. Wang, A.H.-J., Quigley, G.J., Crawford, J.L., van Boom, J.H., van der Marel G., and Rich, A., *Nature, 282,* 680-686 (1979).
40. Crawford, J.L., Kolpak, F.J., Wang, A.H.-J., Quigley, G.J., van Boom, J.H., van der Marel, G. and Rich, A., *Proc. Natl. Acad. Sci., U.S.A., 77,* 4016-4022 (1980).
41. Wang, A.H.-J., Quigley, G.J., Kolpak, F.J., van der Marel, G., van Boom, J.H. and Rich, A., *Science, 211,* 171-176 (1980).
42. Crawford, J.L., Kolpak, F.J., Wang, A.H.-J., Quigley, G.J., van Boom, J.H., van der Marel, G. and Rich, A., *Proc. Natl. Acad. Sci., U.S.A., 77,* 4016 (1980).
43. Drew H., Takano, T., Tanaka, S., Itakura, K. and Dickerson, R.E., *Nature, 286,* 567-571 (1980).
44. Palecek, E., *Prog. Nucl. Acid Res. and Mol. Biol., 18,* 151-213 (1976).

Proceedings of the Second SUNYA Conversation in the Discipline Biomolecular Stereodynamics Volume I, ISBN 0-940030-00-4, Ed., Ramaswamy H. Sarma, Adenine Press, New York, ©Adenine Press

Natural Abundance Carbon-13 NMR Spectroscopic Studies of Native and Denatured DNA

Randolph L. Rill, Peter R. Hilliard, Jr.
Linda F. Levy and George C. Levy
Department of Chemistry
Institute of Molecular Biophysics
Florida State University
Tallahassee, Florida 32306

Introduction

Macroscopically, double stranded DNA behaves as an extremely stiff polymer that approaches the behavior of a rigid rod for lengths of up to a few hundred angstroms, depending on the method of observation. The rigidity of DNA, and factors that affect it, are well known from classic hydrodynamic measurements. Recent advances in time-resolved spectroscopic techniques have led to an appreciation of other biologically significant aspects of the double stranded structure. At the nucleotide level, the DNA structure can no longer be regarded as uniform or static. Relatively long "atypical" sequences containing only one or two of the four nucleotides are fairly common in eukaryotic DNA and are conformationally distinct from the classic B form (e.g., see several articles in this volume). Furthermore, nucleotides in DNA apparently retain sufficient internal degrees of freedom to allow local motions of significant amplitude and frequency. While the overall stiffness of DNA is of consequence in nuclear packaging, local flexibility and adaptability may govern DNA functions involving specific recognition.

Nuclear magnetic resonance studies of different nuclei in DNA potentially can provide information about the relative and absolute motional dynamics of different sites. Certain general features of DNA dynamics are already apparent. High field ^{1}H, ^{31}P and ^{13}C NMR spectra of relatively small (140-300 nucleotide pairs-np), double stranded DNA molecules have now been reported.[1-8] Hogan and Jardetzky[6] have also reported low field (25 MHz) relaxation data and NOE's of protonated sugar and base carbons. All three types of spectra are characterized by relatively narrow lines. For example, sugar carbon resonances are well resolved at 67.9 MHz (excepting the C1′, C4′ doublet), and base carbon resonances are sufficiently resolved to permit reasonable assignments.[4,8] Linewidths at 67.9 MHz are in the range of 100-300 Hz (Ref. 4, see also below). The narrowness of these lines, and the magnitudes of NOE's of protonated carbons reported here (see below) and by others, cannot be explained in terms of any reasonable model of overall molecular motions, but are generally consistent with intramolecular motions with correlation times in the nanosecond range.

There are many potential motional modes in DNA, and strong coupling of these motions is likely. Precise description of DNA motions therefore requires data on the behavior of many DNA sites. Furthermore, simple isotropic models are not sufficient to describe these motions, and need for extensions in present theory and extensive characterization of NMR parameters can be anticipated. We believe that the unique ability of ^{13}C NMR parameters obtained at several magnetic fields to fully probe motional autocorrelation functions for many DNA sites will be particularly useful in this respect. To this end, we are measuring, at multiple fields, carbon-13 resonance linewidths, spin-lattice relaxation times, and Nuclear Overhauser Enhancements of individual sugar and base carbons of carefully fractionated DNA samples. Preliminary comparisons of the NMR parameters of double and single-stranded DNA of different lengths, though insufficient to completely model

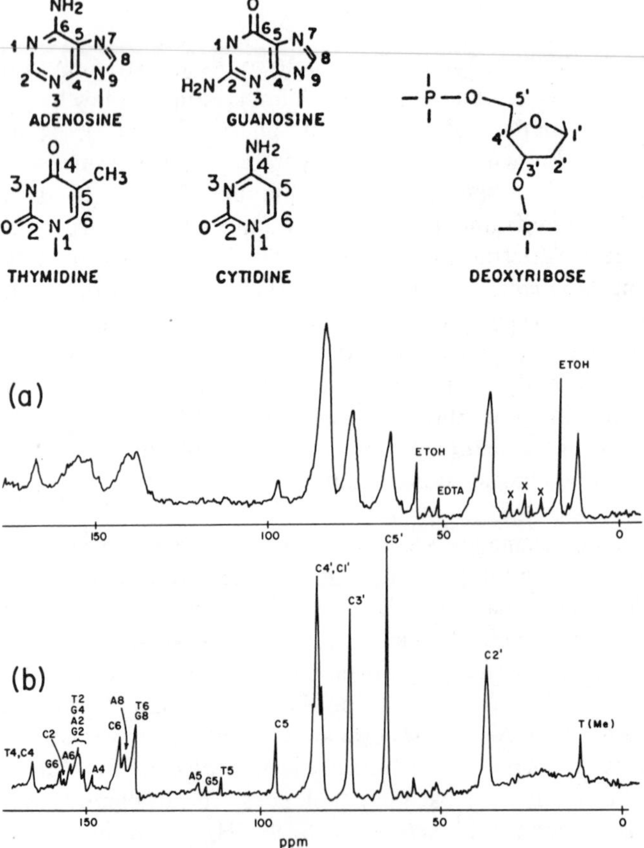

Figure 1. Natural abundance carbon-13 NMR spectra of (a) double stranded and (b) single stranded 120 np DNA at 100.6 MHz (scales approximately identical). Conditions: (a) 80 mg/ml, 32°C, 8000 scans, 20 Hz digital broadening and scan interval 1 sec.; (b) 80 mg/ml, 85°C, 3500 scans, 20 Hz digital broadening and scan interval 1.5 sec. Spectra recorded at the NSF Regional NMR Facility, University of South Carolina.

DNA behavior, suggest some interesting insights into DNA motional dynamics and illustrate some of the problems of interpretation.

Magnetic Field Dependencies of Spectral Linewidths of Double-Stranded DNA

Spectra of fully double-stranded, 160 np DNA have been obtained at four magnetic fields—100.6 MHz, 67.9 MHz, 37.7 MHz, and 22.7 MHz. DNA of this and other well-defined sizes was prepared by digestion of calf thymus chromatin by micrococcal nuclease, followed by chromatography of the digested chromatin on Bio-Gel A5m (Bio-Rad Laboratories).[4,8] As noted previously, linewidths observed at all fields were severalfold smaller than can be accounted for by overall molecular motions. Although spectral resolution was significantly improved at 100.6 MHz (e.g., compare Figure 1 below with Figure 1 of ref. 4), dispersion of shielding for individual types of carbons in different local environments significantly broadens most resonance bands at high fields (Figure 2). Shift dispersions for all sugar carbons are in the range of 1.4-1.9 ppm except for C2′, which shows a lesser increase in linewidth corresponding to a ca. 1.0 ppm dispersion. Thus, the intrinsic resonance

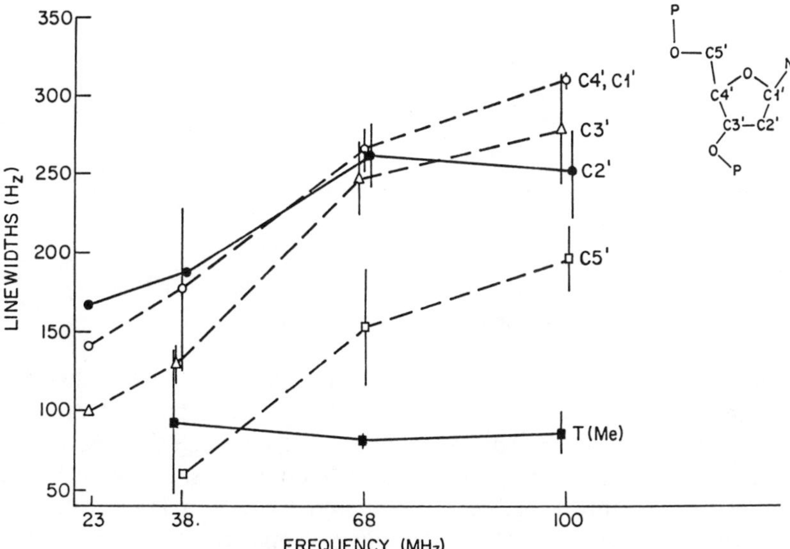

Figure 2. Linewidths of double stranded 160 np DNA (80 mg/ml) as a function of magnetic field from 22.7 to 100.6 MHz: (○), C4′, C1′ band (unresolved); (△), C3′; (□), C5′; (●), C2′; (■), thymidine methyl. Bars indicate the mean deviations of three different spectra at 67.9 MHz and two different spectra at 37.7 and 100.6 MHz. Low field spectra were obtained at 22.7 MHz of FX-90 spectrometers at the University of California at Los Angeles and California Institute of Technology, and at 37.7 MHz on the in-house design, multinuclear spectrometer at Florida State University, SEMINOLE.[12] High field spectra were obtained at 67.9 MHz on a quadrature detection modified Bruker HX-270 at Florida State University and at 100.6 MHz on a Bruker WH-400 at the NSF Regional Facility at the University of South Carolina.

linewidths of DNA carbons are considerably less (~>50%) than measured at 67.9 and 100.6 MHz. Two other features of these spectra are interesting to note. The C5′ sugar resonance linewidth is ca. 40% less than that of other sugar carbons at all fields. This might suggest higher mobility at this site, as expected from model considerations; however, the C5′ NOEF is not higher than that observed for the other sugar carbons (see below). Secondly, the thymidine methyl resonance is little affected by shift dispersion. The reason for this is not obvious since this methyl group is closer to the bases than most sugar carbons.

Length Dependence of the Linewidths of Double-Stranded DNA

Spectra have been obtained for DNA samples with average lengths of 35, 120, 160, 240, 280, and ca. 800 nucleotide pairs (Table I). Virtually no differences in sugar

Table I
Dependence of Sugar Carbon Resonance Linewidths
on the Length of Double Stranded DNA

Carbon	*Linewidth (H_z) for DNA of Length Indicated*[1]					
	25-45 np	*110 np*	*120 np*	*160 np*[2]	*236 np*[3]	*800 np*[4]
C2′	176	219	279	262	344	
C5′	152	(117)[5]	179	154	----	No
C3′	265	270	274	267	(328)	spectrum
C4′,C1′	254	230	270	266	351	observed
T(Me)	54	(127)[5]	97	83	----	

[1]measured at 32°C.
[2]average of 3-4 spectra, precision was ca. 10%.
[3]similar data has been obtained on 280 np DNA contaminated with some mononucleosome length DNA.
[4]whole calf thymus DNA mechanically sheared and S1 nuclease treated, median size is given.
[5]value suspected because of unusual lineshape.

carbon linewidths were observed for DNA 160 np and smaller. This lack of a length dependence of the linewidths is consistent with a dominance of rapid internal motional contributions to spin-spin relaxation times, although shift dispersion could obscure small changes in intrinsic linewidths. In contrast, spectral lines of 240-280 bp DNA were markedly broadened, compared to 160 np DNA, and no useful spectrum was observed for a more polydisperse sample averaging 800 np. (This was prepared from commercial DNA by mechanical shearing, followed by S1 nuclease treatment.[9]) Considering the lack of a length dependence of the linewidths of smaller DNA, this strong length dependence is somewhat paradoxical and may reflect intermolecular ordering in these concentrated solutions.[10,11]

T1's and NOEF's of Double Stranded DNA

Spin-lattice relaxation times (T1's) and Nuclear Overhauser enhancement effects are more sensitive to rapid motions than spin-spin relaxation times, and therefore provide less ambiguous measures of internal motions in polymers than simple linewidth comparisons. Spin-lattice relaxation times for double stranded DNA carbons are relatively short, with NT1's averaging about 0.5 sec at 67.9 MHz for both sugar and protonated ring base carbons (NT1 = T1 times the number of attached protons). NT1's increase with increasing field, reflecting the fact that DNA dynamics fall outside of the extreme narrowing limit (Table II). Significant NOEF's observed

Table II

13C Spin Lattice Relaxation Times in DNA[1]

			160 np DNA[4]		
Carbons		*Native*[2]		*Denatured*[3]	
	37.7 MHz	*67.9 MHz*	*100.6 MHz*	*67.9 MHz*	*100.6 MHz*
C2′	0.20	0.27±.01	0.41±.01	0.28±.03	----
C5′	----	0.23±.01	0.37±.03	0.22±.01	----
C3′	0.21	0.48±.03	0.63±.06	0.50±.01	----
C4′,C1′	0.21	0.54±.04	0.63±.08	0.41±.02	----
T(Me)	----	0.90	1.6 ±.02	----	----
C5	----	----	----	(0.55)	----
			120 np DNA[5]		
		Native[2]		*Denatured*[3]	
C2′	----	0.28	0.45	0.34	0.24
C5′	0.12	0.24	0.33	0.29	0.24
C3′	0.21	0.47	0.57	0.63	0.52
C4′,C1′	0.21	0.49	0.66	0.59	0.47
T(Me)	----	1.0	1.3	----	1.0
C5	----	----	0.65	----	0.65
T4,C4	1.7	----	----	----	----

[1]T_1s, 10-20%. Mean deviations shown are from the average of 2 experiments.
[2]Native DNA measurements at 32°C.
[3]Denatured DNA measurements at 86°C.
[4]160 nP DNA is essentially 100% double stranded by assays described in the text.
[5]120 nP DNA was 15% single stranded by assays described in the text.

for protonated carbons (0.6 to 0.8 at 100.6 MHz) provide the most direct measure of internal motions in DNA (Table III, NOEF = NOE - 1). As discussed by Hogan and Jardetzky,[3,6] NOEF's of this magnitude are consistent with motional components on the order of a few nanoseconds. Resonances of base carbons in double stranded DNA were poorly resolved in general, but the C5 carbon resonance of cytosine was isolated and of sufficient magnitude at 100.6 MHz to provide good estimates of T1 and NOEF. Both parameters for this protonated base carbon are identical, within precision, to the NT1's and NOEF's of protonated sugars, strongly suggesting

Table III
^{13}C Nuclear Overhauser Enhancement Factors
for 120 np DNA at 100.6 MHz

Carbons	*NOEF*	
Sugar	*Native DNA*[1]	*Denatured DNA*[2]
C2′	0.77	1.2
C5′	0.64	1.4
C3′	0.67	1.4
C4′,C1′	0.57	1.3
Base		
T(Me)	0.93	1.2
C5 (protonated)	0.60	1.1
T5 (non-protonated)	0.27	0.56

[1]Measured at 32°C.
[2]Measured at 85°C.

that base motions are coupled with sugar motions occurring on the same, ca. nanosecond time scale. (In this regard, see also the article by Sundaralingam in this volume.)

Effects of Heat Denaturation

Heat denaturation of 160 np DNA decreased resonance linewidths two to four-fold for sugar ring carbons and five-fold for the exocyclic C5′ carbon (Table IV). Base carbon resonances were similarly sharpened, sufficiently to observe reasonable

Table IV
^{13}C NMR Linewidths of Native and Denatured,
160 np DNA at 67.9 MHz

Carbons	*Linewidths*[1]	
Sugar	*Native*[2]	*Denatured*[3]
C2′	262±21	102±11
C5′	154±37	31± 8
C3′	248±23	60±17
C4′,C1′	266±14	125±30
Base		
T(Me)	83± 3	42±14
C5	----	42±14
T4,C4	----	74± 4

[1]In Hz; estimated precision of three experiments (± mean deviation).
[2]Measured at 32°C.
[3]Measured at 85°C.

resolution of several individual resonances at 100.6 MHz (Figure 1). Denaturation also approximately doubled the NOEF's for both sugar and base carbons (Table III), but did not significantly affect any T1's (Table II). The insensitivity of T1's to denaturation is surprising and may reflect three possibilities:

1. The NT1's may correspond to effective correlation times more or less symmetrically disposed about the T1 minimum (Figure 3a).
2. The motional dynamics may reflect a non-exponential autocorrelation function, producing a flattened dependence of NT1 on the rate of molecular reorientation (Figure 3b).
3. The NT1's may have a complex dependence on composite overall and internal motions resulting in little change of the observed parameters.

Changes in NOEF with denaturation argue against (1) acting alone, and both (2) and (3) may contribute to these observations.

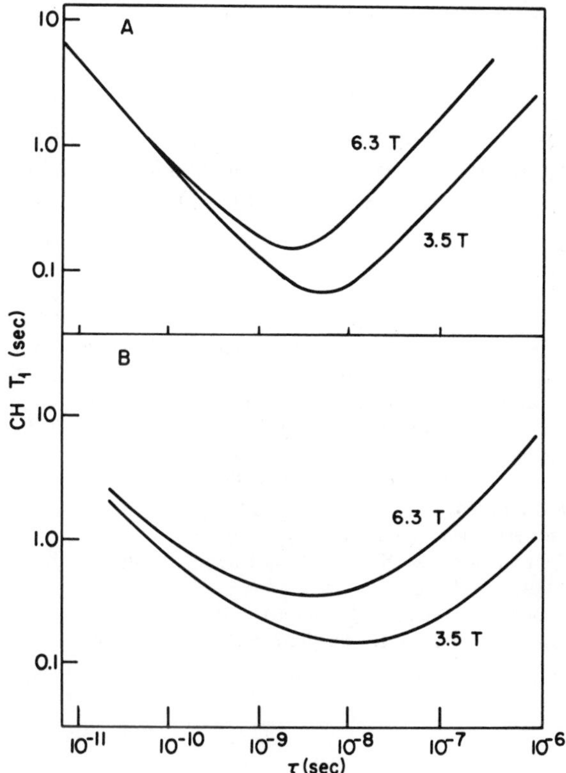

Figure 3. Dipolar T1 for a CH carbon as a function of an effective correlation time for molecular tumbling, τ, at 3.5 and 6.3 Tesla. (a) isotropic or pseudoisotropic tumbling characterized by an exponential autocorrelation function $G(\tau)$; (b) complex behavior resulting from a non-exponential decay of $G(\tau)$. This requires that motion be represented by a set of τ's or a distribution around some mean τ.

Summary

In conclusion, respectable natural abundance carbon-13 NMR spectra, T1's and NOEF's can now be measured at multiple fields for double-stranded DNA of nucleosome core length (140-160 np) and smaller. The relatively narrow resonance linewidths, short T1's, and significant NOEF's of protonated carbons confirm that some sort of rapid internal motions occur in the double stranded structure. Sugar and base motions appear to occur on similar time scales, arguing against major conformational flexibility of the sugar-phosphate backbone independent of the base motions. The more or less uniform decreases in linewidths and increases in NOEF's upon heat denaturation are consistent with increases in motional rates expected from the macroscopically observable decrease in chain stiffness, and again indicate coupling of sugar and base motions. The insensitivity of T1's to denaturation, on the other hand, suggests that the motional dynamics is complex. Proper interpretation of these data requires knowledge of both T1 and NOE parameters, obtained over several magnetic fields, to probe the autocorrelation function at various times. Such experiments are planned, along with theoretical calculations to attempt to fully characterize DNA dynamics.

Acknowledgements

We thank Drs. Ruth Inners, Paul D. Ellis and other members of the staff of the NSF Regional Facility at the University of South Carolina for their assistance. This work was supported by grants from NIH, NSF, and the Department of Energy.

References and Footnotes

1. Davanloo, P., Armitage, I.M., and Crothers, D.M. *Biopolymers 18,* 663-680 (1979).
2. Klevan, L., Armitage, I.M., and Crothers, D.M. *Nucleic Acid Research 6*, 1607-1616 (1979).
3. Hogan, M.E. and Jardetzky, O. *Proc. Natl. Acad. Sci. USA 76*, 6341-6345 (1979).
4. Rill, R.L., Hilliard, Jr., P.R., Bailey, J.T., and Levy, G.C. *J. Amer. Chem. Soc. 102*, 418-420 (1980).
5. Early, T.A. and Kearns, D.R. *Proc. Natl. Acad. Sci. USA 76*, 4165-4169 (1979).
6. Hogan, M.E. and Jardetzky, O. *Biochemistry 19*, 3460-3468 (1980).
7. Bolton, P.H. and James T.L. *Biochemistry 19*, 1388-1392 (1980).
8. Levy, G.C., Hilliard, P.R., Levy, L.F., Rill, R.L., and Inners, R. *J. Biol. Chem.*, in press (1981).
9. Vogt, V.M. *Methods in Enzymology 65*, 248-255 (1980).
10. Ise, N., Okubo, T., Yamamoto, K., Kawai, H., Hashimoto, T., Fujimura, M., and Hiragi, Y. *J. Am. Chem. Soc. 102*, 7901-7906 (1980).
11. Reinert, K.E. and Geller, K. *Stud. Biophys. 45*, 1-11 (1974).
12. Levy, G.C., Rosanske, R.C., Wright, D., and Terpstra, D. Abstracts *Experimental NMR Conference*, Blacksburg, Virginia (1978).

Proceedings of the Second SUNYA Conversation in the Discipline Biomolecular Stereodynamics Volume I, ISBN 0-940030-00-4, Ed., Ramaswamy H. Sarma, Adenine Press, New York,

Fluorescence Decay Studies of Anisotropic Rotations: Internal Motions in DNA

Mary D. Barkley
Department of Biochemistry
University of Kentucky Medical Center
Lexington, KY 40536

and

Andrzej A. Kowalczyk and Ludwig Brand
Department of Biology and the McCollum-Pratt Institute
The Johns Hopkins University
Baltimore, MD 21218

Introduction

Fluorescence polarization is a powerful method for investigating structural and dynamic properties of biological systems. The two sources of depolarization, energy transfer and rotational Brownian motion, may take place on the nanosecond timescale and can be studied by fluorescence. In the absence of energy transfer, fluorescence depolarization provides information about rotational processes of molecules and about the environment in which these occur. The time course of depolarization can be measured by pulse fluorimetry. The incident light flash photoselects a population of molecules whose absorption oscillators are aligned in the direction of the electric field of the exciting light, causing an initial polarization of fluorescence. If the excited molecules reorient due to Brownian motion, then the polarization of the emitted light decays with time. This is conveniently represented by the fluorescence emission anisotropy r,

$$r(t) = \frac{I_V(t) - I_H(t)}{I_V(t) + 2I_H(t)},$$

where the subscripts V and H denote the vertically and horizontally polarized components of the emisson intensity I(t). The time constants for the decay of the emission anisotropy depend only on the rotational behavior of the molecule.

Internal Motions in DNA

Several contributions to this volume are concerned with flexibility of the DNA double helix. The first evidence for nanosecond motions in DNA was obtained in 1970 by Wahl, Paoletti, and LePecq.[1] They detected an ~28 ns relaxation process in fluorescence depolarization experiments of ethidium-DNA complex. Since ethidium is tightly bound to DNA by intercalation and since the relaxation time is too fast for rotation of the entire DNA molecule (~1 ms), they concluded that the observed decay of the emission anisotropy was due to internal rotatory Brownian motion in the DNA helix. Subsequently, theoretical treatments of twisting and rapid bending motions in DNA have been performed which permit quantitative assessment of the amount of flexibility.[2,3] The predicted emission anisotropy for dyes intercalated in DNA with transition dipoles collinear and perpendicular to the helix axis is

$$r(t) = \frac{0.75e^{-\Gamma} + 0.45e^{-(\Delta+\Gamma)} + 0.4e^{-\Delta}}{3 + e^{-\Delta}} \tag{1}$$

where the twisting decay function Γ is given by

$$\Gamma(t) \simeq 4kT(t/\pi C\rho)^{1/2}, \tag{2}$$

and the bending decay function Δ by

$$\Delta(t) \simeq B(t)t^{1/4}. \tag{3}$$

In Equations 2 and 3, k is the Boltzmann constant, T is temperature, C is the torsional stiffness of DNA, ρ is the frictional coefficient per unit length for rotation about the helix axis, and B(t) is a slowly varying function of time that depends on the bending stiffness of DNA. Equation 1 shows that the various twisting and bending motions of DNA collaborate to yield a complicated decay process, with exponentials in $t^{1/2}$ due to twisting and in $\sim t^{1/4}$ due to bending. Except at very early times, relaxation occurs more rapidly from twisting than from bending. For fluorescent probes with lifetimes comparable to ethidium (~23 ns in DNA) the depolarization is due almost entirely to the twisting motions. A similar theory has been developed for EPR experiments of spin-labeled intercalators bound to DNA.[4]

Recently, two laboratories repeated the experiments with ethidium-DNA complex using laser excitation and analyzed their data according to the predicted decay law assuming no contribution from bending.[5,6] In both cases the emission anisotropy exhibited an exponential in $t^{1/2}$ decay. The torsional stiffness of DNA determined from these data is $C = 1.3 \times 10^{-19}$ erg cm, corresponding to an rms fluctuation of about 6° per base pair. The same values were obtained from EPR experiments with an ethidium spin probe.[7] These experimental results provide strong support for the theory described above. However, Genest and Wahl also did fluorescence depolarization measurements of ethidium-DNA complex and fit their data to a biexponential function for the emission anisotropy.[8] Thus, we have a situation in which different models for rotational diffusion are consistent with the fluorescence data. Because there are numerous other examples of complex rotational mechanisms in biological systems, such as segmental flexibility of proteins,[9,10] hindered rotations in membranes,[11,12] and microheterogeneity of rotational behavior, it is necessary to have improved methods which provide enough information to distinguish competing models.

Analysis of Fluorescence Depolarization Data

The difficulty of discriminating between different mechanisms for rotational diffusion from pulse fluorimetry data is illustrated in Figures 1 and 2. Here we have generated computer-simulated data of the type expected for ethidium-DNA complex. Figure 1 shows synthetic data for biexponential decay of the emission anisotropy together with the best fit of these data to a nonexponential function (Equation 1 with $\Delta = 0$). Figure 2 shows the opposite situation, namely synthetic data for nonexponential decay of the emission anisotropy together with the best fit to a biexponential function. In either case the incorrect fitting function provides an acceptable fit to the data, though the shape of the autocorrelation function is not ideal. (If the scatter of the data is random, the autocorrelation function of the weighted residuals exhibits small, high-frequency fluctuations about the zero line.) In principle, these minor deviations in the autocorrelation function as well as other statistical criteria could be used to distinguish between the two decay laws for the emission anisotropy, though in practice systematic instrumental errors would probably obscure such small differences.

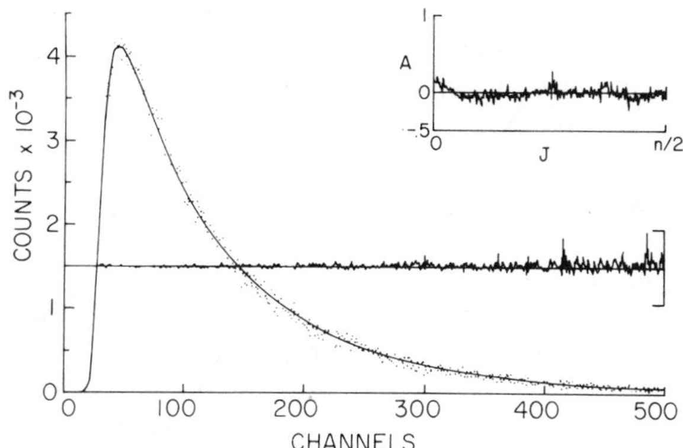

Figure 1. Biexponential decay of the emission anisotropy of ethidium-DNA complex analyzed according to a nonexponential function. Simulation of an experiment with relatively few counts in the difference curve $D(t) = I_V(t) - I_H(t)$. Timing calibration .204 ns/channel, 20°C. The weighted percent residuals and the autocorrelation function of the residuals (inset) are also shown. Points are computer-simulated data for the convolved difference curve with photon-counting noise, $r(t) = 0.11 \exp(-t/14 \text{ ns}) + 0.20 \exp(-t/240 \text{ ns})$, $\tau = 22.5$ ns. Line is best fit to Equation 1 neglecting bending: $C = 1.2 \times 10^{-19}$ erg cm, 17° rms wobble of ethidium in its site on DNA; $\chi_r^2 = 1.1$.

The above example reflects the fact that the information content available from pulse measurements is limited. One can improve the quality of both the data and the analysis procedures. Distortions introduced by the finite width of the excitation pulse and the response time of the detector can be reduced by the use of picosecond laser excitation and faster photomultipliers. Alternatively, various methods

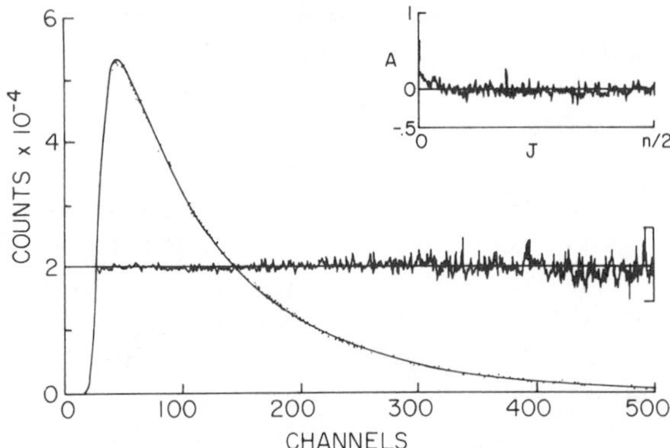

Figure 2. Nonexponential decay of the emisson anisotropy of ethidium-DNA complex analyzed according to a biexponential function. Simulation of an experiment with many counts in the difference curve D(t) = $I_V(t)$ - $I_H(t)$. Timing calibration .204 ns/channel, 20°C. The weighted percent residuals and the autocorrelation function of the residuals (inset) are also shown. Points are computer-simulated data for the convolved difference curve with photon-counting noise, Equation 1 neglecting bending: C = 2 x 10^{-19} erg cm, 16° rms wobble of ethidium in its site on DNA, τ = 23 ns. Line is best fit to a biexponential function, r(t) = 0.08 exp (−t/8 ns) + 0.25 exp (−t/204 ns); χ_r^2 = 1.2.

of deconvolution based on Fourier transforms,[13] nonlinear least-squares,[14] method of moments,[15,16] and Laplace transforms[17] can decrease the uncertainty in the decay parameters recovered from the experimental data. In spite of such technical improvements, it would still probably be difficult to distinguish different models for rotational diffusion and to extract reliable values for the parameters of complicated decay laws. What is really needed, then, are new types of experiments which provide additional information about the relaxation process. One can obtain more information about a rotating system by varying solution conditions, such as solvent viscosity and temperature. This approach gives good results for small molecules in some cases. However, it may not always be feasible for biological molecules, because their structures are often highly dependent on solution variables. Another way to augment the information content of emission anisotropy data is to excite into different electronic transitions of the fluorophore. Excitation of differently oriented absorption oscillators changes the initial distribution of emission dipoles and thus alters the relative contribution of different rotations to the depolarization. In this way one can obtain two or more decay curves for the emission anisotropy which bear a known relationship to each other, thereby decreasing the number of degrees of freedom in the data analysis In the following section we describe the principles of this method and present some results for simple model compounds.

Anisotropic Rotations of Small Molecules

For an asymmetric rigid body, the decay of the enission anisotropy is described by a sum of five exponential terms.[19-21]

$$r(t) = \sum_{i=1}^{5} \beta_i e^{-t/\phi i} \tag{4}$$

However, in practical cases no more than three exponentials will be observed.[22] In the case of ellipsoids of revolution the emission anisotropy is also a sum of three exponentials. The pre-exponentials β_i are geometric factors defined in Figure 3. The values of the pre-exponentials depend on the relative orientation of the absorption and emission dipoles with respect to the diffusion principal axes of the ellipsoid and therefore on the excitation wavelength. On the other hand, the rotational correlation times ϕ_i depend only on the principal diffusion constants of the ellipsoid,

$$\begin{aligned} \phi_1^{-1} &= 6D_\perp, \\ \phi_2^{-1} &= 2D_\perp + 4D_\parallel, \\ \phi_3^{-1} &= 5D_\perp + D_\parallel, \end{aligned} \tag{5}$$

where $D_\perp$ is the rotation rate about an axis perpendicular to the symmetry axis and $D_\parallel$ is the rotation rate about the symmetry axis. The diffusion constants provide important information about the behavior of a rotating molecule, because their values depend on the viscosity η and temperature T of the solvent, the size and shape of the molecule, and the nature of the solute-solvent interaction. For molecules which rotate with the sticking boundary condition of classical hydrodynamics, the diffusion constants are given by the Perrin theory.[23] Thus, depending on the excitation wavelength and the rotational behavior of the molecule, the decay of the emission anisotropy will take different forms. For example, if one of the transition dipoles is parallel to a diffusion principal axis (i.e., $\beta_3 = 0$), the emission anisotropy reduces to a biexponential. If this axis is the symmetry axis of the ellipsoid (i.e., $\beta_2 = \beta_3 = 0$), it reduces further to a monoexponential. The decay is always monoexponential for a sphere (i.e., $D_\perp = D_\parallel$).

To develop the method. we chose two model compounds whose spectral and rotational properties have been previously studied: perylene and 9-aminoacridine. Both fluorophores havc strong $\pi \rightarrow \pi^*$ transitions in the visible and ultraviolet regions, which are positively and negatively polarized; the transition dipole directions are shown in Figure 4. Because fluorescence emission normally occurs from

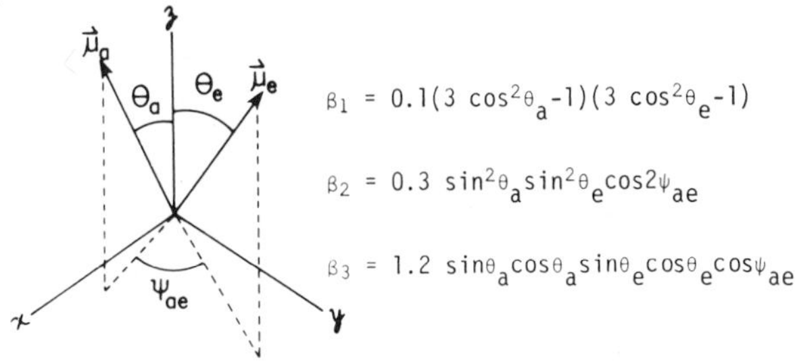

Figure 3. Pre-exponentials for ellipsoid of revolution. Diffusion principal axes of ellipsoid with symmetry axis along z; $\vec{\mu}_a$ absorption dipole; $\vec{\mu}_e$; emission dipole.

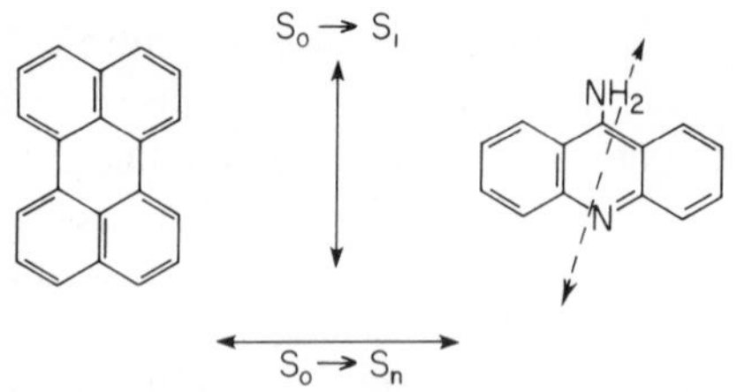

S$_0$ → S$_1$

NH$_2$

N

S$_0$ → S$_n$

Figure 4. Absorption dipole directions of perylene (left) and 9-aminoacridine (right). $S_0 \rightarrow S_1$ at 430 nm; $S_0 \rightarrow S_n$ at 256 or 260 nm. Broken line shows emission dipole direction at 460 nm for 9-aminoacridine. Transition dipoles lie in the plane of the aromatic rings.

the lowest vibronic level of the first excited state, the absorption and emission oscillators are essentially collinear for excitation into the first electronic transition (λ_{ex} = 430 nm) and orthogonal for excitation into the higher electronic transition (λ_{ex} = 256 nm for perylene and λ_{ex} = 260 nm for 9-aminoacridine). Mantulin and Weber found by differential phase fluorimetry that these aromatic compounds represent extremes of rotational behavior, perylene being an anisotropic rotator which slips more rapidly in its plane than it rotates out of plane and 9-aminoacridine being an isotropic rotator.[24]

We measured the fluorescence decay and the decay of the emission anisotropy of perylene and 9-aminoacridine in glycerol by pulsed nanosecond fluorimetry[25] at various excitation wavelengths and temperatures over the range 10°C to 40°C.[26] Under these conditions the fluorescence lifetime of both compounds is essentially constant: about 4.6 ns for perylene and 12.8 ns for 9-aminoacridine. The decay of the emission anisotropy was analyzed according to mono- or biexponential functions. Assuming monoexponential decay, we obtained a single rotational correlation time ϕ at each temperature and excitation wavelength. For both perylene and 9-aminoacridine the value of ϕ depends not only on temperature but also on excitation wavelength (Tables I and II). In the case of perylene, the statistical criteria further indicated that the assumption of monoexponential decay of the emission anisotropy is inadequate. Therefore, we concluded that the wavelength dependence of the single rotational correlation time ϕ is caused by anisotropic rotations. The apparent parameters obtained from the monoexponential analysis are effective averages of the true parameters of a multiexponential decay. Knowing the transition dipole directions, we deduced the location of the diffusion principal axes of the molecules from the wavelength dependence of the apparent single rotational correlation time ϕ. By analyzing the data at two excitation wavelengths for a

Table I
Perylene Decay Parameters

		monoexponential		biexponential			
T°C	λ_{ex}nm	β	ϕ ns	β_1	$\phi_1$ns	β_2	$\phi_2$ns
20	256	−.16	2.5	.10	45	−.24	6.4
	430	.33	10.5	.10	45	.24	6.4
30	256	−.19	0.8	.10	17	−.24	2.7
	430	.32	2.6	.10	17	.24	2.7

Table II
9-Aminoacridine Decay Parameters

		monoexponential		biexponential			
T°C	λ_{ex}nm	β	ϕ ns	β_1	$\phi_1$ns	β_2	$\phi_2$ns
15	260	−.16	52.2	−.16	52		
	317	−.12	59.1	−.19	52	.07	43
	430	.34	44.8	.07	52	.27	43
25	260	−.16	21.9	−.16	22		
	317	−.12	24.6	−.16	22	.05	16
	430	.35	17.1	.07	22	.27	16

consistent set of decay parameters, we were able to obtain accurate values of the parameters for biexponential decay of the emission anisotropy.

Perylene. Using this approach, we determined that perylene rotates as a disk with symmetry axis perpendicular to the plane of the aromatic rings. The decay of the emission anisotropy is expected to be biexponential at both excitation wavelengths. The examples given in Table I show that analysis according to biexponential functions yields pre-exponentials that are independent of temperature but dependent on excitation wavelength and rotational correlation times that are dependent on temperature but independent of excitation wavelength. This contrasts with analysis according to a monoexponential, where both parameters are temperature and wavelength dependent. The highly anisotropic rotation of perylene gives rise to unusual decay curves for the emission anisotropy. Figure 5A shows the experi-

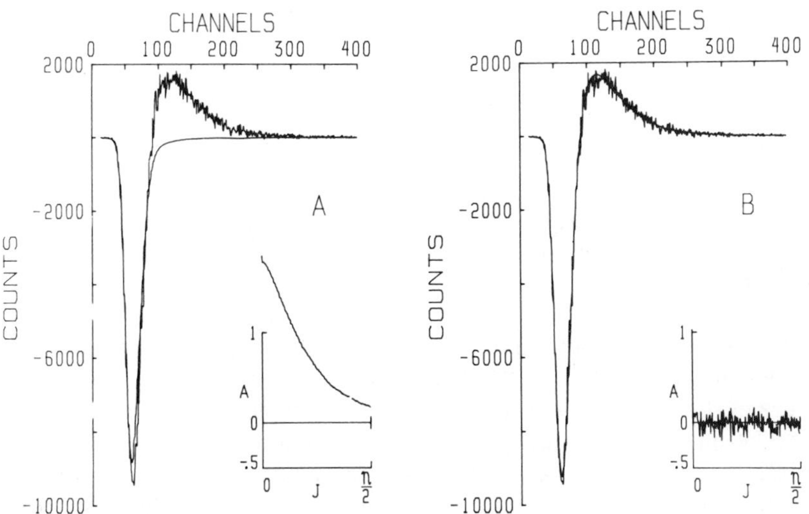

Figure 5. Experimental and computed difference curve of perylene. Timing calibration 0.102 ns/channel, λ_{ex} = 256 nm, λ_{em} = 448 nm, 30°C. The autocorrelation function of the weighted residuals is shown in the inset. A. Monoexponential fit $\beta = 0.19$, $\phi = 0.8$ ns, $\chi_r^2 = 11.7$. B. Biexponential fit $\beta_1 = 0.10$, $\phi_1 = 17$ ns, $\beta_2 = -0.24$, $\phi_2 = 2.7$ ns, $\chi_r^2 = 1.2$.

mental difference curve $D(t) = I_V(t) - I_H(t)$ for perylene at $\lambda_{ex} = 256$ nm and 30°C together with the best fit for a monoexponential decay of the emission anisotropy; the fit is clearly unacceptable. Figure 5B shows the fit for a biexponential; the value of χ_r^2 as well as the autocorrelation function indicate an excellent fit to the data. The impulse responses (deconvolved decay curves) for the emission anisotropy at $\lambda_{ex} = 256$ nm and several temperatures are depicted in Figure 6. The negative emission anisotropy at early times is due to the orthogonality of the absorption and emission dipoles. The fact that the emission anisotropy becomes positive with time prior to decaying to zero demonstrates that rotation within the plane is faster than rotaton out of plane. Such oscillatory behavior is consistent with a biexponential decay law containing a positive and a negative term (Table I). Although oscillating decay functions for the emission anisotropy are expected on theoretical grounds, this is the first case in which they have been observed experimentally. The physical basis of this phenomenon is illustrated in Figure 7. In order for the emission anisotropy to change sign, the relative magnitudes of the vertically and horizontally polarized components of the emission intensity must invert with time. For perylene at $\lambda_{ex} = 256$ nm, those molecules whose aromatic rings are aligned with absorption dipole parallel, and consequently emission dipole perpendicular, to the electric vector of the exciting light are photoselected. The initial horizontal intensity is greater than the vertical intensity so that the emission anisotropy is negative. At early times the fast rotation in the plane of the rings moves the emission dipole in the vertical direction and the emission anisotropy becomes positive. The slow rotation out of plane eventually depolarizes the fluorescence emission at later times.

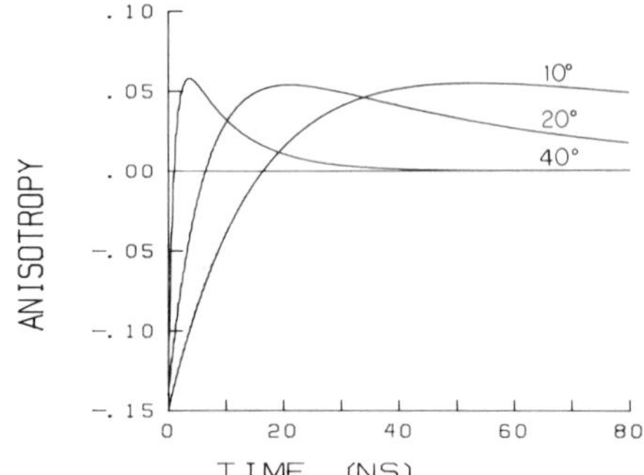

Figure 6. Emission anisotropy of perylene. Impulse response (deconvolved decay) for biexponential decay, $\lambda_{ex} = 256$ nm, $\lambda_{em} = 448$ nm, temperature indicated on curve.

The highly anisotropic rotation of perylene also results in the unusual shape of the Perrin plot shown in Figure 8. While the anisotropic rotation is reflected in slight curvature of the Perrin plot at $\lambda_{ex} = 430$ nm, the steady-state emission anisotropy at $\lambda_{ex} = 256$ nm goes from negative values at low temperatures to positive values at higher temperatures. The steady-state emission anisotropy is the average over all time of the time-resolved emission anisotropy weighted by the fluorescence decay,

$$\langle r \rangle = \int_0^\infty e^{-t/\tau} r(t)dt \Big/ \int_0^\infty e^{-t/\tau} dt,$$

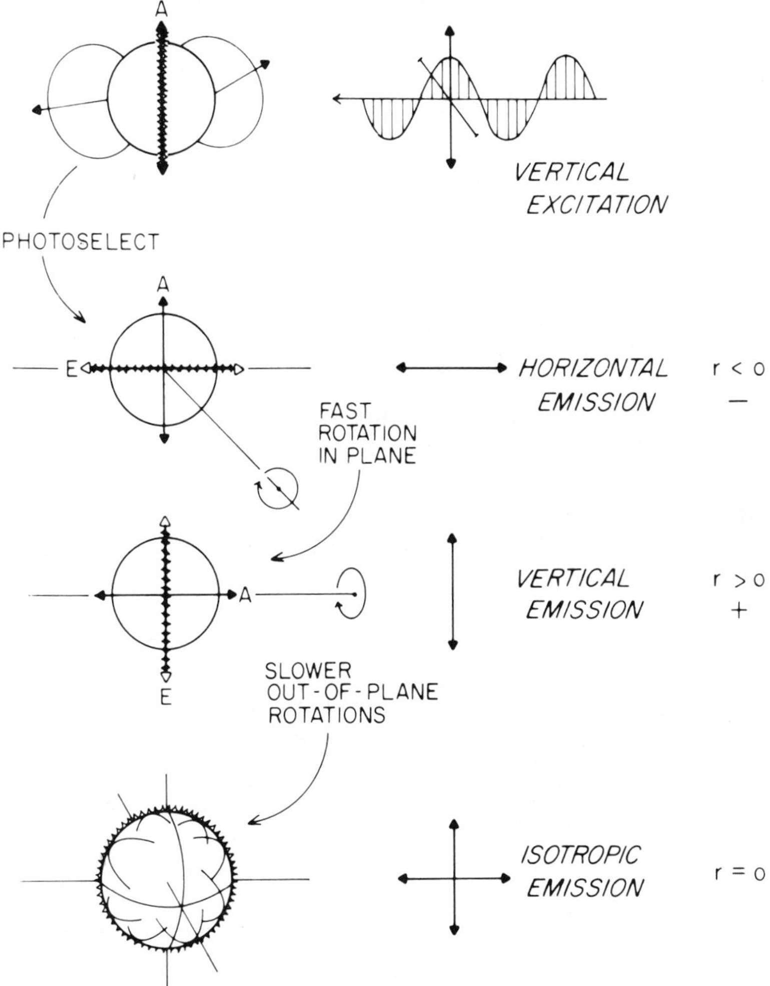

Figure 7. Depolarization of perylene fluorescence. Perylene molecule represented by disk. Absorption dipole A perpendicular to emission dipole E.

where τ is the fluorescence lifetime. For biexponential decay of the emission anisotropy, the steady-state value is given by

$$\langle r \rangle = \beta_1/(1+\tau/\phi_1) + \beta_2/(1+\tau/\phi_2). \tag{6}$$

The steady-state emission anisotropy depends on the relative values of the fluorescence lifetime and the rotational correlation times. For perylene at $\lambda_{ex} = 256$ nm and low temperatures, both rotational correlation times ϕ_1 and ϕ_2 are longer than the fluorescence lifetime so that the fluorescence emission is negatively polarized. The two terms in Equation 6 become comparable in magnitude (though opposite in

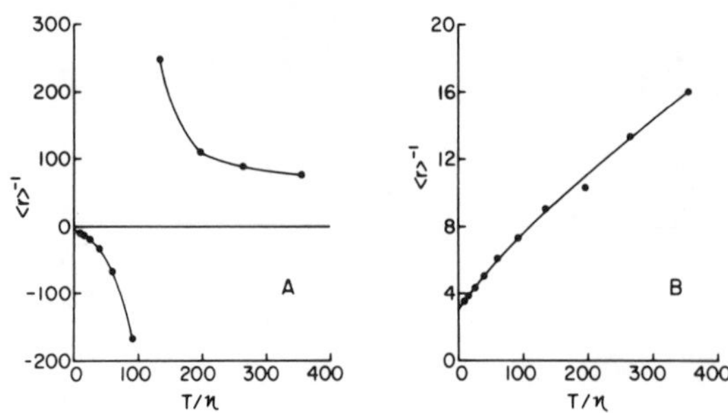

Figure 8. Perrin plots for perylene in glycerol. λ_{em} = 448 nm, viscosity in P. (A.) λ_{ex} = 256 nm, (B.) λ_{ex} = 430 nm.

sign) with increasing temperature until they cancel and $<r> = 0$. At higher temperatures the fast inplane rotation essentially randomizes the emission dipole in the vertical direction during the lifetime of the excited state ($\phi_2 < \tau$), and the fluorescence emission becomes positively polarized. Finally, at very high temperatures (not shown) both rotational correlation times are much shorter than the fluorescence lifetime and $<r> \to 0$.

The principal diffusion constants of perylene in glycerol, which were calculated from the rotational correlation times ϕ_1 and ϕ_2 according to Equations 5, are presented in Figure 9. The ratio of the diffusion constants for the in-plane and out-of-plane rotations $D_{\|}/D_{\perp}$ is 10 ± 1, in excellent agreement with the value of 10 ± 2 obtained previously by pulse fluorimetry for perylene in paraffin over the temperature range −20°C to 50°C.[27] Although the diffusion constants are linear functions of T/η, the in-plane rotation rate $D_{\|}$ of perylene is about 6 times faster than expected from the Perrin theory for an oblate ellipsoid of its size. This supports Mantulin and Weber's conclusion that the rotational behavior of unsubstituted aromatic hydrocarbons approximates a slipping rather than sticking boundary condition.[24]

9-Aminoacridine. We used a similar approach to elucidate the rotational dynamics of 9-aminoacridine. As shown in Table II, the wavelength dependence of the single rotational correlation time ϕ is much less dramatic than in the case of perylene. In addition, the decay of the emission anisotropy at all temperatures and excitation wavelengths gave a good fit to a monoexponential. The slightly anisotropic rotation of 9-aminoacridine is only evident from experiments at different excitation wavelengths. The wavelength dependence of the apparent single rotational correlation time ϕ indicates that 9-aminoacridine rotates as a prolate with symmetry axis parallel to the long axis of the molecule. In this case the decay of the emission anisotropy should be monoexponential at λ_{ex} = 260 nm but biexponential at λ_{ex} = 430 nm. The values of the rotational correlation time ϕ_1 obtained from the data at λ_{ex} = 260 nm were used to recover the parameters from the data at λ_{ex} = 430 nm. The examples given in Table II show that analysis according to a biexponential function at λ_{ex} = 430 nm gives rotational correlation times that are dependent on

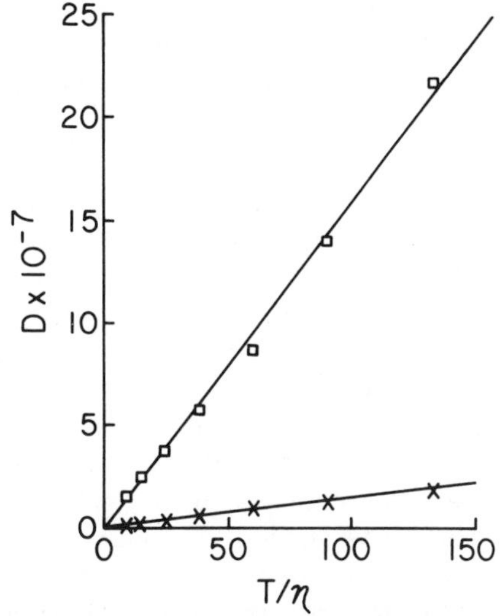

Figure 9. Plots of principal diffusion constants of perylene vs T/η. Lines drawn are least squares fit of the data to the straight line D = const(T/η). (x) $D_\perp$ s^{-1}; (□) $D_\parallel$ s^{-1}.

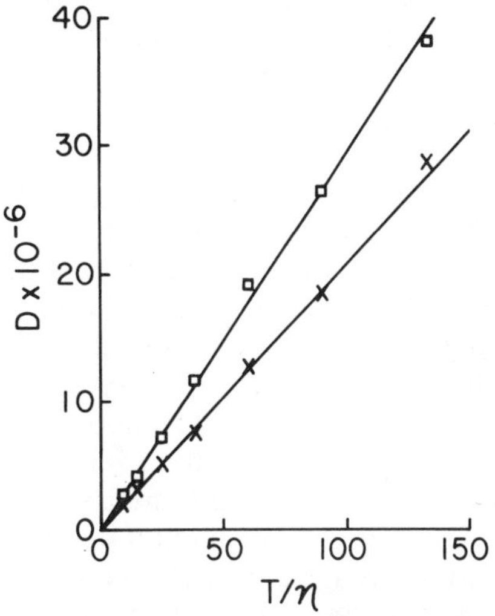

Figure 10. Plots of principal diffusion constants of 9-aminoacridine vs T/η. Lines drawn at least squares fit of the data to the straight line D = const(T/η). (x) $D_\perp$ s^{-1}; (□) $D_\parallel$ s^{-1}.

temperature but independent of excitation wavelength. We also examined the decay of the emission anisotropy at λ_{ex} = 317 nm, which corresponds to a weak, negatively polarized absorption region. The values of the pre-exponentials obtained from the biexponential analysis suggest that the absorption dipole at 317 nm has an out-of-plane component, perhaps due to an n → π* transition.

The principal diffusion constants of 9-aminoacridine in glycerol are summarized in Figure 10. The ratio of the diffusion constants for rotation about the major and minor axes $D_\parallel/D_\perp$ is 1.4 ±0.1. This slightly anisotropic rotation was not observed in the previous study by differential phase fluorimetry.[24] The rotation rates about the major and minor axes $D_\parallel$ and $D_\perp$ are linearly dependent on T/η, and their values are entirely consistent with those calculated from the Perrin theory. Mantulin and Weber likewise concluded that aromatics having substituent groups that hydrogen bond with the solvent rotate with the sticking boundary condition.

Applications

The results described above for simple model compounds demonstrate that pulse fluorimetry is a suitable method for investigating the detailed mechanisms of rotational diffusion. We have shown that excitation of differently oriented oscillators is one way to increase the information available from time-resolved emisson anisotropy data. While the highly anisotropic rotation of perylene is evident from measurements at a single excitation wavelength, the slightly anisotropic rotation of 9-aminoacridine is only revealed by studies at more than one wavelength. This approach is currently being applied to perylene in phospholipid vesicles and to 9-aminoacridine-DNA complex. Preliminary results indicate that both the in-plane and out-of-plane rotations of perylene are hindered in the anisotropic environment of a lipid bilayer.

Excitation into a negatively polarized transition may give rise to oscillating time-dependent emission anisotropy curves. Data of the type shown for perylene may be used as criteria for anisotropic rotational behavior and may offer additional qualitative information. Calculated decay curves based on Equations 4 and 5 and the equations in Figure 3 are shown in Figure 11. In panel A one of the transition dipoles is assumed to be parallel to the symmetry axis of the ellipsoid. Here the decay law reduces to a monoexponential, and both the positive and negative emission anisotropies decay to zero with time as intuitively expected. Panel B represents a situation where the transition dipoles are perpendicular to the symmetry axis of a prolate ellipsoid ($D_{\parallel} > D_{\perp}$). The bottom curve starts at a negative value, passes through zero and becomes positive prior to a final decay toward zero. The initial rapid rise (or rapid drop in the upper curve) reflects the fast rotation about the major axis (or in-plane rotation of a disk with slipping boundary condition), while the slower decay to zero evident after 10 ns comes from rotations about the

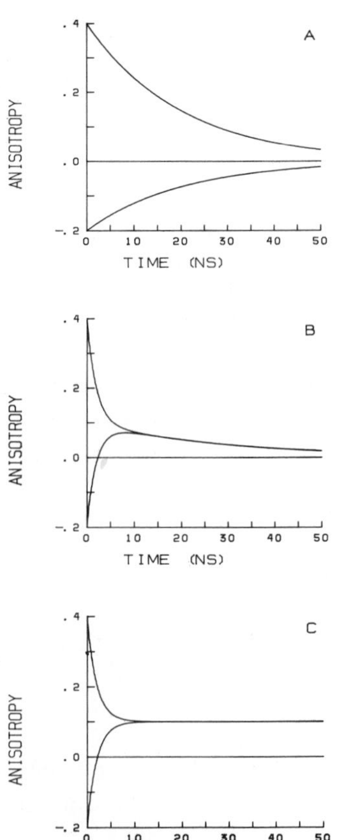

Figure 11. Predicted decay of the emission anisotropy for different orientations of absorption and emission dipoles. In each diagram the upper curve is calculated for collinear transition dipoles and the lower curve for orthogonal dipoles. A. Absorption or emisson oscillator parallel to the symmetry axis of an ellipsoid of revolution. Upper curve $r(t) = 0.4 \exp(-t/20 \text{ ns})$; lower curve $r(t) = -0.2 \exp(-t/20 \text{ ns})$. B. Absorption and emission oscillator perpendicular to the symmetry axis of a prolate ellipsoid ($D_{\parallel} > D_{\perp}$). Upper curve $r(t) = 0.1 \exp(-t/20 \text{ ns}) + 0.3(-t/2 \text{ ns})$; lower curve $r(t) = 0.1 \exp(-t/20 \text{ ns}) - 0.3 \exp(-t/2 \text{ ns})$. C. Similar to B with restricted motion about axes perpendicular to the symmetry axis ($D_{\perp} = 0$).

minor axes (or out-of-plane rotation of a disk). Finally, panel C represents the same system as B but with restricted rotation about the minor axes (or out-of-plane rotation). The asymptotic value of the emission anisotropy is $r_\infty = 0.1$ for both positively and negatively polarized transitions. It is possible that such hindered motion might be approached in biological systems, as in the example of ethidium-DNA complex.

The method presented here requires a fluorophore with two differently oriented absorption dipoles. The molecules used to illustrate this method have one transition in the visible and other transitions in the ultraviolet region. Because most biological macromolecules absorb below 300 nm, the ultraviolet transitions of fluorescent probes are not generally accessible. Therefore, it is desirable to develop probes such as fluorescein derivatives, having both positively and negatively polarized transitions at longer wavelengths.

Acknowledgements

Work in our laboratories is supported by NIH research grants GM22873 (M.D.B.) and GM11632 (L.B.).

References and Footnotes

1. Wahl, P., Paoletti, J., and LePecq, J.-B., *Proc. Natl. Acad. Sci. USA 65,* 417 (1970).
2. Barkley, M.D. and Zimm, B.H., *J. Chem. Phys. 70,* 2991 (1979).
3. Allison, S.A. and Schurr, J.M., *Chem, Phys. 41,* 35 (1979).
4. Robinson, B.H., Forgacs, G., Dalton, L.R., and Frisch, H.L., *J. Chem. Phys. 73,* 4688 (1980).
5. Thomas, J.C., Allison, S.A., Appellof, C.J., and Schurr, J.H., *Biophys. Chem. 12,* (1980).
6. Millar, D.P., Robbins, R.J., and Zewail, A.H., *Proc. Natl. Acad. Sci. USA 77,* 5593 (1980).
7. Hurley, I., Robinson, B.H., Scholes, C.P., and Lerman, L.S. in *Nucleic Acid Geometry and Dynamics,* Ed., Sarma, R.H., Pergamon Press, New York, Oxford, p. 253 (1980).
8. Genest, D. and Wahl, P., *Biochim. Biophys. Acta 521, 502 (1978).*
9. Yguerabide, J., Epstein, H.F., and Stryer, L., *J. Mol. Biol. 51,* 573 (1970).
10. Munroe, I., Pecht, I., and Stryer, L., *Proc. Natl. Acad. Sci. USA 76,* 55 (1979).
11. Wahl, P., *Chem. Phys. 7,* 210 (1975).
12. Lipari, G. and Szabo, A., *Biophys. J. 30,* 489 (1980).
13. Andre, J.C., Vincent, L.M., O'Connor, D., and Ware, W.R., *J. Phys. Chem. 83,* 2285 (1979).
14. Grinvald, A. and Steinberg, I.Z., *Anal. Biochem. 59,* 583 (1974).
15. Yguerabide, *J., Met. Enzym. 26,* 498 (1972).
16. Isenberg, I., *J. Chem. Phys. 59,* 5696 (1973).
17. Gafni, A., Modlin, R.L., and Brand, L., *Biophys. J. 15,* 263 (1975).
18. Witholt, B. and Brand, L., *Biochemistry 9,* 1948 (1970).
19. Belford, G.C., Belford, R.L., and Weber, G., *Proc. Natl. Acad. Sci. USA. 69,* 1392 (1972).
20. Chuang, T.-J. and Eisenthal, K.B., *J. Chem. Phys. 57,* 5094 (1972).
21. Ehrenberg, M. and Rigler, T., *Chem. Phys. Lett. 14,* 539 (1972).
22. Small, E.W. and Isenberg, I., *Biopolymers 16,* 1907 (1977).
23. Perrin, F., *J. Phys. Rad. 7,* 1 (1936).
24. Mantulin, W.W. and Weber, G., *J. Chem. Phys. 66,* 4092 (1977).
25. Badea, M.G. and Brand, L., *Met. Enzym. 61,* 378 (1979).
26. Barkley, M.D., Kowalczyk, A.A., and Brand, L., *J. Chem. Phys.,* in press.
27. Zinsli, P.E., *Chem. Phys. 20,* 299 (1977).
28. Knutson, J.R., Kowalczyk, A.A., and Brand, L., unpublished data.

Proceedings of the Second SUNYA Conversation in the Discipline Biomolecular Stereodynamics
Volume I, ISBN 0-940030-00-4, Ed., Ramaswamy H. Sarma,
Adenine Press, New York, ©Adenine Press

Equilibrium and Kinetic Characteristics of the Low Temperature Open State in Polynucleotide Duplexes.

Richard S. Preisler, Chhabinath Mandal, S. Walter Englander and Neville R. Kallenbach
Departments of Biology, Biochemistry and Biophysics
University of Pennsylvania
Philadelphia, PA 19104

and

Frank B. Howard, Joe Frazier and H. Todd Miles
Laboratory of Molecular Biology
NIAMDD, National Institutes of Health
Bethesda, MD 20205

Introduction

The importance of the regularity and stability of DNA secondary structure for biological processes dependent on complementary base pairing, such as replication, recombination, and repair, is generally recognized. But the conformational flexibility of the double helix recently revealed by a number of physical and chemical probes[1] may also hold significance for biological functions. While a number of motions of bases and backbones occur on a time scale of nanoseconds, the fluctuational base-pair opening in native polynucleotide duplexes under physiological conditions detected by hydrogen-deuterium exchange (HX) has a half-time around one second.[2] This dynamic process could play an important role in the recognition of specific DNA base sequences by proteins such as polymerases, repressors, and restriction enzymes. Local melting appears to occur in the binding of RNA polymerase to form "open" promoter complexes with prokaryotic DNA,[3] an interaction which might depend on prior nucleic acid open states.

A number of small molecules, such as intercalators and diamines, and heavy metal ions also show sequence specificity in their association with DNA. The model for frameshift mutagenesis proposed by Streisinger *et al.*[4] and supported by the experiments of Helfgott and Kallenbach[5] holds that intercalators preferentially associate with mismatched regions. These aspects of ligand-nucleic acid interactions may be related to the effects of nucleotide sequence and defects in the helix on the opening process. We are investigating the properties of the open state in two model double helices of different base sequence, poly(rA)•poly(rU) and poly(rI)•poly(rC). A 1:1 complex of oligo(rA) and poly(rU) provides a model for a DNA molecule with single-strand defects. We have used the methods of hydrogen deuterium exchange (HX) and equilibrium spectroscopy to monitor the dynamics of the open state.

Hydrogen-Deuterium Exchange

The kinetics of hydrogen-deuterium exchange (HX) in poly(rA)•poly(rU) and poly(rI)•poly(rC) can be analyzed according to two first-order processes (Figure 1). An extensive study of HX in poly(rA)•poly(rU) and its component monomers and polymers[2] allowed the assignment of the fast reaction to the uracil N-3 imino proton and the slow reaction to the two exocyclic amino protons of adenine. Since in either complex two of these three protons are involved in Watson-Crick hydrogen bonding (Figure 2), the observed exchange must require base-pair breakage. The adenine amino protons exchange through a pre-equilibrium opening mechanism,

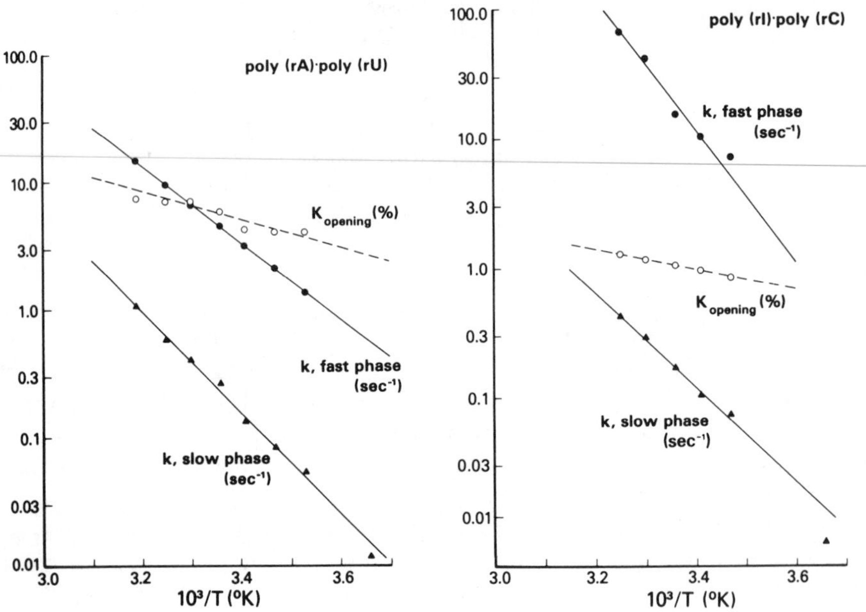

Figure 1. Open state parameters in two polynucleotide duplexes determined by HX at different temperatures. The rate constants (k) yield the following values of $\Delta H^{\neq}$: poly(rA)•poly(rU), fast=15 kcal/mole, slow=18 kcal/mole. Poly(rI)•poly(rC), fast=21 kcal/mole, slow=16 kcal/mole. The equilibrium openings ($K_{opening}$) yield $\Delta H^{\circ}_{opening}$ values of 3.8 kcal/mole for poly(rA)•poly(rU) and 3.7 kcal/mole for poly(rI)•poly(rC).

Polyadenylic (poly(rA)), polyuridylic (poly(rU)), polyinosinic (poly(rI)), and polycytidylic (poly(rC)) acids were purchased from P-L Biochemicals. Solutions of individual polymers were extensively dialyzed to remove small contaminating fragments. Polymer sizes were determined by electrophoresis on 8M urea, 10% acrylamide gels and found to be larger than 200 nucleotides. Concentrations were determined by ultraviolet absorption in a Perkin-Elmer 552 spectrophotometer. Poly(rA) and poly(rU), and poly(rI) and poly(rC), were mixed in equimolar quantities, assayed gravimetrically, to form their respective 1:1 complexes. Each mixture was heated to denaturing temperature and allowed to cool slowly to ensure complete renaturation.

Kinetic properties of base-pair opening reactions were measured in an OLIS stopped-flow spectrophotometer system equipped with a Bausch and Lomb monochromator and cuvette, photomultiplier, and drive syringes from Durrum interfaced with a NOVA2 computer. The polynucleotide in D_2O solution was mixed with nine volumes of H_2O buffer and the resulting small decrease in optical density around 290 nm was monitored on up to three different time scales.[2]

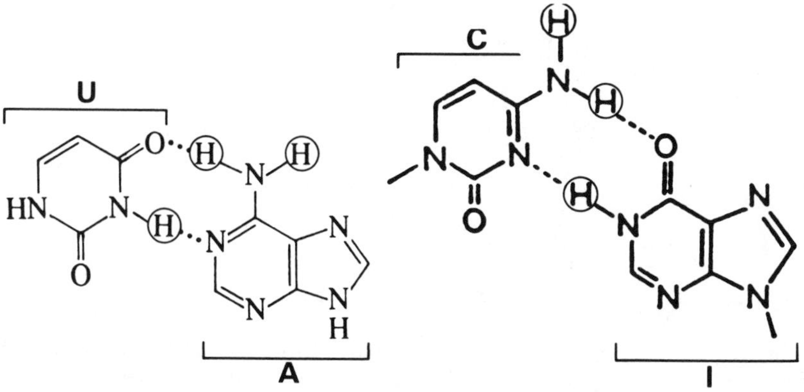

Figure 2. Exchangeable protons in poly(rA)•poly(rU) and poly(rI)•poly(rC). Kinetic experiments measure the exchange of the circled hydrogens.

sensitive to pH and catalysis, and comparison of the rate of this process with the analogous reaction in single-stranded poly(rA) or monomer AMP yields a measure of the equilibrium fraction of open base pairs, $K_{opening}$ (Figure 1). In this article, we have referred HX rates to the poly(rA). On the other hand, exchange of the uracil imino proton is insensitive to pH and catalysis and its rate measures the rate of helix opening.

We have recently performed a similar investigation of HX in poly(rI)•poly(rC), from which we conclude that the fast reaction (Figure 1) corresponds to the N-1 imino proton of hypoxanthine (in inosine) and the slow process involves the two exocyclic amino protons of cytosine (Figure 2). Nakanishi and Tsuboi[8] performed some of the same experiments and reported quantitatively similar results, but they interpreted their data according to a different model, which has since been retracted (M. Nakanishi, personal communication).

The structural characteristics of the open state are not defined by HX measurements. The calculated $\Delta G°$ opening of about 2 kcal/mole for poly(rA)•poly(rU) at 20°C is smaller than the value estimated for formation of a denatured bubble. Two extreme models consistent with the HX data are (1) a purely localized opening, where the uracil (hypoxanthine in poly(rI)•poly(rC)) residue swings out, leaving the adenine (cytosine) stacked[2] and (2) a travelling defect in the form of a coherent solitary wave or soliton.[9]

Equilibrium Spectroscopy

One approach toward refining our understanding of the open state is to look for direct manifestations of opening with equilibrium spectroscopic techniques. Infrared spectrophotometry (IR), while relatively insensitive to base stacking in single-stranded polynucleotides, is a powerful probe of the conformation of base paired nucleic acids. Circular dichroism (CD) provides a sensitive measure of geometri-

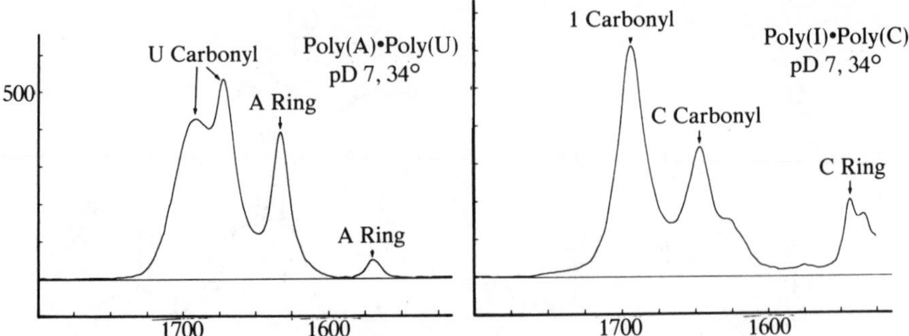

Figure 3. IR spectra of two polynucleotide duplexes (from H.T. Miles in G.D. Fasman, ed. *Handbook of Biochemistry and Molecular Biology,* Third Edition, Nucleic Acids—Vol. I, pp. 604-623, 1975).

cal aspects of interactions between bases, and reflects both stacking as well as base pairing.

Figures 3 and 4 present IR and CD spectra of poly(rA)•poly(rU) and poly(rI)•poly(rC). The amplitudes of the IR and CD bands of both double helices show dramatic changes at temperatures far below the onset of the helix-coil transition defined by UV hyperchromicity. Band frequencies shift to a lesser extent. Figures 5 and 6 present the IR and CD difference spectra generated by subtracting a reference low temperature spectrum from those obtained at higher temperatures.

The Nature of the Open State

Are HX, IR, and CD measuring the same phenomenon? If this were the case, one

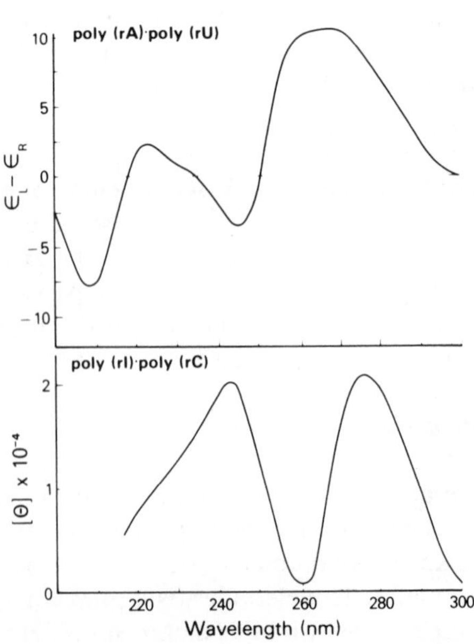

Figure 4. CD spectra of two polynucleotide duplexes (from D.M. Gray, I. Tinoco, and M.J. Chamberlin, *Biopolymers 11* 1235, 1972; and Y. Mitsui, R. Langridge, B.E. Shortle, C.R. Cantor, R.D. Grant, M. Kodama, and R.D. Wells. *Nature 228* 1166, 1970).

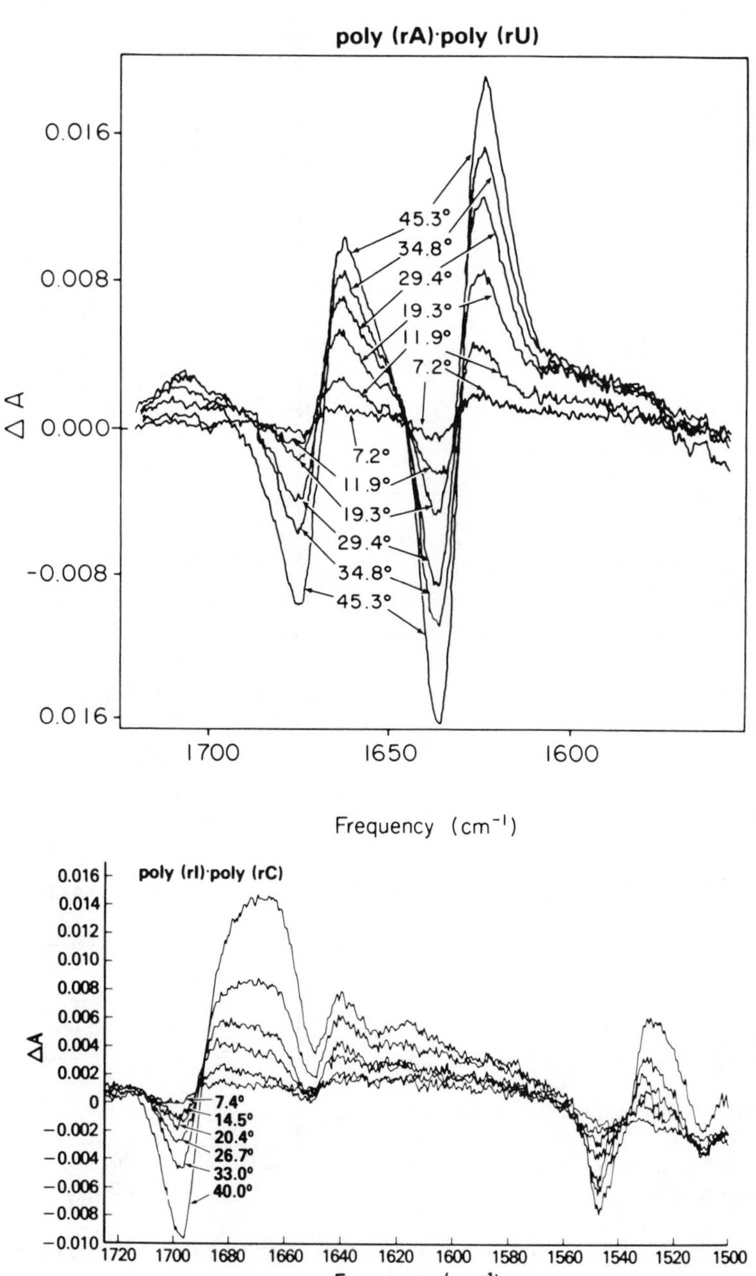

Figure 5. IR difference spectra of two polynucleotide duplexes at different temperatures. The spectrum measured at 4.9°C (poly(rA)•poly(rU)) or 4.5°C (poly(rI)•poly(rC)) was subtracted from all spectra shown. Infrared spectra in D_2O were measured in 26 or 54 micrometer CaF_2 cells in a Perkin-Elmer 580B spectrophotometer, interfaced to a DEC 1103 computer.

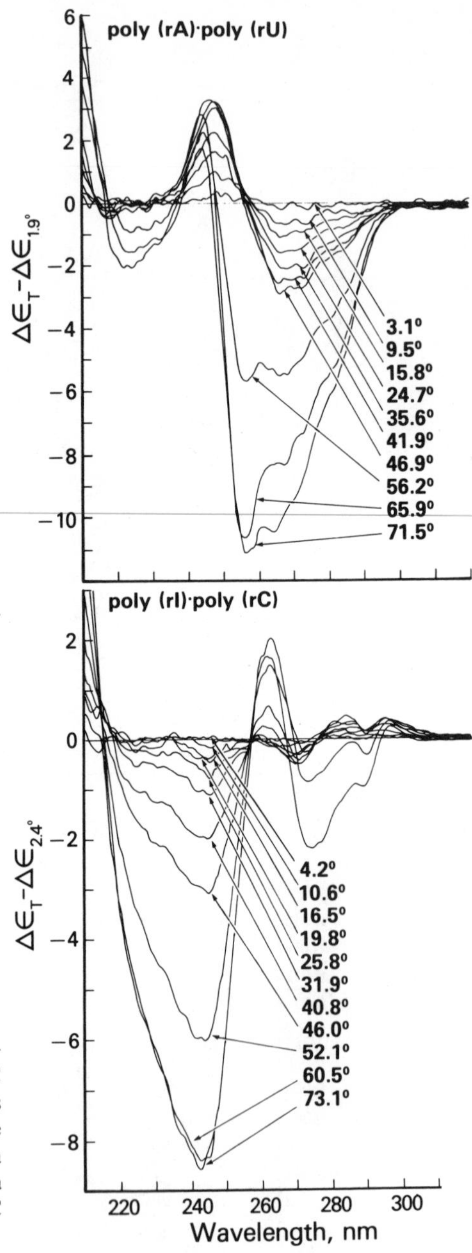

Figure 6. CD difference spectra or two polynucleotide duplexes. The spectrum measured at 1.9°C (poly(rA)•poly(rU) or 2.4°C (poly(rI)•poly(rC)) was subtracted from spectra collected at each of the temperatures shown. Circular dichroic spectra were collected using 1-cm quartz cuvettes with a Jasco J5OOA spectrometer, interfaced to a DEC 1103 computer.

would expect that the temperature dependence, expressed as $\Delta H°$, for these three sets of data would be quantitatively similar. As discussed above, the fraction of base pairs open at equilibrium can be calculated from the rate of the slow HX reaction if referred to the rate of either poly(rA) or AMP. However, the method to use for converting the IR and CD data to fractional openings is not obvious. One could express the difference spectral amplitude at temperature T as a percentage of the difference amplitude in the case of the fully denatured sample's spectrum. But such a procedure assumes that the low-temperature opening reflects the same process as the helix-coil melting transition. The following observations are inconsistent with that assumption: (1) As noted previously, the ΔG° for the opening detected by HX is lower than values determined for denatured bubble formation. (2) In poly(rA)•poly(rU), the opening rate is independent of Na^{+} concentration between 0.1M and 1.0M.[6] (3) Both IR and CD bands for poly(rA)•poly(rU) and poly(rI)• poly(rC) show relatively small frequency shifts in the premelting range, but large shifts near the T_m (Figures 5 and 6).

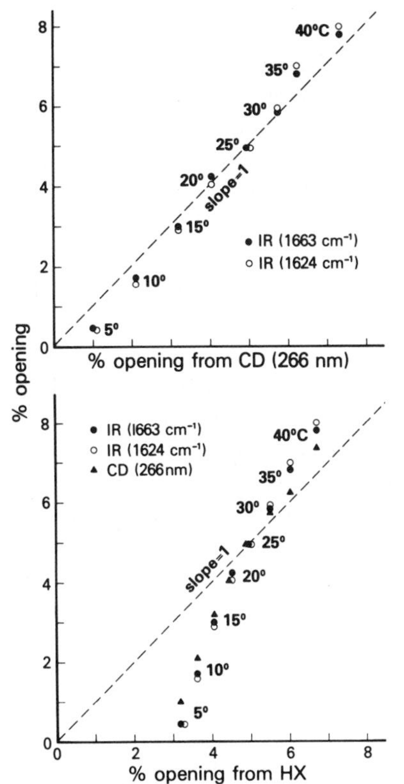

Figure 7. Correlation plots for measurements of opening in poly(rA)•poly(rU). Explanation in text.

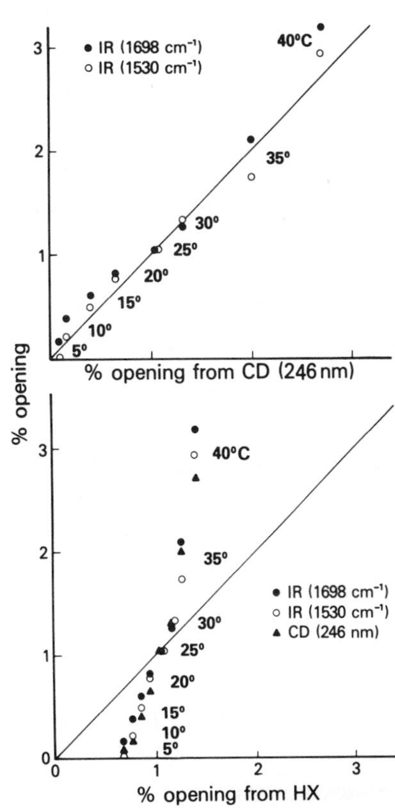

Figure 8. Correlation plots for measurements of opening in poly(rI)•poly(rC). Explanation in text.

The "correlation plots" in Figures 7 and 8 illustrate a method for comparing the temperature effects measured by different techniques that avoids the problem of assigning a 100%-open reference point for the IR and CD data. The difference spectral amplitudes measured at 25°C are set equal to the equilibrium opening values determined from HX at that temperature, 5% for poly(rA)•poly(rU) and 1% for poly(rI)•poly(rC). Amplitudes measured at other temperatures are then normalized to these values.

$\Delta H^{o}_{opening}$ Values from Correlation Plots (kcal/mole)

DUPLEX	HX	IR	CD
poly (rA)·poly (rU)	**3.8**	**7.8** (1663 cm^{-1}) **8.2** (1624 cm^{-1})	**6.7** (266nm)
oligo (rA)·poly (rU)	**3.4**	**9.6** (1663 cm^{-1}) **8.1** (1624 cm^{-1})	
poly (rI)·poly (rC)	**3.7**	**13.8** (1698 cm^{-1}) **13.3** (1530 cm^{-1})	**13.2** (246nm)

It is apparent that the temperature effects on the IR and CD measurements are significantly greater than those reflected by the HX data, especially in the case of poly(rI)•poly(rC). On the other hand, the IR and CD measurements show a striking correlation for both double helices. In the table are listed ΔH° values calculated from the slopes in the correlation plots. It is possible that IR and CD are measuring the same process, but that this is a different phenomenon from the open state revealed by HX. Alternatively, IR and CD may measure not only the opening to which HX is sensitive, but also other modes of conformational change which are either ineffective for, or gratuitous to, HX.

It is interesting to note that the estimated ΔH° for formation of a denatured bubble in poly(rA)•poly(rU) is between 6 and 9 kcal/mole.[2] Comparison of these values with the calculated data in the Table raises the possibility that significant amounts of bubble formation, which is more costly in enthalpy than the minimal base swing-out required for HX, occurs at low temperatures and is detected by IR and CD. If this is true, such openings obviously fail to permit exchange of hydrogens from the bases. Since we have not established the absolute magnitudes of these other open states, it may simply be that the IR and CD detected states exist at lower concentration.

Preliminary Attempts to Differentiate Between a Localized Open State and an Extended One Like a Soliton

Another approach towards elucidating the physical nature of the open state is to compare the opening behavior of oligomer-polymer and polymer-polymer duplexes. A purely localized opening should not be affected by single-strand breaks every 20 nucleotides, but a soliton or other form of extended open state might very well be sensitive to such defects.

Initial HX studies of a 1:1 complex between poly(rU) and crudely fractionated oligo(rA) sample with an average fragment length of 20 nucleotides showed an opening rate 2.5 times faster and an equilibrium opening 1.5 times greater than poly(rA)•poly(rU) at room temperature. However, the significance of the enhancement of opening seen with oligo(rA) is unclear, because the IR spectrum of another oligomer-polymer sample (with a better defined oligo(rA)$_{20}$ from RPC-5 chromatography) was indicative of triple helicity at room temperature.

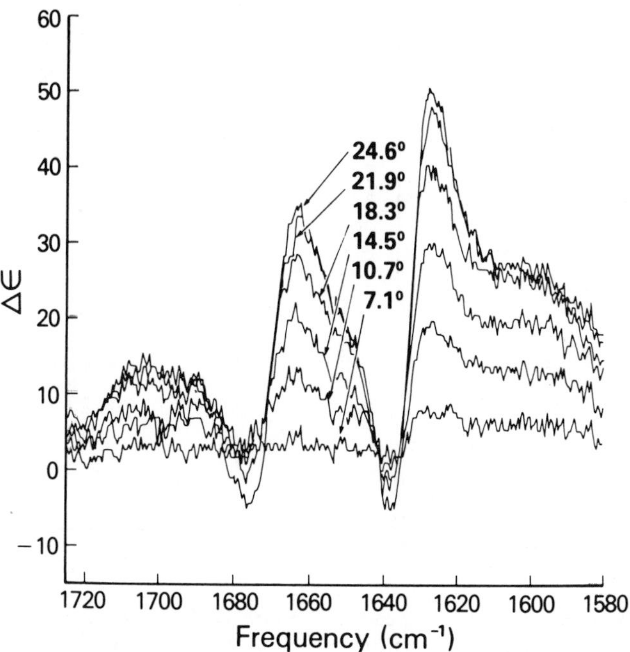

Figure 9. IR difference spectra of oligo(rA)•poly(rU). The spectrum measured at 4.3°C was subtracted from spectra collected at each of the temperatures shown. Oligo(rA) with an average fragment length of 40 nucleotides was prepared by high-pressure chromatography of partially hydrolyzed poly(rA) on an RPC-5 analog (Bethesda Research Laboratories) column. The method used was based on previously published procedures.[6]

Next we prepared oligo(rA)$_{40}$•poly(rU) and found that it was totally double helical below 35°C (the T_m for disproportionation was 45°C). IR difference spectra are presented in Figure 9. Figure 10 plots the IR data for this complex against those for poly(rA)•poly(rU). Calculated $\Delta H°$ values for oligo(rA)$_{40}$poly(rU) are listed in the Table. The temperature effect on opening appears to be somewhat greater for the oligomer-polymer complex measured at 1663 cm^{-1}, but there is no difference from the data at 1624 cm^{-1}. The Table and Figure 11 show HX data for the same complex. The equilibrium opening at room temperature and the $\Delta H°$ determined kinetically do not differ significantly from the values for poly(rA)•-poly(rU). On the other hand, the opening rate (fast phase) is larger at 25°C and more sensitive to temperature in the oligomer-polymer complex. Perhaps a complex containing more frequent interruptions in one strand would show an enhancement of opening at equilibrium. Since disproportionation cannot occur in the IC system, we are planning to compare the opening behavior of complexes between poly(rI) and oligo(rC) or poly(rC).

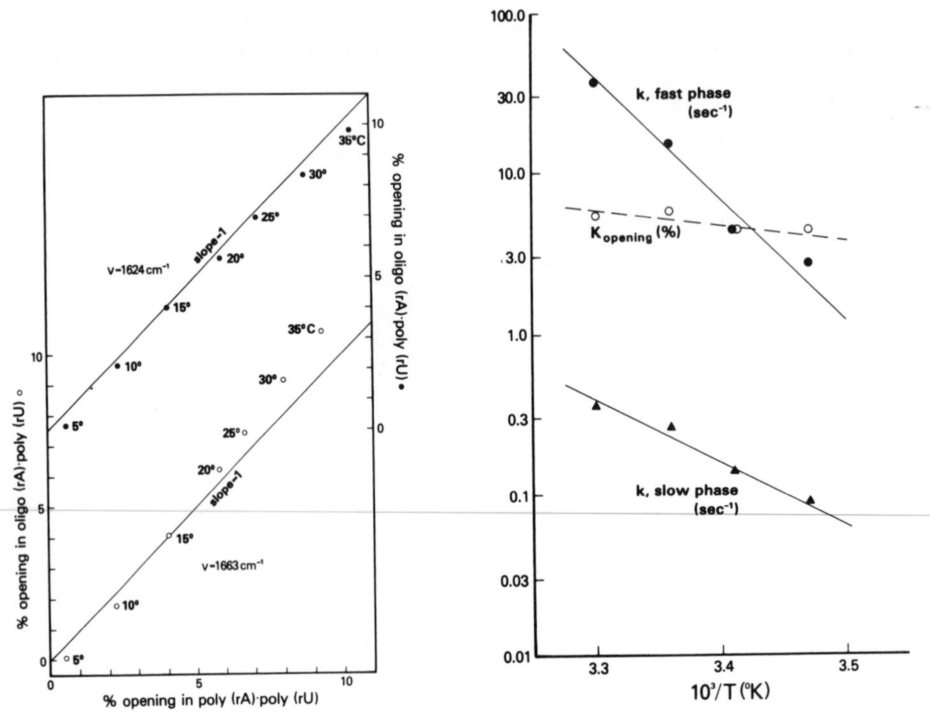

Figure 10. Correlation between IR measurements of opening for oligo(rA)•poly(rU) and poly(rA)•poly(rU). In this case, the measurements at 15°C were set equal to the fraction from HX of poly(rA)•poly(rU), 4.1%.

Figure 11. Open state parameters from HX for oligo(rA)•poly(rU), $\Delta H^{\neq}$ (fast phase) = 28 kcal/mole. $\Delta H^{\neq}$ (slow phase)=17 kcal/mol. $\Delta H^{\circ}_{opening}$ = 3.4 kcal/mole.

Conclusions

The open state measured by HX is qualitatively similar in poly(rA)•poly(rU) and poly(rI)•poly(rC). However, there are considerable quantitative differences—the latter complex opens 15 times faster, but contains only one-fifth as many open base pairs at equilibrium. Infrared spectrophotometry (IR) and circular dichroism (CD) also detect low temperature conformational changes, but their greater apparent ΔH° values make it uncertain whether these methods probe the same phenomenon as the HX open state. A 1:1 complex of oligo$(rA)_{40}$ and poly(rU) shows opening behavior similar to poly(rA)•poly(rU), as measured by HX and IR.

References and Footnotes

1. Sarma, R.H., ed. *Nucleic Acid Geometry and Dynamics.* New York, Pergamon Press (1980).
2. Mandal, C., Kallenbach, N.R., and Englander, S.W. *J. Mol. Biol. 135,* 391-411 (1979).
3. Chamberlin, M. In: Losick, R. and Chamberlin, eds., *RNA Polymerases.* Cold Spring Harbor, New York, pp. 159-192 (1976).
4. Streisinger, G., Okada, Y., Emrich, J., Newton, J., Tsugita, A., Terzaghi, E., and Inouye, M. *Cold Spring Harbor Symp. Quant. Biol. 31,* 77-84 (1966).

5. Helfgott, D.C. and Kallenbach, N.R. *Nuc. Acids Res. 7,* 1011-1017 (1979).
6. Wells, R.D., Hardies, S.C., Horn, G.T., Klein, B., Larson, J.E., Neuendorf, S.K., Panayotatos, N., Patient, R.K., and Selsing, E. *Meth. Enzymol. 65,* 327-347 (1980).
7. Mandal, C., Englander, S.W., and Kallenbach, N.R. *Biochem. 19,* 5819-5825 (1980).
8. Nakanishi, M. and Tsuboi, M. *J. Mol. Biol. 124,* 61-71 (1978).
9. Englander, S.W., Kallenbach, N.R., Heeger, A.J., Krumhansl, J.A., and Litwin, S. *Proc. Natl. Acad. Sci. U.S.A. 77,* 7222-7226 (1980).

Proceedings of the Second SUNYA Conversation in the Discipline Biomolecular Stereodynamics
Volume I, ISBN 0-940030-00-4, Ed., Ramaswamy H. Sarma,
Adenine Press, New York, ©Adenine Press

Short RNA Duplex Stability: Contribution from Non-Base-Paired Residues to the Direction of Stacking

D. Alkema, R.A. Bell, P.A. Hader and T. Neilson
Department of Biochemistry and Chemistry
McMaster University
Hamilton, Ontario L8N 3Z5 Canada

Introduction

Native RNA molecules are long polyribonucleotides which fold uniquely through the formation of hydrogen bonds. Watson-Crick pairs predominate and give rise to a series of duplex regions and single stranded loops.[1-3] If primary RNA sequences is known, prediction of stable RNA conformation should be possible, thus providing some insight into the structure-function relationships of native RNAs. Earlier optical studies[4-12] on factors which control the stability of RNA duplexes, using synthetic oligoribonucleotides, established various thermodynamic parameters[13] now in use, to predict secondary structure from known sequence. Refinement, however, is still necessary as several solutions are possible when applied to long RNA molecules.[14]

Contribution to overall duplex stability from non-base-paired residues has been essentially overlooked, even though these are often stacked and form single helices which are extensions of one of the duplex strands. The three-dimensional structure of transfer RNA[15] is illustrative of this feature. Optical studies have hinted that dangling adenosine residues do increase melting temperature (T_m) of duplexes formed from A_mU_n oligomers[4] and that 3′-dangling residues make a larger contribution than corresponding 5′-residues.[16] Temperature jump studies[17] have demonstrated that extension of stacked non-base-paired sequences enhance stability in short RNA duplex regions. Affinity between the complementary anticodon regions of yeast $tRNA^{Phe}$ and *E. coli* $tRNA^{Glu}$ is stronger, by a factor of 1000, than the interaction between one of these tRNAs and its corresponding complementary triribonucleotide. Increased stability arises from the additional base stacking which occurs in the loop regions.

To investigate further the effects of non-base-paired residues on RNA duplex stability, we are engaged in a varied ^1H NMR program using synthetic oligomers[18,19] of defined sequence. The nature of the NMR experiment provides information on the changing environment of each individual non-exchangeable proton affected by the helix coil transition.[20,21] Optical and relaxation methods are limited in this respect, since only gross conformation change can be observed. Using self-complementary reference duplexes, since their spectra contain fewer signals, we have shown that sequence controls stability even if the same number of G•C and A•U pairs occur.[22] Stacked dangling residues confer additional stability:[21] purines more than pyrimidines and 3′-residues over 5′-residues.[22]

This article reports that the direction of base stacking controls overall stability of duplexes containing terminal non-base-paired residues.

Chemical Synthesis of Oligoribonucleotides

All required oligoribonucleotides were prepared by the general phosphotriester synthesis[18,19] which provides large scale preparation of oligomers necessary for NMR studies (2-3 mg samples). An outline of the synthetic scheme is shown in Figure 1. Stepwise synthesis of a protected oligoribonucleotide starts at the 5′-terminus and proceeds toward the 3′-end of the desired sequence. After coupling each nucleotide residue, the product is deblocked and the spectra of the free intermediate oligomer are used in the assignment of NMR resonances of longer sequences. The spectra also provide an accurate base ratio of the intermediate; the sequence integrity of a longer product is checked at each synthetic step.

Table I

Summary of the Preparation of Protected Oligoribonucleotides[a]

REACTANTS						*PRODUCTS*		
	Quantity			*Quantity*			*Yield*	
Compd.	mg	mmol	Compd.	mg	mmol	Compd.	mg	%
A	1230	1.63	G	770	1.63	AG	1597	69
AG	600	0.42	C	180	0.42	AGC	400	47
AGC	400	0.20	U	80	0.24	AGCU	250	48
AGCU	120	0.05	A	21	0.05	AGCUA	63	43
A	3170	4.20	A	2293	5.04	AA	4937	84
AA	4937	3.51	G	1984	4.21	AAG	3880	54
C	4860	6.65	U	2617	7.98	CU	3778	45
CU	1780	1.42	A	759	1.67	CUA	1900	71
AAG	649	0.31	CU	359	0.38	AAGCU	718	71
AAG	322	0.16	CUA	300	0.19	AAGCUA	380	63
U	550	0.88	G	430	0.88	UG	560	49
UG	500	0.40	C	175	0.41	UGC	480	63
UGC	350	0.19	A	95	0.21	UGCA	360	77
UGCA	300	0.12	A	60	0.13	UGCAA	180	47
G	2500	3.24	C	1677	3.89	GC	1860	41
GC	1860	1.32	A	721	1.58	GCA	1862	69
GCA	400	0.20	A	106	0.23	GCAA	270	51

[a]Column 1 contains the 5′-trityloxacetyl reactants and A stands for TracA^{bz}tOH; column 4 contains the incoming nucleosides and G stands for HOG^{bz}tOH; column 7 contains the trityloxacetyl product and AC stands for TracA^{bz}tpG^{bz}tOH. Two equivalents of pyridinium mono-2, 2, 2-trichloroethyl phosphate activated by 4 equiv. of MST in anhydrous pyridine is used in each phosphorylation step. The coupling step to the incoming nucleoside is driven by 1.2 equiv. of MST.

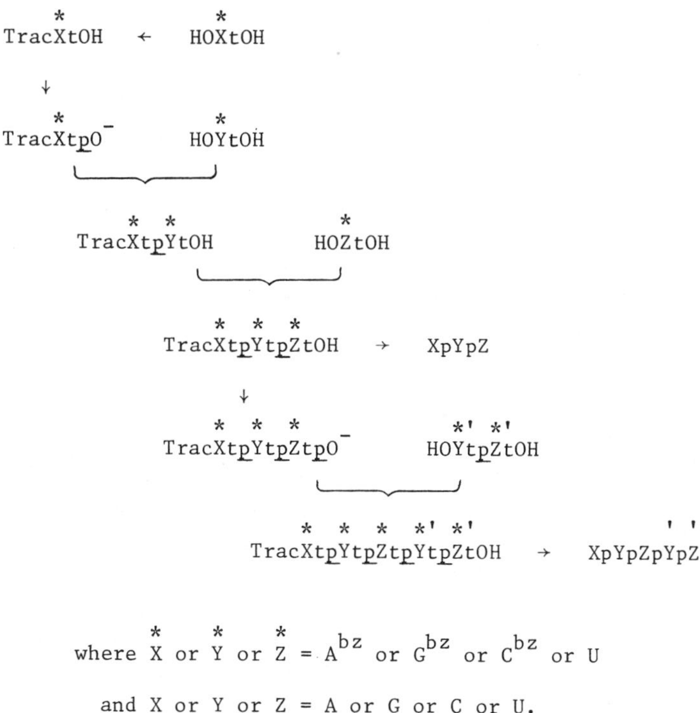

Figure 1. Chemical synthesis of oligonucleotides of defined sequence.
Abbreviations used: AGCU, tetranucleotide triphosphate ApGpCpU. In addition to the abbreviations recommended by IUPAC-IUB Commission (1970), the following are also used: trac, triphenylmethoxyacetyl; t, tetrahydropyranyl; pO$^-$, 3′-0-(2, 2, 2-trichloroethyl) phosphate; p between two characters, 3′, 5′-(2,2,2-trichloroethyl) phosphotriester; MST, mesitylenesulfonyl 1,2,4-triazolide; DSS, 2,2-dimethyl-2-silapentane-5-sulfonate; NMR, nuclear magnetic resonance; T_m, melting temperature (°C). All oligoribonucleotides are written in the 5′→3′ sequence, and the bases are numbered from the 5′ end. For double helices, the base pairs are numbered from left to right:

12345
→
AGCUA
••••
AUCGA

The following free oligoribonucleotides were used in this study: AGCU, AGCUA, AAGCU, AAGCUA, UGCA, UGCAA, GCA, GCAA and AGC. Table I contains the preparative data for the protected oligomers. Complete deblocking of the sequences was carried out using the three-step procedure.[18] Free oligoribonucleotides were isolated using paper chromatography (Table II).

Variable Temperature NMR Studies of Duplex Formation

The ^{1}H NMR spectra were obtained in the Fourier transform mode of Bruker WH-90, WM-250 and WM-400 spectrometers equipped with quadrature detection.

Table II
Experimental Data of Free Oligoribonucleotides

Compd.	R_F[a]	% Yield[b]	Compd.	R_F[a]	% Yield[b]
AGC	0.36	60	UGC	0.49	57
AGCU	0.31	33	UGCA	0.24	37
AGCUA	0.12	40	UGCAA	0.14	40
AAGCU	0.12	34	GCA	0.35	45
AAGCUA	0.08	41	GCAA	0.35	38

[a]Chromatography system: 1 M ammonium acetate - ethanol (50/50) on Whatman 40.
[b]Calculated from UV spectrophotometric data assuming a 90% hypochromicity factor.

Probe temperatures are maintained to within ±1°C by a Bruker variable temperature unit. Samples were lyophilized twice from D_2O and then dissolved in 100% D_2O (Aldrich) which contained 0.01 M sodium phosphate buffer (pD = 7.0) and 1.0 M sodium chloride. t-Butanol-OD was used as an internal reference and the chemical shifts are reported in ppm relative to DSS. Chemical shift values for each aromatic and anomeric proton were plotted against temperature in the range 70°-10° (Figure 2 for AGCUA). Cooperative behaviour of duplex formation was indicated by the sigmoidal nature of most curves. Melting temperatures (T_m's) were determined by graphic method.[9] Table III contains the average T_m values for all duplexes studied.

Chemical Shift Assignments of Free Oligomers

The procedure of incremental analysis[20] was used to make the chemical shift assignments summarized in Tables IV, V and VI. Irradiation of pyrimidine H-6

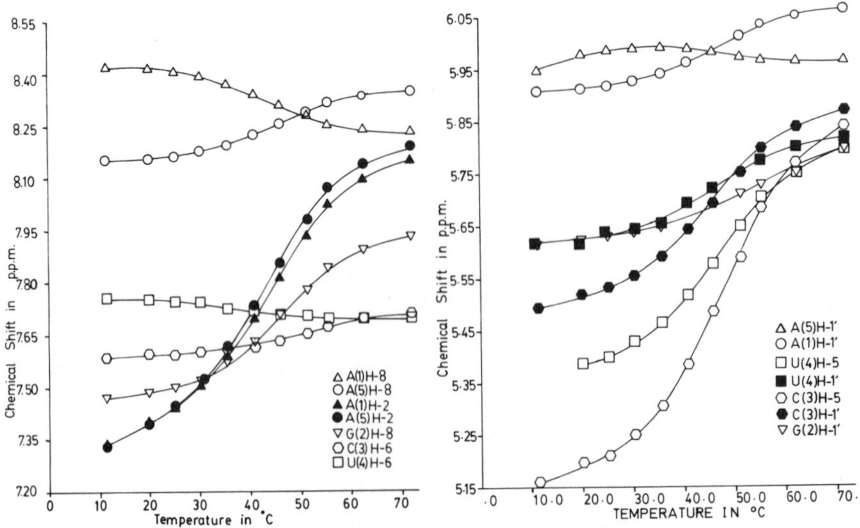

Figure 2. Chemical shift vs temperature for AGCUA.

Table III
Melting Temperatures of Base Paired Duplexes

Duplex	Average T_m (°C)	Duplex	Average T_m (°C)
AGC	ca. 0	GCA	33
AGCU	34	GCAA	34
AGCUA	45	UGCA	33
AAGCU	30	UGCAA	46
AAGCUA	48		

resonances and the subsequent collapse of corresponding pyrimidine H-5 resonances aided in the assignment of these resonances as well as the assignment of anomeric (H-1′) ribose resonances. Often UH-1′ and CH-1′ resonances were equivalent at 70°, but separated at lower temperatures. At low temperatures (<30°C),

Table IV
Chemical Shifts of AGCU Series at 70°C in D_2O[a,b]

Proton	AG	AGC	AGCU	AGCUA	AAGCU	AAGCUA
A(1)H-8	8.238	8.242	8.248	8.229	8.201	8.182
A(2)H-8					8.224	8.224
A(5)H-8				8.343		
A(6)H-8						8.323
A(1)H-2	8.186	8.180	8.176	8.144	8.084	8.038
A(2)H-2					8.119	8.051
A(5)H-2				8.186		
A(6)H-2						8.114
G(2)H-8	7.942	7.926	7.929	7.929		
G(3)H-8					7.844	7.810
C(3)H-6		7.744	7.737	7.704		
C(4)H-6					7.713	7.667
U(4)H-6			7.768	7.698		
U(5)H-6					7.767	7.686
A(1)H-1′	5.967	5.974	5.978	5.965	5.891	5.888
A(2)H-1′					5.930	5.922
A(5)H-1′				6.063		
A(6)H-1′						6.040
G(2)H-1′	5.842	5.812	5.791	5.798		
G(3)H-1′					5.710	5.695
C(3)H-1′		5.886	5.903	5.869		
C(4)H-1′					5.869	5.812
C(3)H-5		5.876	5.850	5.839		
C(4)H-5					5.788	5.730
U(4)H-1′			5.892	5.817		
U(5)H-1′					5.869	5.790
U(4)H-5			5.821	5.793		
U(5)H-5					5.806	5.712

[a]pD = 7.0; Concentrations: 9-12 mM.
[b]Chemical shifts are in ppm relative to DSS using *t*-butyl alcohol-OD as an internal reference and are accurate to ±0.005 ppm.

ribose H-1′ proton signals were observed as broad singlets ($J_{1'2'}$ < 1.0 Hz). Such a reduction in the coupling constant reflects a high degree of base stacking.[23]

AGCU Series (Table IV). The resonances of AG were assigned by comparison with published data.[24] Extension of the sequence to AGC resulted in the appearance of three additional doublets in the 70°C spectrum: C(3)H-6 (7.744 ppm), C(3)H-5 (5.876 ppm) and C(3)H-1′ (5.886 ppm).

Aromatic base resonances of the AGCU spectrum were readily assigned by observing two additional doublets: U(4)H-6 (7.768 ppm), U(4)H-5 (5.821 ppm), whereas the anomeric ribose U(4)H-1′ resonance coincided with the C(3)H-1′ at 5.892 ppm.

Comparison of the 70°C spectrum of AGCUA with that of AGCU shows the appearance of three new signals: A(5)H-8 (8.343 ppm), A(5)H-2 (8.186 ppm) and A(5)H-1′ (6.063 ppm). Addition of the terminal A resulted in a general shielding of most of the resonances. The only difficulty this shielding caused was in the assignment of the AH-2 resonances. The more shielded resonance was assigned to A(1)H-2, which reflects the general shielding trend throughout the sequence.

The resonances attributable to the additional 5′-adenosine in AAGCU were routinely assigned by reference to the AGCU 70°C spectrum: A(1)H-8 (8.204 ppm), A(1)H-2 (8.084 ppm) and A(1)H-1′ (5.891 ppm). This 5′-dangling adenosine results in some shielding of the neighbouring residues but is not as extensive as with the 3′-dangling adenosine in AGCUA.

The spectrum for AAGCUA which contains both an additional 3′ and 5′ adenosines over AGCU was assigned by comparison to the spectra for AAGCU and AGCUA.

Table V

Chemical Shifts of UGCA Series at 7O°C in $D_2O^{a,b}$

Proton	UG	UGC	UGCA	UGCAA
A(4)H-8			8.362	8.270
A(5)H-8				8.278
A(4)H-2			8.202	8.069
A(5)H-2				8.164
G(2)H-8	8.004	8.015	7.975	7.966
C(3)H-6		7.809	7.711	7.692
U(1)H-6	7.723	7.731	7.716	7.710
A(4)H-1′			6.068	5.948
A(5)H-1′				6.020
G(2)H-1′	5.904	5.896	5.794	
C(3)H-1′		5.932	5.875	
C(3)H-5		5.958	5.853	5.838
U(1)H-1′	5.816	5.812	5.811	
U(1)H-5	5.828	5.832	5.811	5.807

Footnotes: See Table IV.

The resonances corresponding to the A(1)H-8, A(1)H-2 and A(1)H-1′ were assigned as 8.182 ppm, 8.053 ppm and 5.888 ppm respectively. The signals appearing at 8.232 ppm, 8.114 ppm and 6.040 ppm were assigned to A(6)H-8, A(6)H-2 and A(6)H-1′, respectively.

UGCA Series (Table V). The resonances of UG were assigned by comparison with published data.[25] Extension to UGC resulted in the appearance of three additional doublets in the 70°C spectrum: C(3)H-6 (7.809 ppm), C(3)H-5 (5.958 ppm) and C(3)H-1′ (5.832 ppm).

The adenosine resonances of the UGCA spectrum were easily identified by the appearance of two additional singlets: A(4)H-8 (8.362 ppm) and A(4)H-2 (8.202 ppm) and a ribose doublet, A(4)H-1′ (6.068 ppm). A general shielding trend extended throughout the sequence and agrees with the findings for AGCUA.

Comparison of the 70°C spectrum of UGCAA with its precursor UGCA shows the appearance of another set of adenine signals: A(5)H-8 (8.278 ppm), A(5)H-2 (8.164 ppm) and A(5)H-1′ (6.020 ppm). Again a general shielding occurs.

The assignments for GCA and GCAA were obtained in similar fashion and can be found in Table VI.

Table VI
Chemical Shifts of GCA Series at 70°C in $D_2O^{a,b}$

Proton	GC	GCA	GCAA
A(3)H-8		8.348	8.313
A(4)H-8			8.313
A(3)H-2		8.191	8.101
A(4)H-2			8.181
G(1)H-8	7.966	7.920	7.909
C(2)H-6	7.780	7.909	7.717
A(3)H-1′		6.061	5.977
A(4)H-1′			6.038
G(1)H-1′	5.884	5.812	5.809
C(2)H-1′	5.913	5.883	5.867
C(2)H-5	5.910	5.837	5.849

Footnotes: See Table IV.

Effects on Duplex Stability

Secondary RNA structure results from more than mere Watson-Crick base-pairing. Base stacking is controlled by length and nearest neighbour interactions (i.e. sequence). Non-complementary base-pairs, dangling bases, loop size and terminal fraying effects, also contribute to general stability. Optical studies can provide values for overall conformation change but fail to evaluate the algebraic contribution of each feature. Variable temperature ¹H NMR spectroscopy has that capability, and already has been used to identify various non-bonded interactions, G-U base-pairs[26] and terminal effects[22] as perturbations within short duplexes. The role of stacking direction in short RNA duplex stability will now be considered.

Direction of Base Stacking

A series of synthetic oligoribonucleotides of the type $(M)_mGC(N)_n$ was prepared.[18,19] Each contains a self-complementary GC core, and additional residues (M or N = A or U), destined to occupy base paired regions or, 3′ or 5′-dangling positions upon duplex formation (see Figure 3).

Sequences of the type, $(M)_mGCA(N)_n$, in particular GCA, GCAA, UGCA and UGCAA, clearly demonstrate the increased helix stability derived from the contribution of a non-base paired adenosine residue. Comparison of GCA with UGCA,[27] both of which exhibit a T_m of 33°C, dramatically illustrates that the stability in GCA is primarily due to the presence of 3′ well-stacked, non-base-paired adenosine residues which are equivalent to A•U pairs in their stabilizing effect. On the other hand, the corresponding duplex from trinucleotide AGC, also contains a GC core but has 5′-dangling adenosines and exhibits an estimated $T_m \simeq 0$°C. These adenosine residues also non-base-paired, must contribute less to overall duplex stability. Presumably the greater helical overlap of a 3′-base residue generates a greater

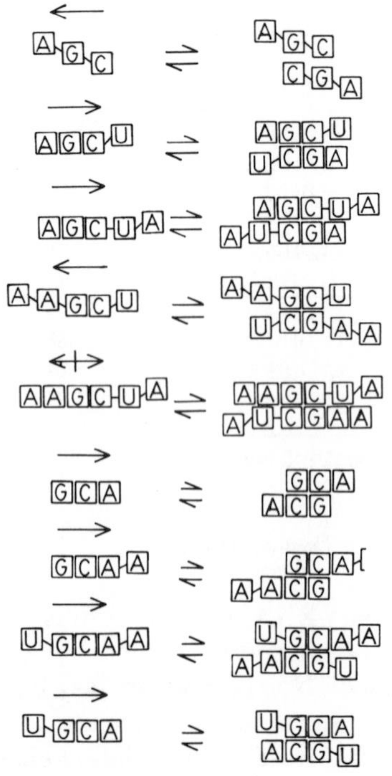

Figure 3. Diagramatic representation of how base stacking can effect helix stability in short oligoribonucleotides of the type $(M)_mGC(N)_n$. The degree of stacking is illustrated by deviations from the linear array of boxes which represent a fully stacked helix. Arrows above represented sequences indicate the direction of stacking as determined from $J_{1',2'}$ variable temperature plots. The degree of base stacking in the individual helices are for temperatures at the T_m's of the various duplexes.

aromatic ring-current interaction within a strand, enhancing overall base stacking which in turn, strengthens duplex formation. This observation clearly contrasts those noted earlier in the case of oligomer, CAUG ($T_m \simeq 25°C$) where dangling adenosine residues increased stability ($T_m \simeq +11°C$), irrespective of whether the base was 3′ or 5′.[22] It should be noted, however, that the 3′-dangling adenosine flanks a G residue in CAUGA and the 5′-dangling adenosine flanks a C residue in ACAUG.

GCA vs. AGC. The $J_{1'2'}$ coupling constants observed at all the H-1′ signals in the $(M)_mGC(N)_n$ series of compounds showed a gradual, sometimes erratic, decrease in magnitude as the temperature was lowered. Figure 4 shows the plots of $J_{1'2'}$ against temperature for the two representative cases GCA and AGC. While the trends in the behaviour of the coupling constants were similar (the $J_{1'2'}$ values decreased in the order A G, C at all temperatures), the position of the adenosine residue is opposite (5′ as opposed to 3′) in the two sequences. This residue causes the direction of base stacking in AGC (C before G before A) to be reversed in comparison to GCA where the stacking is G,C before A.

In the dinucleoside monophosphate series, low values of the $J_{1'2'}$ coupling constant

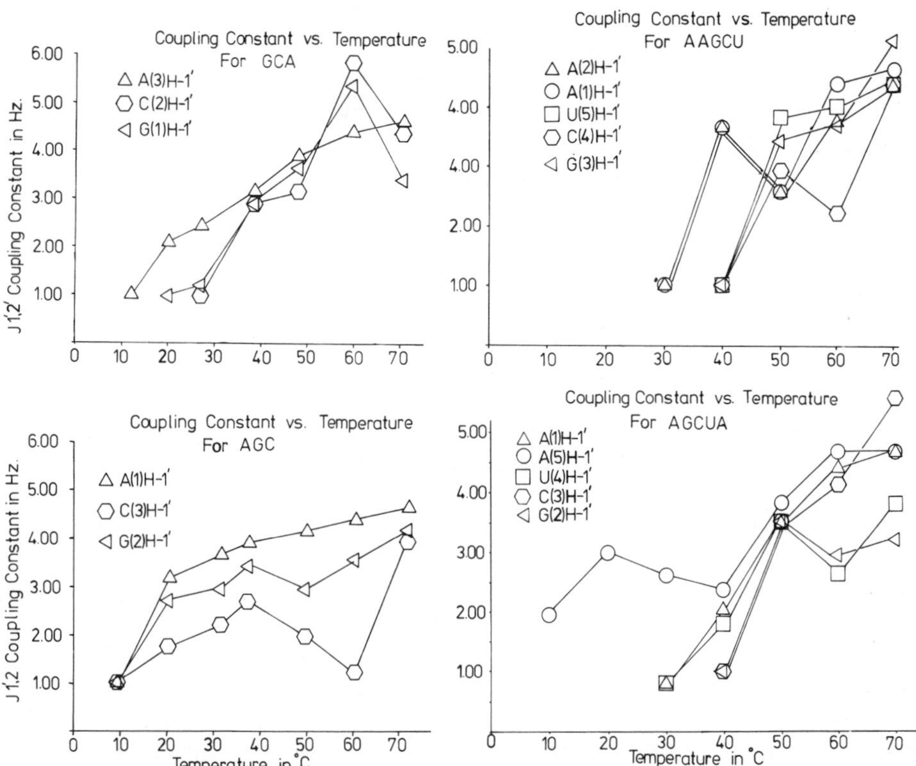

Figure 4. Coupling constant vs temperature for GCA, AGC, AAGCU and AGCUA.

have been associated with a *3′-endo* ribose ring conformation and a high degree of base stacking.[24,25] If a similar relationship exists between the magnitude of $J_{1'2'}$ and base stacking for trimers as for dimers, then the present data, imply that the trimer AGC stacks from the 3′-end to the 5′-end, while the GCA oligomer stacks from the 5′-end to the 3′-end.[28]

5′-to 3′-directional stacking is found throughout the $(M)_mGCA(N)_n$ oligomers so far investigated, and also applies in the $(M)_mAGC(N)_n$, (m = 0 or 1) series of compounds that exhibit greater duplex stability as reflected in their T_m's. Dangling adenosine residues appear to enhance duplex stability in $(M)_mGC(N)_n$ compounds when present at the 3′-end of the molecule, where their presence does not disrupt the 5′-end to 3′-end base stacking. This contrasts the effect in the corresponding duplexes with 5′dangling adenosine residues which apparently alter the base stacking direction to 3′ to 5′, with an accompanying decrease in stability. Trimer AGC where the 5′-dangling adenosine residue disrupts the GC core base stacking is a case in point.

AAGCU vs. AGCUA. Duplex formed from oligomer, AAGCU, which contains a 5′-dangling adenosine residue also exhibits an overall 3′-end to 5′-end base stacking (Figure 4) while the duplex from oligomer, AGCUA, which contains a 3′-dangling adenosine residue, exhibits the more stabilizing 5′-end to 3′-end base stacking (Figure 4), and has a T_m = 45°C. Duplex, AAGCU, has a T_m even lower ($T_m \simeq$ 30°C) than that of the parent duplex AGCU (T_m = 34°C). 5′-Dangling adenosine residues apparently destabilize the duplex formed by AAGCU by altering the base stacking direction, subsequently lowering the helical overlap of base residues and increasing fraying in the AAGCU duplex. In contrast, duplexes from oligomers, AGCUA and UGCAA, where the 3′-dangling adenosine residue extends the 5′-end to 3′-end base stacking, strengthens the 3′-terminal A•U base pairing, and increases the overall stability.

AAGCUA. Duplex from AAGCUA which contains *two* 3′- and *two* 5′-dangling adenosine residues, exhibits a slightly higher T_m (48°) than duplex AGCUA. Apparently the effect of the 3′-dangling adenosines predominates and directs base stacking to adopt the 5′-to 3′-direction. The additional 5′-dangling adenosines become an extension to this established stacking and so enhances overall duplex stability.

Terminal A•A Pairs

Comparison of the duplex from AAGCUA which has two terminal A•A non-base-paired pairs with its parent tetramer duplex AGCU, is significant (T_m's 48° vs 33°). Internal A•A pairs are centres of instability within duplexes, for example duplex, CAAUG:CAAUG cannot be detected.[26] However, in the duplex with terminal A•A pairs, sufficient geometric flexibility must exist to tolerate the adenosine residues in opposing positions and still maintain extension of base stacking to these same adenosine residues. A partially "unwound" hexamer duplex appears to be more stable than a "tight" tetramer duplex with a similar Watson-Crick base-paired core.

This finding has direct bearing on the junction of duplex and non-duplex regions within native RNA molecule. Purine residues in the immediate non-paired positions may contribute to overall stability despite their apparent steric interaction.

Conclusion

In our search for factors which stabilize native RNA molecules, the present article introduces two additional effects which control secondary structure, namely the nature of bases adjacent to any duplex region and the direction of base stacking. Minor perturbations (an additional A residue) can change this direction and so duplex stability. It is tempting to speculate on the biological significance of possible changes in duplex stability promoted by reverse base stacking as a result of interaction with proteins or other nucleic acid molecules, for example, in polymerase recognition sites.

References and Footnotes

1. Gross, H.J., Domdey, H., Lossow, C., Jank, P., Raba, M., Alberty, H. and Sänger, F. *Nature 273,* 203-208 (1978).
2. Fiers, W., Contreras, R., Duerinck, F., Haegeman, G., Tserentant, D., Merregaert, J., Min Jou, W., Molemans, F., Raeymaekers, A., van der Berghe, A., Volckaert, G. and Ysebaert, M. *Nature 260,* 500-507. (1976).
3. Fiers, W., Contreras, R., Duerinck, F., Haegeman, G., Merregaert, J., Min Jou, W., Raeymaekers, A., Volckaert, G., Ysebaert, M., Van de Kerckhove, J., Nofl, F. and Van Montagu, M. *Nature 256,* 273-278 (1975).
4. Martin, F.H., Uhlenbeck, O.C. and Doty, P. *J. Mol. Biol. 57,* 201-215 (1971).
5. Uhlenbeck, O.C., Martin, F.H. and Doty, P. *J. Mol. Biol. 57,* 217-229 (1971).
6. Craig, M.E., Crothers, D.M. and Doty, P. *J. Mol. Biol. 62,* 383-401 (1971).
7. Pörschke, D., Uhlenbeck, O.C. and Martin, F.H. *Biopolymers 12,* 1313-1335 (1973).
8. Gralla, J. and Crothers, D.M. *J. Mol. Biol. 73,* 497-511 (1973).
9. Gralla, J. and Crothers, D.M. *J. Mol. Biol. 78,* 301-319 (1973).
10. Uhlenbeck, O.C., Borer, P.N., Dengler, B. and Tinoco Jr., I. *J. Mol. Biol. 73,* 483-496 (1973).
11. Wickstrom, E. and Tinoco Jr., I. *Biopolymers 13,* 2367-2383 (1974).
12. Borer, P.N., Dengler, B., Tinoco Jr., I. and Uhlenbeck, O.C. *J. Mol. Biol. 86,* 843-853 (1974).
13. Tinoco Jr., I., Borer, P.N., Dengler, B., Levine, M.D., Uhlenbeck, O.C., Crothers, D.M. and Gralla, *J. Nature New Biol. 246,* 40-41 (1973).
14. Studnicka, G.M., Rahn, G.M.,Cummings I.W. and Salser, W.A. *Nucl. Acids Res. 5,* 3365-3386 (1978).
15. Rich, A. *Accts. of Chemical Research 10,* 388-396 (1977).
16. Gennis, R.B. and Cantor, C.R. *Biochemistry 9,* 4714 (1970).
17. Grosjean, H., Soll, D.G. and Crothers, D.M. *J. Mol. Biol. 103,* 499-509 (1976).
18. England, T.E. and Neilson, T. *Can. J. Chem. 54,* 1714-1721 (1976).
19. Werstiuk, E.S. and Neilson, T. *Can. J. Chem. 54,* 2689-2696 (1976).
20. Borer, P.N., Kan, L.S. and Ts'o, P.O.P. *Biochemistry 14,* 4847-4863 (1975).
21. Romaniuk, P.J., Hughes, D.W., Gregoire, R.J., Neilson, T. and Bell, R.A. *J. Am. Chem. Soc. 100,* 3971-3972 (1978).
22. Neilson, T., Romaniuk, P.J., Alkema, D., Hughes, D.W., Everett, J.R. and Bell, R.A. *Nucl. Acids Res., Symp. Series No. 7,* 293 (1980).
23. Altona, C. in, *Structure and Conformation of Nucleic Acids and Protein-Nucleic Acid Interactions.* Sundaralingam, M. and Rao, S.T., eds. p. 613 (1975). Baltimore, University Park Press.
24. Lee, C.-H., Ezra, F.S., Kondo, N.S., Sarma, R.H. and Danyluk, S.S. *Biochemistry 15,* 3627-3639 (1976).

25. Ezra, F.S., Lee, C.-H., Kondo, N.S., Danyluk, S.S. and Sarma, R.H. *Biochemistry 16,* 1977-1987 (1977).
26. Romaniuk, P.J., Hughes, D.W., Gregoire, R.J., Bell, R.A. and Neilson, T. *Biochemistry 18,* 5109-5116 (1979).
27. Alkema, D., Bell, R.A., Hader, P.A. and Neilson, T. *J. Amer. Chem. Soc. 103,* 2866-2868 (1981).
28. Everett, J.R., Hughes, D.W., Bell, R.A., Alkema, D., Neilson, T. and Romaniuk, P.J. *Biopolymers 19,* 557-573 (1980).

Proceedings of the Second SUNYA Conversation in the Discipline Biomolecular Stereodynamics
Volume I, ISBN 0-940030-00-4, Ed., Ramaswamy H. Sarma,
Adenine Press, New York, ©Adenine Press

Why Do Nucleic Acids Form Helices?

Douglas H. Turner, Matthew Petersheim, Diane DePrisco Albergo,
T. G. Dewey, and Susan M. Freier
Department of Chemistry
University of Rochester
Rochester, N.Y. 14627

Introduction

The importance of nucleic acid structure and dynamics for the process of life is evident. Nevertheless, there is no clear understanding of the fundamental forces that determine these structures and dynamics.[1-4] Interactions that have been proposed to be important include: (1) vertical electronic interactions between the bases,[5-7] (2) hydrogen bonding, (3) hydrophobic bonding,[8-10] and (4) solvophobic bonding due to the energy required for cavity formation in water.[11-15] We are attempting to estimate the relative contributions of these forces to helix formation in both single and double strand nucleic acids, by measuring thermodynamic, kinetic, and spectroscopic properties of model systems.[16-22] Since helix formation is driven by the enthalpy change, we have concentrated on the origin of this favorable enthalpy term. In the following, we review our progress thus far.

Single stranded polynucleotides such as polyadenylic acid (poly A) and polycytidylic acid (poly C) are known to undergo a transition at neutral pH from a random coil conformation at high temperature to a helical, stacked conformation at low temperature:[23-28]

$$\text{poly X (random coil)} \underset{k_{-1}}{\overset{k_1}{\rightleftharpoons}} \text{poly X (stacked)}$$

Presumably, this transition does not involve any hydrogen bonding of the bases. Thus the thermodynamics provide an estimate for the contribution of intrastrand "stacking" effects to helix formation. Unfortunately, there is considerable disagreement over these thermodynamics. For example, the reported enthalpy change associated with poly A stacking ranges from −3 to −13 kcal/mole.[1,24-26,29-32] These values can be compared with the ΔH of about −7 kcal/mole base pair associated with forming fully stacked, double helical poly A-poly U from approximately half stacked poly A and completely unstacked poly U at 40°C.[31-33] It is reasonable to assume that poly U stacking is associated with a negligible ΔH since it is essentially a random coil above 15°C.[34,35] Thus, depending on which value is correct for the ΔH of stacking in poly A, it is possible to conclude that base-base stacking is a negligible or dominant contributor to the ΔH of double helix formation.

The major problem in measuring the thermodynamics for single strand stacking is that the transitions are very broad so the properties of the pure random coil and helical states cannot be directly measured. For example, in analyzing absorbance vs. temperature melting curves of single strand stacking, it is not possible to measure the extinction coefficients, ϵ_U and ϵ_S, of the completely unstacked and stacked states (see Figure 2). This makes it difficult to determine the fraction of each state present at each temperature, and therefore the thermodynamics. In principle, this problem can be overcome by curve fitting, if there is a suitable model describing the transition. The best model for single strand stacking is the one dimensional Ising model.[26,36] Unfortunately, there are five parameters in the theory: ϵ_U, ϵ_S, ΔH, ΔS, and σ, the cooperativity parameter. Unique determination of all five parameters is not possible from melting curves alone. We have recently overcome this limitation by combining melting curve data with differential scanning calorimetry (DSC).[20] In essence, DSC provides additional experimental values that make possible the determination of all five parameters. In practice, the curvature of the excess heat capacity vs. temperature (DSC) curve is compared with that predicted from the ΔH and ΔS derived from fitting melting curve data assuming a particular σ. The σ providing the best congruence of predicted and measured curvature is the cooperativity parameter for the transition. It is necessary to focus on the curvature of the DSC curve because there is a linear background heat capacity that is not associated with stacking. The spectroscopically predicted and calorimetrically measured excess heat capacity curves that give best agreement for poly C are shown in Figure 1. The parameters derived are $\sigma = 1$ (i.e. the transition is not cooperative) and $\Delta H = -9$ kcal/mol stack.[20] Using previously published calorimetric[31,33] and spectroscopic[18] data for poly A, we obtain $\sigma = 0.3$[31] or 0.6[33] corresponding to $\Delta H = -5.6$ or -8.0 kcal/mol stack, respectively. The magnitudes of these values suggest base-base stacking is a major contributor to the stability of double helices.

As pointed out above, the thermodynamics of single strand stacking are difficult to determine. It is therefore desirable to have an independent measure of the contribution of base stacking to helix stability. One approach is to add a nucleotide that cannot base pair onto the end of a fully formed double helix. Such "dangling ends" are known to increase the stability of the double helix,[37-40] and are thought to be important for codon-anticodon interaction.[38,39] To investigate this effect, we have preliminarily measured the thermodynamics of double helix formation for the self-complementary ribo-oligonucleotides CCGG, ACCGG, and CCGGA.[41] One advantage of this approach is that it couples the stacking interaction to a very cooperative helix-coil transition. This makes it easier to derive the thermodynamics. For example, Figure 2 shows melting curves for the non-cooperative poly C and the cooperative ACCGG transitions. Previously published NMR melting curves for CCGG indicate that all four base pairs melt together.[42] Our own results confirm this, and indicate that ACCGG behaves similarly (see Figure 3). This strong cooperativity results in a sharp melting transition that can be analyzed with a two-state model. It has been demonstrated that reliable thermodynamics for this type of transition can be obtained from spectrophotometric melting curves by plotting inverse melting temperature vs. log (concentration).[21,43] The thermodynamics

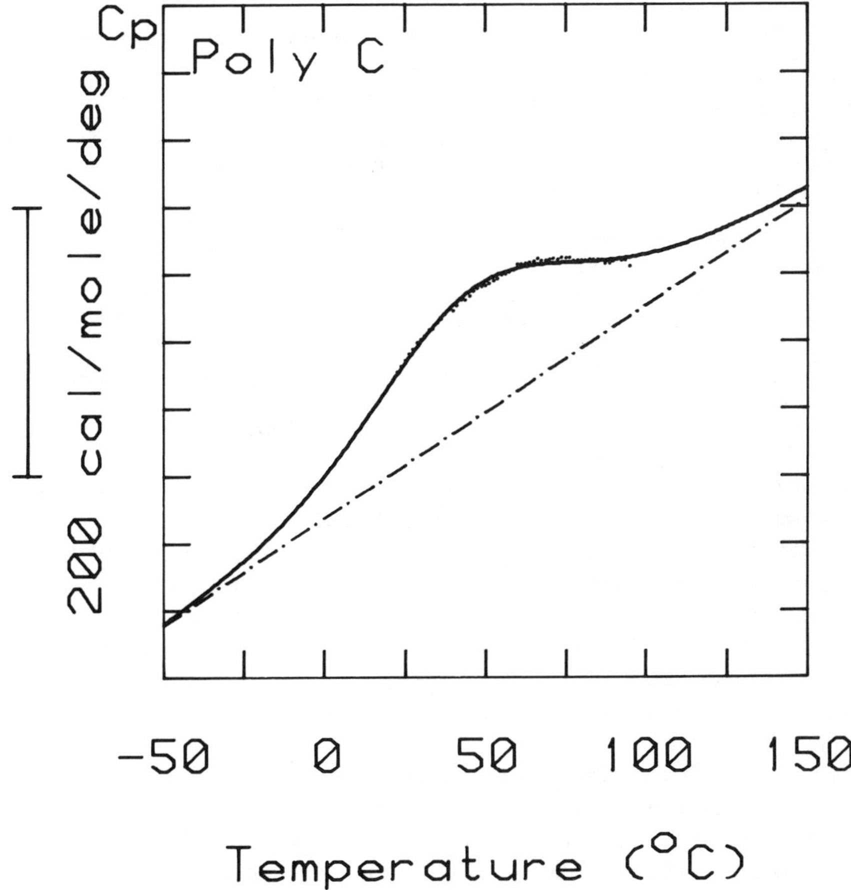

Figure 1. Experimental and calculated heat capacity curves for the coil to helix transition of poly C. ΔH = −9.04 kcal/mol-stack; ΔS = −27.6 cal/(mol-stack deg); $\sigma = 1.0$. The solid curve (——) is the sum of the calculated excess heat capacity and a linear background heat capacity (— • —). The experimental data are represented as dots superimposed on the solid curve. The scale on the ordinate is 50 cal/(mol deg) per division.

obtained from this analysis are listed in Table I. Significantly, the extra adenine stacks increase the melting temperature by more than 10°C, and result in a more favorable enthalpy change for helix formation. The magnitude of the enthalpy effect is quite different for ACCGG and CCGGA. Future work will provide more details on the sequence dependence of the stacking contribution of dangling ends. In general, however, the results support the idea that base-base stacking is a major contributor to helix stability, and that it is associated with a favorable enthalpy change.

Two origins have been suggested for the favorable enthalpy of stacking. The first is vertical electronic interactions between the bases.[5-7] These include dipole-dipole, dipole-induced dipole, and induced dipole-induced dipole forces. The second potential contributor is solvophobic bonding, as proposed by Sinanoglu.[11-15] In this clever

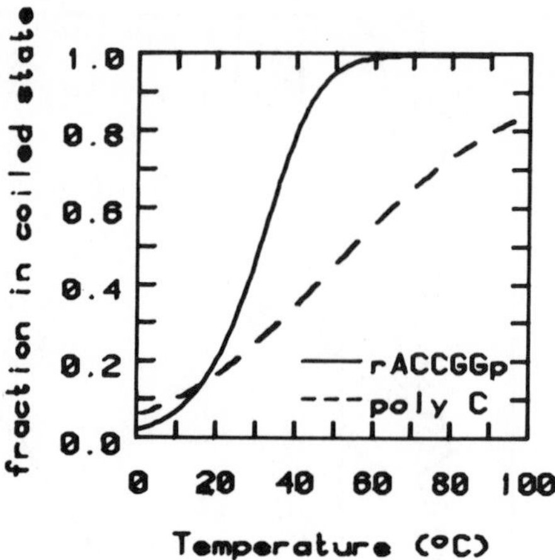

Figure 2. Fraction of molecules in the coil state for r-ACCGGp at a strand concentration of 5.0 x 10^{-5} M (——) and single stranded poly C (- - -) as a function of temperature. The fraction in the coil state was calculated using the two-state model assuming $\Delta H = -39$ kcal/mol and $\Delta S = -108$ cal/(mol deg) for r-ACCGGp and $\Delta H = -9.04$ kcal/mol stack and $\Delta S = -27.6$ cal/(mol-stack deg) for poly C. The thermodynamic parameters for r-ACCGG were obtained from a plot of log (concentration) vs. the spectrophotometrically measured melting temperature in 1 M NaCl, 10 mM sodium cacodylate, 0.1 mM EDTA, pH 7. The poly C thermodynamics were obtained from a fit of absorbance vs. temperature profiles in 0.05 M sodium cacodylate, pH 7 to the two-state model.

theory, the solvent has to form a cavity in order to accommodate the dissolved nucleic acid. The energy required for cavity formatiom is proportional to its surface area. Water has an unusually high surface tension so that cavity formation is quite unfavorable. Thus there is a free energy derived from the solvent that favors conformations with minimal surface area. For example, this would favor stacked single strands over random coils, and double helices over single strands. Since the surface tension of water results from a large surface enthalpy, solvophobic bonding can provide a driving enthalpy term for solute associations.

To test for solvent contributions to stacking, we have measured the effects of solvent perturbation on several reactions, including the dimerizations of thionine[16] and proflavin[17] dyes, single strand stacking in poly A[19] and poly C,[20] and double helix formation in $(dG\text{-}dC)_3$.[22] Some of the solvent mixtures used in these studies are listed in Table II, along with several of their properties. It is seen that cosolvents can drastically alter the properties of water. The effects of these perturbations on several nucleic acid reactions are reviewed below.

Table II indicates that addition of alcohols substantially lowers the surface tension of water which is a free energy term. If solvent cavity forces were important for single strand stacking, then presumably this would decrease the lifetime of the

Table I
Coil to Helix Thermodynamics for Ribo-oligonucleotides
in 1 M NaCl, 0.01 M Cacodylate, 0.0001 M EDTA, pH 7

Oligomer	T_m(°C) (5x10^{-5} M)	$-\Delta H$ (kcal/mol)	$-\Delta S$ (cal/deg-mol)
CCGGp	15	29	81
CCGGAp	43	47	129
ACCGGp	32	39	108

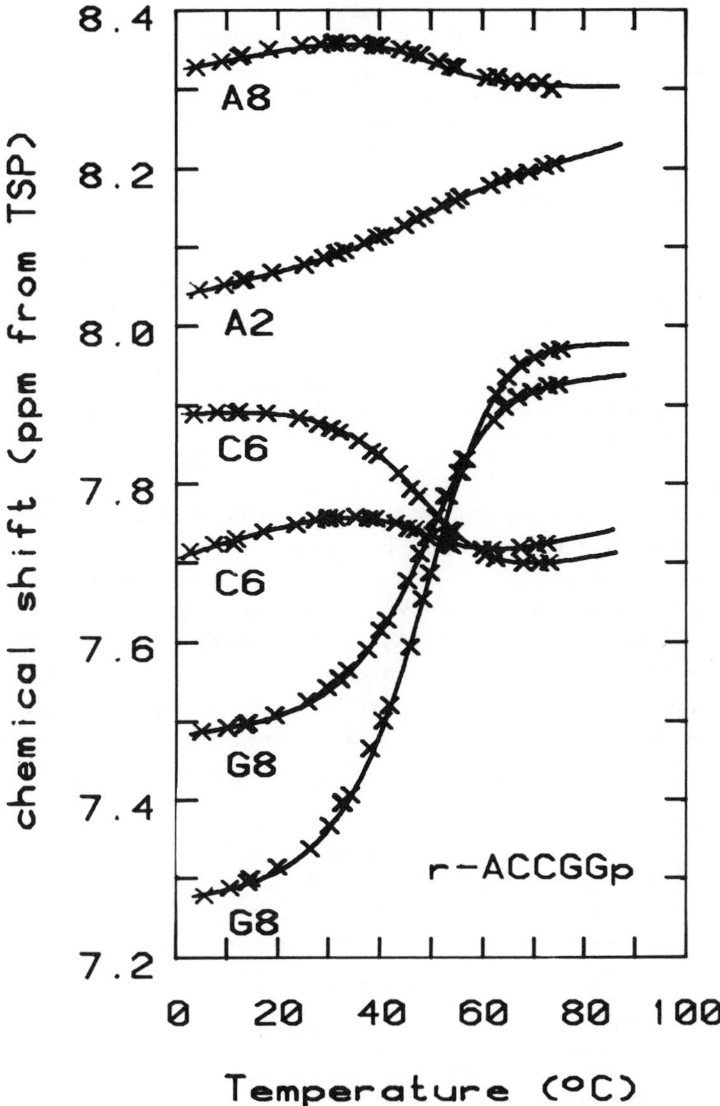

Figure 3. Chemical shift vs. temperature for the guanine (H-8), cytosine (H-6), adenine (H-8) and adenine (H-2) protons of 0.77 mM r-ACCGGp in 1.0 M NaCl, 10 mM cacodylate, 1 mM EDTA, pH 7 (99.8% D_2O). The spectra were obtained with a Bruker WH-400 MHz spectrometer; chemical shifts are referenced to the internal standard, TSP. The solid lines are calculated fits of the data to a two-state model with sloping baselines. All six protons have T_m's between 44 and 47°C.

stacked state. The implicit assumption is that the cavity required for the transition state is larger than that for the stacked species. Using the laser temperature jump method, we have measured the rate constants for stacking, k_1, and unstacking, k_{-1}, in poly A and poly C as a function of solvent. These are listed in Table III.[19,20] For poly A, the rates derived assuming $\sigma = 0.5$ or 1 are both listed. The absolute

Table II
Physical Constants of Aqueous Solvent Systems at 25°C

Cosolvent	Mol %	Viscosity cP	Cosolvent dipole moment (debye)	Surface tension dyn/cm	*Surface enthalpy dyn/cm
water	100	0.89	1.9	72.0	118
methanol	10	1.24	1.7	50.0	84
ethanol	5	1.40	1.7	46.4	92
ethanol	10	1.90	1.7	36.6	75
n-propanol	10	2.15	1.7	26.4	52
urea	10	1.15	4.6		
formamide	10	1.04	3.4		

*estimated from surface tension data of Teitelbaum, B.R., Gortolova, T.A., and Sidorova, E.E., *J. Phys. Chem.* (Russian) *25,* 911 (1951).

magnitudes of the rates depend on σ. However, the qualitative trends are independent of σ, as long as it is solvent independent. The cooperativity parameters measured for poly A and poly C range from 0.3 to 1. This small cooperativity, if present, can be attributed to an entropy term arising from restriction of rotation about the base-sugar bond,[26] and should be solvent independent. The lifetime of the stacked state is the inverse of k_{-1}. Comparison of Tables II and III indicates there is no correlation between k_{-1} and solvent surface tension. In fact, most of the solvent perturbations have little or no effect on k_{-1}. Interestingly, there is a correlation between the rate of stacking, k_1, and inverse solvent viscosity. This is shown in

Table III
Rate Constants for Single Strand Stacking in 0.05 M Sodium Cacodylate
All Rates x $10^{-6}(s^{-1})$

Solvent (mole%)	Poly A at 25°C				Poly C at 30°C	
	k_1 ($\sigma = 1$)	$k_1(\sigma = 0.5)$	k_{-1} ($\sigma = 1$)	k_{-1} ($\sigma = 0.5$)	k_1 ($\sigma = 1$)	k_{-1} ($\sigma = 1$)
H_2O	7.0	10.6	3.2	5.5	17	8
5% MeOH	5.0	7.7	3.3	6.4		
5% EtOH	4.4	7.5	3.7	6.2	15	9.9
10% EtOH	3.7	6.4	4.1	7.1	10	8
5% PrOH	4.4	7.4	3.4	5.7		
10% PrOH					10	7
5% CH_3CN	8.3	14.4	6.9	11.1	18	13
5% urea	8.0	13.1	6.7	11.9	14	9
5% formamide	5.3	9.0	4.8	8.2	16	12
10% glycerol	1.4	2.9	3.4	4.8	6	5
15% glycerol					2.3	2.8
2.9% sucrose					6.2	3.2

Figure 4 for both poly A and poly C. The result suggests the process is controlled by a rotational diffusion mechanism. Surprisingly, the unstacking rate does not show a similar dependence. While the solvent dependence of the unstacking rates are difficult to interpret, they clearly provide no direct evidence implicating solvophobic bonding.

In search of solvophobic effects, we have also measured the thermodynamics of double helix formation with the oligomer $(dG\text{-}dC)_3$.[22] The salt conditions were 1 M NaCl, 45 mM cacodylate, pH 7. Circular dichroism spectra indicate that at low temperature, the oligomer forms a right handed double helix in all the solvents employed. Typically, a two-state model is used to derive the thermodynamics of double helix formation with oligonucleotides.[21,37] NMR results indicate this is a reasonable treatment for $(dG\text{-}dC)_3$.[44] In particular, Figure 5 shows NMR melting curves for the C(H-6) protons of this oligomer. The data indicates all six base pairs melt cooperatively. Moreover, integration of the exchangeable protons at low temperature indicates all six base pairs are fully formed. Further support for two-state behavior has been obtained by measuring the ΔH for $(dG\text{-}dC)_3$ helix formation by both calorimetry and spectrophotometry. The values are −59.6 and −56.9 kcal/mole, respectively.[21] The close agreement again indicates the two-state model is adequate. Thus derivation of thermodynamic parameters from spectrophotometric melting curves is reasonably straight-forward for $(dG\text{-}dC)_3$, and it provides a convenient model system for studying solvent effects.

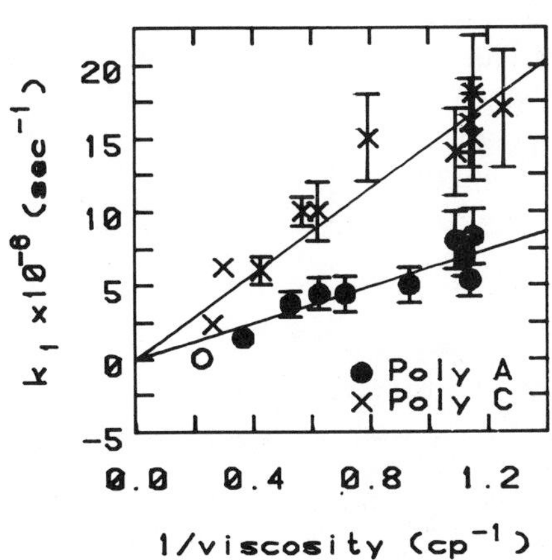

Figure 4. Coil to helix rate vs. reciprocal solvent viscosity for poly C at 30°C (x) and poly A at 25°C (•) in 0.05 M sodium cacodylate, pH 7. Two-state model. The open circle is the value for poly A in 15 mol % glycerol as estimated in reference 19.

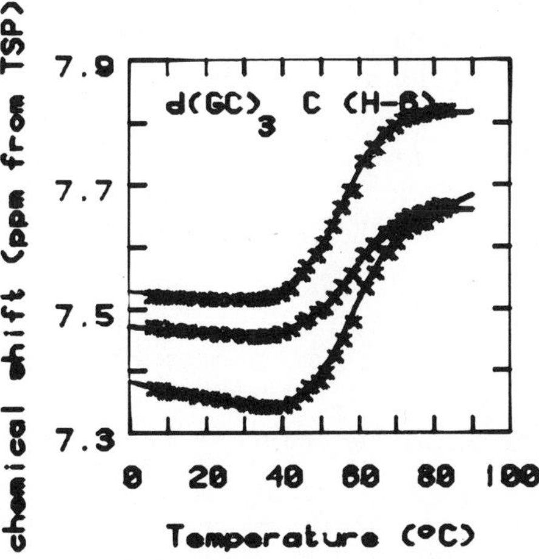

Figure 5. Chemical shift vs. temperature for the cytosine (H-6) protons of 1.64 mM $d(GC)_3$ in 0.01 M sodium cacodylate, 1 mM EDTA, pH 7 (99.8% D_2O). The spectra were obtained on a Bruker WH-400 MHz spectrometer. The chemical shifts are referenced to the internal standard, TSP. The solid lines are calculated fits of the data to a two-state model with sloping baselines. The T_m's of all three protons are between 54 and 57°C.

The potential contribution of solvent cavity terms to the thermodynamics of helix formation with $(dG\text{-}dC)_3$ can be readily estimated. The cavity around random coil, single strand guanines and cytosines can be approximated by spheres with radii of 3.47 and 3.88Å, respectively.[15] If the double helix is a cylinder with radius 7.5Å and length 17Å, then the helix formation is accompanied by a decrease in solute-solvent surface area of 889Å^2/helix. Since the surface tension and surface enthalpy of water at 25°C are about 72 and 118 dyn/cm, this corresponds to a ΔG and ΔH of −46 and −75 kcal/mol strand, respectively. The surface tension quoted is for an air-water interface. This may not be appropriate for a nucleic acid-water interface, and a lower value may be more reasonable. For example, the surface tension and enthalpy for a benzene-water interface are about 34 and 45 dyn/cm, respectively.[45] On the other hand, recent calculations of solvent accessible surfaces of nucleic acids by Alden and Kim indicate the decrease in area may be greater than the above estimate.[46] While the approximations are crude, they clearly suggest solvent cavity terms could make a major contribution to helix formation. If they did, then solvent perturbations should affect the ΔG and ΔH of helix formation. In particular, inspection of Table II indicates the addition of alcohols at 10 mole % to water lowers both the surface tension and enthalpy by roughly a factor of two. Thus this perturbation should make the ΔG and ΔH of helix formation less favorable.

Using a two-state model, we have analyzed optical melting curves to determine the thermodynamics of $(dG\text{-}dC)_3$ helix formation in various solvent mixtures.[22] The results are listed in Table IV. While alcohols at 10 mole % make the ΔG of helix formation less favorable, this is not the result of a less favorable ΔH. The magnitudes of the thermodynamic changes are also rather modest considering the substantial changes in solvent properties listed in Table II. A larger effect on $(dG\text{-}dC)_3$ thermodynamics is observed at 20 mole % ethanol. However, CD spectra indicate this is associated with a conformational change in the double helix.[22] Similar conformational changes in DNA have been attributed to dehydration.[47] Thus the thermodynamic effect at 20 mole % ethanol may reflect an enthalpy contribution

Table IV
Thermodynamics of Double Helix Formation
by $(dG\text{-}dC)_3$ in 1 M NaCl, 0.045 M Cacodylate, pH 7

Solvent	Mole %	$(dG\text{-}dC)_3$ $-\Delta H$(kcal/mole)	$-\Delta S$(cal/mol deg)
H_2O	100	56.9	155.5
Methanol	10	58.4	162.9
Ethanol	10	58.5	166.5
Propanol	10	62.0	175.9
Formamide	10	47.3	130.5
N,N-Dimethyl-Formamide	10	38.6	107.5
Urea	10	49.5	138.9
Ethanol	20	43.4	121.1

from tightly bound water. Once again, the data provide no direct evidence for solvent cavity contributions to the favorable enthalpy of helix formation.

Table IV also lists the thermodynamics for $(dG\text{-}dC)_3$ helix formation in 10 mole % urea, formamide, and dimethylformamide. In all three cases, the ΔH of helix formation is less favorable than in water, suggesting these cosolvents compete with the interactions providing the driving enthalpy of helix formation. These three cosolvents have dipole moments greater than water (see Table II). Thus they should compete with vertical base-base electronic interactions and/or hydrogen bonding more effectively than water.

The results of the above studies can be summarized as follows:

(1) Most of the favorable enthalpy driving helix formation can be attributed to base "stacking" interactions.
(2) We find no evidence that solvent cavity terms contribute to the favorable enthalpy of stacking.
(3) Specific solvation may affect the thermodynamics of helix formation.

These results provide working hypotheses for future studies of the origins of nucleic acid stability. Additional work, in progress, should provide more detailed insights into the forces involved.

Acknowledgments

We thank Dr. Frank Buff for several stimulating discussions. This work was supported by National Institutes of Health grant GM 22939. D.H.T. is an Alfred P. Sloan Fellow.

References and Footnotes

1. Cantor, C.R. and Schimmel, P.R., *Biophysical Chemistry Part I: The Conformation of Biological Macromolecules,* W.H. Freeman, San Francisco, Chapter 6 (1980).
2. Bloomfield, V.A., Crothers, D.M., and Tinoco, I., *Physical Chemistry of Nucleic Acids,* Harper and Row, New York, Chapter 6 (1974).
3. Pullman, B., ed., *Molecular Associations in Biology,* Academic Press, New York (1968).
4. Crothers, D.M. and Zimm, B.H., *J. Mol. Biol. 9,* 1 (1964).
5. DeVoe, H. and Tinoco, I., *J. Mol. Biol. 4,* 500 (1962).
6. Pullman, A. and Pullman, B., *Adv. Quantum Chem. 4,* 267 (1968).
7. Pullman. B. and Pullman, A., *Prog. Nucl. Acid Res. and Mol. Biol. 9,* 327 (1969).
8. Herskovits, T.T. *Biochemistry 2,* 335 (1963).
9. Herskovits, T.T. *Arch. Biochem. Biophys. 97,* 474 (1962).
10. Kauzmann, W., Adv. *Protein Chem. 14,* 34 (1959).
11. Sinanoglu, O. in *Molecular Associations in Biology,* B. Pulman, Ed., Academic Press, New York, pp. 427-445 (1968).
12. Sinanoglu, O. and Abdulnur, S., *Fed. Proc. Amer. Soc. Exp. Biol. 24,* Pt III, S-12 (1965).
13. Sinanoglu, O. and Abdulnur, S., Photochem. Photobiol. 3, 333 (1964).
14. Sinanoglu, O., *Int. J. Quant. Chem. 18,* 381 (1980).
15. Abdulnur, S., Ph.D. thesis, Yale University (1966).

16. Dewey, T.G., Wilson, P.S. and Turner, D.H., *J. Am. Chem. Soc. 100,* 4550 (1978).
17. Dewey, T.G., Raymond, D.A. and Turner, D.H., *J. Am. Chem. Soc. 101,* 5822 (1979).
18. Dewey, T.G. and Turner, D.H., *Biochemistry 18,* 5757 (1979).
19. Dewey, T.G. and Turner, D.H., *Biochemistry 19,* 1681 (1980).
20. Freier, S.M., Hill, K.O., Dewey, T.G., Marky, L.A., Breslauer, K.J., and Turner, D.H., *Biochemistry 20,* 1419 (1981).
21. Albergo, D.D., Marky, L., Breslauer, K.J., and Turner, D.H., *Biochemistry 20,* 1409 (1981).
22. Albergo, D.D. and Turner. D.H., *Biochemistry 20,* 1413 (1981).
23. Holcomb, D.N. and Tinoco, I., *Biopolymers 3,* 121 (1965).
24. Leng, M. and Felsenfeld, G., *J. Mol. Biol. 15,* 455 (1966).
25. Stannard, B.S. and Felsenfeld, G., *Biopolymers 14,* 299 (1975).
26. Applequist, J. and Damle, V., *J. Am. Chem. Soc. 88,* 3895 (1966).
27. Adler, A., Grossman, L., and Fasman, G.D., *Proc. Natl. Acad. Sci. USA 57,* 423 (1967).
28. Brahms, J., Maurizot, J.C., and Michelson, A.M., *J. Mol. Biol. 25,* 465 (1967).
29. Neumann, E. and Ackermann, T., *J. Phys. Chem. 73,* 2170 (1969).
30. Breslauer, K.J. and Sturtevant, J.M., *Biophysical Chem. 7,* 205 (1977).
31. Filimonov, V.V. and Privalov, P.L., *J. Mol. Biol. 122,* 465 (1978).
32. Rawitscher, M.A., Ross, P.D., and Sturtevant, J.M., *J. Am. Chem. Soc. 85,* 1915 (1963).
33. Suurkuusk, J., Alvarez, J., Freire, E., and Biltonen, R., *Biopolymers 16,* 2641 (1977).
34. Lipsett, M.N., *Proc. Natl. Acad. Sci. 46,* 445 (1960).
35. Young, P.R. and Kallenbach, N.R., *J. Mol. Biol. 126,* 467 (1978).
36. Zimm, B.H. and Bragg, J.K., *J. Chem. Phys. 31,* 526 (1959).
37. Martin, F.H., Uhlenbeck, O.C., and Doty, P., *J. Mol. Biol. 57,* 201 (1971).
38. Yoon, K., Turner, D.H., Tinoco, I., von er Haar, F., and Cramer. F., *Nucl. Acids Res. 3,* 2233 (1976).
39. Grosjean, J., Söll, D.G. and Crothers, D.M., *J. Mol. Biol. 103,* 499 (1976).
40. Romaniuk, P.J. Hughes, D.W., Gregoire, R.J., Neilson, T., and Bell, R.A., *J. Am. Chem. Soc. 100,* 3971 (1978).
41. Petersheim, M. and Turner. D.H., unpublished results.
42. Arter, D.B., Walker, G.C., Uhlenbeck, O.C., and Schmidt, P.G., *Biochem. Biophys. Res. Commun. 61,* 1089 (1974).
43. Breslauer, K.J., Sturtevant, J.M. and Tinoco, I., Jr., *J. Mol. Biol. 99,* 549 (1975).
44. Freier, S.M. and Turner, D.H., unpublished results.
45. Good, R.J. and Elbing, E., *Ind. Eng. Chem. 62,* #3, 54 (1970).
46. Alden, C.J. and Kim, S.H., *J. Mol. Biol. 132,* 411 (1979).
47. Girod, J.C., Johnson, W.C., Huntington, S.K., and Maestre, M.F., *Biochemistry 12,* 5092 (1973).

Proceedings of the Second SUNYA Conversation in the Discipline Biomolecular Stereodynamics
Volume I, ISBN 0-940030-00-4, Ed., Ramaswamy H. Sarma,
Adenine Press, New York, ©Adenine Press

High Resolution Thermal Dispersion Profiles of DNA

R.D. Blake,* F. Vosman† and C.E. Tarr†
Departments of *Biochemistry and †Physics
University of Maine
Orono, Maine 04469

Introduction

Some of the most distinctive properties of DNA are due to a difference in energy of only 10-12% between the stability of the G-C and A-T base pair (bp). The relationship between base sequence and helix stability has attracted interest ever since the pioneering studies of Marmur and Doty[1] but perhaps not as intensely as it has recently with the rapid development of sequence libraries. The principal experimental approach is through melting curves, obtained by monitoring changes in UV absorbance associated with the melting process. The conformational state of DNA correlates with its energy level at any temperature within the range of melting. Because it is stronger than A-T the random incorporation of G-C among A-T base pairs in a helix leads to regions that behave as what may best be described as phase boundaries between helical and coil regions during melting.[2-4] Rather than a single sharp transition as seen with synthetic homopolymer systems capable of many degenerate energy levels, natural DNAs exhibit discrete regions of order-disorder. This difference between simple synthetic systems and the (quasi-)random heteropolymer DNA is demonstrated in the melting profile of Figure 1 obtained from a composite specimen of both synthetic and natural helixes. The homopoly — rA•rU, copoly — d(A-T)•-d(T-A) and homopoly — dA•dT helixes melt sharply within a range of only 2-3°, with single t_m at 44, 47 and 54°C. The contrasting profile for λDNA, consisting of almost equivalent numbers of A-T and G-C base pairs ($F_{G\text{-}C}$ = 0.49), melts at a much higher temperature (75.5°C), over a wide 20° range and exhibits numerous subtransitions for the dissociation of different domains in the molecule. Short segments enriched with G-C base pairs form discrete interfacial boundaries between helix and coil phases during melting since they require more energy than that normally needed to dissociate a given base pair. Given the specimen is both short and of homogeneous sequence within the population, the melting curve will exhibit a fine structure like that in Figure 1, reflecting considerable detail about local sequence domains. The subtransitional fine structure arises from the simultaneous dissociation of discrete domains in all molecules of the population. With suitable analysis it is possible to construct denaturation maps of DNAs up to 50,000 bp in length to a resolution of only 200-300 bp.

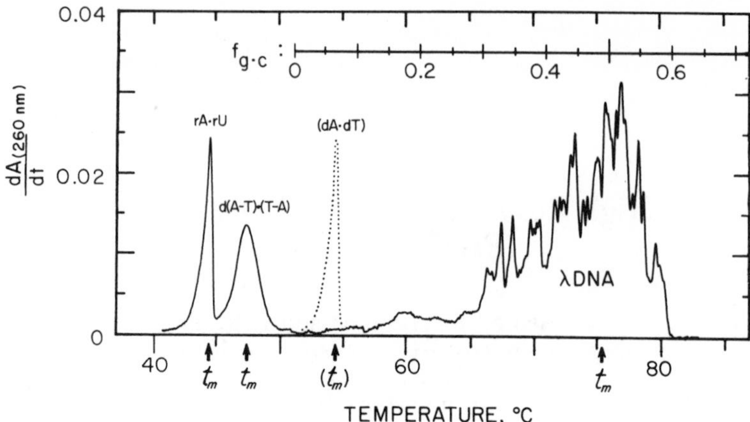

Figure 1. A derivative-reconstructed differential-absorbance melting curve of lambda ($cI_{857}S_7$) DNA with rA•rU and d(A-T)•d(T-A) markers in 0.018M-Na^+ obtained at a constant 6.75°C/hr rate of heating and with a temperature difference, $\Delta t = 0.23 \pm 0.004°$ between samples. The melting temperatures were determined by one-half the integrated (numerically) area under the profile. The inset Marmur-Doty scale for the variation of $t_{m,i}$ for each subtransition with the fractional G-C base composition of the domain, $f_{gc,i}$, corresponds to the expression $t_{m,i} = t_{at} + \Delta t \bullet f_{gc,i}$, where t_{at} is the mean t_m for d(A-T)•d(T-A) and dA•dT (not shown), and $\Delta t = 48.98°$.

Evidence for melting by discrete domains originated with electron microscopic mapping of denaturation pioneered in Inman's laboratory,[5,6] and later from melting curves of the change in absorbance with temperature showing complex fine structure.[7-15] Recent studies from this laboratory demonstrated that each subtransition in the optical profile behaves independently in response to the local counterion, again indicating the independence of domains during melting. Excellent correspondence between electron micrographs of partially denatured Col E1 DNA and a high resolution optical profile has been demonstrated recently by Borovik et al.[16] Electron microscopy is one essential means of establishing the loci of domains undergoing dissociation, while the much greater resolution of the optical profile is essential for quantitative thermodynamic analysis. Still further evidence for domain melting is found in the recent results of electrophoresis of DNA fragments through urea-formamide gradient gels at high temperatures.[17] When the fragment reaches the concentration of denaturant in the gel that is sufficient to dissociate the weakest domain, the electrophoretic mobility of the fragment abruptly stops. The stopping point in the gel for each fragment is determined by the base composition of the weakest domain and is independent of the size of the fragment.[17]

Our interest in this phenomenon has been primarily focused on the specific factors that determine the stability of domains. In recent work we have examined the thermodynamic effects of base composition, locus and size of domains on stability,[14,15,18] while other studies have examined the effects of counterion and solvent on stability.[19,20] Still other studies have focused on the effects of mispairs, lesions and ligands.[21] Concomitant with these studies has been a continuing effort to increase the sensitivity and resolution of the absorbance profiles for more unambiguous deconvolution of subtransitions and for the detection of subtle energetic perturbations.

Melting Curves

The sensitivity of melting curves obtained by recording the loss of hypochromicity near 260 nm has been improved dramatically in the past decade, facilitating the numerical derivitization of the change in absorbance. That derivitization enhances fine structure is well known,[22] making complex profiles amenable to direct analysis. The usual practice is to generate the derivative numerically at each temperature from smoothed absorbance data obtained at high densities. Depending on the length of the specimen and the distribution of G-C bp, the numerical method will generally require an on-line computer to handle the large number of computations involved. Profiles obtained in this fashion generally exhibit a remarkably high sensitivity to subtle changes in melting with minimal distortion from numerical manipulations. The result is an impressively detailed record of the melting process. The performance characteristics of a number of instrumental systems has been reviewed by Wada.[23]

An alternative method for obtaining derivative profiles is based on a finite difference approximation of the derivative. The difference in absorbance between two identical DNA solutions raised colinearly in temperature is measured continuously with a dual-beam ratio-recording spectrophotometer.[14] A constant difference of 0.1-0.3 (±0.001)°C between samples throughout the temperature range of the exper-

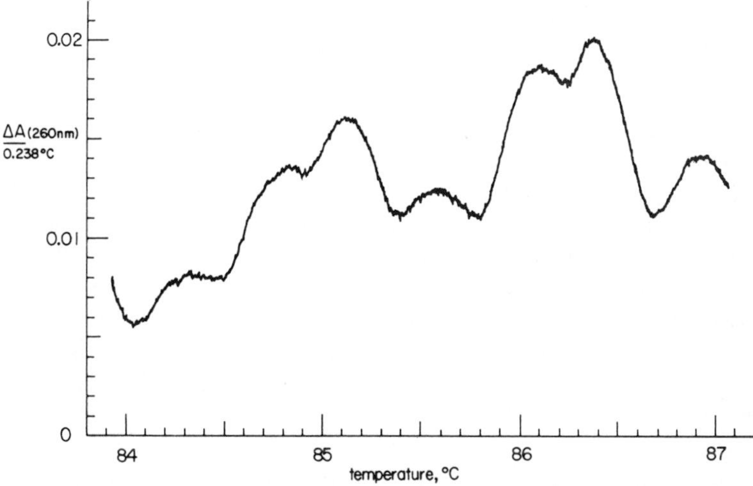

Figure 2. An expanded three-degree segment of a differential-absorbance melting curve of lambda DNA in 0.10M-Na^+, obtained at 6.75°C/hr.[15] The absorbance of the DNA was 2.493 at 260 nm in a volume of 280 μl/cell; and the experiment measured in a modified Cary 118c double-beam ratio-recording prism spectrophotometer.

iment has been shown to be sufficient to resolve most subtransitions in a DNA of moderate length, e.g. ~10,000 bp.[15] It has been our experience that the difference approximation method gives a higher signal-to-noise sensitivity, and is generally more convenient than the numerical approach, and therefore is the principal method in use in this laboratory. An example of a melting profile obtained by this method is shown in Figure 1. An expanded three-degree segment of the associated analog signal is shown in Figure 2. The reproducibility of such a profile is excellent, to within about twice the noise level seen in Figure 2. The specimen in both figures is the $CI_{857}S_7$ lysogenic strain of λDNA with a fractional G-C composition, $F_{G\text{-}C}$ = 0.49, over a length of 48,000 bp. Normally the profile is digitized at 0.01°C intervals and stored on mag-tape for further processing and analysis. Processing includes a reconstruction of the true derivative from the experimental differential data using a Taylor series expansion routine.[15] Application of this routine to simulated melting curve produced by a nine component Gaussian error function indicates the maximum error in the reconstructed derivative is rarely more than 0.1% when the Δt between samples is 0.2° or smaller,[15] which is well within the capabilities of the system.

Spectral Analysis

A spectral approach is necessary to evaluate the fractional G-C composition of domains undergoing dissociation because the familiar Marmur-Doty relationship is only applicable to the macroscopic melting behavior of a heterogeneous population of DNAs. Marmur and Doty[1] described the relationship between the overall melting temperature, t_m, of a mixed population of DNAs and average fractional G-C composition, F_{gc}, by the linear function

$$t_m = t_{at} + \Delta t \cdot F_{gc} \qquad (1)$$

In this expression t_{at} represents the melting temperature of a DNA containing only A-T base pairs in random nearest neighbor arrangement, while Δt is the increase in melting temperature expected for the incorporation of the stronger G-C base pair. The linear inset scale in Figure 1 represents the Marmur-Doty relationship for the variation of $t_{m,i}$ with the fractional G-C base composition of each subtransition in the lambda profile. This scale was established from the mean value for the t_ms of d(A-T)•d(T-A) and dA•dT and the overall t_m of λDNA, for which F_{gc} = 0.49. Thus, t_{at} = 51.25°C and Δt = 48.98°C, for the ionic conditions of this experiment. For the $t_{m,i}$ of many of the subtransitions in the lambda profile, however, it has been found that the $f_{gc,i}$ predicted by this scale may vary from known values by ±4%.[19]

Theoretical considerations indicate $t_{m,i}$ of individual domains may deviate from that predicted by the Marmur-Doty relationship.[3,4,29] This expectation has been qualitatively substantiated by experimental results,[19] necessitating a spectral analysis for the determination of domain base composition. The basis of this approach is a dispersion of the hypochromic effect with wavelength. Felsenfeld and Fresco and their respective co-workers[24-27] pioneered the spectral approach for the determination of base composition from melting curves, which has been modified for the analysis of derivative profiles.[14,15,28] Melting curves have conventionally been monitored at 260 nm, and while both A-T and G-C base pairs contribute to the loss of hypochromicity at that wavelength, they do so to different extents. On the other hand, the conformational transition of an A-T base pair is invisible at 282 nm, while the change for G-C base pairs is quite large. In principal, therefore, a determination of the base composition of each domain giving rise to a discernible subtransition can be made from experiments conducted at just two wavelengths. The sensitivity of the method will be greatest when the derivative absorbance values at the two wavelengths show the largest difference for contributions from the dissociation of both A-T and G-C pairs. Ideally, one wavelength should be transparent (isosbestic) for either A-T or G-C, while the other should correspond to the wavelength with the largest value for the change in dA/dT representing the sum of contributions from both pairs. In this way it is possible to determine not only the G-C composition of domains exhibiting a subtransition but their lengths, in base pairs, as well.

We have found 282 nm to be the mean isosbestic wavelength for all seven nearest neighbors of the A-T bp, while the largest change in derivative absorbance for both pairs is near 260 nm.[14] In our earlier work it was shown that

$$f_{gc}(T) = \Delta\varepsilon_{at}(260) \Big/ \frac{\Delta\varepsilon_{gc}(282)}{[dA(282)/dA(260)]} - [\Delta\varepsilon_{gc}(260) - \Delta\varepsilon_{at}(260)] \quad (2)$$

relating the ratio of derivative melting curves obtained at 282 nm and 260 nm, at any temperature, to the fractional G-C base composition of the DNA dissociating at that instant. There are four constants in this expression for the derivative extinction coefficients, $\Delta\epsilon(\lambda)$, of the A-T and G-C base pair at 260 and 282 nm. The value for $\Delta\epsilon$ (282) = 0, while $\Delta\epsilon_{at}$ (260) was determined with synthetic polynucleotides. The other two $\Delta\epsilon$ were evaluated by least squares fit of equation 2 to the ratio of melting curves on a series of bacterial DNAs, illustrated in Figure 3. From the spectral analysis of melting thirteen standard DNAs of known base composition, $\Delta\epsilon$ were found to have the following values:

$$\Delta\epsilon_{at}(260) = 2765 \pm 95 \text{ liter (mole}\bullet\text{cm)}^{-1}$$
$$\Delta\epsilon_{at}(282) = 0$$
$$\Delta\epsilon_{gc}(260) = 1014 \pm 138$$
$$\Delta\epsilon_{gc}(282) = 2009 \pm 106$$

With these values in equation 2 the f_{gc} of domains undergoing a conformational transition can be determined from the ratio of experiments at two wavelengths, as illustrated in Figure 4. Lengths of domains are determined from their base composition and the fractional areas under the derivative melting curve contributed by the corresponding subtransition. The latter requires the profile be decomposed into

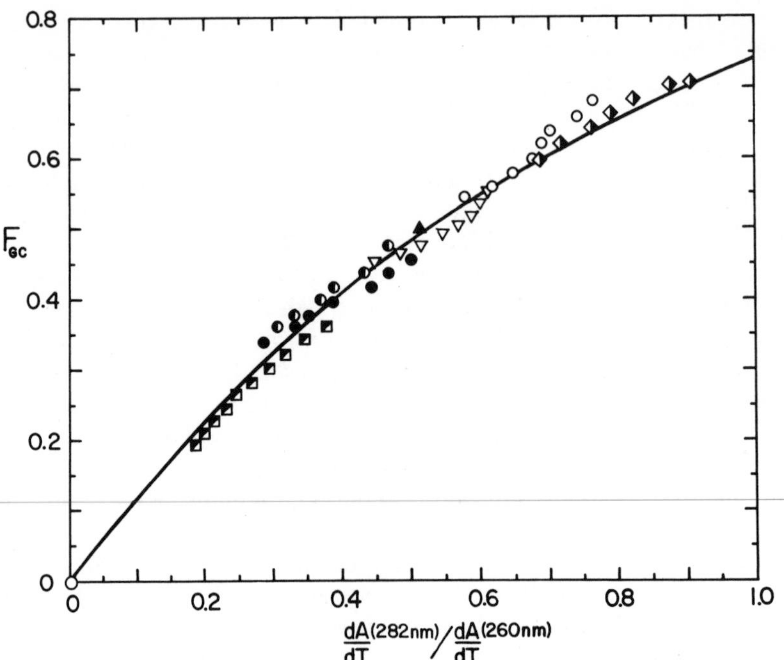

Figure 3. Variation of the fractional G-C composition, with the ratio of melting curves obtained at 282 nm and 260 nm. Data are from melting curves of DNA from *Cl. perfringens* (□), *Pseudomonas* BAL 31 (•), *B. subtilis* (○), d(A-C)•d(G-T) (▲), *E. coli* (△), *Ps. fluorescens* (○), and *M. lysodeikticus* (◇). The line through data is given by expression (2).

subtransitional components according to a suitable thermodynamic analysis for melting by domains.

Thermodynamic Considerations

The usual approach to the analysis of such curves involves comparison with theoretical curves, calculated from known sequences on the basis of existing helix-coil transition theory. The virtues of this approach are clear, however, due to simplifying assumptions, e.g. the parameterization of stacking energies, it suffers an inability to account for local variations in energetic details and, of course, cannot be used for the analysis of curves from unknown sequences. Azbel has advocated a direct analysis of melting profiles through considerations of the primary sequence for abrupt variations in the local energy level. The direct approach can be extended to the analysis of curves from unknown sequences, given a suitable model for melting of domains. As a first approximation we consider the complex oscillatory fine structure in melting curves as the result of a collection of two-state helix-coil subtransitions for discrete domains. The contribution of each domain to the observed derivative absorbance profile will then be given by[29]

$$\frac{dA(T)}{dT}_i = \frac{\Delta A_i \Delta H_i}{4RT^2_i} \left\{ \cosh^2 \left[(\Delta H_i / 2RTT_{m,i} \; (T - T_{m,i}) \right] \right\}^{-1} \tag{3}$$

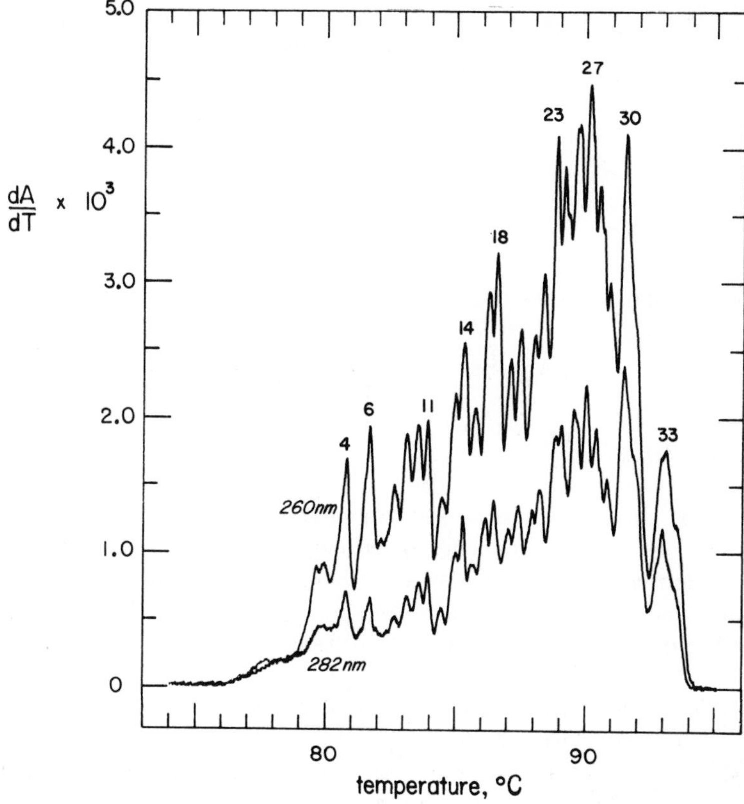

Figure 4. Derivative-reconstructed melting curves of lambda DNA obtained at 260 nm and 282 nm in 0.0955 M NaCl, 5 mM cacodylate, 0.2 mM Na - EDTA, pH 6.8. The absorbance of the DNA was 2.493 at 260 nm and 1.173 at 282 nm, the Δt was 0.238, and the heating rate 6.750°C/hr. The numbers refer to features where $dA^2/d^2T = 0$ and not to subtransitions. There are 87±5 subtransitions under this profile.[21]

where ΔA_i is the change in absorbance, and ΔH_i is the change in enthalpy for the melting of the ith domain. The unit enthalpy is then given by

$$\overline{\Delta H_i} = \Delta H_i / \ell_i \tag{4}$$

where l_i is the length of the domain. The domain length is calculated from the relationship

$$\ell_i = [(\Delta A_i(282)/f_{gc,i}) / (\Delta A_{tot}(282)/F_{gc})]\ell_{tot} \tag{5}$$

where, as before, $f_{gc,i}$ is the fractional G-C composition of the domain and F_{gc} the overall base composition of the molecule. The total change in absorbance in equation 5 is obtained by numerical integration of the profile, while ΔA_i is evaluated from a Gaussian surrogate for equation 3. This equation has been shown to be very closely approximated by the Gaussian probability function

$$\frac{dA(T)_i}{dT} = \frac{dA}{dT}\bigg|_{T_{m,i}} \exp\left\{-[T-T_{m,i}]^2/2\,\sigma_i^2\right\} \qquad (6)$$

so that

$$\Delta A_i = \frac{dA}{dT}\bigg|_{T_{m,i}} \sigma_i\left(\frac{\pi}{2}\right)^{\frac{1}{2}} \qquad (7)$$

and from Gaussian deconvolution of the profile we obtain

$$\Delta H_i = 4RT_{m,i}^{\ 2}/\sigma_i\left(\frac{\pi}{2}\right)^{\frac{1}{2}}. \qquad (8)$$

Profile Deconvolution

Even the most subtle features in these complex thermal dispersion profiles are reproducible, provided the replicate experiments are conducted at the same ionic strength.[19] Such constancy indicates a nonrandom relationship between putative subtransitions and sequences of corresponding domains. It is therefore of interest to resolve the profile into its constituent subtransitions.

Deconvolution is accomplished by numerical methods, and begins with the enumeration of the subtransitions, together with a good approximation of their t_m, peak heights and widths at half maximum. The enumeration process is initiated by placing a Gaussian band shape (equation 6), with the three adjustable parameters constrained by rigid thermodynamic considerations, under the principal maximum in the experimental profile. The constraints involve fixing the transition enthalpy, ΔH_i, for each domain (cf equations 3, 8) to a value that corresponds to that expected by calorimetric measurements. The latter can be estimated for domains of different fractional G-C content with good precision by the expression

$$\overline{\Delta H}^{cal} = 11250 + (2.6-F_{gc})(450 \log[Na^+] - 580), \text{ cal } (\text{mol}\cdot\text{bp})^{-1} \qquad (9)$$

obtained by fitting a large number of independent values for the unit calorimetric transition enthalpy from the literature.[19] This first subtransition is then subtracted from the experimental profile, and the process iterated until the entire area under the melting curve is accounted for. Parameters defining the line shape of subtransitions are then refined by a least squares method that optimizes all three adjustable parameters of the Gaussian function by minimizing the error between the sum of components and the experimental profile. The optimization procedure is based on a maximum neighbor method that combines the precision and speed of gradient-search and linear-interpolation methods.[30] The rigid constraints that were placed on the unit transition enthalpy during the initial enumeration stage of deconvolution are relaxed during the optimization procedure, although the *overall* unit enthalpy remains fixed to that indicated by equation 9 for the estimated calorimetric value.

Because of this relaxation the unit enthalpies for domains show some variation after the optimization procedure, although they do not drift much beyond ±5% of the overall (calorimetric) value for the molecule. The flexibility of this approach, where the enthalpy is recognized explicitly as an adjustable parameter, is suited for the analysis of profiles from unknown DNA sequences, and for the evaluation of extraordinary variations in free energy, such as that associated with ligand site binding, lesions, heterogeneity in base composition for each domain, contributions from loop enthalpy or experimental uncertainties.

Analysis of SV40 Fragment, ϕx174-RF and Col E1 DNA Melting Curves

As a test of the two-state assumption (equation 3) we have analyzed a number of melting curves of small DNAs. The analysis was performed on these curves without consideration of the sequence. The first profile (Figure 5) is of a small 1164 bp restriction fragment of SV40 DNA, and was taken from the recent work of Gabbarro-Arpa et al.[31] The deconvolution pattern contains eight subtransitions as shown in Figure 5, with an average $\overline{\Delta H_i}^{app} = 7950 \pm 480$ cal. (mol•bp)$^{-1}$, compared to $\overline{\Delta H}^{cal} = 7942$ from equation (9). The difference between experimental and deconvoluted profiles of SV40 fragment are shown in the lower half of Figure 5. It can be seen that the fit over the first degree or so of melting is not as good as it is over the remaining three degrees. While the constraints on $\overline{\Delta H_i}^{app}$ are relaxed during the optimization stage of deconvolution, they were maintained, in this case, to within ±10% of $\overline{\Delta H}^{cal}$, otherwise the apparent enthalpy for the first two subtransitions would have assumed very large values of no physical significance. We suspect our

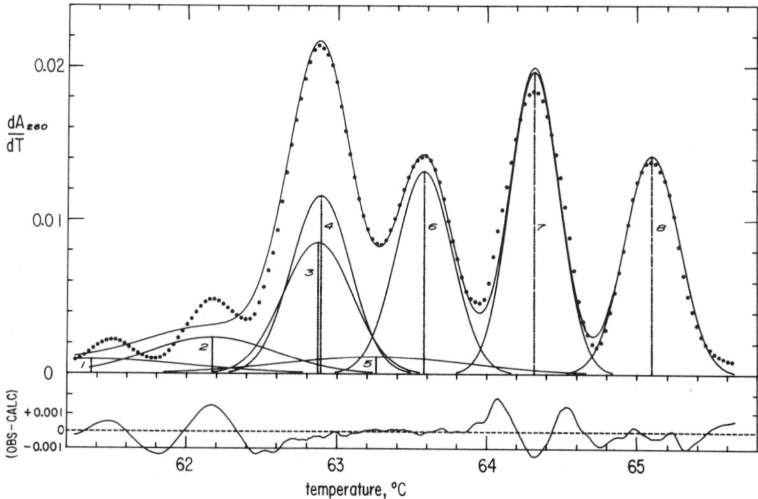

Figure 5. Melting curve of a 1164 bp restriction fragment of SV40 DNA.[31] The experimental curve is represented by the solid circle; conditions: 10 mM Na-citrate, 0.1 mM Na-EDTA, with a heating rate of 4°C/hr. The solid line through data points represents the sum of absorbance changes for the subtransitions shown.[29] The difference between observed and calculated melting curves is plotted beneath the melting curve.

inability to fit the early part of melting reflects conditions of non-equilibria due to the low ionic conditions of the experiment. By contrast, the last three large sub-transitions in Figure 5 yield quite reasonable values for $\overline{\Delta H_i}^{app}$ when treated as single, well-resolved transitions. The average variation of $\overline{\Delta H_i}^{app}$ from $\overline{\Delta H}^{cal}$ is only ±5%, which we take as strong support of the two-state assumption.

Figure 6 shows the results of a similar deconvolution analysis of a melting curve from ϕx174-RF DNA, taken from the work of Vizard et al.[32] This experiment was obtained at the moderately low ionic condition of 0.03M-Na^+, while Figure 7 shows results of a second experiment in 0.195M-Na^+. An excellent fit of deconvoluted to observed profile is achieved with 19 subtransitions under both profiles (Tables I and II). Despite obvious differences in the appearance of the profiles obtained at the

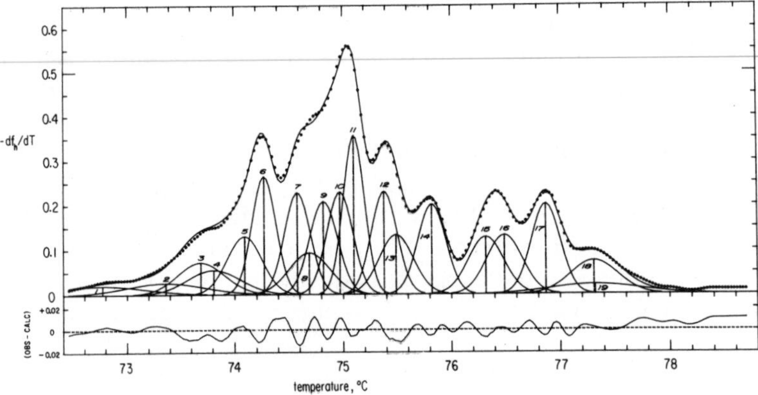

Figure 6. Observed and deconvoluted melting curves of ϕx174-RF DNA in 30mM-Na^+.[32] Experimental conditions and characteristics of domain are given in Table I.

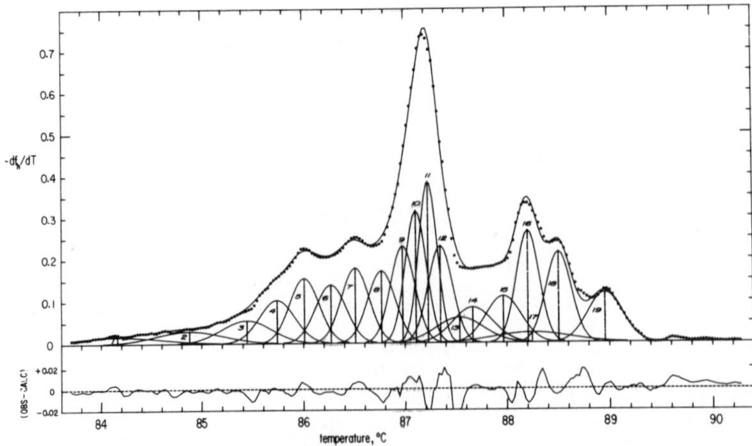

Figure 7. Observed and deconvoluted melting curves of ϕx174-RF DNA in 195 mM-Na^+.[32] Details summarized in Table II.

Table I

Domain Characteristics of ϕx174-RF DNA[a]

Domain	$t_{m,i}$ (°C)	$\frac{df}{dt}\Big\vert_{t_{m,i}}$ × 10^{-1}	σ_i, (°C)	ΔA_i, × 10^{-2}	Length (bp)	$\overline{\Delta H}^{app}$ (bp^{-1})
1	72.77	0.196	0.402	1.97	106	8913
2	73.36	0.257	0.363	2.34	128	8320
3	73.69	0.715	0.218	3.91	212	8336
4	73.81	0.547	0.244	3.34	180	8708
5	74.09	1.30	0.158	5.15	277	8731
6	74.27	2.65	0.111	7.38	397	8679
7	74.58	2.28	0.120	6.84	370	8709
8	74.69	0.930	0.188	4.38	238	8648
9	74.82	2.08	0.128	6.68	360	8321
10	74.97	2.29	0.120	6.87	370	8682
11	75.10	3.57	0.098	8.79	474	8256
12	75.38	2.30	0.120	6.90	371	8679
13	75.49	1.33	0.158	5.27	284	8585
14	75.82	2.01	0.128	6.46	348	8657
15	76.32	1.28	0.162	5.21	280	8519
16	76.49	1.32	0.162	5.37	389	8262
17	76.87	2.02	0.133	6.71	363	8125
18	77.31	0.743	0.214	3.98	126	8519
19	77.32	0.212	0.406	2.16	118	8272

Total number of base pairs = 5386
Average number of base pairs per domain = 282
Total enthalpy from all subtransitions = 4.570 × 10^7 cal/mol of ϕx174 DNA
Mean calorimetric enthalpy estimated by Eq. (9) = 8517 cal/mol bp
Apparent overall unit enthalpy (total enthalpy/ total bp) = 8503 cal/mol bp

[a] From Vizard et al. (Ref.32). Conditions: 29 m*M* NaCl (0.03*M* Na^+), 1 m*M* sodium cacodylate, 10^{-4} *M* EDTA, pH 7.

two ionic strengths, as well as significant differences in expected calorimetric enthalpies, we find these curves require the same number of domains for optimum fit. Examination of Tables I and II shows the underlying correspondence between domains, with match-ups that are generally within ±10 bp. It would appear that most phase boundaries are conserved over the ionic range of these experiments and that the differences in these profiles are due to differences in the dependencies of $t_{m,i}$ on salt concentration.[19]

Figure 8 shows the results of the deconvolution of a melting curve from the 6594 (±20)bp Col E1 Plasmid DNA taken from the recent work of Borovik et al.[16] This DNA has an extraordinarily broad distribution of base composition throughout the entire molecule, as indicated by the broad dispersion of subtransitions. Deconvolution analysis indicates the first, completely isolated band near 64.28°, arises from the dissociation of two domains of 245 and 108 bp, with $\overline{\Delta H}^{app}$ that are slightly less than the expected $\overline{\Delta H}^{cal}$ for Col E1 DNA. The fit of those subtransitions

Table II

Domain Characteristics of ϕ174-RF DNA[a]

Domain	$t_{m,i}$ (°C)	$\left.\frac{df}{dt}\right|_{t_{m,i}}$ × 10^{-1}	σ_i (°C)	ΔA_i, × 10^{-2}	Length (bp)	$\overline{\Delta H}^{app}$ (bp^{-1})
1	84.16	0.162	0.444	1.80	97	9392
2	84.88	0.293	0.333	2.45	132	9240
3	85.44	0.547	0.244	3.34	180	9301
4	85.74	1.03	0.175	4.52	244	9555
5	86.01	1.55	0.145	5.64	805	9232
6	86.27	1.39	0.150	5.21	282	9714
7	86.51	1.79	0.137	6.14	332	9036
8	86.77	1.72	0.137	5.90	319	9418
9	86.98	2.31	0.120	6.93	375	9167
10	87.11	3.16	0.103	8.12	439	9142
11	87.23	3.86	0.090	8.68	469	9787
12	87.35	2.31	0.120	6.93	375	9186
13	87.54	0.604	0.235	3.56	192	9143
14	87.67	0.846	0.197	4.17	225	9336
15	87.97	1.11	0.171	4.76	257	9415
16	88.21	2.67	0.111	7.44	402	9272
17	88.25	0.240	0.372	2.24	121	9208
18	88.51	2.16	0.124	6.71	363	9221
19	88.96	1.79	0.167	4.97	269	9276

Total number of base pairs	= 5386
Average number of base pairs per domain	= 282
Total enthalpy from all transitions	= 5.014 × 10^7 cal/mol of ϕx174 DNA
Mean calorimetric enthalpy estimated by Eq. (9)	= 9307 cal/mol bp
Apparent overall unit enthalpy (total enthalpy/ total bp)	= 9310 cal/mol bp

[a] From Vizard et al. (Ref. 32). Conditions: 194 m*M* NaCl (0.195*M* Na^+), 1 m*M* sodium cacodylate, 0.1 m*M* EDTA, pH 7.

as well as several of the other almost completely resolved transitions is excellent, again giving strong support to the two-state assumption.

A variation of ±4% is found in $\overline{\Delta H}_i^{app}$ for the four experiments analyzed. Much of this variation undoubtedly reflects experimental error, however, some is due to variations in base composition and to the particular process by which a domain melts.[3,4,29]

Melting Processes

As pointed out by Azbel[3,4] there are five distinct processes for the piecewise dissociation of the two strands:

1. Formation of an internal loop. Such a loop results in the formation of two phase boundaries. Following Azbel $n_b = +2$, where n_b is the net change in phase boundaries.

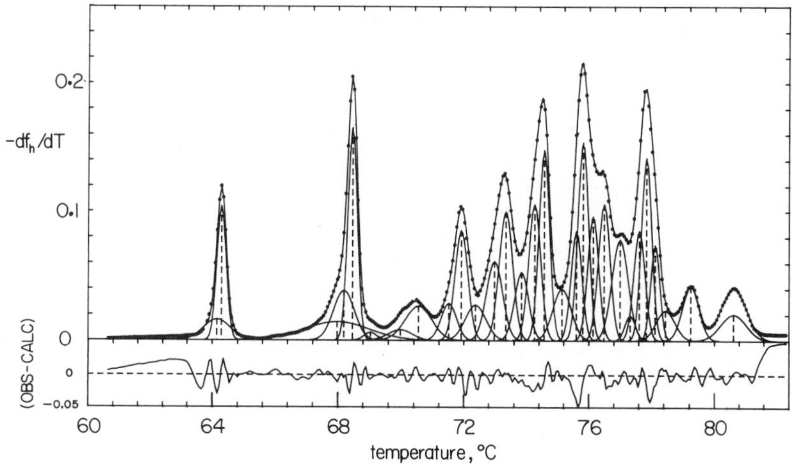

Figure 8. Observed and deconvoluted melting curves of Col E1 DNA in 0.1 x SSC[16] (19.5 mM-Na^+).

2. Melting from the ends ($n_b = +1$).
3. Enlargement of an internal loop ($n_b = 0$).
4. Enlargement of an internal loop into free ends ($n_b = -1$).
5. Coalescence of two internal loops ($n_b = -2$).

These processes are summarized in Table III. Each process brings a different contribution to melting. The formation of loops produces the greatest effect, primarily through an increase in loop entropy. However, some effect of loop formation may also involve additional enthalpy for initiation of the loop.[33] Assuming a small positive loop initiation enthalpy, ΔH_{in}, the Marmur-Doty relationship, equation 1, is amended to

$$T_m = (T_{AT} + \Delta t \cdot f_{GC}) - T_{m\infty}^2 \, \Delta S_{in}/\Delta H_{in} \tag{10}$$

where $T_{m\infty}$ is the melting temperature for all processes averaged over the entire length of the molecule. Thus, if loop formation results in $\Delta S_{in} < O$ and $\Delta H_{in} > O$, we see that melting by the first process (B in Table III) will lead to a $T_{m,i}$ higher than predicted by the Marmur-Doty relationship (equation 1). Other processes lead to $T_{m,i}$ greater, less than or equal to that predicted by equation 1 as summarized in Table III. Predictions by the model used to derive equation 10 for the different energetic effects of five distinct processes of local melting, and summarized in Table III, are in excellent agreement with those made previously by Azbel on quite different grounds.

Effect of Sodium Ion on Domain Stability

In addition to base composition and sequence we expect the melting of domains to show a dependence on counterion concentration in relation to their size and proximity to previously melted domains.[34-36] Accordingly, a study of the effects of Na^+

Table III
Summary of Melting Processes

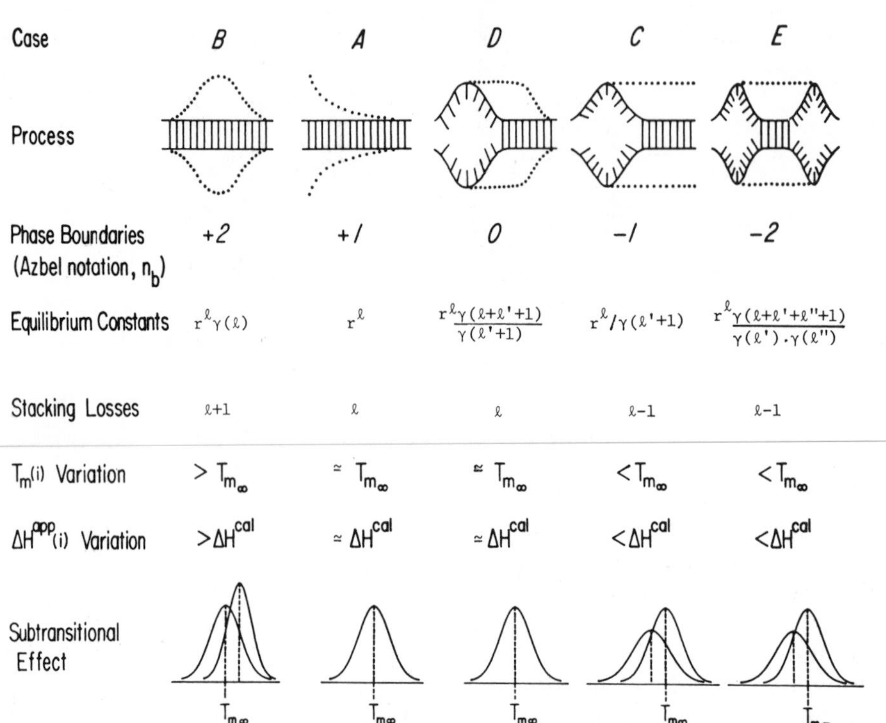

Case	B	A	D	C	E
Process					
Phase Boundaries (Azbel notation, n_b)	+2	+1	0	−1	−2
Equilibrium Constants	$r^{\ell}\gamma(\ell)$	r^{ℓ}	$\frac{r^{\ell}\gamma(\ell+\ell'+1)}{\gamma(\ell'+1)}$	$r^{\ell}/\gamma(\ell'+1)$	$\frac{r^{\ell}\gamma(\ell+\ell'+\ell''+1)}{\gamma(\ell')\cdot\gamma(\ell'')}$
Stacking Losses	$\ell+1$	ℓ	ℓ	$\ell-1$	$\ell-1$
$T_m(i)$ Variation	$> T_{m_\infty}$	$\approx T_{m_\infty}$	$\approx T_{m_\infty}$	$< T_{m_\infty}$	$< T_{m_\infty}$
$\Delta H^{app}(i)$ Variation	$> \Delta H^{cal}$	$\approx \Delta H^{cal}$	$\approx \Delta H^{cal}$	$< \Delta H^{cal}$	$< \Delta H^{cal}$
Subtransitional Effect	$T_{m\infty}$	$T_{m\infty}$	$T_{m\infty}$	$T_{m\infty}$	$T_{m\infty}$

was carried out on the melting of lambda ($cI_{857}S_7$) DNA. Lambda is a particularly good specimen because the distribution of base composition is exceptionally broad favoring a broad dispersion of subtransitions, as can be seen in Figures 1 and 4. It is also a convenient specimen because of the large number of distinct subtransitions emanating from the same DNA molecule. At low Na^+, where long range electrostatic effects are most significant, changes in the dispersion profile are quite prevalent.[19] Changes below 0.04M-Na^+ generally occur in an abrupt fashion indicating the balance between alternative phase boundaries in the establishment of a domain is sometimes delicate, reflecting long-range perturbations by weakly screened electrostatic forces. Also, we have good evidence now[21] that many variations in the thermal dispersion profiles obtained in less than ~0.015M-Na reflect nonequilibria at the 6.75°C/hour rate of heating used. The rates of both denaturation and renaturation decrease dramatically with counterion concentrations below 0.02M.[37-39]

At $[Na^+] > \sim 0.06$M the dispersion profiles are more conserved, indicating that screening of long-range forces is effective only above that value. A number of the better resolved subtransitions remain apparently unchanged over a wide range of $[Na^+]$. Figure 9 illustrates the variation of $t_{m,i}$ with log $[Na^+]$ for nine of these subtransitions, corresponding to numbered peaks in Figure 4. (The numbers in Figure 4 refer to graphical features in the melting curve, not to subtransitions,

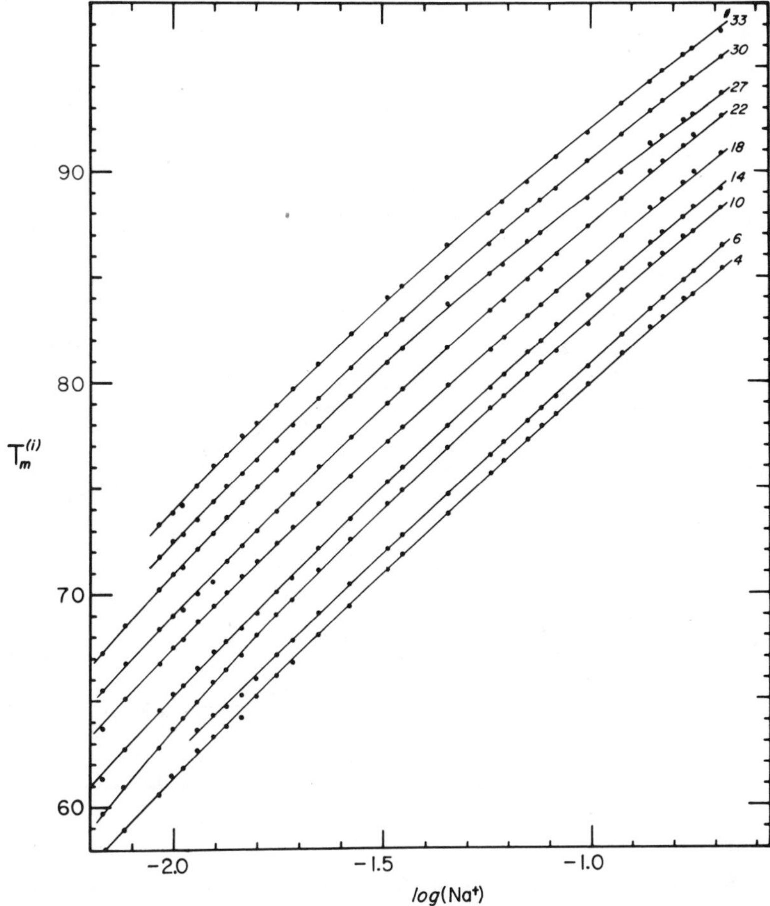

Figure 9. Variation of melting temperature with $[Na^+]$ for representative subtransitions in lambda DNA[19] corresponding to numbered peaks in Figure 4.

which number close to 90 (ref.21).) Due to a reciprocal relationship between the slopes of the curves in Figure 4 and subtransition enthalpy,[40,41] the early melting subtransitions of more A-T rich domains, e.g. numbers 4,6..., vary more strongly with $[Na^+]$ than do the late melting subtransitions. The consequence is that the dispersion between early and late transitions is compressed with increasing $[Na^+]$. This convergence is seen clearly in Figure 10 where the difference in $t_{m,i}$ between selected pairs of subtransitions is plotted as a function of $[Na^+]$. The Δt_m of two very distant transitions, numbers 33 and 4, is a constant 12, 81°C up to 0.03M-Na^+ but then falls rapidly to 11.30°C at 0.20M-Na^+. Differences between $t_{m,i}$ of other transitions indicate that convergence is a reflection of the sodium ion-dependent properties of the individual domains and not to any change in average helix or correlation lengths, as has been frequently suggested.

The electrostatic behavior of these more conserved domains can be analyzed in

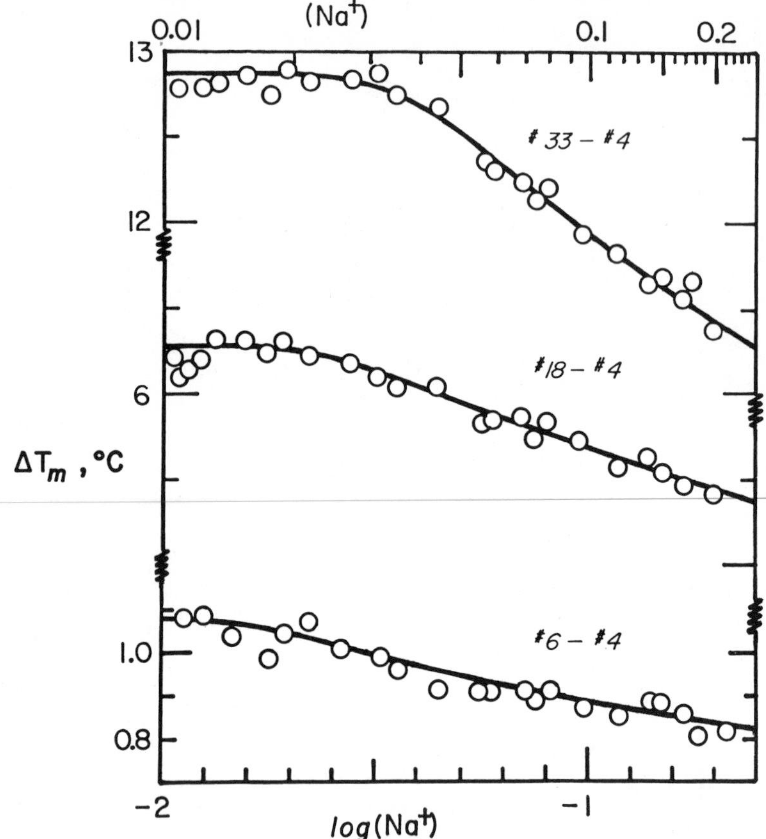

Figure 10. The difference in melting temperature with $[Na^+]$ for representative pairs of subtransitions from lambda DNA,[19] numbered in Figure 4. Note different scales for the three curves.

thermodynamic terms in the familiar way by the variation of $dt_{m,i}/d\log[Na^+]$ with the difference in territorially bound Na^+ to the helix and coil states.[40,41] Since the G-C composition of each domain can be readily determined by spectral analysis (cf. Figure 4) the analysis can be extended to a determination of the amount of Na^+ bound to A-T and G-C residues in the coil state. Given a relatively uniform linear charge density over the *ith* domain, unperturbed by neighboring domains, we can adopt the relationship[40,41]

$$dt_{m,i}/d\log[Na^+] = 2.3\alpha' R(T_{m,i})^2(\Psi_h-\Psi_c)/\Delta H'_m \quad (11)$$

where α' is the activity coefficient of Na^+ with a value of 0.94 ±0.025 over the limited concentration range of this study, and $\Delta H'_m$ is the enthalpic change per mole of nucleotide residue. The quantity $(\Psi_h$-$\Psi_c)$ represents the fractional loss of condensed Na^+ per mole of nucleotide residues upon going from the helix to coil state and typically has a value close to 0.17 near 0.1M Na^+ (refs. 40, 41). The ion association parameter Ψ is given by

$$\Psi = 1-(2\xi)^{-1} \tag{12}$$

where

$$\xi \equiv q^2/\epsilon kTb \tag{13}$$

is a charge density parameter governing the condensation process. Here q is the charge on the phosphate, k is Boltzmann's constant, ϵ is the bulk dielectric constant and b is the mean contour axis univalent charge spacing (Å) in the helix or coil. These expressions therefore relate measured slopes, $dt_{m,i}/d\log[Na^+]$, to differences in molecular charge spacing in the helix and coil, so that such analysis can be used as a probe of differences or perturbations of detailed structure features in the molecule.

Slopes of $t_{m,i}$ with log $[Na^+]$ in Figure 9 were determined over narrow ranges of $[Na^+]$ by a moving 10-point linear least-squares analysis. Results for three subtransitions of representative behavior appearing at different regions of melting are illustrated in Figure 11. All slopes, regardless of the G-C content of the domain, approach a 'constant' value of ~20.5°C per decade change of $[Na^+]$ at low salt. However, as the $[Na^+]$ is increased the slopes, $dt_{m,i}/d\log[Na^+]$, decrease rapidly, and in direct proportion to the G-C composition of the domain. Thus, number 4, an early melting comparatively low G-C containing domain, falls from 20.5° at 0.012M-Na^+ to only 17.4°C at 0.15M-Na^+, whereas number 33 falls to 15.8°C over the same range of Na^+. All three subtransitions exhibit a distinct break in the variation of $dt_{m,i}/d\log[Na^+]$ with $[Na^+]$ near 0.06M-Na^+. This is approximately the same ionic condition where subtransitional detail in the melting profile becomes conserved, reflecting a more uniform electrostatic environment for all domains above this value. At the same time the $dt_{m,i}/d\log[Na^+]$ slopes for over two-dozen subtransitions in the lambda profile now show a clear dependence on the G-C composition of the domain, determined by the dual wavelength method for base analysis. At 0.14M-Na^+, the highest $[Na^+]$ for which we have a 10-point least-squares slope, this variation can be described by the relationship

$$(dt_{m,i}/d\log[Na^+])_{0.14M\text{-}Na^+} = 19.96-6.65f_{g \cdot c} \tag{14}$$

which compares very well to the relationship

$$dt_m/d\log[Na^+] = 18.30-7.04F_{gc} \tag{15}$$

for a heterogeneous population of DNAs.[42]

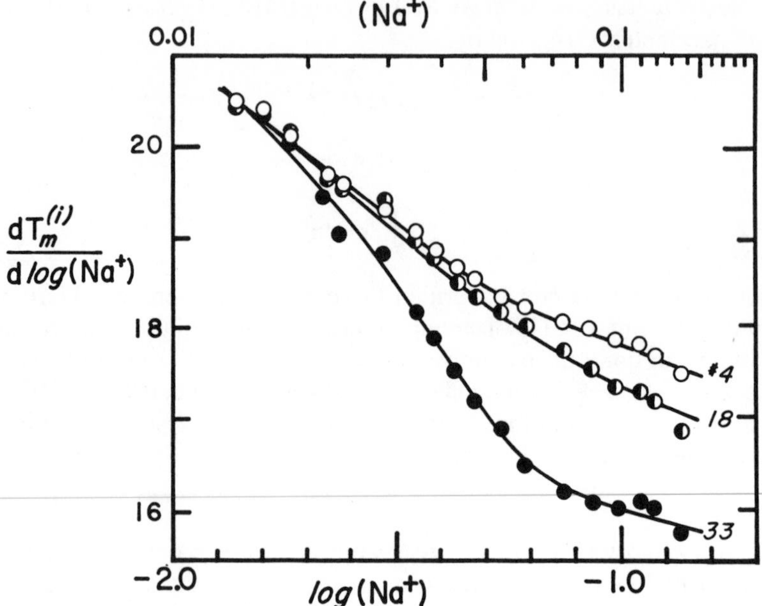

Figure 11. Variation of $dt_{m,i}/d\log[Na^+]$ with $[Na^+]$ for representative subtransitions in lambda DNA,[19] illustrated in Figure 4.

Above ~0.06M-Na^+ the charge density approaches uniformity over the entire length of the molecule, regardless of its conformational state. Under these conditions values for the measured slopes, $dt_{m,i}/d\log[Na^+]$, indicate directly the extent of liberation of Na^+ brought about by the precipitous fall in polynucleotide charge density during the helix→coil transition, as described in equation (11). Values for the transition enthalpy from expression (9), obtained from compiled calorimetric

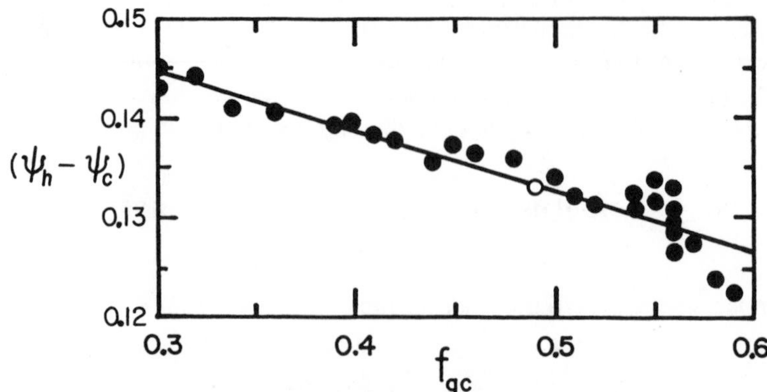

Figure 12. Variation of the fractional difference in Na^+ bound 'territorially' to the helix and coil with the fraction G-C composition of domains in lambda DNA producing a very distinct subtransition.[19] The open circle represents the value from the *overall* t_m and $dt_m/d\log[Na^+]$ for lambda DNA.

measurements,[19] and the measured slopes were used in equation (11) to determine the fractional loss of Na^+, Ψ_h-Ψ_c, during the helix→coil subtransitions of more than two-dozen domains in lambda DNA. The results are illustrated in Figure 12. The amount of Na liberated from the helix during the transition is significantly lower for the G-C base pair. Substitution of expressions (14) and (9) into (11) will confirm that the decrease in $dt_{m,i}/d\log[Na^+]$ is clearly not compensated by the increase in $\Delta H_m'$ over the range of f_{gc} for domains in lambda. Moreover, the extrapolated value of (Ψ_h-Ψ_c) at $f_{gc} = 0$ is in excellent agreement with that determined with the synthetic homopoly(dA•dT) helix. These results indicate that the A-T bp liberates 40% more Na^+ than the G-C bp. Either the A-T bp binds more Na^+ than the G-C bp in the helical state, or A and T residues bind less than G and C in the coil state. Since the former interpretation is ruled out by results of Shapiro et al[43] on the basis of differential equilibrium dialysis experiments, the results above indicate a persistent high level of Na^+ bound to dissociated G and C residues. A higher mean charge density of the single stranded G-C residue over that of A-T is consistent with the extraordinary proclivity of the former, particularly G residues to stack, preserving a shorter axial spacing between phosphates. From equation (13) we calculate the mean axial spacing of the single-stranded A-T residue is 3.98Å, but only 3.10Å for the G-C residue.

Acknowledgements

We are grateful to our many friends and colleagues who contributed to different aspects of this work, and especially to C. Hathaway, P.V. Haydock, S. Lefoley, R.M. Stephens and W.S. Yen. This work was supported by N.I.H. Grant GM22827 and MAES Project No. 08402.

References and Footnotes

1. Marmur, J. and Doty, P., *J. Mol. Biol 5,* 109-118 (1962).
2. Azbel, M. Ya., *Biopolymers 12,* 1591-1609 (1973).
3. Azbel, M. Ya., *Proc. Natl. Acad. Sci. USA, 76,* 101-105 (1979).
4. Azbel, M. Ya., *Biopolymers 19,* 61-80; 80-95; 95-109 (1980).
5. Inman, R.B., *J. Mol. Biol. 18,* 464-476 (1966).
6. Inman, R.B., *J. Mol. Biol. 28,* 103-116 (1967).
7. Reiss C., Michel, F. and Gabarro, J., *Anal. Biochem. 62,* 499-508 (1974).
8. Yubuki, S., Gotoh, O., and Wada, A., *Biochim. Biophys. Acta 395,* 258-273 (1975).
9. Vizard, D.L. and Ansevin, A.T., *Biochemistry 15,* 741-750 (1976).
10. Ansevin, A.T., Vizard, D.L., Brown, B.W. and McConathy, J., *Biopolymers 15,* 153-174(1976).
11. Gotoh, O., Husimi, Y., Yubuki, S. and Wada, A., *Biopolymers, 15,* 655-670 (1976).
12. Lyubchenko, Y.L., Frank-Kamenetskii, M.D., Vologodskii, A.V., Lazurkin, Y.S. and Gauze, G.C., *Biopolymers 15,* 1019-1036 (1976).
13. Reiss, C. and Arpa-Gabbarró, T., *Prog. Mol. Subcell. Biol. 5,* 1-30 (1977).
14. Blake, R.D. and Lefoley, S.G., *Biochim. Biophys. Acta 578,* 233-246 (1978).
15. Yen W.S. and Blake, R.D., *Biopolymers 19,* 681-700 (1980).
16. Borovik, A.S., Kalambot, Yu.A., Lyubchenko, Yu.L., Shitov, V.T. and Golovanov, Eu.I., *Nucl. Acids Res. 8,* 4165-4184 (1980).
17. Fischer, S.G. and Lerman, L.S., *Cell, 19,* 191-200 (1979).
18. Blake, R.D. Stephens, R., Fischer, S.G., and Lerman, L.S. (in preparation).

19. Blake, R.D. and Haydock, P.V., *Biopolymers 18,* 3089-3109 (1979).
20. Blake, R.D., *Biochemistry* (in press, 1981).
21. Blake, R.D., Stephens, R., Haydock, P.V., Yen. W.S., Hathaway, C., Vosman, F., and Tarr, C. (in preparation).
22. Talsky, G., Mayring, L. and Kreuger, H., *Angew. Chem. Int. Ed. Engl. 17,* 785-799 (1978).
23. Wada, A., Yabuki, S. and Husumi, Y., Crit. Rev. Biochem. CRC Press Inc. (1980).
24. Felsenfeld, G. and Sandeen, G., *J. Mol. Biol. 5,* 587-610 (1962).
25. Felsenfeld, G. and Hirschman, S.Z., *J. Mol. Biol. 13,* 407-427 (1965).
26. Fresco, J.R., Klotz, L.C. and Richards, E.G., *Cold Spring Harbor Symp. Quant. Biol. 28,* 83-90 (1963).
27. Fresco, J.R., in "Informational Macromolecules" (Vogel, H.J., Bryson, V. and Lampon, J.D., eds.), p.121, Academic Press, New York (1963).
28. Coutts, S.M., *Biochim. Biophys. Acta, 232,* 94-106 (1971).
29. Yen, W.S. and Blake, R.D., *Biopolymers 20,* 1161-1181 (1981).
30. Marguardt, D.M., *J. Soc. Ind. Appl. Math. 11,* 431-441 (1963).
31. Gabbarro-Arpa, J., Tougart, P. and Reiss, C., *Nature, 280,* 515-517 (1979).
32. Vizard, D.L., White, R.A. and Ansevin, A.T., *Nature 275,* 250-251 (1978).
34. Schildkraut, C. and Lifson, S., *Biopolymers 3,* 195-208 (1965).
35. Elson, E.L., Scheffler, I.E. and Baldwin, R.L., *J. Mol. Biol.* 54, 401-415 (1970).
36. Baldwin, R.L., *Accts. Chem. Res. 4,* 265-272 (1971).
37. Massie, H. and Zimm, B.H., *Biopolymers 7,* 475-493 (1969).
38. Ross, P.D. and Sturtevant, J.M., *J. Amer. Chem. Soc. 84,* 4503-4511 (1962).
39. Blake, R.D. and Fresco, J.R., *J. Mol. Biol. 19,* 145-160) (1966).
40. Manning, G., *Q. Revs. Biophys. 11,* 179-246 (1978).
41. Record, M.T., Jr., Anderson, D.F. and Lohman, T.M., *Q. Revs. Biophys. 11,* 103-178 (1978).
42. Frank-Kamenetskii, M.D. *Biopolymers 10,* 2623-2624 (1971).
43. Shapiro, J.T., Leng, M. and Felsenfeld, G., *Biochemistry 8,* 3233-3241 (1969).

Proceedings of the Second SUNYA Conversation in the Discipline Biomolecular Stereodynamics
Volume I, ISBN 0-940030-00-4, Ed., Ramaswamy H. Sarma,
Adenine Press, New York,

Base Sequence and Melting Thermodynamics Determine the Position of pBR322 Fragments in Two-Dimensional DNA Gel Electrophoresis

L.S. Lerman, S.G. Fischer, D.B. Bregman, and K.J. Silverstein
Center for Biological Macromolecules
Department of Biological Sciences
State University of New York
Albany, New York 12222

Introduction

We have previously described a new system in which the electrophoretic transport of DNA in a gel is strongly and sharply regulated by each molecule's susceptibility to denaturing conditions (Fischer and Lerman[1]). Molecules are driven into an ascending gradient of a denaturing solvent at an elevated temperature. Each molecule undergoes an abrupt, drastic dimunition of its electrophoretic mobility at a critical depth into the gradient. The residual mobility after the transition is so much less than the original mobility that the patterns become nearly independent of time after the last molecule has experienced its transition. The location of DNA at the conclusion of the experiment represents a position in the gradient, reflecting the denaturing potential due to a combination of both temperature and solvent, rather than a time point according to the velocity of migration. We have combined separations based on this principle with separations based on conventional, velocity-determined, length separations in the perpendicular direction. Systematic fragmentation of a monodisperse DNA sample results in reproducible two-dimensional patterns. The physical meaning of the coordinate for velocity separation, which we conventionally orient along the abcissa, is reasonably clear; it is our purpose here to see what meaning can be attributed to the coordinate along the gradient direction.

Our exploration of this system is still at a very early stage. We have been more concerned with qualitative characteristics of the DNA distribution patterns than with close control of operating details to permit close quantitative comparisons. Nevertheless, the intimate involvement of our results with the theory of melting of the double helix leads to a number of interesting inferences and guidelines for further work.

Techniques

The mobility transition is induced by a combination of temperature elevation and an organic solvent mixture, adjusted such that the most easily melted, naturally occurring DNA is expected to have the mid-point of its helix-random chain transition near the beginning of a linearly increasing gradient and the least meltable natural DNA near the end. The solvent mixture consists of urea and formamide in constant ratio, such that a concentration of 7 M urea implies also 40% (v/v) formamide. The total cation concentration is 0.044 M, of which 0.02 M is ionized tris. The migration is carried out with the gel submerged in a stirred water bath at 60.0 degrees, but the actual gel temperature is very slightly higher because of ohmic heating during electrophoresis. Greater detail is presented in Fischer and Lerman.[2] In all of the pBR322 experiments the polyacrylamide concentration was 65 mg/ml.

Solvent-Temperature Relations

To determine the interdependence of solvent concentration and temperature in defining the transition point for the abrupt drop in electrophoretic mobility, which we shall call the retardation point, we have examined the distance migrated by the six EcoRI restriction endonuclease fragments of lambda DNA in constant solvent concentration at various temperatures for a fixed length of time. The measurements are based on gels in which a solvent gradient was perpendicular to the direction of migration, rather than the standard configuration, in which the gradient and migration are parallel. Representative patterns of this sort have been

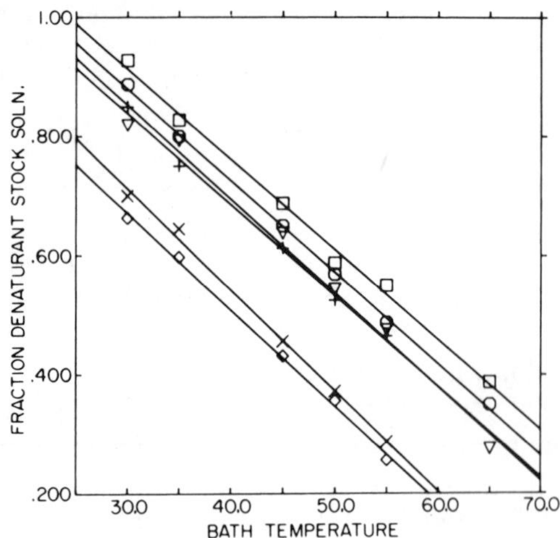

Figure 1. *Retardation position vs. temperature.* The urea-formamide concentration at which the electrophoretic mobility of each EcoRI fragment of lambda DNA in a polyacrylamide gel has dropped to half its value in purely aqueous solution is shown as a function of the temperature of the bath in which the gel was submerged during electrophoresis. Since the gels were prepared with a gradient of urea and formamide in constant ratio perpendicular to the direction of DNA migration, each molecule traveled in constant urea and formamide composition (except for loss of solvent by diffusion in a narrow band near the starting level of the gel). Concentration is indicated as fraction of the maximum in the gel, which was 7.0 M urea and 40% (v/v) formamide. The solvent also contained 0.04 M Tris, 0.020 M sodium acetate, and 1 mM EDTA at pH 8, (TAE). Representative fluorescence photographs and other details are presented in Fischer and Lerman.[1] The lambda fragments are indicated as follows: A, 21.8 kb (◇), E, 5.9 kb (X), D, 7.5 kb (△) F, 3.4 kb (+), C, 5.5 kb (○), B, 4.8 kb (□). The least squares regression lines shown have a mean slope of .0158 deg C.

presented previously (Fischer and Lerman[1]). A plot of the mid-points of the mobility transitions for each of the six fragments as a function of temperature is shown in Figure 1, together with least-squares lines. It will be seen that the solvent concentration required for retardation follows a simple linear trade-off against temperature, and that the slope is essentially the same for all fragments. The uniformity of slope in Figure 1 implies that the effect of temperature on lowering the retarding urea-formamide concentration is independent of base composition of the fragments. The equivalent conclusion has been reached calorimetrically for the effect of urea on the melting of calf thymus and salmon sperm DNA (Klump and Burkart[3]). We will use the average slope, degree scale unit, from these results in the calculations to be presented below. We may note that the transition mid-point does not necessarily represent the concentration at which molecules moving into an ascending gradient drop to their final mobility, but the difference is small and perhaps uniform.

What Happens at Retardation?

The appearance of a systematic correspondence between electrophoretic retardation and the helix-random chain transition of DNA strongly suggests that melting itself constitutes the basis of retardation. We would like to consider whether the distribution of molecules in the gradient can be related to their expected melting behavior in any simple way. Some of our previous results provide a significant rule (Fischer and Lerman[4]). Each of the six fragments of lambda DNA produced by the action of EcoRI was subjected separately to fragmentation by hydrodynamic shear. The pattern of retardation positions of the sheared material was compared with the retardation positions of each of the intact fragments. The sheared samples provide broad distributions with a number of peaks through the gradient. There are at least three obvious maxima or shoulders in the shearate from fragments A, B, and C, and more than six each from D, E, and F. The fragments are labeled in the conventional way, in order of descending length. The resolution and reproducibility of the patterns suggest that a more searching analysis would indicate a much larger number of distinct components in each pattern. In every case the retardation point of the intact restriction fragment can be identified with the very beginning of the complex distribution given by the same fragment after shearing; the intact restriction fragment is retarded at the same depth as the most easily retarded subfragment snipped from it by the shearing. It appears irrelevant that some or most of the subfragments penetrate very much deeper into the gradient, or nearly to the bottom, before retardation.

If retardation is to be attributed to melting, the melting of only a small part of the fragment is sufficient. The difficulties in detecting partial melting of an extremely dilute DNA sample in a gel at 60 degrees necessitates a less direct approach to the melting hypothesis. As a first approximation we can attempt to correlate the fractional base composition of a uniform arbitrary sample of each molecule with its retardation point. Although the relation between base sequence and melting temperature represents an intricate problem in statistical mechanics, the approximate correlation of fractional AT or GC content with melting temperature for relatively long molecules is well known and can serve as the basis for a simple comparison.

Arrays of Fragment Families

It is useful to consider two-dimensional patterns prepared in two different ways. A relatively simple DNA, that of the plasmid pBR322, can be fragmented by restriction endonucleases to provide a set of well-defined smaller molecules which may contain common sub-sequences within shorter or longer segments. The sequence of each fragment can be examined to find the region of highest adenine/thymine density, which will presumably represent the most easily melted part of the fragment.

A different sort of fragmentation is achieved with hydrodynamic shear. Applied marginally in excess of the threshhold turbulence needed for one double strand scission, it gives a mixture of sequences differing by very small length increments, mostly products of a scission near the center of the original molecule (Burgi and Hershey[5]). The series of fragments related by very slight differences at the ends is useful in identifying features of the sequences that affect melting properties. If the material were indeed randomly sheared so that all possible beginning and end points were represented, the position of each restriction fragment in the two-dimensional gel would match part of the shear pattern. Because of the control of shearing in the present experiments, most restriction fragments will not have counterparts in the shear distribution; virtually all of the shear fragments carry one or the other end of the linearized plasmid molecule, and fragments representing only central sections of pBR322 will not be generated. The two-dimensional pattern indicates, nevertheless, the retardation levels of many of the high melting priority sequences and serves to stake out reference points in the coordinate system against which all restriction fragment positions can be fixed.

The pBR322 Shear Pattern

A typical pattern given by the shear fragments of pBR322 after linearization with EcoRI is shown in Figure 2. The sample also included a mixture of restriction fragments of the replicative form of ϕX 174 as molecular length markers.

The conspicuous spot in the top right corner corresponds to unfragmented linearized plasmid DNA. There is an intense zone covering a broad span of fragment lengths at essentially the same depth immediately to the left of the plasmid spot. Continuing to the left, there are two short, relatively weak zones higher in the gradient at the short molecule end of the first zone and then another long zone just slightly deeper in the gradient than the first. The series of spots immediately below the top series of zones derive entirely from the fragments of ϕX DNA. The lower zones offer a relatively complex pattern beginning with a series of short, closely associated streaks at the long fragment end, a numerous series of spots in an inverted V, which we will call a stile because of the double staircase effect, a conspicuous series of zones over a long span of fragment lengths at almost but not

Figure 2. *Two-dimensional shear of pBR322.* pBR322, linearized with EcoRI, was sheared in a Virtis homogenizer Model 60K at 19,000 rpm for 15 min at a concentration of 5 μg/ml in 10 ml TAE in a 50 ml cup. The DNA was ethanol precipitated and resuspended to a concentration of 200 μg/ml in TAE, containing also 10% (v/v) glycerol, 40 mM EDTA and bromophenol blue. To this sample Hae III, Hpa I and Hpa II restriction endonuclease digests of φX 174 rf DNA were added for length calibration. The sizes of these fragments seen on the gel (base pairs) are 3730, 2748, 1697, 1353, 1264, 1078, 872, 603, and are designated φX A-H. A total of 5 μg DNA including about 0.35 μg φX 174 DNA was loaded onto a 14 mg/ml agarose gel and run for 17 hours at 2 v/cm. The ethidium-stained strip was sealed across the top of a 65 mg/ml acrylamide gel containing an ascending denaturing gradient and run as described in the text.

quite the same depth at the bottom of the gradient, and two fainter zones at intermediate depths. The zones are conspicuously narrower in the ordinate direction than in the length separation as a result of focusing due to the mobility transition.

The Loci of pBR322 Restriction Fragments

We have been able to establish correspondence between restriction fragment positions and shear fragment's zones by mixing restriction endonuclease fragments with sheared DNA before electrophoresis. The restriction fragments make compact spots that can be discerned easily if the concentration of sheared DNA is not too high. An outline summarizing the results of a series of gels including restriction fragments is shown in Figure 3. Data were assembled into a composite by assigning an abcissa value determined by the ϕX174 calibration and the pBR322 fragment length, together with an ordinate scaled according to the shear pattern.

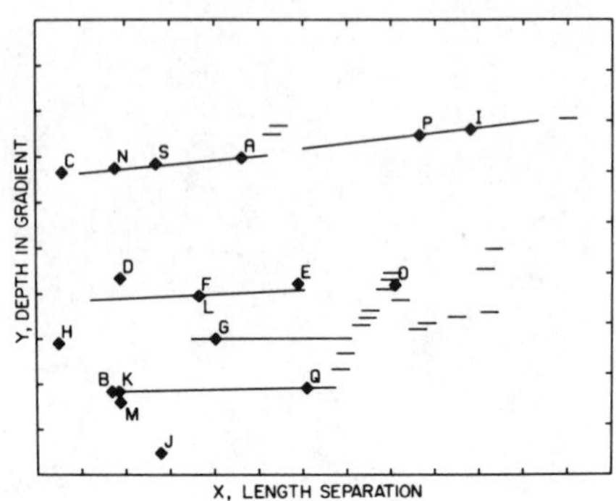

Figure 3. *Composite diagram of two-dimensional gel of positions of pBR322 restriction fragments superimposed on shear patterns.* pBR322 was cut with EcoRI and then with various restriction endonucleases to produce a set of fragments with defined ends shown in Figure 4. The cleavage products of different digests were mixed with 2 μg sheared pBR322 and ϕX 174 length standards and run in two dimensions as described in Figure 2. The coordinates of the pBR322 restriction fragments with respect to the shear pattern were obtained from several gels and summarized. The ϕX 174 length standards are not shown.

Recognition of Local Sequences

It is helpful to consider the character of the fragments and their coordinates in the gel in terms of a compositional map of the plasmid. We have calculated the local base composition as a sliding average within a frame of 51 bp using the nucleotide sequence determined by Sutcliffe.[6] The average is plotted in Figure 4 at the position of the central base pair. The numbering proceeds from the EcoRI site, following Sutcliffe's convention, and we shall refer to the polarity of the sequence as the high or low direction according to that numbering. Figure 4 also indicates the span of each of the restriction fragments that have been located on the shear pattern, with an arbitrary letter identification. A fragment of pBR322 which extends from base 1424 to the right-hand end falls near the center of the strong zone in the top right, and the fragment which extends from 2067 to the end falls in the same zone near its long end. Since both these, as well as the whole linearized plasmid include the most conspicuously AT-dense region near 3,200 and the AT-dense region at the high end, we infer that the strong zone at the top right is defined by melting in one or the other of these regions.

The strong zone at the upper left, which might be regarded as a continuation from the top right, corresponds to the loci of fragments A and N, both of which contain the high end of the linearized plasmid, but exclude the peak of high AT density at 3200. None of the fragments retarded deeper in the gradient include either of these regions.

A group of fragments, E, F, L, and O, substantially overlap the region between 1500 and 2100. That region exhibits a relatively uniform, slightly elevated AT density. The fragments are retarded at a middling level in the gradient. Fragment G, which lies along the rough high ground between the broad and narrow valleys is retarded further down. Since the low end of the plasmid appears to be AT rich, it might be

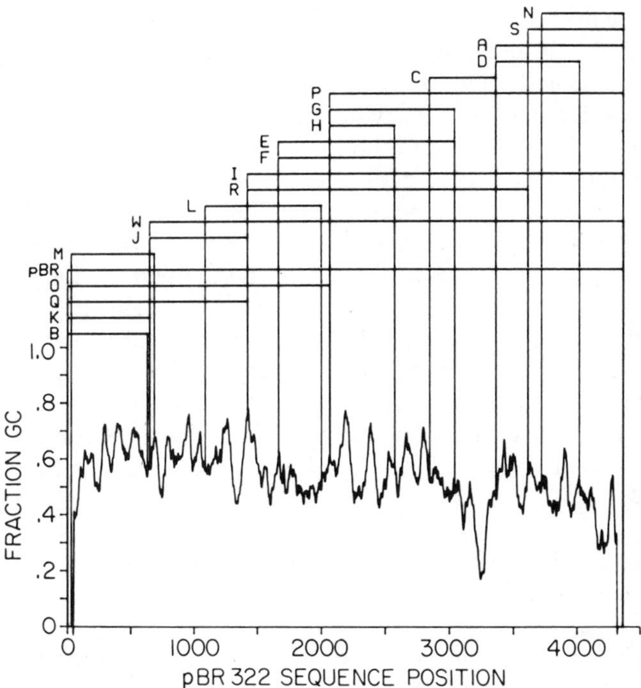

Figure 4. *Span of pBR322 restriction fragments displayed on fractional GC map of linear pBR322.* The fractional GC content of pBR322, numbered from the EcoRI site at 0 bp is plotted as a sliding average within a frame of 51 bp in the lower half of the figure. The sequences spanned by each of the pBR322 restriction fragments which were analyzed by two-dimensional electrophoresis are represented as bars above the GC map.

expected that the small fragments from that end, B, K, M and Q, would lie high in the gradient. Actually, the AT rich region is much shorter than a domain length adequate for retardation. Nevertheless, the position of these fragments is somewhat problematic as will be seen below.

Retardation Depth and Base Composition

In searching for the possibility of a simple relation between retardation and composition of the fragments, we need to consider a number of characteristics of the system: the length of the region that must melt to effect the mobility transition, the position of the melting region relative to the ends of the fragment, and the presence of other regions in the same fragment with similar melting properties. It is plausible to suppose that the retarding length is related to the cooperativity of melting in that it must consist of at least one domain (see Blake et al[7]) or a sufficient number of small domains the total length of which exceeds some critical number of nucleotide pairs. Although the total of the domain lengths will be expected to vary considerably among the various fragments we will assume that they can be approximated by a single value.

The small slope of the zones defined by similar families of shear fragments suggests that each fragment in the gradient continues to increase slowly after retardation; the larger the fragment, the less the residual velocity. Although retardation depths may be identical for fragments carrying the same region of highest melting priority, residual velocities may differ and the final depth may be a little greater for short than for long fragments. The effect would be expected to be most pronounced for fragments that are retarded earliest, since they remain in the electric field the longest after retardation. That expectation corresponds to the features of Figure 3, where it is seen that the zones at the top have the greatest slope. We have introduced some compensation for the effect of varying fragment lengths in a simple way. The slopes of the three clearly defined zones in Figure 3 vary as the square of the height above the lowest zone (the strongest zone at the top right deviates a little). The parameters of that variation have been applied to all of the fragments in Figure 3 to extrapolate the position of each to the depth it would have directly under the spot given by intact pBR322. The largest correction, that for fragment C, is less than 12% of full scale.

A search by means of a simple computer program identified the section of specified length with the lowest average GC density in each fragment.

A correlation diagram for the corrected gradient depth and the base composition of the 301 bp section richest in AT in each of the fragments is shown in Figure 5. A least squares regression line has been drawn based on 16 points, excluding the four fragments that terminate at 0 or 31 in the pBR322 sequence. The correlation coefficient is 0.992; the standard deviation of the scatter around the regression line is 0.29, or about 2.5% of full scale. The line is given by the relation

$$y = 3.01\, X_{GC} - 1.154 \qquad (1)$$

where X_{GC} is the fractional composition with respect to the sum of A and T. The 95% confidence limits for the slope are 2.79 and 3.23. (1) By introducing the coefficient of proportionality between temperature and gradient composition derived from the data of Figure 1 and the scale factor equation 1 becomes

$$T_m = (44.7 \pm 3.3) X_{GC} \qquad (2)$$

including the 95% confidence limits. Although the value of the intercept is not meaningful because of some arbitrariness in details of the system, the coefficient of the compositional factor should be identifiable with the temperature difference between the extremes of composition for DNA melting as observed by more direct procedures. The difference in melting temperatures between (hypothetically) pure AT and pure GC DNAs depends on cation concentration, and has not been examined for mixed solvents under conditions similar to ours. Nevertheless, the low salt values given by Blake et al.[7] may be appropriate, since urea, as noted above, has no substantial base specificity. Their value is 48.98 degrees C, in reasonable agreement with our slope. The retardation depths can be compared also with the average base

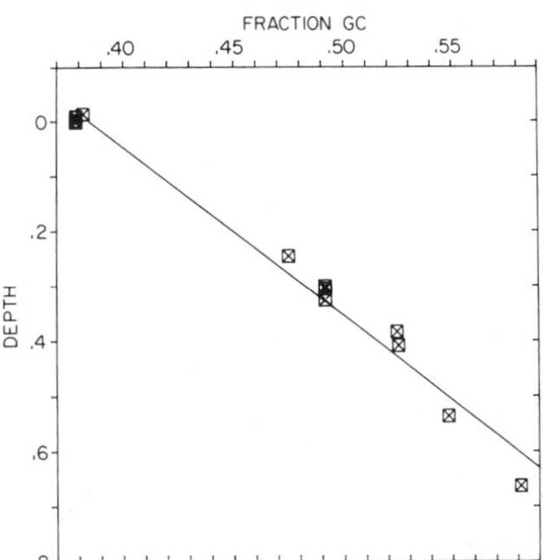

Figure 5. Retardation depth of pBR322 restriction fragments plotted against the fraction GC of the most AT-rich 301 bp sequence in each fragment. Fragment depths are those shown in Figure 3 corrected for residual velocity.

composition of the entire fragment; these values are included in Figure 5. Since the 301 bp region is a large part of every fragment and its composition is included in that of the entire fragment, overall composition cannot be totally uncorrelated with depth. Nevertheless, the correlation is seen to be much weaker (r = .82).

Fraying vs. Bubbles

It is clear from thermodynamic considerations that the temperature required for the helix-random chain transition varies according to position of the domain with respect to the presence or absence of neighboring helix at the transition temperature. Simplified descriptions of the neighboring domain effects have been presented by Azbel[8,9] and by Yen and Blake,[10] who both suggest (somewhat differently) that the T_m expected for an indefinitely long polymer is perturbed upward or downward in discrete steps if there is adjacent helix remaining unmelted at one or both sides or if there is already melted helix on one or both sides (see also Blake, et al., this symposium). Azbel proposes that the melting temperature, T_m, of a domain will be given by

$$T_m = (T_{GC} - T_{AT})X_{GC} + \Delta n T_b/d \tag{3}$$

where $(T_{GC} - T_{AT})$ is the difference in melting temperature of infinitely long polymers at the extremes of composition, d is the length of the domain, and Δn is the change in the number of boundaries between melted and unmelted regions brought about by melting of the specified domain. Following Blake, we regard the effect of a fragment end as the same as that of a boundary with a melted section. The perturbations are generally expected to be less than about 5 deg C.

We find that one end of the most AT-dense 301 bp region of each fragment, with two exceptions, lies within 70 bp of the fragment end, and mostly much closer. The exceptions are fragments D and G that have their lowest 301 bp average conent of 0.475 and .525 separated from an end by 128 and 251 bp, respectively. However, the

GC compositions of the end regions are only slightly higher, 0.488 and 0.535. The handicap for interior melting of 301 bp given by the second term of equation 3, using the parameters presented in the following section, is 0.67 deg C. This values exceeds the T_m advantage of the slightly lower interior GC content, given by the first term of equation 3, (0.58 and 0.45 deg, respectively), and higher melting priority must be assigned to the terminal regions of both fragments. We infer that our selection of restriction fragments correlates reasonably well because retardation is attributable to end melting for the whole set.

Although the terminal region of 301 bp has the lowest GC density in all fragments extending to the high end and in the intact linearized plasmid, the region near 3200 contains a higher AT density if the average is taken over a span of 251 bp or fewer. We cannot yet defend the necessity of melting 300 bp rather than merely 250, but the handicap for interior melting again applies and confirms the end melting priority.

The Shear Pattern and Middle Melting

Although much of the pattern of fragments of pBR322 produced by one shear scission is closely identified with the loci of restriction fragments that appear to be end melters, certain features of the shear pattern depart significantly from the positions expected for end melting. Consider the two major upper zones in Figures 2 and 3. Both include shear fragments that match loci of restriction fragments whose retardation is determined by melting at the high end, i.e., beyond 4000 bp. These are fragments A, I, N, and P. It might be reasonable to infer that the region at the high end will represent highest melting priority in any molecule which includes this region after shearing regardless of the point of scission. Why then, is there an interruption at 1300 bp, instead of continuity between the two zones? The possibility of a substantial shift in T_m if an end is introduced near a meltable region offers an explanation; we may examine the properties of fragments sheared near 3200. Although the 301 bp region centered near the high end has a lower expected T_m than the 3200 region as infinite polymers, the difference is slight, only 0.30 degree. If our estimate of the melted length necessary for retardation is too large, the order would be inverted. The T_m difference for 250 bp is 0.35 deg C in favor of the 3200 region. Even so, the thermodynamic advantage of end melting implies that the 4100 region retains priority. The temperature elevation for melting a middle 250 bp domain, relative to one that is bounded by helix on only one side after melting can be estimated from equation 3 using Azbel's[9] value for the perturbation constant, T_b, which may differ for a mixed solvent and appropriate values of Δn. For middle melting, Δn is +2, since there are no relevant helix-random chain boundaries before melting, and there are two after the domain melts. For end melting $\Delta n = 0$; the virtual helix-chain boundary at the end of the fragment is merely translated inward by melting of the end domain. The difference in T_m values for any sequence at an end and the same sequence surrounded by high melting helices is then 200/deg C, or 0.8 deg for a 250 bp domain. Thus the highest melting priority for fragments terminating near 3200 would shift from the 4100 region to the 3300 region, since end melting of a sequence denser in AT would be possible in those fragments. The easier melting of high end fragments of this particular length is expressed in the pattern by the spots just above the interruption between the upper zones.

Average base composition for a length of 301 bp was used in the melting calculation presented here because it provided the best correlation with retardation depth for restriction fragments, but the correlations for somewhat smaller lengths are not substantially inferior. As noted above, we do not think that the actual retardation length is uniform for all fragments.

The converse situation appears in the lower stile, the part of the pattern near restriction fragment 0. The depths of spots in the stile are replotted in Figure 6

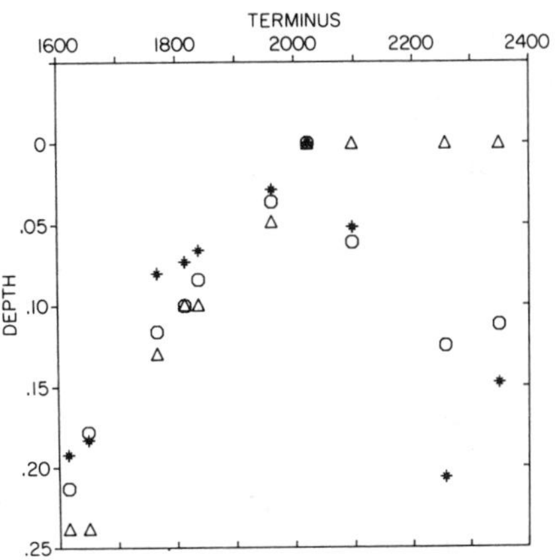

Figure 6. *The retardation pattern of pBR322 fragments extending from 0 to lengths between 1600 and 2400.* Position measurements from Figure 2 have been converted to a linear length scale. Gradient depth is shown in scale units representing distance on the fluorescence photograph. Data points, O; depth expected from the base composition of the region densest in AT, △; depth expected from the base composition of the end region.*

against the lengths estimated for the center of each small streak from the ϕX174 calibration. It will be seen that the stile is formed by low end fragments terminating in the 1400-2400 region, a broad, slightly AT-dense basin. Their high end, complementary fragments will have been retarded in the top zones. Termination near 2022 (the top fragment in the stile) provides a long region of high AT density, obviously favorable to end melting. Termination further toward the high end introduces more GC near the end of the fragment, hence descent along the stile to the right. Termination further toward the low end requires melting into the region denser in GC to reach critical retardation length, hence descent along the stile to the left. Figure 6 includes also a calculated pattern, represented by Δ, of expected depths (as T_m) if the difference between end and middle melting were disregarded, using the base composition term of equation 3 only. It will be seen that retardation of all of the larger fragments at the same depth would be expected, since the highest AT density would be the same for all, the basin region around 1800-2000. Expectation on the basis of end melting (here calculated for a 241 bp region at the end of each fragment) shown as *, is in much closer correspondence with the data, except for a conspicuous overestimate of T_m for the 2256 and 2348 bp fragments. It can be seen from Figure 4 that the ends of these fragments are GC dense and well removed from the AT-dense basin. The assumption of end melting is inappropriate, and middle melting of the 1800-2000 basin is favored, despite the penalty. Thus middle melting with 0.486 GC (fragments 0-2256 and 0-2348) retards at roughly the same depth as end melting with .530 GC (fragment 0-1767).

Conclusions

The good agreement between the data and a simplified analysis of the relation between sequence and melting supports the hypothesis that the large change in electrophoretic mobility is the consequence of melting of only a few hundred contiguous base pairs. Gradient position is interpretable in terms of base composition of that region. The patterns show that a particular sequence melts more easily when it is at the end of a more GC dense molecule rather than embedded in the interior, as expected theoretically, and provide an estimate of the temperature difference, perhaps the first test of this effect. The considerations demonstrate the utility of the two-dimensional gel system as a means for discerning the coarse-grained features of nucleotide sequence of substantial lengths of DNA, and suggest that even small variations in sequence are detectable. Since retardation characteristics constitute a more robust reflection of sequence than does restriction enzyme susceptibility, which is controlled by single base changes at relatively sparse loci, it can be expected to serve as a particularly useful tool in certain types of genetic analysis.

Acknowledgements

This work has been supported by a grant from the National Institutes of Health, GM-24030. We are grateful to Dr. Nadrian Seeman for helpful discussions and assistance with programming, to Dr. Richard Orville for offering use of the false color video densitometer for measurement of pattern coordinates, and to Pamela Laird for preparation of the manuscript.

References and Footnotes

1. Fischer, S. G., and Lerman, L.S., *Cell 16*, 191-200 (1979)
2. Fischer, S. G., and Lerman, L.S., *Methods in Enzymology 68*, 183-191 (1979).
3. Klump, H.,and Burkart, W., *Biochem. Biophys. Acta 475*, 601-604 (1977).
4. Fischer, S. G., and Lerman, L.S., *Proc. Natl. Acad. Sci. USA 77* 4420-4424 (1980).
5. Burgi, E., and Hershey, A.D., *Biophys. J. 3*,309-321 (1963).
6. Sutcliffe, J.G.,*Cold Spring Harbor Symposium 46*, 77-90 (1977).
7. Blake, R. D., Vossman, F., and Tarr, C.E., in *Biomolecular Stereodynamics, Volume I*, Ed., Sarma, Ramaswamy H., Adenine Press, N.Y. (in press).
8. Azbel, M.Ya, *Proc. Natl. Acad. Sci. USA. 76*,101-105 (1979).
9. Azbel, M.Ya, *Biopolymers 19*,61-80 (1980).
10. Yen, W.S., and Blake, R.D. *Biopolymers 20*, 1161-1181 (1981).

SUBJECT INDEX